Dictionary of Global Bioethics

Henk ten Have ·
Maria do Céu Patrão Neves

Dictionary of Global Bioethics

Springer

Henk ten Have
Center for Healthcare Ethics
Duquesne University
Pittsburgh, PA, USA

Maria do Céu Patrão Neves
University of the Azores
Ponta Delgada, Portugal

ISBN 978-3-030-54160-6 ISBN 978-3-030-54161-3 (eBook)
https://doi.org/10.1007/978-3-030-54161-3

This Springer imprint is published by the registered company Springer Nature Switzerland AG
The registered company address is: Gewerbestrasse 11, 6330 Cham, Switzerland

Preface

The *Dictionary of Global Bioethics* seeks to clarify terms, concepts, arguments, perspectives, and theories used in the emerging field of global bioethics. Bioethics emerged in North America as a new discipline in the 1970s and quickly expanded to other continents. Many new ideas and original approaches were used as a result of this discipline becoming increasingly applied not only within various healthcare contexts, but also in policies, regulations, and legislations. Later and especially since 2000 it became clear that the scope of bioethics was genuinely global following the globalization of medical research, medical technologies, and healthcare practices and in the face of global challenges such as pandemics and climate change. Viewed in this light bioethics is increasingly redefined as "global bioethics." This book aims to assist those who want to understand the most commonly used concepts in this emerging field of global bioethics.

This dictionary is a reference work that follows and supplements two previous reference volumes. The *Handbook of Global Bioethics* (2014) elaborates the global ethics principles included in the Universal Declaration on Bioethics and Human Rights that was unanimously adopted by the member states of UNESCO in 2005. It also elaborates several relevant global problems. The *Encyclopedia of Global Bioethics* (2016) systematically discusses topics and themes that are relevant in the contemporary global bioethics debate. This dictionary is different in that it provides brief clarifications of relevant topics. The presentations are succinct much as is the case with regular dictionaries and aim to help the reader to get an initial feel for meaning, content, relevance, and associated issues. Although it does not always provide final definitions (which is sometimes difficult because the field is emerging and developing) the topics do provide dynamic conceptualizations and thematic problematizations that elucidate what is at stake and spell out the various perspectives and related challenges as briefly as possible.

The entries presented are arranged in four categories that can be broken down into four different parts: "organizations" (Part I), "documents" (Part II), "persons" (Part III), and "topics" (Part IV). The selection of entries has been based on just how frequently they occur in the relevant literature dealing with global bioethics. We have scanned the keywords and indexes of reference publications, books on global bioethical issues, and journal publications in this area. However, the selection is ours and therefore is open to criticism.

Nonetheless, we are convinced that this dictionary covers the most relevant issues in today's global bioethics.

Part I deals with those organizations that actively work in the field of global bioethics. We have included those that have a global outreach and mission in addition to actively working within the field of global bioethics. Although some organizations are regional or based in a particular country, those included here have as part of their mission to make contributions to the global debate. We have not included organizations with a more limited scope or those that do not contribute to the global ethical debate. Some of the organizations selected have been around for some time and during the past few decades have increasingly focused on global challenges. Some have deliberately promoted ethical approaches and contributed to the growth of global bioethics. Others are relatively new and owe their establishment to the processes of globalization. All the organizations presented play active roles in global bioethics despite having different natures. They range from professional associations, international agencies, international ethics committees, and scientific and humanitarian bodies.

Part II discusses relevant documents. We have identified and listed those that play a significant role in global bioethics and can be used as reference points for global debate. Most of these documents have been produced by international organizations. Since many topics in Part IV of this book refer to these documents further explanation is warranted in Part II. We have not included documents that have been developed at the national level despite possibly having a global impact. Declarations and policy statements have been adopted over the years by international bodies in response to global challenges. The first documents were focused on medical research. Later documents dealt with other specific issues such as environmental concerns, international exchange and trade, and genetics and life sciences. A few documents pertain in particular to global bioethics since they formulate ethical frameworks and principles for the ethical application of medicine, healthcare, and medical research around the world.

Part III briefly introduces a small band of people who have played substantial roles in promoting global bioethics or have introduced concepts that have significantly contributed to amplifying the global bioethical debate. Some worked for organizations and institutes that promoted ethical reflection on global issues. Others promoted fundamental ideas that were constitutive of current global ethical discourse. We are aware of many people around the globe who have enriched bioethics with their intellectual and practical activities. Although many have been working in the field of bioethics, not all have explicitly argued that bioethics should be a global endeavor. To lessen any risk for biases we have chosen to present only those people who have made a recognizable contribution during their lifetime but are sadly no longer among us.

Part IV comprises the bulk of the dictionary. It presents brief descriptions of topics in alphabetical order. Choosing which topics to include was relatively easy, although it can never be complete since global bioethics is constantly changing with new issues and topics arising all the time. An example is the COVID-19 pandemic that was unknown when we started to work on this book. The selection of topics was based on the systematic perusal of current literature relating to global bioethics. Some topics are familiar to bioethical discourse and included in other dictionaries. However, the emphasis will be on their relevance to global bioethical discourse. Many other topics are new. The processes of globalization have generated a range of new challenges facing bioethics. Furthermore, medical innovation and new biotechnologies pose new moral questions that need to be addressed often in a global context.

The *Dictionary of Global Bioethics* has been designed to be a helpful resource, especially for students and young scholars. It offers a broad introduction to global bioethics and gives readers a quick feel for the main concepts and relevant knowledge. It should also be a helpful resource for different academic scholars and professionals (healthcare professionals, researchers, educators, and lawyers) who, armed with deep knowledge and vast experience in their own fields of expertise, can benefit from a fast and straightforward look at other associated fields. Anyone wanting to know more about global developments in bioethics will hopefully find this book a useful resource.

Since this is the first dictionary covering the rapidly evolving field of global bioethics we realize this edition will need constant improvement and continuous updating. Any feedback on the overall composition or on particular entries would therefore be appreciated. Suggestions for future entries would also be welcome.

We would like to thank Floor Oosting (Springer/Nature) for her encouragement and support in developing the dictionary. We hope the book will be of use to those interested in global bioethics and will inspire them to explore the new ethical panoramas that continuously present themselves across the globe.

Pittsburgh, USA — Henk ten Have
Ponta Delgada, Portugal — Maria do Céu Patrão Neves

Contents

Persons

Topics

About the Authors

Henk ten Have studied medicine and philosophy at Leiden University (the Netherlands) (MD 1976; Ph.D. 1983). He worked as a researcher in the Pathology Laboratory, University of Leiden; as a practicing physician in the Municipal Health Services of Rotterdam before being appointed as Professor of Philosophy in the Faculty of Medicine and Faculty of Health Sciences, University of Limburg, Maastricht; and subsequently as Professor of Medical Ethics and Director of the Department of Ethics, Philosophy and History of Medicine in the University Medical Centre Nijmegen (the Netherlands). In 2003 he joined UNESCO as Director of the Division of Ethics of Science and Technology. From 2010 until 2019 he was Director of the Center for Healthcare Ethics at Duquesne University in Pittsburgh (United States). In July 2019 he retired and was made Emeritus Professor at Duquesne University. His latest books are *Global Bioethics: An Introduction* (2016), *Vulnerability: Challenging Bioethics* (2016), and *Wounded Planet: How Declining Biodiversity Endangers Health and How Bioethics Can Help* (2019). He is editor of the *Encyclopedia of Global Bioethics* (3 volumes, 2016) and (with Bert Gordijn) editor of the *Handbook of Global Bioethics* (4 volumes, 2014).

Maria do Céu Patrão Neves studied philosophy at Lisbon University (Portugal) and Louvain-la-Neuve (Belgium) (Ph.D. 1991) and bioethics at The Kennedy Center, Georgetown University, Washington DC, where she was a visiting scholar. She is currently Full Professor of Ethics at the University of the Azores (Portugal). She has taught applied ethics and biomedical ethics in several Portuguese universities. She was consultant on Ethics of Life for the President of the Portuguese Republic, a member of the National Ethics Committee, and sat on the Board of Directors of the International Association of Bioethics. She is a member of several ethics committees and advisory boards, a member of UNESCO's Global Ethics Observatory (a system of databases relating to the ethics of science and technology), and an expert on ethics for the European Commission. Her latest book is *TheBrave World of Bioethics: The Origin of Bioethics in Portugal through Its Pioneers* (2016). She is editor of *Applied Ethics* (12 volumes, 2017–2018) and *Ethics, Science, and Society: Challenges for BioPolitics* (2019). She is coordinator of the project "Biomedical Ethics and Regulatory Capacity Building Partnership for Portuguese Speaking African Countries (BERC-Luso)" financed by the European Development Clinical Trials Partnership/European Commission (2018–2021).

Part I

Organizations

ASBH

The American Society of Bioethics and Humanities (ASBH) was established in 1998 through a merger of three existing organizations in the United States: the Society for Health and Human Values (SHHV), the Society for Bioethics Consultation, and the American Association of Bioethics. The oldest organization in the field of what later became bioethics was the SHHV, which was founded in 1969 as a professional association of persons committed to human values in medicine. It undertook several initiatives, notably in the field of medical ethics education.

The goals of the ASBH are to promote the exchange of ideas, to foster multidisciplinary, interdisciplinary, and interprofessional scholarship, research, teaching, policy development, professional development, and collegiality among people engaged in clinical and academic bioethics and the medical humanities. Such people engage in the following activities: encouraging consideration of issues related to values with regard to health services; the education of healthcare professionals; conducting educational meetings relative to the issues in healthcare ethics; encouraging research related to the ethical issues at hand; and contributing to public discussion of these endeavors and interests including how they relate to public policy (http://asbh.org/). The governing body of the ASBH is the Board of Directors. This board consists of a five-member executive committee, nine directors at large, and one student director. The ASHB meets annually in a different US city every October. The society also has several committees. Each of these consist of members with similar interests (e.g., on public health or rural bioethics) who meet and communicate with one another throughout the year. Another committee is the Clinical Ethics Consultation Affairs Committee. This committee supports the goals of healthcare ethics consultation (including the certification of consultants) and is currently developing standards of accreditation for clinical ethics programs. The ASBH consists of approximately 1,800 members who are affiliated with a range of disciplines such as medicine, nursing, social work, theology, philosophy, education, medical humanities, and research studies. Many of these members work in a variety of healthcare settings such as hospitals, government agencies, and universities.

H. ten Have and M. do C. Patrão Neves, *Dictionary of Global Bioethics*,
https://doi.org/10.1007/978-3-030-54161-3_1

CIOMS

The Council for International Organizations of Medical Science (CIOMS) was established in 1949 in Brussels by the WHO and UNESCO (https://cioms.ch). The two founding organizations were concerned with facilitating the exchange of views and scientific information in the medical sciences. They achieved this goal by securing continuity and coordination between international organizations of medical sciences and by furnishing them with material and other forms of aid to keep them fully updated about their mission. The exchange of information and the provision of substantial and financial assistance to conferences and their participants helped them to achieve this endeavor. CIOMS is an international, non-governmental, and non-profit organization that brings together member organizations from the biomedical scientific community such as medical research councils and academies of sciences. The mission of CIOMS is to advance public health through guidance on health research including ethics, medical product development, and safety. It is especially interested in promoting international activities and thus has taken a global perspective from its inception. Initially focused on coordinating international medical conferences CIOMS became more engaged with health policy, ethics, and human values, especially after the adoption by the World Health Assembly in 1977 of the goal of health for all. An early area of interest was medical research and clinical trials. CIOMS has emerged as a leading agency in exploring and clarifying the ethical issues involved in the use of human subjects in drug and vaccine research, especially for developing countries. Under the leadership of its Secretary-General Zbigniew Bankowski it organized international dialogue and issued ethical guidelines. In 1989 CIOMS published the *International Guiding Principles for Biomedical Research Involving Animals*. In 1993 it published the *International Ethical Guidelines for Biomedical Research Involving Human Subjects*. These guidelines were widely used in developing countries. CIOMS was also one of the first to articulate the role of vulnerability in medical research. The guidelines were revised and updated in 2002 and 2016. CIOMS was also one of the first international organizations to draw attention to global bioethics. Following a conference in Mexico in 1994 it adopted the Declaration of Ixtapa that emphasized a global agenda for bioethics. The governing body of CIOMS is the General Assembly who represent its international and national membership. The Executive Committee has 16 members including a permanent secretariat located in Geneva and housed in offices made available by the WHO.

H. ten Have and M. do C. Patrão Neves, *Dictionary of Global Bioethics*,
https://doi.org/10.1007/978-3-030-54161-3_2

Council of Europe/CoE (*See also* Oviedo Convention)

The Council of Europe (CoE) is the oldest European organization still active and the continent's leading human rights organization (https://www.coe.int/en/web/portal). It was founded in the aftermath of the Second World War (1949) in Strasbourg (France) by 8 countries to promote human rights in Europe. Today it comprises 47 member states and 6 observer members. Before becoming members of the CoE all countries have to sign the European Convention on Human Rights, a treaty designed to protect human rights, democracy, and the rule of law. The CoE has two main bodies: the Committee of Ministers, composed of the Ministers for Foreign Affairs of the member states, is the council's decision-making body; and the Parliamentary Assembly, composed of members of the national parliaments of each member state, seeking to spread human rights and democratic ideals in its neighborhood. The European Court of Human Rights and the European Directorate for the Quality of Medicines are two other very important bodies both aiming to promote common standards, charters, and conventions streamlining cooperation among European countries. Regarding bioethics the CoE has long been concerned about important human rights issues raised by the application of biology and medicine. Its first recommendations on the issue date from 1976. In 1985 the Resolution "On human rights and scientific progress in the field of biology, medicine and biochemistry" recommended that council intensify its work in relation to these problems from the standpoint of human rights and consider practices and legislation from both national and international contexts. The CoE was the first international institution to acknowledge the need for an international body to address bioethical issues. The Committee of Experts on Bioethics/CAHBI, an ad hoc pluri-disciplinary body, was set up in 1985 under the direct authority of the Committee of Ministers to intensify work on the impact of the progress of biomedical sciences on human beings, later to be recognized as the fourth generation of human rights (i.e., biological rights) or rights related to the application of biology and medicine to human beings. In 1992 the CAHBI became permanent and its name changed to Steering Committee on Bioethics/CDBI. In 2012, following the reorganization of intergovernmental bodies at the CoE, the Committee on Bioethics/DH-BIO took over the responsibilities of the CDBI. The main concerns of the bioethics committee prevailed such as guaranteeing a person's integrity, respecting human dignity, and drawing up common guidelines for all member states so that they can deal with the new situations created by rapid development of the life sciences. The most prominent achievement of the CoE's Ethics Committee was the Convention on Human Rights and Biomedicine (1997) and its additional protocols. The CoE also developed important resolutions on different topics such as Xenotransplantation (2003), Protection of the Human Rights and Dignity of Persons with Mental Disorders (2004), and Research on Biological Materials of Human Origin (2006, revised 2016). In 2014 the major topic was the total ban on any form of trading

H. ten Have and M. do C. Patrão Neves, *Dictionary of Global Bioethics*,
https://doi.org/10.1007/978-3-030-54161-3_3

in human organs. In 2015 the DH-BIO adopted a Statement on Genome Editing Technologies. Some other works are in progress such as predictivity, genetic testing and insurance, medical treatment in end-of-life situations, and emerging technologies. Although the CoE's influence is restricted to European member states, it has an impact worldwide because its documents and the rulings of the European Court establish soft law and powerful precedents.

EGE (*See also* International committees)

The European Group on Ethics in Science and New Technologies (EGE) is a neutral, independent, pluralist, and multidisciplinary body (gathering a broad range of professional competencies such as biology, genetics, medicine, pharmacology, agricultural sciences, information and communications technology, law, ethics, philosophy, and theology). It is composed of 15 experts tasked with advising the European Commission on ethical aspects of science and new technologies in connection with the preparation and implementation of community legislation or policies (https://ec.europa.eu/research/ege/index.cfm). In 1991 the President of the European Commission (EC) Jacques Delors established the Group of Advisers to the European Commission on the Ethical Implications of Biotechnology (GAEIB) to promote ethical discussions about the development of biotechnology and to reflect ethically on the decision-making process for community research and technological development policies. At that time it was a 6-member ad hoc committee appointed by the European Commission's president. In 1997 (under Jacques Santer's Presidency) the GAEIB was restructured: its remit widened to cover not only biotechnology but also communication and information technologies; it was opened up to broader consultation; it adopted the current designation; and the number of members increased (9 members in 1994; 12 members in 1998). Although the group could take the initiative and write Opinions, the European Parliament and Council of Ministers could now also request Opinions from the group. Members were chosen following a call for expression of interest (CEI). The group became a permanent body and its role was strengthened such that it could then consider the ethical aspects of science and technology in all activities of the European Commission, especially those connected with community legislation and policies. The EGE continued to strengthen and widen the scope of its ethical analyses by producing Opinions on patenting inventions involving human stem cells; clinical research in developing countries; genetic testing in the workplace; umbilical cord blood banking; and ICT implants in the human body (during its second mandate 2000–2005). It produced Opinions on the ethical review of human embryonic stem cell research under the EU Research Programme and on nanomedicine, agriculture technologies, and animal cloning for food supply (during its third mandate 2005–2010). It also produced Opinions on the ethics of information and communication technologies; the ethics of new health technologies and citizen participation; an ethical framework for assessing research; production and use of energy; and the ethics of security and surveillance technologies. It also made Statements on clinical trials, research integrity, and gene editing. Opinions are regularly preceded by roundtables gathering together representatives of European institutions, experts in specific fields, and stakeholders representing different interests. Until 2014 the EGE reported directly to the President of the European Commission. Since then under Jean-

H. ten Have and M. do C. Patrão Neves, *Dictionary of Global Bioethics*,
https://doi.org/10.1007/978-3-030-54161-3_4

Claude Juncker's presidency the group reports to the European Commissioner for Research, Science and Innovation and continues to provide high-quality advice for European policies and legislation on issues that have evident ethical, societal, and fundamental rights dimensions. The EGE's positions hold a lot of sway over other national and international bioethical institutions.

ESPMH

The European Society for Philosophy and Medicine (ESPMH) was established in 1987 by a group of philosophers, physicians, ethicists, and other professionals interested in ethics. They were especially motivated to reflect critically on the role of medicine and health in present-day societies. These concerns were not only focused on ethics, but on the intersection of healthcare and society. Examples of such concerns included the functioning of social security and welfare systems, or the impact of the demands of healthcare on the national budget of most Western countries. They were critical about the tendency in Western cultures to approach resulting problems by searching exclusively for technocratic and economic solutions neglecting philosophical analysis and ethical evaluations. At that time new teaching programs and research projects on meta-medical subjects were introduced in many European countries but often met with strong opposition, if not straightforward obstruction. Thus it was imperative to the founders that these efforts be strengthened through contacts and cooperation at the European level. The society also seeks to articulate particular European values and approaches that often motivate policies and practices (http://www.espmh.org/). The governing body of ESPMH is its board of directors that currently consists of 9 members. A General Assembly of the membership occurs at annual conferences. The ESPMH has approximately 330 members from 47 countries. The goals of ESPMH are to stimulate and promote development and methodology in the field of philosophy of medicine and healthcare, to be a center of contact for European scholars in this field, and to promote international contact between members. Two activities are undertaken to accomplish these goals. One is the annual conference that takes place in different cities in various countries—Lisbon (Portugal) in 2018; Oslo (Norway) in 2019. The other is publication of the European journal *Medicine, Health Care and Philosophy*, which was first produced in 1988 and consists primarily of original contributions. The specific goals of the journal are to focus on European perspectives and the interrelation between philosophy and medicine. Subscription to the journal is included in the membership fee of ESPMH. Although ESPMH is a European organization, its influence in the global bioethics debate is significant since it articulates perspectives that often differ from mainstream bioethics discourse.

H. ten Have and M. do C. Patrão Neves, *Dictionary of Global Bioethics*,
https://doi.org/10.1007/978-3-030-54161-3_5

FAO

The Food and Agriculture Organization (FAO) was founded in 1945 in Quebec (Canada). It has 194 member states with offices in 130 countries. Its headquarters were established in Rome (Italy) in 1951 (www.fao.org). The goals of the FAO are the eradication of hunger, food insecurity and malnutrition; the elimination of poverty; and the sustainable management and utilization of natural resources including land, water, air, climate, and genetic resources for the benefit of present and future generations. To accomplish these goals the FAO adopted a new strategic framework in 2013. Its priority is to help eliminate hunger. The world can produce enough food to feed everyone adequately, yet nearly 800 million people still suffer hunger. Another goal is to make agriculture, forestry, and fisheries more productive and sustainable. The FAO is engaged in activities in several areas. First, it serves as a knowledge network. Its staff (e.g., agronomists; specialists in forestry, fisheries, and livestock; nutritionists; social scientists; economists) collect, analyze, and disseminate data to support the transition to sustainable agriculture. It also supports members in the development of agricultural policy and legislation designed to alleviate hunger. Second, since it is a neutral forum the FAO provides a setting where rich and poor nations can come together to build common understanding. It also engages with the food industry and non-profits to provide support and services to farmers and to facilitate greater public and private investments in strengthening the food sector. Finally, the FAO develops mechanisms to monitor and warn about risks posed by a wide range of hazards as well as threats to agriculture, food, and nutrition. It informs countries about successful risk reduction measures that they can include in all policies related to agriculture. When need arises, it makes sure disaster response plans are coordinated at all levels. Over the years the FAO has issued many reports and developed standards. In cooperation with the WHO it published international food safety standards (*Codex Alimentarius*). In 2000 an ethics panel of 8 independent experts was established to examine ethical issues regarding food and agriculture, to advise the FAO on key ethical issues, and to raise the level of public awareness on ethical considerations. The panel published its first report in 2000 whose emphasis was on biotechnology and genetically modified organisms. The FAO also started a new series of ethics publications: *Ethical Issues in Food and Agriculture* (2001), *Genetically Modified Organisms, Consumers, Food Safety and the Environment* (2001), *The Ethics of Sustainable Agricultural Intensification* (2004), and *Ethical Issues in Fisheries* (2005). It was argued that ethical considerations are inherent in the program of the organization since at least five values are involved: the value of food (as essential to human survival), the value of enhanced well-being (advancing human dignity), the value of human health (unsafe and insufficient food deteriorates health), the value

H. ten Have and M. do C. Patrão Neves, *Dictionary of Global Bioethics*,
https://doi.org/10.1007/978-3-030-54161-3_6

of natural resources (biodiversity is the source of food), and the value of nature (nature itself must be valued). Nevertheless, there has been no sign of activity from its ethics panel since 2006. In 2009 the FAO established an ethics office with a primary focus on integrity and an ethics code of conduct. In 2012 it founded the Ethics Committee whose mandate was to operate as an advisory panel on all matters pertaining to ethics within the organization. However, ethics was deemed to be related to compliance and integrity issues, and the mandate expired in 2016.

HUGO

At the first meeting on genome mapping and sequencing that took place at Cold Spring Harbor Laboratory, New York (USA) in 1988 the initiative was taken to establish the Human Genome Organization (HUGO). It was established as an international coordinating scientific body in the rapidly developing field of genomics as a kind of United Nations for the human genome. HUGO was founded later in 1988 in Montreux (Switzerland) and its head office is located in Seoul (South Korea) (http://www.hugo-international.org/). Its mission is to bring the benefits of genomic sciences to humanity by promoting fundamental genomic research within nations and throughout the world. It fosters scientific exchange in genomics with a particular emphasis on countries experiencing scientific growth and supports discourse in the ethics of genetics and genomics with a global perspective. For the latter objective HUGO established the Committee on Ethics, Law and Society (CELS). The purpose of this committee includes the promotion of discussion and understanding of social, legal, and ethical issues as they relate to the conduct of and the use of knowledge derived from human genome research. It acts as an interface between the scientific community, policy-makers, educators, and the public. It fosters greater public understanding of human variation and complexity and collaborates with other international bodies in genetics, health, and society with the goal of disseminating information. It deliberates about policy issues to provide advice to the HUGO Council and to issue Statements where appropriate. Since 1995 it has published 8 Statements (i.e., on cloning, 1999; benefit sharing, 2000; gene therapy research, 2001; human genomic databases, 2002; stem cells, 2004; and patentability of genes, 2013). HUGO reports annually on its activities to the HUGO Council (a committee with 9 members). HUGO also acts on any matter related to genomics and genomics research as needed or requested.

H. ten Have and M. do C. Patrão Neves, *Dictionary of Global Bioethics*,
https://doi.org/10.1007/978-3-030-54161-3_7

IAB (*See also* SIBI)

The International Association of Bioethics (IAB) was established in 1992 at the initiative of three philosophers—Australians Peter Singer and Helga Kuhse and the American Dan Wikler. It gathers together a membership from various disciplines made up mainly of philosophers, physicians, and lawyers from 34 countries sharing an interest in bioethics. Its main goal is to link those working in bioethics all over the world facilitating the exchange of information to promote not only an interdisciplinary or transdisciplinary dialogue, but also a free and open cross-cultural discussion smoothing the way for international contact (http://www.bioethics-international.org/). The association specifies its educational and scientific objectives as organizing and promoting periodic international conferences on bioethics and encouraging the development of research and teaching in bioethics. IAB has no physical head office. IAB is managed by a board of directors elected by all members following a proportional representation system. It has encouraged the formation of networks dedicated to specific fields within bioethics such as arts and bioethics; bioethics education; environmental bioethics; feminist approaches to bioethics, genetics and bioethics; international bioethics for philosophy and bioethics; public health ethics; and an Ibero-American network. The association is the organizer of the World Congress of Bioethics that has taken place every 2 years since 1992 in a different continent in cooperation with a local institution focusing on a variety of topics such as "Global Bioethics: Dream or Nightmare" (1998), "Bioethics and Public Policy in the Next Millennium" (2000), "Power and Injustice" (2002), "Bioethics in a Globalized World: Creating Space for Flourishing Human Relationships" (2010), "Contextual, Social, Critical: How We Ought to Think about the Future of Bioethics" (2012), "Individuals, Public Interests and Public Goods: What Is the Contribution of Bioethics?" (2016) and "Health for All in an Unequal World: Obligations of Global Bioethics" (2018). In 1997 a group of scholars criticizing what they considered to be the Anglo-American orientation of the IAB decided to create another global bioethics organization that they called the International Society of Bioethics (Sociedad Internacional de Bioética/SIBI). *Bioethics* is the official journal of IAB and *Developing World Bioethics* is its companion journal.

H. ten Have and M. do C. Patrão Neves, *Dictionary of Global Bioethics*,
https://doi.org/10.1007/978-3-030-54161-3_8

IAEE

The International Association for Education in Ethics (IAEE) was established as a result of an initiative by UNESCO in March 2010 to bring together a group of experts to act as a preparatory group for the establishment of such an organization. The goal of this initiative was to promote the teaching of ethics in various areas of science and technology and to enhance exchange of experiences among experts from different parts of the world. The association was officially incorporated as a non-profit organization in 2011. Its first General Assembly took place in 2012 in the United States in Pittsburgh, Pennsylvania (https://www.ethicsassociation.org). The goals of the IAEE are to enhance and expand the teaching of ethics at national, regional, and international levels; to exchange and analyze experiences with the teaching of ethics in various educational settings; to promote the development of knowledge and methods of ethics education; and to function as a global center of contact for experts in this field; and to promote contacts between members from countries around the world. These goals have been achieved by organizing international conferences and other scholarly meetings (with a maximum interval of 2 years), promoting the establishment of regional and national chapters, publishing and disseminating reports of its scholarly meetings and other suitable material, cooperating and collaborating with other organizations, making use of other valid means at the association's disposal, and maintaining close cooperation with UNESCO. The governing body of the IAEE is its board of directors that currently consists of 9 members. A General Assembly takes place during international conferences. The IAEE has approximately 100 members from 35 countries. The IAEE organized international conferences on education and ethics in Pittsburgh (USA) in 2012, Ankara (Turkey) in 2014, Curitiba (Brazil) in 2015, Logroño (Spain) in 2016, Mangalore (India) in 2017, Stellenbosch (South Africa) in 2018, and Porto (Portugal) in 2019. The IAEE is also associated with the *International Journal for Ethics Education* that was launched in 2016. The aim of the journal is to provide a global platform for the exchange of research data, theories, experiences, reports, and opinions on ethics education in a broad range of areas of applied ethics. The journal is particularly interested in contributions about teaching programs in developing countries giving detailed information concerning the content of teaching programs and the objectives of the programs, teaching methods, and tactics needed to evaluate the impact of ethics education. It also addresses general issues regarding ethics education such as policy, implementation, international trends, and philosophy of education. The scope of the journal is broad. It covers education in the major areas of applied ethics including bioethics, business

H. ten Have and M. do C. Patrão Neves, *Dictionary of Global Bioethics*,
https://doi.org/10.1007/978-3-030-54161-3_9

ethics, engineering ethics, environmental ethics, law and ethics, medical ethics, nursing ethics, science ethics, and social sciences and ethics. The journal also provides a forum for interdisciplinary studies from diverse cultural and religious contexts. Subscription to the electronic version of the new journal is included in the IAEE membership fee.

IBC

The International Bioethics Committee (IBC) was established in 1993 by Federico Mayor, the Director-General of UNESCO (http://www.unesco.org/new/en/social-and-human-sciences/themes/bioethics/international-bioethics-committee/). It is a body of 36 independent experts appointed by the Director-General to provide him or her with the best possible advice. The experts represent a variety of disciplines including genetics, medicine, law, philosophy, ethics, history, and social sciences. The committee is enriched by individuals who come from different regions and who bring with them a variety of backgrounds—professional, cultural, and moral. They do not represent their member states and are not expected to follow any instructions from their governments. They represent themselves and are members of the committee because of their scientific expertise and qualifications. The main mission of the IBC is to provide recommendations to the Director-General that reflect the best possible expertise in the scientific and ethical domain. The committee reports directly to the Director-General. The IBC meets at least once a year. It can also create subsidiary bodies such as working groups tasked with elaborating particular issues. In the past the IBC has produced a series of reports covering issues such as genetic screening and testing (1994), the use of embryonic stem cells in therapeutic research (2001), and pre-implantation genetic diagnosis and germline interventions (2003). The IBC played a major role in the development of UNESCO's Universal Declaration on Bioethics and Human Rights. In 2001, when the General Conference of UNESCO invited the Director-General to undertake studies to explore the possibility of elaborating universal norms in bioethics, the IBC was tasked with making a feasibility study. In 2003 the General Conference gave the Director-General a mandate to develop a declaration on universal norms in bioethics. The IBC was once again entrusted with drafting the text. The IBC set up a drafting group, organized a survey, engaged in consultations with various stakeholders, and organized hearings with other intergovernmental (e.g., WHO and FAO) and international organizations (e.g., HUGO and WMA) and national bioethics committees. It also initiated a series of rotating conferences in 10 countries to discuss drafts of the declaration. The final IBC draft text was the basis for negotiations among member states that ultimately adopted the text in 2005.

H. ten Have and M. do C. Patrão Neves, *Dictionary of Global Bioethics*,
https://doi.org/10.1007/978-3-030-54161-3_10

MSF

Médecins Sans Frontières (MSF—Doctors Without Borders) was founded in 1971. A group of young physicians from France volunteered to provide humanitarian assistance by working with the Red Cross in the civil war in Biafra (Southern Nigeria). Shocked by the cruelties of the war they wanted to speak up on behalf of the victims. Since the Red Cross maintained the principle of neutrality they accused the organization of complicity with war criminals. The young doctors established a new organization based on the belief that all people have the right to medical care regardless of gender, race, religion, creed, or political affiliation and that the needs of these people outweigh respect for national boundaries (http://www.doctorswithoutborders.org). MSF is an independent volunteer organization consisting mainly of doctors and other healthcare workers. Since 1980 it has opened offices in 28 countries and employs more than 30,000 people globally. Since its founding MSF has treated over 100 million patients—with 8.25 million outpatient consultations carried out in 2014 alone. Its charter advocates four principles: assistance to populations in distress, to victims of natural disasters, to victims of man-made disasters, and to victims of armed conflict. They do so irrespective of race, religion, creed, or political convictions. MSF adheres to the principles of neutrality and impartiality and does so in the name of universal medical ethics and the right to humanitarian assistance. It claims the right to full and unhindered freedom in the exercise of its functions. It respects the professional code of ethics and maintains complete independence from all political, economic, or religious powers. Since its members are volunteers they understand the risks and dangers of the missions they carry out and make no claim for themselves for any form of compensation. However, the association has been known to pay compensation. MSF emphasizes that its actions are guided by the rules of medical ethics, particularly the duty to provide care without causing harm to individuals or groups. MSF workers respect patient autonomy, confidentiality, and their right to informed consent. They treat patients with dignity and with respect for their cultural and religious beliefs. MSF furthermore underlines that impartiality and neutrality are not synonymous with silence. They will bear witness if they notice violence and atrocities. They are known to speak out publicly on such matters. The organization seeks to bring attention to extreme need and unacceptable suffering when access to lifesaving medical care is hindered, when medical facilities come under threat, when crises are neglected, or when the provision of aid is inadequate or abused. For their humanitarian work across the globe MSF was awarded the Nobel Peace Prize in 1999.

H. ten Have and M. do C. Patrão Neves, *Dictionary of Global Bioethics*,
https://doi.org/10.1007/978-3-030-54161-3_11

SIBI

The International Society of Bioethics (Sociedad Internacional de Bioética or SIBI) was founded in 1997 in Gijón (Spain) (http://www.sibi.org). The aim of SIBI is to promote, support, disseminate, and consolidate knowledge about bioethics to attain its full application in a number of fields such as the medical, biological, environmental, and food science and technology fields. Its mission is also to encourage legal and pedagogical developments regarding bioethics at the national and international level and to foster a legal, social, and ethical analysis of any problem posed by scientific advances, especially in medicine and biotechnology. One of its main tasks is to disseminate the Convention on Human Rights and Biomedicine—also known as the Oviedo Convention because it was opened for signatures in the city of Oviedo (Spain)—as issued by the Council of Europe. SIBI organizes congresses throughout the world. The global focus is evident in the themes of subsequent conferences. Examples of such a focus include themes related to water problems (2004), biological weapons (2005), hunger and poverty (2009), the environment and sustainable development (2011), and violence against women (2016). SIBI also publishes a journal (*Revista de la Sociedad Internacional de Bioética*) that has been published since 1998. It is published bi-annually and is available in both Spanish and English. The society furthermore awards the SIBI Prize for Bioethics. The bi-annual prize recognizes the most relevant work, publication, or teaching in the field of bioethics, or relevant work that has most contributed to building the language of bioethics. The SIBI Prize was first awarded in 2000 to Van Rensselaer Potter. In 2002 it was awarded to UNESCO and the Council of Europe. SIBI is legally, economically, and administratively supported by a foundation bearing the same name. The SIBI Foundation is managed by its board of trustees composed of the founders of the society including the town council of Gijón and the University of Oviedo.

H. ten Have and M. do C. Patrão Neves, *Dictionary of Global Bioethics*,
https://doi.org/10.1007/978-3-030-54161-3_12

UNESCO

The United Nations established the United Nations Educational, Scientific and Cultural Organization (UNESCO) in 1945 (http://en.unesco.org/). It is a special agency within the United Nations that is headquartered in Paris (France). Since its foundation it has worked in four areas: education, culture, science, and communication. Established shortly after World War II it aims to prevent future wars and to promote peace and security through international cooperation in these areas to increase respect for justice, human rights, fundamental freedoms, and the rule of law. The Preamble to its Constitution famously states that "since wars begin in the minds of men, it is in the minds of men that the defenses of peace must be constructed." Peace must be founded on the intellectual and moral solidarity of humanity. Julian Huxley, the first Director-General of UNESCO, pointed out that science should develop for the benefit of humanity implying a restatement of morality. This emphasis explains the involvement of UNESCO in ethics associated with science, culture, and education. With the emergence of the life sciences in the 1970s UNESCO began to organize activities in the field of bioethics. It established the International Bioethics Committee (IBC) in 1993 in connection with a work program and budget (with a separate administrative unit in the social sciences sector). The main objective of UNESCO in bioethics is the development of international normative standards. UNESCO has 193 member states and therefore is the only existing global platform for all nations to explore and discuss the values and principles that they share and to negotiate on normative instruments. So far the General Conference has adopted three normative instruments in the field of global bioethics: the Universal Declaration on the Human Genome and Human Rights (1997), the International Declaration on Human Genetic Data (2003), and the Universal Declaration on Bioethics and Human Rights (2005). Besides standard-setting UNESCO also functions as a laboratory of ideas, a clearinghouse, capacity builder, and catalyst for international cooperation. Since the adoption of the declarations emphasis has been put on their implementation and application. The bioethics program initiated specific activities in countries and regions such as programs to establish and train national bioethics committees, the Ethics Education Program (to train ethics teachers), and the Global Ethics Observatory (to collect data and information about ethics infrastructure, experts, teaching programs, and ethics resources in specific countries).

H. ten Have and M. do C. Patrão Neves, *Dictionary of Global Bioethics*,
https://doi.org/10.1007/978-3-030-54161-3_13

WHO

The United Nations established the World Health Organization (WHO) in Geneva (Switzerland) in 1948. It is a special agency within the United Nations that is focused on international public health (www.who.int). The objective of the WHO is the attainment by all people of the highest possible level of health. According to its constitution the organization will do this through acting as the directing and coordinating authority on international health work, international collaboration, assistance to governments in strengthening health services, providing technical assistance, and in emergencies necessary aid upon the request or with the acceptance of governments. The WHO also seeks to stimulate and advance work to eradicate epidemic, endemic, and other diseases. The WHO Constitution also formulates basic principles. One of the best known is that health is considered a state of complete physical, mental, and social well-being and not merely the absence of disease or infirmity. Another principle is that the enjoyment of the highest attainable standard of health is one of the fundamental rights of every human being regardless of race, religion, political belief, and economic or social condition. It articulates the importance of social justice underlining that unequal development in different countries in the promotion of health and control of diseases, especially communicable disease, is a common danger. The WHO has had its successes such as the global eradication of smallpox, control of SARS, and ongoing campaigns against polio and other diseases. It has also addressed the broader determinants of health with campaigns on essential drugs, tobacco control, and diet and nutrition. However, one criticism of the WHO relates to its lack of global leadership, especially in connection with the Ebola epidemic (2013–2016). In 2002 the WHO established an ethics team in what is now called the Global Health Ethics Unit. The aim is to provide a focal point for the examination of ethical issues throughout the organization. The unit also supports member states in addressing ethical issues. The unit has several functions foremost among which is developing normative guidance such as issuing guidelines for patient safety, epidemic response, and vaccine research (e.g., *WHO Guidelines on Ethical Issues in Public Health Surveillance*, 2017). It also produces ethics tools and resources (e.g., *Global Health Ethics: Key Issues*, 2015) and engages in capacity building (training on public health and research ethics). The unit furthermore reviews WHO research ethics (hosting the secretariat of the WHO Research Ethics Review Committee) and partners with international leadership (hosting the secretariat of the biennial Global Summit of National Ethics/Bioethics Committees; and hosting the Global Network of Collaborating Centers).

H. ten Have and M. do C. Patrão Neves, *Dictionary of Global Bioethics*,
https://doi.org/10.1007/978-3-030-54161-3_14

WIPO

The World Intellectual Property Organization (WIPO) was established in 1967 and is headquartered in Geneva (Switzerland). WIPO has 189 member states (www.wipo.int). The objectives of WIPO are to promote the protection of intellectual property throughout the world through cooperation among states and where appropriate in collaboration with any other international organization and to ensure administrative cooperation among the relevant bodies. "Intellectual property" is defined as rights related to literary, artistic, and scientific works; performances of performing artists; phonograms and broadcasts; inventions in all fields of human endeavor; scientific discoveries; industrial designs, trademarks, service marks, commercial names, and designations; protection against unfair competition; and all other rights resulting from intellectual activity in the industrial, scientific, literary, or artistic fields. WIPO defines its mission as leading the development of a balanced and effective international intellectual property system that enables innovation and creativity for the benefit of all.

WIPO established its Ethics Office in 2010. This reports annually to the WIPO Coordination Committee. The office undertakes promotional activities including ethics and integrity training. This training is mandatory for all staff at every level of the organization. It focuses on ethical conduct and the culture of the organization. It also developed the WIPO Code of Ethics. The Chief Ethics Officer provides confidential advice to senior management, managers, and all staff members. Another role of the Ethics Office is norm-setting and policy development. It recently issued a new financial disclosure policy and a whistle-blower protection policy. All ethical activities have an internal function as a mechanism to safeguard integrity. WIPO does not engage in broader ethical debates concerning intellectual property, patenting, or indigenous knowledge systems. While there are many debates in global bioethics about the status of knowledge and information (as private or common goods or as commodities) and about the balance between protecting intellectual property and public health, WIPO focuses on strict IPR regulations that are primarily important for economic growth.

H. ten Have and M. do C. Patrão Neves, *Dictionary of Global Bioethics*,
https://doi.org/10.1007/978-3-030-54161-3_15

WMA

Established in Paris in 1947 the World Medical Association (WMA) is an association representing 112 national medical societies (www.wma.net). First located in New York the WMA eventually moved its headquarters to Ferney-Voltaire (France) in 1974. The official languages of the WMA are English, French, and Spanish. The organization was created out of concerns over the medical profession during the Second World War. Its mission is to ensure the independence of physicians and to work for the highest possible standards of ethical behavior and care by physicians. The WMA endeavors to provide a forum for its member associations to communicate freely, to cooperate actively, to achieve consensus on high standards of medical ethics and professional competence, and to promote the professional freedom of physicians worldwide. The WMA developed to become a platform for global consensus on medical ethics. It first prepared its Charter of Medicine to serve as an oath or promise for doctors upon receiving their medical degrees. This text became known as the *Declaration of Geneva* after its adoption in 1948. A subsequent report on war crimes and medicine led to the development of the International Code of Medical Ethics that was adopted in 1949. The WMA established its permanent Committee on Medical Ethics in 1952. The organization is active in the areas of medical ethics, human rights, public health, health systems, and education. To promote the highest possible standards of ethical behavior and care by physicians the WMA adopted global policy statements on a range of ethical issues related to medical professionalism, patient care, research on human subjects, and public health. The best known is the *Declaration of Helsinki* on medical research involving human subjects that was first adopted in 1964 and amended seven times since. Most recently the WMA issued the *Declaration of Taipei* on ethical considerations regarding health databases and biobanks (2016). The WMA also serves as a clearinghouse of ethics information resources for its members. To achieve this goal the WMA cooperates with academic institutions, global organizations, and individual experts in the field of medical ethics. In 2005 it launched its *Medical Ethics Manual*, a brief introduction to medical ethics that is used in medical education programs. It also developed a medical ethics course containing 8 hours of postgraduate continuing education. Training modules were developed for health research ethics.

H. ten Have and M. do C. Patrão Neves, *Dictionary of Global Bioethics*,
https://doi.org/10.1007/978-3-030-54161-3_16

WTO

The World Trade Organization (WTO) was established in 1995 and was the outcome of the Uruguay Development Round of trade negotiations replacing the General Agreement on Tariffs and Trade (GATT). The WTO is based in Geneva (Switzerland) and consists of 164 member countries (www.wto.org). The WTO is the global international organization dealing with the rules of trade between nations. It is the only forum that can create binding and enforceable international trade rules. All major decisions are made by the membership as a whole with each member having an equal voice. In practice, however, emphasis is on consensus decision-making. The highest decision-making body is the Ministerial Conference that meets at least once every 2 years. The goal of the WTO is to ensure that trade flows as smoothly, predictably, and freely as possible. The functions of the WTO are the administration of WTO trade agreements, acting as a forum for trade negotiations, the handling of trade disputes, monitoring national trade policies, and providing technical assistance and training for developing countries. The WTO is the only UN body that can strictly enforce its rules through a dispute settlement mechanism. Countries bring disputes to the WTO if they think their rights under agreements are being infringed. Judgments by specially appointed independent experts are binding. Although developing countries are a majority in the WTO, they have found it difficult to influence decision-making that is often determined by informal processes. They frequently have no delegation in Geneva. They often lack expertise and matters are made worse by the WTO secretariat being weak and failing to provide much assistance. In practice, the most powerful members are those best equipped and able to negotiate deals to their advantage. The decisions of the WTO are primarily based on technical and economic matters. Global economic governance as demonstrated by the WTO, however, has become increasingly criticized. WTO Ministerial Conferences have been the site of significant global contests (most notably in Seattle, 1999). It is argued that trade should benefit people—not just markets and economies. Economic activities must be pursued within a broader context of human development by promoting human rights, overarching policies, and targets aimed at eliminating poverty. Therefore international trade policy should take into consideration public health, access to medication, food security, human rights, and environmental impacts. The WTO is receptive to criticism and responded to it by starting the Doha Development Round of negotiations in 2001. Its focus was on the developing countries and making globalization more inclusive. The Doha Ministerial Conference produced two declarations that address the needs of poorer

H. ten Have and M. do C. Patrão Neves, *Dictionary of Global Bioethics*,
https://doi.org/10.1007/978-3-030-54161-3_17

countries and their peoples. The future of these negotiations nonetheless is uncertain. The agreement reached in 2003 is important for global bioethics. This agreement addressed issues related to access to patented medicines ultimately giving priority to public health considerations over protection of patents under certain conditions (*see* Doha Declaration).

Part II
Documents

CBD

The Convention on Biological Diversity (CBD) was adopted during the Earth Summit in 1992 in Rio de Janeiro and signed by 150 governments. It entered into force in December 1993. As a convention it is an internationally binding treaty. The CBD affirms that the conservation of biological diversity is a common concern of humankind. Biological diversity is defined as "the variability among living organisms from all sources." It includes species, ecosystems, and genetic resources. Concerned about the increasing loss of biodiversity across the globe and the negative impacts such degradation has on human well-being the CBD emphasizes the need for conservation and sustainable use of biological diversity. This is demonstrated in its three objectives: conservation of biological diversity, sustainable use of its components, and fair and equitable sharing of benefits arising from the utilization of genetic resources. Article 3 formulates the basic principle of the CBD: states have the sovereign right to exploit their resources pursuant to their own environmental policies. Rather than postulating that biodiversity is the common heritage of humankind, it is regarded as the property of states. This position came about as a result of pressure applied by developing countries that had experienced this notion of common heritage. Such a notion opened up their biodiversity resources to exploitation by (usually) foreign agents. Arguing that these resources were common property (i.e., in fact nobody's property) such states felt they could be appropriated by others. Developing countries regarded this as biopiracy. A good example involved pharmaceutical companies removing natural products from developing countries and then transforming them into drugs that are subsequently patented and unaffordable for people from the originating developing countries. To limit these exploitative practices the CBD took the approach of allocating ownership to states. If states are the legitimate owners of biological diversity, then this would allow states to control commercial transactions. Biodiversity is primarily regarded as a resource that can be exchanged, traded, and commercialized. The CBD assumes that this compromise between trade and protection will promote conservation and sustainable use. Subsequent meetings of parties have regulated access to genetic resources and traditional knowledge. The Nagoya Protocol, adopted in 2010 and entered into force in 2014, stipulates rules for the fair and equitable sharing of benefits from the use of genetic resources. Although there have been successes (e.g., the growth of protected areas), the targets to reduce biodiversity are often not met. The overwhelming majority of countries do not meet the self-identified objectives of halting habitat loss, saving endangered species, reducing pollution, and making fishing and agriculture more sustainable. Loss of biodiversity is increasing—not diminishing.

H. ten Have and M. do C. Patrão Neves, *Dictionary of Global Bioethics*,
https://doi.org/10.1007/978-3-030-54161-3_18

DoH

The World Medical Association developed the Declaration of Helsinki (DoH) (*see* WMA). The Nuremberg Code was the first international framework of ethical principles for medical research (*see* Nuremberg Code). Shocked by the atrocities and human rights violations in Nazi Germany the code formulated such basic principles as the necessity for consent to be voluntary, competent, informed, and understanding. However, the code was often regarded as a marginal document applicable only to barbarian physicians such as Nazi doctors. However, the thalidomide scandal in 1962 and increasing confusion between experimentation and therapy led to growing concern over the ethical aspects of medical research, especially clinical trials for new drugs. It was here that the WMA became concerned about the ethical framework of medicine and medical science. It had adopted the Declaration of Geneva in 1948 and the International Code of Medical Ethics in 1949 thus advancing basic ethical principles for healthcare. Since 1953 it has become increasingly involved in human experimentation. After lengthy debates the WMA adopted the Declaration of Helsinki in 1964. While the Nuremberg Code was primarily concerned with the rights of patients, the DoH focused on the conduct and responsibilities of the physician-researcher and the need to balance the production of therapeutic knowledge with the need to protect the health and interests of patients. The 1964 declaration also made a distinction between therapeutic and non-therapeutic research. It gave higher priority to the assessment of risk than to obtaining informed consent. The DoH was revised 11 years later. In the meantime shocking revelations of unethical research had been disseminated (e.g., the studies of Henry Beecher in 1966). There was a clear need to strengthen the declaration, especially regarding vulnerable subjects and populations. The revision placed more emphasis on the interests of subjects of research than on the duties of doctors. It also introduced the famous statement that concern for the interest of the human subject must always prevail over the interests of science and society. The declaration was revised again in 1983, 1989, and 1996. The fifth revision in 2000 created controversy. It dropped the distinction between therapeutic and non-therapeutic research. However, the most controversial issue was the use of placebos in clinical trials despite proven effective treatment existing. This issue raises the question of what the standard of care in resource-poor countries includes: the best treatment available or the best proven treatment? The revised text does not rule out placebo controls even when a proven method of treatment exists. At the same time it clearly states the principle of post-trial access in which research participants should continue to receive the medication when the trial is over and the results are beneficial. The 2000 revision of the DoH politicized the document. The FDA in the United States announced that it rejected the revised DoH and that it would no longer refer to it. The WMA undertook further revisions in 2008 and 2013. As to the use of placebos the latest

H. ten Have and M. do C. Patrão Neves, *Dictionary of Global Bioethics*,
https://doi.org/10.1007/978-3-030-54161-3_19

revision now formulates that the benefits, risks, burdens, and effectiveness of a new intervention must be tested against those of the best proven intervention(s). At the same time it mentions several notable exceptions. The statement on post-trial provisions in the revision has been weakened. It mentions that provisions for post-trial access should be made in advance of a clinical trial—not that they should be implemented.

Doha Declaration

The Doha Declaration was adopted in 2001 by members of the World Trade Organization (*see* WTO) as a result of negotiations about the Agreement on Trade-Related Aspects of Intellectual Property Rights (*see* TRIPS). The main criticism of TRIPS was that by exclusively prioritizing trade important public health and environmental concerns were neglected. Emphasis on the protection of property rights made access to medication unaffordable for many, particularly in the developing world. The Doha Declaration states that the TRIPS agreement does not and should not prevent members from taking measures to protect public health. It allows the use of compulsory licensing. Government authorities are thus able to license the use of a patented invention to a third party or government agency with the consent of the patent-holder. The declaration also affirms that parallel importation cannot be challenged under the WTO dispute settlement system. Parallel importation enables governments to import patent products for lower prices than those on the international market. Such imports take place without the consent of patent-holders and therefore are often challenged. In principle the Doha Declaration gives countries flexibility regarding patent rights to promote better access to medication. However, in practice its impact has been rather limited. Very few countries have taken up opportunities for compulsory licensing and parallel importation. Major countries such as the United States and international organizations such as the European Union now prefer bilateral and regional trade agreements circumventing WTO arrangements. In these agreements they can often pose more stringent requirements. Such requirements extend the lifetime of patents, relax environmental restrictions, and give multinational corporations the ability to sue governments. Negotiations over such agreements are secret in that there is no possibility of public debate and democratic input fueling fears that standards of public health and environmental protection are lowered when the focus is only on trade.

H. ten Have and M. do C. Patrão Neves, *Dictionary of Global Bioethics*,
https://doi.org/10.1007/978-3-030-54161-3_20

Earth Charter

In 2000 an international group of organizations and individuals adopted the Earth Charter at the headquarters of UNESCO in Paris. The charter is a declaration of principles for a just, sustainable, and peaceful global society. At the initiative of personalities such as Maurice Strong and Mikhail Gorbachev a worldwide consultation process was started to draft a text. The Earth Charter Commission approved the draft during a meeting held at the headquarters of UNESCO in March 2000 and officially launched the charter in June 2000 with a ceremony at the Peace Palace in The Hague (the Netherlands). The charter assumes that current trends are not sustainable. Patterns of production and consumption are causing environmental devastation, depletion of resources, and massive extinction of species. The benefits of development are not shared equitably with the gap between rich and poor widening. Injustice, poverty, ignorance, and violent conflict are widespread and the cause of great suffering. Emphasizing that the Earth is everybody's home the charter argues that human beings share responsibility for present and future generations. The charter identifies four fundamental principles: respect and care for the community of life; ecological integrity; social and economic justice; and democracy, non-violence, and peace. The initial statement of the first fundamental principle is to respect Earth and life in all its diversity. This implies that everybody recognize all beings are interdependent and every form of life has value regardless of its worth to human beings. It further implies that the inherent dignity of all human beings and the intellectual, artistic, ethical, and spiritual potential of humanity be affirmed. As an ethical framework the charter is tasked with guiding any transition to a sustainable future. It is the result of a global bottom-up approach driven by civil society to formulate a global ethics, in contrast to the Universal Declaration of Human Rights and the Universal Declaration on Bioethics and Human Rights. It is an aspirational document that can easily be neglected by governments and policy-makers since it is not the result of intergovernmental negotiations. On the other hand. it shows how people across the world can agree on several fundamental principles to address the environmental crisis.

H. ten Have and M. do C. Patrão Neves, *Dictionary of Global Bioethics*,
https://doi.org/10.1007/978-3-030-54161-3_21

Nuremberg Code

Following the end of the Nuremberg trials of Nazi doctors after the Second World War in 1947 judges adopted the Nuremberg Code. This so-called doctors' trial proceeded over the course of 2 years (1946–1947) and resulted in the condemnation of German physicians involved in experiments in concentration camps. Of the 16 found guilty 7 were hanged. The judges based their verdict on universal human rights principles. The final judgment presented the Nuremberg Code that comprised 10 ethical principles for medical research with human beings. The very first principle asserts the essential nature of voluntary consent of people taking part in the experiment. The second principle states that the experiment should be such as to yield fruitful results for the good of society, unprocurable by other methods or means of study, and not random and unnecessary in nature. The third principle affirms that the experiment should be so designed and based on the results of animal experimentation and a knowledge of the natural history of the disease or other problem under study and that the anticipated results will justify performance of the experiment. According to the fourth principle the experiment should be so conducted as to avoid all unnecessary physical and mental suffering and injury. The strength of the code lies in its emphasis on universal application of its principles. Furthermore, it strongly articulates the importance of informed consent. Nonetheless, since the code originated in an American military tribunal it did not gain general acceptance. Even within the United States the code was regarded as an aberration and not applicable to civilized physicians.

H. ten Have and M. do C. Patrão Neves, *Dictionary of Global Bioethics*,
https://doi.org/10.1007/978-3-030-54161-3_22

Oviedo Convention (*See also* Council of Europe)

The Convention for the Protection of Human Rights and Dignity of the Human Being with regard to the Application of Biology and Medicine: Convention on Human Rights and Biomedicine was prepared by the Committee of Experts on Bioethics/CAHBI and issued by the Council of Europe (CoE). It is known as the Oviedo Convention because it was opened for signature in Oviedo (Spain) in 1997. This legally binding international document in the field of bioethics was requested by the Steering Committee on Bioethics (CDBI). Its first draft was presented in 1992 and subsequently adopted by the Committee of Ministers in 1996. It entered into force in 1999 and was ratified by 29 of the 47 member states of the Council of Europe (2017). The Oviedo Convention aims to protect human dignity and human rights in the face of the growing number of new biomedical technologies and thus contributes to the fourth generation of human rights in the biomedical field (see: Human Rights). The Oviedo Convention defines a minimum common standard that should apply to all member states. The fundamental principle is that the interests of human beings must come before the interests of science or society (Article 2). The convention relates to those issues in the biomedical realm where it is possible to build consensus such as consent (free and fully informed consent as a requirement for all medical interventions and special provisions for those unable to give consent or in emergency situations—Chap. 2); rights to privacy and keeping personal information private (the right to know or not to know medical information—Chap. 3); matters relating to the human genome (modifying the human genome with no effect on descendants, genetic testing only for health-related purposes, and there should be no discrimination based on genetic characteristics—Chap. 4); scientific research (the importance of freedom of scientific research and the need to comply with rules set up by the convention to perform research on human beings; prohibiting the creation of embryos for scientific research—Chap. 5); organ transplantation (organ removal from living donors can only be performed for therapeutic reasons and without financial gain—Chaps. 6 and 7). Some of the Convention's consensual provisions have been elaborated and supplemented by additional protocols (Article 31) such as Prohibition of Cloning Human Beings (1998); Transplantation of Organs and Tissues of Human Origin (2002); Biomedical Research (2005); and Genetic Testing for Health Purposes (2008). A protocol on the Protection of Human Rights and Dignity of Persons with Mental Disorders is under elaboration. Legal questions concerning interpretation of the convention are addressed in the European Court of Human Rights (Article 27).

H. ten Have and M. do C. Patrão Neves, *Dictionary of Global Bioethics*,
https://doi.org/10.1007/978-3-030-54161-3_23

Rio Declaration on Environment and Development

The Rio Declaration adopted in 1992 at the United Nations Conference on Environment and Development is widely known as the Earth Summit. The conference in Rio de Janeiro was a follow-up to the United Nations Conference on the Human Environment that took place in Stockholm in 1972. The Earth Summit was a significant global event. It was the largest gathering of world leaders in history. Its purpose was to reconcile protection of the environment with economic development. The summit agreed on several legally binding agreements and opened them for signature such as the Convention on Biological Diversity, the Framework Convention on Climate Change, and the United Nations Convention to Combat Desertification. The conference also established the Commission for Sustainable Development. The Rio Declaration proclaimed 27 principles for sustainable development. The first principle emphasizes that human beings are at the center of concerns for sustainable development. They are entitled to a healthy and productive life in harmony with nature. The declaration also promotes the notion of future generations. Principle 3 states that the right to development must be fulfilled so as to equitably meet development and environmental needs of present and future generations. The basic precondition for sustainable development, however, is the eradication of poverty. This will entail international cooperation and true global partnerships to restore the health and integrity of the Earth's ecosystem (Principle 7) because without reducing and eliminating unsustainable patterns of consumption a higher quality of life for all and sustainable development cannot be achieved. Principle 10 emphasizes the importance of the participation of all all concerned citizens regarding environmental issues. Another important principle that is explicitly mentioned is precaution. Principle 15 advocates a precautionary approach: "Where there are threats of serious or irreversible damage, lack of full scientific certainty shall not be used as a reason for postponing cost-effective measures to prevent environmental degradation." The Rio Declaration is also vital because it stresses the crucial role of women and indigenous peoples in environmental management and development (Principles 20 and 22). The Rio Declaration has been central to the development of international environmental law. The declaration also formulated ethical principles that would play an important role in subsequent bioethical debates. However, progress has been slow, especially regarding climate change and biodiversity erosion. Since 1992 the momentum toward a sustainable future has been lost.

H. ten Have and M. do C. Patrão Neves, *Dictionary of Global Bioethics*,
https://doi.org/10.1007/978-3-030-54161-3_24

TRIPS

The Agreement on Trade-Related Aspects of Intellectual Property Rights (TRIPS) is an international legal agreement between members of the World Trade Organization (WTO) and was enacted in 1995 (https://www.wto.org/english/tratop_e/trips_e/intel2_e.htm). It covers all areas of intellectual property from copyright, trademarks, geographical indications, industrial designs, to patents. It sets the standards for protection of these property rights, identifies mechanisms of enforcement, and submits disputes to the WTO's dispute settlement procedures. Adoption of the TRIPS agreement during the Uruguay Development Round of trade negotiations imposed the intellectual property rights regime developed in the United States on the rest of the world. A powerful lobby group of American companies succeeded in convincing Western governments and pressuring developing countries into compliance. The agreement was severely criticized, particularly by developing countries. One reason is that it imposes a notion of property rights as private rights, while in many countries traditional knowledge is frequently regarded as community or group rights. Natural products that are the basis of new medications are patented by foreign companies despite having been used in indigenous healing systems for centuries. Another reason is that the agreement (as well as the WTO) often gives exclusive priority to trade while neglecting considerations of public health or environmental protection. Such neglect became apparent when new medication became available for HIV/AIDS. Patenting this medication created monopolies that resulted in high prices making it unaffordable for many populations in developing countries. Criticism of the narrow interpretation of the TRIPS agreement finally led to new negotiations that resulted in the Doha Declaration in 2001. It accepted that the agreement should not prevent countries from taking measures to protect public health. Access to medication should be compatible with the provisions of TRIPS. Whether this is the case in practice is the subject of ethical dispute. Although the flexibilities of TRIPS are difficult to implement for developing countries, developed countries nowadays prefer to make bilateral trade agreements with even more stringent regulations about property rights than TRIPS and the WTO.

H. ten Have and M. do C. Patrão Neves, *Dictionary of Global Bioethics*,
https://doi.org/10.1007/978-3-030-54161-3_25

UDBHR

The Universal Declaration on Bioethics and Human Rights (UDBHR) was unanimously adopted by the member states of UNESCO in October 2005. It is the first international document in the field of bioethics adopted by all governments and as such marks the emergence of global bioethics. Member states entrust the organization with the development of a broad bioethics framework since it succeeded earlier in drafting a document in the specific area of the human genome (*see* UDHGHR). Several important principles have therefore already been formulated and accepted by the international community. Although the declaration does not define bioethics, it does provide a description of its scope. It addresses ethical issues related to medicine, life sciences, and associated technologies as applied to human beings considering their social, legal, and environmental dimensions. The declaration has multiple aims the most important of which is to provide a universal framework of principles and procedures to guide states in the formulation of legislation, policies, and other instruments in the field of bioethics. At the same time, since the bioethical principles identified in the text are founded on human rights and fundamental freedoms, bioethics concerns every individual. Thus the declaration also aims to guide the actions of individuals, groups, communities, institutions, and corporations (public and private). The core of the declaration is its 15 principles. They express different obligations and responsibilities of moral agents concerning different categories of moral objects (moral patients). Although there is no hierarchy among the principles identified, they are arranged according to a gradual widening of the range of moral objects such as individual human beings (with principles of human dignity, benefits and harms, and autonomy), other human beings (with principles of consent, privacy, and equality), human communities (with principles of respect for cultural diversity), humankind as a whole (principles of solidarity, social responsibility, and sharing of benefits), and its environment (with principles of protecting future generations, the environment, the biosphere, and biodiversity). Principles in the declaration strike a balance between individualist and communitarian perspectives. They present a broad view of bioethics, as advocated by Potter, including social and environmental concerns. These principles are anchored in the rules governing human dignity, human rights, and fundamental freedoms. The declaration also refers to the application of principles. It calls for professionalism, honesty, integrity, and transparency in the decision-making process. It encourages the establishment of independent, multidisciplinary, and pluralist ethics committees by states. Finally, it emphasizes the need for states to foster bioethics education and training at all levels and underlines the importance of encouraging information and knowledge dissemination programs about bioethics.

H. ten Have and M. do C. Patrão Neves, *Dictionary of Global Bioethics*,
https://doi.org/10.1007/978-3-030-54161-3_26

UDHGHR

The Universal Declaration on the Human Genome and Human Rights (UDHGHR) was adopted by acclamation by the General Conference of UNESCO in November 1997. One year later it was endorsed by the General Assembly of the United Nations. It is the first normative instrument elaborated and adopted by UNESCO. It also initiated the ethics program of the organization. In June 1992 the Director-General of UNESCO decided to set up the International Bioethics Committee (*see* IBC). The task of the committee was to explore how to draft an international instrument to protect the human genome. The IBC started with preliminary studies in ethics and genetics. It took into account the need for a universal approach and the significance of cultural diversity. The committee developed an ethical framework that instigated intense debate and that finally resulted in the document adopted as the UDHGHR. The declaration strongly emphasizes international human rights discourse. The first articles relate to human dignity. They state that the human genome inspires the fundamental unity of all members of the human family and recognition of their inherent dignity and diversity. The declaration also articulates that the human genome is the heritage of humanity, at least in a symbolic sense. Furthermore, it states that everyone has a right for their dignity and their rights to be respected regardless of their genetic characteristics. Dignity makes it imperative not to reduce individuals to their genetic characteristics and to respect their uniqueness and diversity. Human dignity also implies that the human genome in its natural state shall not give rise to financial gains. Subsequently, the declaration specifies the rights of persons concerned. Free and informed consent is a prerequisite for research. Research protocols should be submitted for prior review. A rigorous assessment of potential risks and benefits should also be undertaken before research. Each individual has the right to decide whether or not to be informed of the results of genetic examination and the resulting consequences. No one shall be subjected to discrimination based on genetic characteristics that are intended to infringe or have the effect of infringing human rights, fundamental freedoms, and human dignity. The declaration also states that practices contrary to human dignity such as reproductive cloning of human beings shall not be permitted. As a global ethics document it stresses the need for international cooperation, sharing of benefits, freedom of research, and solidarity towards individuals, families and population groups who are particularly vulnerable to or affected by disease or disability of a genetic character.

H. ten Have and M. do C. Patrão Neves, *Dictionary of Global Bioethics*,
https://doi.org/10.1007/978-3-030-54161-3_27

UDHR

The Universal Declaration of Human Rights (UDHR) was adopted in December 1948 by the United Nations General Assembly. At that time the United Nations had 58 member states of which 48 members voted in favor of the declaration, 8 abstained, and 2 did not vote. The declaration is regarded as a milestone in the history of human rights. Although the Charter of the United Nations mentions respect for and observance of human rights, this was the first time since its establishment in 1945 that the international community specified which human rights were at stake. The atrocities committed in the Second World War reinforced the need to define human rights. A drafting committee chaired by Eleanor Roosevelt submitted its work in May 1948. In its preamble the declaration recognizes the inherent dignity and the equal and inalienable rights of all members of the human family. The first two articles establish the basic concepts of dignity, liberty, and equality. Article 1 states that all human beings are born free and equal in dignity and rights. They should act toward one another in a spirit of kinship. Article 2 reiterates that everyone is entitled to all the rights and freedoms outlined in the declaration without distinction of any kind such as race, color, sex, language, religion, political or other opinion, national or social origin, property, birth, or other status. Articles 3–11 establish other individual rights such as the right to life, liberty, and security of persons (Article 3) and the right to recognition as a person (Article 6). Slavery, torture, cruel treatment, arbitrary arrest and detention, and exile are firmly rejected. Articles 12–17 establish the rights of the individual toward the community such as freedom of movement, seeking asylum from persecution, the right to a nationality, the right to marry and found a family, and the right to own property. Article 18 declares that everyone has the right to freedom of thought, conscience, and religion. Article 19 affirms the right to freedom of opinion and expression; Article 20 the right to freedom of peaceful assembly and association; Article 21 the right to take part in the government of his or her country. Articles 22–27 articulate the economic, social, and cultural rights of individuals such as the right to social security (Article 22), the right to work (Article 23), the right to rest and leisure (Article 24), the right to a standard of living adequate for the health and well-being of individuals and their families—this right includes food, clothing, housing, medical care, and necessary social services (Article 25), the right to education (Article 26), the right to participate in the cultural life of the community, and the right to share in scientific advancement and its benefits (Article 27). Adoption of the declaration is celebrated each year on December 10, known as International Human Rights Day.

H. ten Have and M. do C. Patrão Neves, *Dictionary of Global Bioethics*,
https://doi.org/10.1007/978-3-030-54161-3_28

Part III
Persons

Bankowski

Zbigniew Bankowski (died 2010) was a medical doctor and experimental pathologist born in Poland. From 1965 to 1975 he served as a staff member of the WHO in Switzerland where he was responsible for research coordination and the program for research training and grants. Subsequently, he was WHO representative in Tunisia combining this post with a visiting professorship in histology at the Faculty of Medicine in Tunis. In 1975 he was elected Secretary-General of CIOMS (Council for International Organizations of Medical Sciences) where he served until 1999. In this position he made significant contributions to global bioethics. Bankowski took the initiative to reorient the activities of CIOMS to the social and ethical implications of scientific and technological aspects of biomedicine, whereas until that time the focus was primarily on biomedical science. In 1976 CIOMS appointed its standing Advisory Committee on Bioethics. The annual international conferences of the organization also explicitly addressed ethical concerns. Researchers, policy-makers, and ethics experts from around the world were brought together in roundtables (e.g., in Portugal, 1978 and the Philippines, 1981). This started an intensive global dialogue and gradually produced a consensus to develop guidelines that could be internationally recognized. Together with the WHO a joint project was launched to elaborate the principles of the Helsinki Declaration into a set of global guidelines. In these guidelines CIOMS was one of the first organizations to emphasize the concept of vulnerability as an important ethical notion at the global level. Bankowski edited and produced a series of books including *Poverty, Vulnerability, the Value of Human Life, and the Emergence of Bioethics* (1994), *Ethics and Health* (1995), and *Ethics and Human Values in Health Policy* (1996). These books disseminated bioethical discourse, especially in the developing world. His colleagues and friends created the Zbigniew Bankowski Lecture Fund to support an annual lecture at CIOMS on the ethical aspects of health policy.

H. ten Have and M. do C. Patrão Neves, *Dictionary of Global Bioethics*,
https://doi.org/10.1007/978-3-030-54161-3_29

Hellegers

Andre Hellegers (1926–1979) was the founder and first director of the Kennedy Institute of Ethics at Georgetown University in the United States. Born in the Netherlands he studied medicine in Edinburgh (United Kingdom). He specialized in fetal physiology in the United States. In 1967 he was appointed professor of obstetrics and gynecology at Georgetown University. In 1971 Hellegers took the initiative to establish the first bioethics institute with financial support from the Kennedy family. He not only used the new term of bioethics and made it internationally known, but also created an academic infrastructure for future research and teaching bringing together a group of prestigious scholars who inspired bioethical activities. Hellegers had a wider view on bioethics than the Georgetown perspective for which his institute became known. Although the institute was initially focused on the ethics of human development and reproduction, he advocated a view of bioethics that not only addressed the challenges of medical technology but also included issues such as poverty, immigration, disability, and vulnerability. He also strongly promoted an interdisciplinary approach arguing that various disciplines could enrich each other. Although he was a pioneer of bioethics, he finally considered bioethics a global endeavor. He died in his home country of the Netherlands when he was supporting the establishment of a new bioethics institute there.

H. ten Have and M. do C. Patrão Neves, *Dictionary of Global Bioethics*,
https://doi.org/10.1007/978-3-030-54161-3_30

Jahr

Fritz Jahr (1895–1953) was a Protestant pastor and teacher in the city of Halle an der Saale in central Germany. He studied theology, philosophy, music, and history. Jahr published 18 short papers between 1927 and 1934 most of which can be found in the journal *Ethik: Sexual- und Gesellschafts-Ethik* (Ethics: Sexual and Societal Ethics). In 1927 Jahr published *Bio-Ethik: Eine Umschau über die Ethischen Beziehungen des Menschen zu Tier und Pflanze* (Bioethics: A Panorama of the Human Being's Ethical Relations with Animals and Plants) in the magazine *Kosmos*. In the article he argues that animals and plants are moral partners of human beings. Recent biopsychological research has shown that there is an equivalence between humans and other animals since they all have "souls." Animals and plants must therefore be treated as partners. According to Jahr this indicates the need for a new ethics that he termed *Bio-Ethik*. People not only have moral duties toward humans, but also toward all living beings. This is what Jahr calls the "bioethical imperative." He suggests that we "… respect every living being in principle as an end in itself and treat it, if possible, as such." Similar ideas were proposed by Albert Schweitzer but his contributions are not found in Jahr's literature. Jahr's work was forgotten primarily because all his publications were printed in German. It is not clear whether his work had any influence on medicine, especially during a time where the Nazi ideology was rapidly developing. Furthermore, after the war Halle fell within the borders of the German Democratic Republic that housed a communist ideology in which the exchange of information became difficult. Jahr's work was rediscovered in 1997. He is now regarded, especially in Europe, as the creator of the term and concept of bioethics and has equal standing with Potter and Hellegers who introduced the term "bioethics" in English discourse in the 1970s. Bioethicists argue that Jahr promoted a comprehensive approach to bioethics: human beings should be considered relative to the surrounding world. Jahr takes a broader approach to bioethics than the commonly used concept since it encompasses not only individual human beings, but the entire world of life.

H. ten Have and M. do C. Patrão Neves, *Dictionary of Global Bioethics*,
https://doi.org/10.1007/978-3-030-54161-3_31

Pardo

Arvid Pardo (1914–1999) was the first Permanent Representative of Malta to the United Nations. He held the position from 1964 to 1971. In November 1967 he made a prophetic speech to the General Assembly claiming that ocean resources should be regarded as the common heritage of humanity. Arguing that the seabed and the ocean floor are common property he felt they should be used and exploited for peaceful purposes and to the benefit of humankind as a whole. Pardo initiated an international negotiation process that resulted in 1982 in the Convention on the Law of the Sea. Article 136 of this convention states: "The Area [the deep sea bed and the ocean floor] and its resources are the common heritage of mankind." Articulating the concept of common heritage Pardo introduced an important notion in the discourse of global bioethics. Many bioethical discussions at the global level concern whether particular resources can be owned at all and whether they can become private property. Examples are water, genes, and indigenous knowledge. The notion of common heritage includes such factors as non-appropriation; management by all people; international sharing of benefits; freedom of scientific research for the benefit of all people; and conservation for future generations. Although the notion of common heritage limits certain human activities, is outside the commercial sphere, and cannot be commercially exploited, it should be made available for common use and common benefit. There are also limits to scientific and technological interventions. Pardo feared that since the most developed countries are technologically much more advanced than most other countries they could exploit, for example, minerals at the bottom of oceans and harvest the benefits of what should be regarded as a common good. This should apply not only to the oceans but also outer space and the Arctic.

H. ten Have and M. do C. Patrão Neves, *Dictionary of Global Bioethics*,
https://doi.org/10.1007/978-3-030-54161-3_32

Pellegrino

Edmund Pellegrino (1920–2013) is one of the founders of present-day philosophy of medicine and bioethics. Educated in medicine, especially renal medicine, he became the President of the Catholic University of America. In 1978 he took up a teaching post at Georgetown University where in 1983 he became the Director of the Kennedy Institute of Ethics. He founded the Center for Clinical Bioethics at Georgetown University Medical Center (renamed the Edmund D. Pellegrino Center for Clinical Bioethics in 2013). He remained at Georgetown until his retirement in 2013. Pellegrino was a productive author with 600 published articles and chapters to his name. He wrote 23 books, predominantly in philosophy and ethics. In *A Philosophical Basis of Medical Practice* published in 1981 (with David Thomasma) he presented a philosophical theory of medicine that grounded the growth of bioethics. The book inspired a resurgence of interest in the philosophy of medicine. Previously Pellegrino founded the *Journal of Medicine and Philosophy* (1976) and was its first editor-in-chief. Pellegrino was strongly convinced that medicine is a special moral enterprise. In another well-known publication, *The Virtues in Medical Practice* (published in 1993 with David Thomasma), he argued that medicine is a special kind of human activity that is intrinsically related to specific virtues such as trust, compassion, prudence, and justice. Pellegrino was also a pioneer in the teaching of humanities and ethics in medical schools. He published many studies and reports about ethics-teaching programs and made recommendations to improve and expand ethics education. In 2004 Pellegrino was appointed as a member of the International Bioethics Committee of UNESCO. He was instrumental in the drafting and adoption of the Universal Declaration on Bioethics and Human Rights in 2005. The same year he was appointed chair of the President's Council on Bioethics in the United States (a position he held until 2009).

H. ten Have and M. do C. Patrão Neves, *Dictionary of Global Bioethics*,
https://doi.org/10.1007/978-3-030-54161-3_33

Potter

Van Rensselaer Potter (1911–2001) is regarded the first to use the term "bioethics" in publications (in 1970 in the journal *Perspectives in Biology and Medicine*). Born in South Dakota in the United States Potter was educated in biochemistry at the University of Wisconsin-Madison. After obtaining a Ph.D. in 1938 he received a postdoctoral fellowship and traveled to Sweden to work in the Biokemiska Institutet in Stockholm. On his return to the United States he was appointed to the faculty of the University of Wisconsin (in 1940). He worked for more than 50 years as a professor of oncology at the McArdle Laboratory for Cancer Research at the University of Wisconsin in Madison. Potter was an enthusiastic scientific researcher. In the 1960s he began to publish on issues outside his initial field of cancer research. Oncology for Potter was essentially interdisciplinary and could not merely focus on individual and medical perspectives. He believed that the same dimension of social responsibility characterized other areas of research. His earlier publications on the concept of human progress, the interrelation between science and society, and the role of the individual in modern society were included as chapters in his first book on bioethics *Bioethics: Bridge to the Future* published in 1971. To address the challenges of contemporary science and technology Potter argued that a new discipline was needed that combined science and philosophy, biological knowledge, and knowledge of human value systems. He called this new discipline "bioethics" and argued it should be oriented toward the future as a science of survival. Although the new term was quickly and widely used, Potter was disappointed since its content did not differ from the usual medical ethical approaches. To emphasize that bioethics implies a broad approach—not merely one focused on the individual person but also taking into account the social and environmental context—Potter coined a new term "global bioethics." In his second monograph *Global Bioethics* (1988) he argued that bioethics should not only be comprehensive but also worldwide in scope. With this new term Potter anticipated the later development of global bioethical issues. This required a broader focus than clinical patient problems and an ethical perspective on the social and environmental settings in which such problems arise.

H. ten Have and M. do C. Patrão Neves, *Dictionary of Global Bioethics*,
https://doi.org/10.1007/978-3-030-54161-3_34

Part IV
Topics

Ableism

Ableism refers to individual discrimination or social prejudice against people with physical, intellectual, or psychiatric disabilities. It is based on a concept of normality that dictates the superiority of those who comply with this standard (i.e., the concept of normality) in relation to others who are disabled and do not comply with it and hence are considered inferior. The concept of ableism was forged mainly in the second half of the 20th century in Western countries in the wake of the human rights movement and its statement of the absolute and equal dignity of all people regardless of their particular characteristics such as gender (sexism), ethnicity (racism), or disability (ableism). The idea of normality as a unique and single pattern for all human beings is criticized and denounced as artificial. It is stressed that all people are different and have different abilities. Moreover, it is common for people with some disabilities to develop and enhance other abilities. Therefore the term "disabled person" was substituted by "person with a disability." The latter term now tends to be substituted by "person differently abled." Ableism is still a reality in many developing countries where disability is stigmatizing. People with disabilities are considered inferior and excluded from access to education and from the world of work ending up living on handouts. Although ableism is generally condemned in developed countries, in practice many barriers remain and hindering persons differently abled from enjoying the same rights as all other citizens.

H. ten Have and M. do C. Patrão Neves, *Dictionary of Global Bioethics*,
https://doi.org/10.1007/978-3-030-54161-3_35

Abortion

Abortion is the termination of a pregnancy and results in the death of the embryo or fetus. Termination can be spontaneous or artificial. Spontaneous abortion is called "miscarriage." Deliberate termination of a pregnancy that would normally result in a live birth is morally problematic. Such an abortion can be the result of various methods used in different stages of gestation such as surgical intervention (suction evacuation) and medication. A common method today is the abortion pill (mifepristone) that first became available in the 1980s. There is a debate about whether free access to this pill should be allowed since it can lead to its use as contraceptive or whether it is better to distribute it on medical prescription. Many countries have legalized abortion. When it is not legal, many abortions are unsafe. The ethical debate on abortion primarily concerns the moral status of the embryo. If the embryo is regarded as having the same moral standing as an adult human being from the moment of conception, then abortion will not be permissible. It is killing a human being in its earliest stages of development and denying the embryo or fetus the right to life. On the other hand, there is the view that the embryo has no moral standing. That means the pregnant woman may decide to terminate the pregnancy for any reason that is important to her. There are also intermediate views. They articulate that the moral standing of the embryo is developing gradually and argue that a substantial change occurred when the blastocyst forms since that means a biological individual has originated. Another substantial change begins at implantation or at the end of ontogenesis around the 12th week. On the basis of such arguments a moral status comparable with that ascribed to human beings is accomplished later in embryonic development such that abortion may be permissible in earlier stages but not later in the development. Another contested ethical issue is the reason the pregnancy may be terminated. The right to life position rejects all reasons. Sometimes an exception is made when the life of the mother is at stake. The pro-choice position assumes that all reasons are valid as long as the pregnant woman has made a free and deliberate choice. New reproductive and genetic technologies have made abortion practices more complicated. In countries such as China and India prenatal diagnostics is used to determine the sex of the embryo. Parents decide to terminate the pregnancy more often when the embryo is female. This has led to important imbalances between the sexes in society. In Western countries prenatal screening for Down's syndrome has been simplified with the introduction of a non-invasive test. Several countries have made testing available in public health programs. If Down's syndrome is detected, then the result in most cases is abortion. In countries such as Iceland and Denmark the syndrome has almost disappeared. This begs the question as to what conditions are sufficient to make abortion morally permissible. Serious genetic conditions that will cause a life of

H. ten Have and M. do C. Patrão Neves, *Dictionary of Global Bioethics*,
https://doi.org/10.1007/978-3-030-54161-3_36

suffering without the possibility of cure or treatment will weigh more heavily in moral considerations than conditions that create less suffering. It is not only up to the individual to decide to terminate life in such circumstances since individual decisions have social consequences as gender imbalances demonstrate. Moreover, in many cases societies have created care arrangements for people with disabilities to prevent stigma and discrimination. Contemporary abortion practices may encourage new eugenics policies based on the desires and wishes of individual people. Although the current debate on abortion is little different from when it started in the 1960s, it should be acknowledged that contraception today is much more successful, accessible, and cheaper. There are also many more contraceptive methods. This new reality should contribute to more responsible contraception and fewer abortions. Furthermore, knowledge about gestational life is much more advanced than it was. Today the clear perception of the human life to be terminated could lead to new arguments in the debate on abortion.

Abuse, Child

Children have been regarded for a long time as the property of their parents, especially the father. In all cultures parents are responsible for their children and take decisions in their best interests. The adoption of the United Nations Convention on the Right of the Child in 1989 was a turning point. The convention came into force in 1990. It has been ratified by all members of the United Nations except the United States. The convention formulates a human rights perspective applied to children. The basic idea is that all members of the human family have inherent dignity and equal and inalienable rights. Article 2 states that appropriate measures should be taken to "ensure that the child is protected against all forms of discrimination or punishment on the basis of the status, activities, expressed opinions, or beliefs of the child's parents, legal guardians, or family members." Article 19 explicitly refers to the possibilities for abuse stating that "States Parties shall take all appropriate legislative, administrative, social and educational measures to protect the child from all forms of physical or mental violence, injury or abuse, neglect or negligent treatment, maltreatment or exploitation, including sexual abuse, while in the care of parent(s), legal guardian(s) or any other person who has the care of the child." Protecting the child from all forms of sexual exploitation and sexual abuse is articulated in Article 34. Ethical debate of child abuse is concentrated on two issues. One is clarification of abuse. Identifying actions as abuse, maltreatment, and neglect can be problematic given that there are many differences in how children are treated in various cultures. The defining characteristic is avoidable harm that leads to serious suffering and that impairs the development of children. Harm can be physical and emotional and is caused intentionally. It can also be caused not only by commission but also omission by neglecting or ignoring the needs of children. The other ethical issue is what to do when child abuse is detected or suspected. Since parental liberty and family privacy are highly regarded in all cultures health professionals should operate very carefully. They should be aware of the risk factors. If they suspect abuse, then they should monitor the situation of the child without interfering with the privacy of home life. They should be fully cognizant that the most extreme alternative is to bring the child under external protection mechanisms.

H. ten Have and M. do C. Patrão Neves, *Dictionary of Global Bioethics*,
https://doi.org/10.1007/978-3-030-54161-3_37

Abuse, Concept

According to its etymological roots abuse means excess of use where excess is in itself contrary to harmony or to what is convenient. Abuse has two meanings. One is misuse or the use of something for a bad purpose (e.g., alcohol abuse). The second is maltreatment such as the cruel and violent treatment of a human or animal (e.g., physical abuse). Both meanings emphasize excess. Although drinking alcohol is not necessarily bad, it is if someone drinks too much. Although being firm toward someone else or an animal is not necessarily bad, it is when it is excessive. Abuse can have many reasons that are often related to differences in status, authority, and power; to discrimination and stigmatization; to harassment and hate; and to institutional and legal neglect. All people can be victims of abuse because they are intrinsically vulnerable and the more vulnerable they are (e.g., children and elderly) the more they may become victims of abuse. People do not generally live as isolated individuals but as social beings in relationships. This means that the actions of others can control, intimidate, threaten, or injure them. Even in relations involving intimacy, dependency, and trust, power can be misused. The examples of child abuse and elder abuse illustrate this. Although it is obvious to all that abuse is morally wrong, there are at least three main reasons. First, abuse is harmful and in many cases has long-term consequences for victims. Second, it is a betrayal of trust because in many cases abuse takes place within relationships of trust thus violating a basic condition necessary for the security of people and for them to flourish. Third, abuse dehumanizes and shows a lack of respect for other people and disregard for human dignity.

H. ten Have and M. do C. Patrão Neves, *Dictionary of Global Bioethics*,
https://doi.org/10.1007/978-3-030-54161-3_38

Abuse, Elder

A relatively new phenomenon is elder abuse. The significant increase in life expectancy in many societies has led to aging of the population. Although instances of elder abuse have been reported in history, it has become an increasing concern since the 1970s. The WHO defines elder abuse as "a single or repeated act, or lack of appropriate action, occurring within any relationship where there is an expectation of trust, which causes harm or distress to an older person." The WHO reports (in 2018) that 1 in 6 people 60 years and older have experienced some form of abuse in community settings during the past year. Elder abuse is particularly more frequent in institutions such as nursing homes and long-term care facilities with 2 in 3 staff in the United States reporting that they have committed abuse in the past year. There are different manifestations of abuse. Although abuse is often related to violent behavior, in a wider sense it includes neglect, abandonment, lack of respect, and depriving people of living with dignity. Elder abuse can be physical, psychological, sexual, and financial. Abuse and neglect can happen not only within the family setting but also in institutional settings. Cases are underreported. A prevalent form of abuse is misuse or misappropriation of financial assets by someone who is related to the older person. Aging is often associated with diminishing autonomy such that elderly persons are not always able to react properly to manipulation and abuse.

H. ten Have and M. do C. Patrão Neves, *Dictionary of Global Bioethics*,
https://doi.org/10.1007/978-3-030-54161-3_39

Access to Healthcare

The human right to health as outlined in the 1948 Universal Declaration of Human Rights states that everyone has the right to the highest attainable standard of physical and mental health. This includes access to all medical services and means that medical services and healthcare institutions must be accessible, equally available, acceptable, and of good quality where and when needed. The declaration mentions health as part of the right to an adequate standard of living (Article 25). The right to health was again recognized as a human right in the 1966 International Covenant on Economic, Social and Cultural Rights. Since then numerous international human rights documents have recognized such a right but until recently it has had little effect on governments. Since 2000 this has changed. General Comment No. 14 of the Committee on Economic, Social and Cultural Rights (2000) specifies the normative implications of the right to health as follows. A healthcare system should have the following characteristics: (1) Availability. Functioning public health and healthcare facilities, goods and services should be available in sufficient quantity; (2) Accessibility. Health facilities, goods and services should be accessible to everyone without discrimination. This has four dimensions: non-discrimination; physical accessibility; economic accessibility; and information accessibility; (3) Acceptability. All health facilities, goods and services must be respectful of medical ethics and culturally appropriate, i.e. respectful of the culture of individuals, minorities, peoples and communities; (4) Quality. Health facilities, goods and services must be scientifically and medically appropriate and of good quality. This requires skilled medical personnel, scientifically approved medication and hospital equipment, safe and potable water, and adequate sanitation. Access to healthcare therefore is a universal right of all human beings. Everyone should have access to healthcare on an equitable basis. Healthcare should be affordable, comprehensive, and physically accessible. In contemporary global bioethics the major challenge is how this right can be realized in practice. The question is what services should be prioritized and which healthcare needs should be considered first, especially when resources are limited. The ethical challenge of access is also related to insufficient basic health insurance as one barrier to access. Many people, especially in the United States, have poor access to healthcare because they do not have insurance coverage. Already existing disparities in health status are therefore reinforced. In many other countries, however, the problem of access to healthcare is even more serious. It is not simply a case of lack of financial resources or insurance, healthcare is often not accessible due to a lack of infrastructure and the absence of healthcare services and health professionals. A recent report of the World Bank and the WHO states that at least half the population of the world cannot obtain basic health services.

H. ten Have and M. do C. Patrão Neves, *Dictionary of Global Bioethics*,
https://doi.org/10.1007/978-3-030-54161-3_40

Access to Medication

Almost 2 billion people have no access to basic medicines and hence do not benefit from the advances of medical science. Lack of access to medication is a complex problem. Although affordability is an important determinant, deficient infrastructures play major roles. Drugs need to be safe and of good quality. This means there should be a system of pharmacovigilance that has secure supply chains to ensure substandard or suspect medical products do not enter the market. Access to pain medication is limited in many countries and international conventions for the control of narcotic drugs are often a barrier to access. An important barrier to access to medication is the current system of intellectual property rights and patents. Since the development and introduction of new drugs is market driven this system has not led to investments in new products for poor populations who do not have the requisite purchasing power. This means that research into and development of medication are not driven by the health needs of poor populations. Moreover, access to medication is also limited in developed countries today as a result of the prices of new drugs increasing. Thus the issue of access is no longer only a problem for developing countries. The increasing costs of medication have become a concern for patients, prescribers, insurers, and policy-makers in most countries. High drug prices in the United States are the result of government-protected monopolies granted to drug manufacturers. Unlike other countries the prices of new prescription drugs are not regulated or negotiated in the United States when they come onto the market as borne out by a number of recent scandals that shocked the public. For example, the company Mylan who produces the epinephrine autoinjector EpiPen used by 3.6 million Americans as a life-saving treatment for anaphylactic reactions increased its price by 500% in 7 years. Scandals like this reveal the myth that drug prices are high because pharmaceutical companies reinvest profits in research and development of innovative and better products. In reality, major companies spend only 10–20% of their income on research. The WHO has developed a list of essential medicines that they have updated every 2 years since 1977. Essential medicines are those that satisfy the priority healthcare needs of the population. They are selected with due regard to public health relevance, evidence of efficacy and safety, and comparative cost-effectiveness. The basic idea behind the list is that such medicines should be available within the context of functioning health systems at all times in adequate amounts, in appropriate dosage forms, with assured quality and adequate information, and at a price the individual and the community can afford. In many Western countries it is argued that access to medication is sometimes too easy resulting in some drugs such as sleeping pills or painkillers being abused. An example is the abuse of antibiotics that are not appropriately prescribed thus being responsible for the growing global problem of immunity against effective medication.

H. ten Have and M. do C. Patrão Neves, *Dictionary of Global Bioethics*,
https://doi.org/10.1007/978-3-030-54161-3_41

Addiction (*See* Substance Abuse)

Addiction is considered a condition in which an individual is unable psychologically and physically to stop consuming a chemical, drug, or substance or engaging in an activity, although it is causing psychological and physical harm. Addiction can refer to dependence on substances such as cocaine or alcohol. It can also involve the inability to stop certain activities such as eating and gambling. The individual concerned cannot control the use of a substance or engaging in an activity. Addiction is often regarded as a chronic disease characterized by compulsive behavior that is difficult to control. The decision to take drugs is in most cases voluntary. Repeated drug use can lead to changes in the brain that interfere with the ability to resist urges to take drugs. Such brain changes can be persistent making recovery from drug use disorders fragile and relapses common. It is estimated that 40 million Americans 12 years and older abuse or are addicted to nicotine, alcohol, or other drugs. Other countries such as Iran, the Russian Federation, and Afghanistan have rates of illicit drug use and drug addiction that are considerably higher. In several European countries such as Spain and Portugal drug use has been decriminalized. The United Kingdom is considered the world's most drug-addicted country. It is estimated that around 2.8 million people took illicit drugs in 2014, mostly cannabis, cocaine, and ecstasy, while a huge number of people were addicted to alcohol. Ethical issues concerning addiction are related to autonomy, harm, and vulnerability. Addiction threatens the autonomy of the individual. If it leads to compulsive behavior, then the notion of personal choices no longer applies. Moreover, addiction has serious harmful consequences not only for the addicted individual but also for society because addiction may be associated with crime and prostitution. Vulnerability to addiction exists for particular populations such as patients with severe pain who need prescription medication like morphine that may lead to addiction.

H. ten Have and M. do C. Patrão Neves, *Dictionary of Global Bioethics*,
https://doi.org/10.1007/978-3-030-54161-3_42

Adoption

Adoption is the transfer of legal parenting rights from one parent to someone else. Related adoption is when a relative takes over the parenting of another person, usually a child, from the biological or legal parents of that person. In non-related adoption all rights and responsibilities are assumed by someone who has no genetic links to the adopted person. Adoption has been a common practice historically in diverse cultures. Today it is a global practice. The country with the largest number of children adopted per 100 live births is the United States. Each year 135,000 children are adopted, which means 1 of every 25 US families with children has an adopted child. According to the US Census about half of these families have both biological and adopted children. In other countries the numbers are much lower. For example, in Italy there were 3,158 adoptions in 2006 and 1,044 in Sweden in 2002. Although England and Wales passed adoption laws in 1926, most other countries have done so recently: the Netherlands in 1956, Sweden in 1959, and West Germany in 1977. In recent decades reproductive technologies have helped infertile couples have biological children. Thus they frequently choose such an option instead of adoption. Adoption has an inherent ethical conflict when it comes to priority interests. Are the needs of children or the needs of adoptive parents prior? The primary purpose of adoption is to meet the needs of children. Adoption itself is controversial since it may permanently separate children from their biological parents. While some children have lost their parents (are orphans) or have been abandoned, some biological parents may not be able to adequately care for their children such that the children concerned are neglected and abused. Another ethical issue is confidentiality. It used to be common practice not to inform children that they have been adopted. It is commonly assumed today that children need to know not only because they have the right to know but also because it is beneficial for their psychological well-being. The result is many adoptees end up searching for their birth parents. Biological parents may also be identified by chance through genetic testing. An area of ethical concern is international adoption. This type of adoption increased after the Second World War and grew steadily for several decades. In the 1990s many children were adopted from Russia after the breakup of the Soviet Union. This was also the case with China whose one-child policy resulted in many abandoned children. Between 2000 and 2010 citizens in 27 countries adopted 410,000 children. After decades of growth the number of international adoptions has decreased by 50% since 2004. This is not the result of a smaller number of orphans worldwide or a decrease in demand from potential parents. It is down because of an increase in regulations concerning foreign adoptions and growing reluctance in some countries to send children abroad. For example, the Russian Federation banned the adoption of its children by families in the United

H. ten Have and M. do C. Patrão Neves, *Dictionary of Global Bioethics*,
https://doi.org/10.1007/978-3-030-54161-3_43

States in 2014, while countries such as Ghana and Guatemala banned all adoptions by foreigners. A related challenge is child trafficking. There is also an intermediate situation between legal adoption and child trafficking. This is legal adoption involving the payment of a fee. In other words, the child is bought (the more you are able to pay the faster you get the child). It is sometimes unclear how adoptions are arranged in emergency situations, natural disasters, and civil wars. The Hague Convention on Protection of Children and Co-operation in Respect of Intercountry Adoption has regulated international adoption and child trafficking since 1995.

Advance Care Planning (*See* Advance Directive)

Advance care planning (ACP) (a.k.a. end-of-life care) is the process in which patients discuss their goals and preferences for future treatment and care with their relatives and healthcare providers. This process is important since patients may receive treatments such as resuscitation that they preferred not to receive. Sometimes patients are acutely hospitalized while their preference was to stay at home in the last phase of life. Timely consideration and discussion about what a patient wishes to be done in such circumstances means treatment and care can be adjusted to follow patients' values. It also makes relatives more comfortable that care is indeed following patients' wishes if they are no longer able to express them. Since ACP is not a one-time conversation but a process, it needs to be repeated. Values and preferences can change over time. Patient preferences can be documented in an advance care directive that can be used if patients are unable to express their preferences themselves. Moreover, such documents need updating. In practice there are many uncertainties surrounding ACP. What should be discussed? At what moment in time? Does it apply to everyone? Whose responsibility is it? Although the importance of ACP is generally recognized, especially for end-of-life care, it is not often used in practice. ACP primarily involves communication. This will take time that is often not reimbursed. It also assumes openness to discuss values and goals in relation to life and death. In a cultural context in which human finiteness and death are not easily acknowledged and an ideology of rescue is prevalent such discussions will be difficult. This is the reason ACP has been blamed for the appearance of "death panels" in the United States. Even the suggestion that medical intervention could be limited in view of patients' values is ethically unacceptable to some people. Physicians are often not trained in communicating with patients. They also tend to avoid end-of-life discussions while patients and their loved ones generally appreciate such conversations.

H. ten Have and M. do C. Patrão Neves, *Dictionary of Global Bioethics*,
https://doi.org/10.1007/978-3-030-54161-3_44

Advance Directive (*See* Advance Care Planning; Living Will)

Advance directives are written documents that specify the medical preferences of competent people. They are prepared to provide instructions for future treatment in case such people become incompetent. There are two types of advance directives: living wills that specify the preferred treatment options and durable power of attorney that designates a surrogate who can make decisions for the incompetent person without specifying the type of decisions. Although the two types are not mutually exclusive, both options can be combined. Advance directives are regarded as the expression of the values of a competent person. They are a means to exercizing individual autonomy. Advance directives are now recognized as legally valid documents in the United States, although legislation varies from state to state. An advance directive becomes legally valid as soon as it is signed in front of the witnesses required. Healthcare institutions are obliged to have a policy for keeping records of patients' directives. However, advance directives do not have legal status in most countries across the world. From an ethical point of view they provide significant indications about patients' wishes irrespective of whether they are binding or not. One problem with advance directives is that not many people make them. If they are written, then the instructions are often not very specific. When surrogate decision-makers are designated, they are not always available or competent to make decisions. They also might have conflicts of interests. Another problem is that advance directions are not often updated. Although they may well have been applicable at the time they were made, patients might have changed preferences but not have changed their directives. This means that the legal duration of advance directives might be an ethical problem. Specific moral problems arise when the instructions in the living will are in conflict with the best interests of the patient. Circumstances may have changed that could not be anticipated by the patient when he or she was competent and drafted the living will. Advance directives are controversial in countries that have legalized euthanasia and assisted suicide. Commonly, directives set limits to medical interventions so that patients may request the termination of life-saving treatment. However, in some countries (e.g., Belgium and the Netherlands) they can now request active intervention to terminate life such as the injection of lethal drugs. Such requests are not valid if legislation only applies to competent patients (as is the case in several US states that have legalized assisted suicide). For the proper use of an advanced directive it is important to distinguish between product and process. The written document is a product that results from a longer term effort of communication about patients' values. It is the outcome of advance care planning. The role of physicians therefore is to encourage patients to write an advance directive documenting the communication that has taken place.

H. ten Have and M. do C. Patrão Neves, *Dictionary of Global Bioethics*,
https://doi.org/10.1007/978-3-030-54161-3_45

Advocacy

Advocacy means taking action on behalf of other people. Healthcare professionals are used to doing just that. An example is the pediatrician who discovered lead poisoning in drinking water in the city of Flint in Michigan and went to the press when the authorities did not undertake any remedial action. Another example is the general practitioner who repeatedly discovered bruises and injuries in a child, suspected abuse, and reported this to the authorities. Although witnessing harm, suffering, injustice, and threats to health will instigate a moral judgment, in the case of advocacy it will determine that action is undertaken to remediate the harm or injustice. Advocacy has many ethical challenges. It broadens the role of the health provider. It also goes beyond the usual clinical domain and is often related to issues of justice. People question whether advocacy is therefore appropriate for healthcare providers. Healthcare practitioners, mostly nurses, often find themselves as patients' advocates. Such a situation is more common in Anglo-Saxon countries. Nurses claim that advocacy within a healthcare setting can bring about distancing among healthcare professionals and introduce or enhance a legal dimension to healthcare. At the same time, if health is often more influenced by social determinants than the provision of healthcare, healthcare providers cannot neglect the impact of these determinants on the health of their patients and have to take on tasks beyond their usual clinical activity. There are also different levels of advocacy. Health professionals can advocate for individuals (e.g., in disputes with health insurance companies). They can also act on behalf of the community (e.g., pediatric societies arguing that pediatricians should engage with the poverty of patients). Advocacy is furthermore directed at structural sources of injustice and in such cases it will require collective action. Finally, advocacy is important at the global level. NGOs and international organizations are often the only ones who can act on behalf of vulnerable populations such as displaced people and refugees.

H. ten Have and M. do C. Patrão Neves, *Dictionary of Global Bioethics*,
https://doi.org/10.1007/978-3-030-54161-3_46

Aesthetic Medicine (*See* Cosmetic Surgery)

Aesthetic medicine reflects a new trend in medicine. It is the application of medical procedures to improve the physical appearance of patients to their satisfaction. It uses non-invasive or minimally invasive cosmetic procedures. The patients concerned are not suffering any illness and are usually in good health. They want to be fit and to manage the effects of normal aging or want to change their appearance. The purpose of aesthetic medicine is enhancement and embellishment. The practitioners of aesthetic medicine include dermatologists, plastic surgeons, and all manner of specialists. It is a lucrative business. The reason it is booming is because so many people are dissatisfied with their bodies and want to change some part of it by such procedures as liposuction, breast augmentation, facelifts, or cosmetic dentistry. Among the favorite procedures are botox injections (making you look younger), dermal fillers (for fuller and softer lips), laser hair removal (getting rid of facial and body hair), microdermabrasion (rejuvenating the skin and removing dark pigmentation), and chemical peels (improving skin tone and texture). There is a risk that the use of aesthetic medicine can become addictive. Some people use it so frequently that psychological support is needed in some extreme cases. The rise of aesthetic medicine reflects the emphasis on consumer demand in healthcare. People regard aesthetic medicine as a potential panacea for personal and relationship difficulties. However, treatment may lead to adverse effects on body function and health. The individual autonomy of the patient should then be balanced by the principle of non-maleficence. Practitioners should make sure that patients have realistic expectations, that they know the benefits and harms, alternative options, and the risks of surgery and anesthesia. Another ethical concern is that aesthetic medicine is openly and sometimes aggressively advertised in some countries. If this new discipline is primarily driven by economic concerns, then it could be argued that it is striving to create demands that would not exist without advertising. It then also reinforces the market ideology of consumption and performance with the primary aim to profit from it thus deliberately exaggerating contemporary worries about body image and physical appearance. One phenomenon is that the population demanding aesthetic medicine (other than for any medical reason) is becoming younger (often involving minors). This begs the question as to whether there should be an age limit for interventions already established. It is against this background that aesthetic medicine practitioners have developed codes of ethics. For example, the American Society for Aesthetic Plastic Surgery in its *Code of Ethics* (2013) defines the ethical responsibilities practitioners have to patients, to colleagues, in practice settings, and toward the profession. The first set of

H. ten Have and M. do C. Patrão Neves, *Dictionary of Global Bioethics*,
https://doi.org/10.1007/978-3-030-54161-3_47

responsibilities includes patient safety, competence, informed consent, patient assessment, confidentiality and privacy, professional fees, conflicts of interest, sexual relationships, and sexual harassment. Aesthetic medicine can also have a therapeutic goal in specific cases (e.g., for people who have suffered severe burns). In such cases it contributes to self-esteem.

Ageism

Ageism is discrimination on the basis of age. In healthcare it refers to providing less favorable treatment to persons because of their chronological age. Like racism or sexism negative discrimination on the basis of age is unjustifiable. In principle it can be directed against young people but in practice it is usually directed against old people. Unfortunately, the reaction against racism, sexism, and other forms of negative discrimination is much stronger than against ageism. In fact, society commonly accepts ageism without fully recognizing it as negative discrimination. Different forms of ageism can be distinguished. Chronological age itself can be used for discrimination. The mere fact that people have reached a specific age is used to justify differential treatment. Another form is to assume that people of a particular age group have characteristics that in fact they do not possess, but still use this as a reason for discrimination. A third form is to attribute a characteristic that some people possess and use it as justification for discrimination. Compulsory retirement is a case here, especially when it is justified by loss of competence with aging. Age plays a specific role in bioethical discussions regarding treatment and health policy. Since the elderly often have particular needs different care and treatment may be required. This is not discrimination, but the fine-tuning of healthcare according to needs identified. The growing population of elderly also requires substantially more care than younger populations. This discrepancy has advanced arguments for differential treatment of the elderly. One is the argument regarding triage. It assumes that the elderly use more than their fair share of healthcare resources despite possibly not having the greatest need. This is because they are often close to death irrespective of the medical interventions attempted. More resources should therefore be made available for conditions that affect younger people. Another argument is the fair innings argument. The assumption is that persons who have reached the normal span of human life have lesser claims on resources than persons who have still a lifetime before them. In allocating medical resources priority should therefore be given to conditions that affect the young. Both arguments are contested. The elderly have contributed to the healthcare system that exists, have earned a right to treatment that the younger generations have not yet accomplished, and hence should be entitled to the same treatment as other generations. It can also be argued that life in the recent past was more difficult and harder than it is now for the young so that at least some elderly persons have not had a fair innings.

H. ten Have and M. do C. Patrão Neves, *Dictionary of Global Bioethics*,
https://doi.org/10.1007/978-3-030-54161-3_48

Agricultural Ethics

Agricultural ethics is the area of practical ethics that examines agriculture. Although it is part of the wider area of food ethics, it concentrates on how food is produced. Agriculture is regarded as a major step in civilization that took place 10,000 years ago. It led to human settlement, the founding of cities and civilizations, and population growth. Estimates are that currently 40% of the land surface of the Earth is used for agriculture. Ethical queries concerning agriculture refer to a basic ethical dilemma. First, it is argued that to prevent hunger and malnutrition contemporary agriculture should produce more food because the world population is growing. The second argument is that expansion of modern agriculture will further increase loss of biodiversity. The third argument holds that biodiversity is necessary for human health. Such arguments lead to the conclusion that to keep people alive and healthy and to feed future generations modern agriculture is damaging biodiversity, the sustainable production of food, and global health, although in principle it can provide sufficient nutrition for an increasing world population. It is commonly held today that hunger and undernourishment are not the result of shortage of food, but of poverty and unequal global distribution. For the ethical debate on hunger this means that the focus should move from food production to better access. It is also assumed that it is not agriculture as such but the way in which it has developed and been practiced that is problematic. This begs the fundamental question as to whether it is possible to have an agricultural production system that respects the value of biodiversity and does not reduce biological diversity. Answers to this question are diametrically opposite. One is the agro-ecological approach that combines food production and biodiversity protection in a sustainable way. This approach balances the interests of humans and ecosystems. It combines a number of alternative ways of agriculture such as traditional practices, ecological farming, and organic food production. The best solution to current problems is widespread conversion to alternative agriculture. In many countries the relation between society and the food industry is transforming and farmers are willing to change their production methods. The other approach emphasizes that science and technology are the best way to deliver more food. Genetically modified crops promise to feed growing populations through their higher yields while promoting sustainable practices. It is argued that genetic modification (GM) can provide crops that are resistant to insects and diseases and have higher nutrient quality. This can reduce the level of pesticides and herbicides used thus leading to less exposure to toxic chemicals and improving human health. Such potential advantages mean biotechnology is increasingly used in agriculture. About 11% of the cultivated area in the world is covered by GM crops, particularly in the United States, Canada, Argentina, Brazil, and India. However, in the European Union 57% of the

H. ten Have and M. do C. Patrão Neves, *Dictionary of Global Bioethics*,
https://doi.org/10.1007/978-3-030-54161-3_49

population are not willing to support GM food. They regard GM food as not offering benefits, as unsafe, as inequitable, and as worrying. Major concerns exist regarding human health and biodiversity. Research findings differ and the evidence is often ambiguous. Since data are mostly provided by the industry it is unclear how objective and reliable such data are. The consequence of all this is that many people are going over to different kinds of foods (e.g., organic or vegan). It is argued that precision agriculture will be the future: a farm management approach using information technology and a wide array of items such as GPS guidance, control systems, sensors, robotics, drones, autonomous vehicles, and variable rate technology such that crops can be specifically managed and returns maximized.

AIDS (*See* HIV)

AIDS is acquired immune deficiency syndrome. It is the result of infection with human immunodeficiency virus (HIV). It is the last stage of HIV infection and will lead to death if not treated. After infection it can take 2–15 years to develop. The infection destroys the immune system. When AIDS occurs, the immune system has already been severely damaged. Patients develop opportunistic infections or cancers. They may have recurring fever, chronic diarrhea, persistent fatigue, weight loss, and skin rashes. HIV can be suppressed by retroviral drugs. They do not cure the infection but suppress viral replication within the body of the patient and help to strengthen the immune system. AIDS and HIV infection have become a major public health threat. More than 35 million have died as of the end of 2017. The WHO estimated there were approximately 36.9 million people living with HIV at the end of 2017 and that 1.8 million people across the world became newly infected in that year. Although AIDS has become a chronic disease (i.e., a disease that can be lived with) in developed countries, it is still fatal in the rest of the world.

H. ten Have and M. do C. Patrão Neves, *Dictionary of Global Bioethics*,
https://doi.org/10.1007/978-3-030-54161-3_50

Alternative Medicine

Alternative medicine refers to medical systems and interventions that are outside the established domain of conventional or regular medicine and outside the domain of scientific evidence. It is also called complementary medicine. It is widely used. More than 40% of adults in the United States use some form of alternative medicine. It is not clear exactly what is covered by the term "alternative." Treatments and interventions range from acupuncture, chiropractic medicine, energy therapies, homeopathy, Chinese medicine, yoga, massage, herbal medicine, meditation, biofeedback, and osteopathy. The use of alternative medicine in many countries is growing. The ethical concerns regarding alternative medicine focus on the status of knowledge and evidence since alternative treatments lack scientific evidence regarding safety and efficacy. They are introduced without the rigorous testing that is required for conventional treatments. Alternative practitioners however argue that they are operating on the basis of empirical evidence that is usually acquired during a long period of time. Because reliable data on efficacy and safety are unavailable it is also questioned whether the choice of alternative treatment really is an informed choice. People can be easily deceived, especially when treatments are advertised through the internet. There are also risks for the patients when they forego conventional treatment and substitute alternative remedies for conventional drugs. Another risk is that some alternative practitioners in the absence of a formal education and legal requirements can exploit personal vulnerabilities and make an easy living without seriously considering the benefits and interests of patients. The WHO in its strategy concerning traditional medicine advocates integrating alternative and conventional medicine thus enabling consumers to have a wider choice of potential treatments. It is argued that the two approaches to medicine are not incompatible but complementary. Although there is agreement that further research is necessary, there are disputes about the appropriate methodology. It is furthermore argued that more and better regulation is needed. However, integration and cooperation as advocated by the WHO will be difficult since alternative medicine is not just a collection of tools and products. It represents a worldview, philosophy of life, and perspective on health and illness that often differs substantially from orthodox medicine. Alternative practitioners frequently have a different approach to the doctor–patient relationship rejecting reductionism and emphasizing a subjective interaction with the patient as an active participant rather than a passive recipient of treatment.

H. ten Have and M. do C. Patrão Neves, *Dictionary of Global Bioethics*,
https://doi.org/10.1007/978-3-030-54161-3_51

Altruism (*See* Authenticity)

The concept of altruism derives from the Latin term *alter* meaning "other." It refers to a perspective centered on the other as opposed to one centered on the "I." It was in its opposition to (the Hobbesian) individual selfishness that the philosopher August Comte introduced the notion of altruism to characterize voluntary actions toward others and on their behalf. Altruism is traditionally regarded as a virtue meaning that despite not expressing an obligation (it is not a principle) it constitutes a desirable trace of character. Therefore, it should be encouraged as it has been throughout human history first by example and always by education. Altruism relates to other virtues such as generosity (the willingness to uninterestedly share with others what is one's own), philanthropy (the willingness to help others), or solidarity (the willingness to a reciprocity of feelings or responsibilities). The most prominent normative expression of altruism is the so-called Golden Rule "Don't do unto others what you don't want others to do unto you" (negative injunction) or "Do unto others as you would have them do unto you" (positive injunction), which is common to most religions and cultures being an important rule of our common morality. Nevertheless, it was powerfully challenged under a libertarian perspective that inverted the Golden Rule into "Do to others what they perceive as good," what they want to be done to them in what can be interpreted as reinforcing altruism where the "other" is not a reflection or projection of the "I" or an alter ego (at the intersubjective level), but another *alter* in itself (at the alterity level). The libertarian version of the Golden Rule can be particularly relevant in the healthcare sector denying the possibility of someone—healthcare professionals or family and friends—taking decisions for the patient or for loved ones as if they were deciding for themselves (a criterion universally considered as good assuming that no one wants for themselves what is bad). This can be right, but what must be stressed is the possibility that the other's autonomy to want something is different from what I would choose for myself. Treating the other as myself is refusing the other to be other and reducing him or her to an alter ego. The traditional Golden Rule is also well established in the idiomatic expression "put oneself in someone's shoes." The effort to be "in someone's shoes" is methodologically desirable reinforcing empathy, but cannot be an end in itself because no one can really be in the other's shoes in the other's circumstances (a hypothetically similar factual experience would always be differently lived by two persons who have different genetic predispositions; cultural, social, and economic environments; stories of life; relationships; projects; fears; and hopes). Although altruism is highly admired as a moral virtue, it should be noted that it is often pointed out that its

H. ten Have and M. do C. Patrão Neves, *Dictionary of Global Bioethics*,
https://doi.org/10.1007/978-3-030-54161-3_52

opposition to selfishness is from biological and psychological perspectives arguable. Besides, some would also argue that altruism is common to animals; others, however, deny this considering that virtues require voluntary acts and willingness, which is not compatible with the instinctive behavior of animals. Altruism has been considered a virtue throughout time and space and can be said to be a transcultural (and ecumenical) virtue.

Animal Cloning (*See* Animal Ethics; Animal Research; Cloning)

Cloning in the animal world is achieved naturally in several ways. Asexual reproduction is when an organism creates a copy of itself without any contribution of genetic material from another individual. It is the most elementary form of (plant and) animal cloning and happens in nature through fragmentation (a new organism grows from a fragment of the progenitor), gemmulation (aggregates of cells mostly archaeocytes become isolated), and parthenogenesis (an unfertilized egg develops into a new individual). Although not involving genetic material from a second source, parthenogenesis can be considered sexual reproduction because it involves gametes. However, animal cloning today refers to artificial cloning (i.e., laboratory techniques developed due to advances of genetics that produce cells, tissues, organs, or offspring genetically identical to the donor). Only artificial cloning is under ethical scrutiny. There are two techniques for animal cloning: artificial twinning in which embryonic cells are separated at an early stage of development; and somatic cell nuclear transfer (SCNT) in which the nucleus of a somatic cell is removed and inserted into an enucleated egg. There are two variations of SCNT: the Roslin Technique that successfully cloned the first mammal in 1996 (the sheep Dolly); and the Honolulu Technique developed in 1998. Each of these procedures targets different goals such as reproduction of an individual or reprogramming of cells to produce specific substances—the major achievement of cloning techniques. Meanwhile, the health problems that affected Dolly and led to the animal's euthanasia have been overcome, although the inefficiency of SCNT is still considerable and the abnormal development of embryos remains a challenge. Nevertheless, animal cloning has extended to other species such as pigs, horses, and bulls. In 2017 the first primates (monkeys) were successfully cloned by SCNT in China. Animal cloning today tends to be more frequently applied for a variety of purposes in a wide range of fields each one raising different ethical issues. In biomedical research cloning techniques can be used to genetically modify animals so that their cells and organs can be transplanted into humans or to produce therapeutic proteins in both cases threatening animal welfare and disposing of their lives for human benefit only. Animal cloning created new opportunities for research and new human uses for animals, which has led to further emphasis on ethical matters regarding animals such as reducing the number of animals involved in research and using them only when it is impossible to replace them and when the outcome of the research is highly important. Stock breeding can be highly beneficial for poor societies by creating identical copies (clones) of economically valuable specimens, refining some species with economically desirable characteristics, or adapting them to more demanding environmental conditions and specific social situations. However, these interventions entail the denaturalization of animals that may be

H. ten Have and M. do C. Patrão Neves, *Dictionary of Global Bioethics*,
https://doi.org/10.1007/978-3-030-54161-3_53

perceived as disrespectful. Cloning is used in sports to create clones of champions such as racehorses, while developing an artificial process of species selection. It also intensifies the use of animals for human entertainment violating their inherent value. The cloning of pets as a way to overcome their short life expectancy is a growing business reinforcing the ethical criticism of human ownership of animals and their commodification. In endangered species and in the future also extinct species cloning opens new horizons for the preservation of biodiversity. This is the only field where cloning is applied on behalf of animal life.

Animal Ethics (*See* Animal Welfare; Animal Rights; Animal Research; Vegetarianism; Zoocentrism)

Animal ethics broadly refers to ethical theories and to moral and legal practices concerning the relationships that humans ought to have toward animals. It is a growing field that covers many different human activities involving animals mainly at the scientific, industrial, and utilitarian level. Animals are used by science in fundamental research for the development of knowledge in specific domains (i.e., embryology, cytology, genetics, even psychology); in applied research for the resolution of practical problems (i.e., biomedical experimentation); and in healthcare as a clinical resource (i.e., production of specific substances such as insulin or used for xenotransplantation). Animal ethics has been successful in reducing the number of animals used and in refining their use to situations where they cannot be replaced and where the expected benefit for humans is very significant. Industries also use animals as raw material for the production of an almost endless range of objects for profit. There are many ethical problems at this level and perhaps the most serious are related to intensive livestock farming where animals are forced to grow fast, to live in confined space, and are transported and slaughtered under poor conditions. Animal body parts are furthermore used for futile human desires such as animal skin for bags or horn for aphrodisiac substances. It is at this level that animal ethics has led to the establishment of requirements to reduce the suffering and promote the well-being of animals prohibiting some uses. Moreover, animals are used by humans as convenient utilities for cooperation and entertainment purposes. Animals collaborate with humans in several working tasks (i.e., in agriculture or in policing), as a means of transportation (i.e., horses, elephants, and camels), as companions (i.e., for blind people or elderly), and therapeutic strategies (i.e., horses or dogs for children with disabilities). They are also used to entertain humans in displays (zoos, aquariums, and contests), in exhibits (circuses, fights, and shows), and in sports (e.g., hunting and fishing). The main concerns of animal ethics regarding these activities are focused on prohibiting what causes suffering, promoting the well-being of animals, rejecting denaturalization, and advocating for animal freedom. It is common to date the beginning of animal ethics back to the work of the English philosopher Jeremy Bentham. In his *An Introduction to the Principles of Morals and Legislation* (1789) he asked concerning the distinction between man and animals: "What else is it that should trace the insuperable line? Is it the faculty of reason, or perhaps, the faculty for discourse? … the question is not, Can they reason? nor, Can they talk? but, Can they suffer?" Nevertheless, it is possible to find in human history different cultures and authors advocating for the protection of animal life: in the East there is Hinduism, Buddhism, and Jainism and in the West there are the likes of Pythagoras, Plutarch, and Porphyry, and Montaigne, Montesquieu, and Condillac. Many have claimed that animals should not have the status of things and that humans have a moral duty not to kill and eat them and thus be vegetarians. Bentham brought

H. ten Have and M. do C. Patrão Neves, *Dictionary of Global Bioethics*,
https://doi.org/10.1007/978-3-030-54161-3_54

animal issues into the public sphere promoting a debate that still continues. He also established a philosophical (utilitarian) and legal (rights based) foundation for animal ethics. Animal ethics today has developed two fundamental orientations. One is the animal liberation movement instigated by Peter Singer who condemns human prejudice against animals that has led to *speciesism* (or discrimination based on the species). The other is the emphasis on animal rights promoted by Thomas Regan who claims the existence of natural rights for all beings with interests of their own. If animals are capable of suffering, then they have interests and hence moral claims and ought to be part of the moral community. Consequently, their moral status and their inherent value (and thus the rejection of the point of view that they only have instrumental value) as sentient beings needs to be acknowledged. Animal ethics has been concerned directly with the moral status and the welfare of animals from an individualistic perspective (zoocentrism). This can conflict with the view that it is important to preserve ecological wholes such as populations or species (environmentalism).

Animal Research (*See* Animal Ethics; Animal Welfare; Animal Cloning)

The use of animals in scientific research appears to be almost as old as the beginning of medicine. For example, vivisection is already reported in the *Corpus Hippocraticum* (4th to 1st century BC) and by Galen (2nd century). The use of animals was strongly revived in the Renaissance. In the 19th century Claude Bernard advocated animal experimentation including vivisection as indispensable to the progress of medicine. Following the utilitarian English philosopher Jeremy Bentham who was a pioneer of the contemporary movement of animal ethics and stressed animals' capacity for suffering and the human moral duty to protect them, several initiatives took place regarding animal experimentation and animal welfare. The Victoria Street Society was the first antivivisection society created in England in 1875. Animals were no longer considered objects (things) at man's disposal, but sentient beings with their own interests. Nevertheless, from the end of the 19th century until the mid-20th century the diversity of species and the number of individuals used in research continued to grow together with a multiplicity of field experiments (in physics and astronomy) due to increasing knowledge about animals and the progress of science. The turning point in animal research was the work of William Russell and Rex Burch in 1959 when they presented major principles of humane animal lab experimentation known as the three Rs (3Rs): replacement (animals should be substituted by alternative methods of research such as non-living models whenever possible without affecting achievement of the proposed scientific goals); reduction (in the absence of an alternative method the number of animals involved in research should be reduced, while respecting the statistical requirement to assure the validity of the study); and refinement (during animal experimentation procedures to minimize pain, suffering, and discomfort should be in place and the experiment should stop when the pain crosses the maximum threshold established). Although the 3Rs are not yet fully observed, the creation of animal ethics committees (beginning in the 1970s, in Canada, Sweden, and Australia) in institutions performing animal experimentation to review the ethical acceptability of animal research and of a regulatory framework for animal research (from the 1980s on) have been crucial for the replacement, reduction, and refinement of animal research as well as for the establishment of welfare conditions during and after the experiments. Mention should be made of the *International Guiding Principles for Biomedical Research Involving Animals* (1985) published by the Council for International Organizations of Medical Sciences (CIOMS); the European Union's Directive (EU86/609) to protect animals used in scientific procedures in 1986; and the Directive 2010/63/EU that substitutes the former, strengthens the legislation, and improves the welfare of those animals still needed to be used. This Directive is the most restrictive worldwide and refers not only to vertebrates, but also to invertebrates capable of

H. ten Have and M. do C. Patrão Neves, *Dictionary of Global Bioethics*,
https://doi.org/10.1007/978-3-030-54161-3_55

feeling pain (such as cephalopods); animals bred for research; mammal fetuses in their last trimester of gestation; animals used for fundamental research, education, and training (but not behavioral studies in zoo animals or animals used in military experiments). The Directive totally prohibits using higher primates and endangered species. It requires establishment of an ethical committee in institutions and ethical review of all research using animals. The Directive tries to reconcile two ethical principles that conflict in animal research: on the one hand, freedom of research that is needed for scientific progress and is willing to use animals whenever considered scientifically pertinent; on the other hand, animal welfare that demands respect for animal life and its inherent value and that often advocates for a total ban of the use of animals in research.

Animal Rights (*See* Animal Ethics; Animal Research; Animal Welfare; Vegetarianism; Zoocentrism)

The Animal Rights Movement advocates that animals as subjects of a life (i.e., having a life of their own and their own story of life) also have interests that can only be adequately protected if they are considered as rights. Animals are then also said to be subjects of rights. The claim that animals have rights is quite recent. It was introduced, developed, and extensively advertised by the North American utilitarian philosopher Tom Regan in his work *The Case for Animal Rights* (1983). For Regan an action is moral if it tends to achieve the greatest good for the greatest number of individuals. Since animals are capable of suffering and having their own needs they should also be considered with a view to their welfare. Influenced by Kant, Regan attributes an inherent worth to all subjects of a life in their uniqueness. He thus assumes the obligation to protect these subjects. Regan pushes his advocacy of animal welfare as far as possible stressing that respect for and the protection of animals cannot be left to the goodwill of people, but must be legally established. It is not enough to have prudential rules that attribute duties to man such as considering the protection of animals as human duties. The recognition of animal interests as animal rights is the only way to switch animal protection from a human moral duty to an animal legal right. It is under the rubric of animal rights that Regan advocates a total ban on the use of animals as a means for human ends (abolitionism). This includes the consumption of animals and by-products, animal experimentation in scientific research, and all other uses such as for clothing or entertainment (in contrast to Peter Singer who balances competing claims according to the greatest good). Moral claims are absolute under an animal rights framework. Countries worldwide currently tend to regulate the human use of animals—livestock, research, pets, and entertainment—thus progressively attributing rights to them and revising their moral status.

H. ten Have and M. do C. Patrão Neves, *Dictionary of Global Bioethics*,
https://doi.org/10.1007/978-3-030-54161-3_56

Animal Welfare (*See* Animal Ethics; Animal Research; Animal Rights)

Animal welfare refers to the quality of life of animals that humans relate to and have moral obligations. The state of an animal's welfare indicates how the animal is coping (physiologically, behaviorally, cognitively, and emotionally) with the conditions in which it lives. The requisite quality of life or standards of care largely refer to the protection of each animal's interests such as adequate living conditions according to the species in question (nourishment, comfort, safety, health, and longevity). Particular requirements are also specified for animals such as pets (not to be abandoned), livestock (stress-free transportation and slaughter), laboratory animals (reducing pain and discomfort as much as possible), and zoo animals (reproducing conditions found in the wild), and for entertainment animals such as those in circuses or bullfighting (to be progressively abolished). Concerns about animal welfare started with and are still mainly focused on farm animals. The United Kingdom was the first country to implement laws protecting animals. The Act to Prevent the Cruel and Improper Treatment of Cattle became law in 1822 and the Protection of Animals Act (the first general animal protection law) in 1911. The European Union is a global leader in the field of animal welfare and has been producing important legislation to promote animal welfare. It too started with farm animals. The European Convention for the Protection of Animals Kept for Farming Purposes (1976) strengthened the five freedoms first presented in the Brambell Report (1965). Such freedoms were from hunger and thirst; discomfort; pain, injury, and disease; and from fear and distress while expressing their normal behavior. Since then the European Union has issued legislation in several specific situations such as improving the lot of sows and laying hens, banning animal testing for cosmetics, and improving animal transportation. It adopted a strategy for the Protection and Welfare of Animals (2012–2015) and funded training initiatives and workshops for professionals both within and outside the European Union to ensure that animal welfare legislation is enforced. In 2018 the European Union created the European Union Reference Centre for Animal Welfare with the aim of providing coordinated assistance to member states in carrying out official controls in the field of animal welfare. A label related to animal welfare (beyond that existing for table eggs) is being considered to help consumers buying food of animal origin to make informed choices. The European Union is also moving to harmonize standards to protect animals used in laboratory tests and to protect wild animals. Currently, a great deal of investment has gone into practical methods to scientifically assess animal welfare (mostly farm and laboratory animals and a few related to companion animals) following a variety of indicators under different

H. ten Have and M. do C. Patrão Neves, *Dictionary of Global Bioethics*,
https://doi.org/10.1007/978-3-030-54161-3_57

categories: behavioral, physical, physiological, and production oriented. Moreover, there is a growing awareness to change from the dominant environmentally centered assessment to a more challenging and more accurate animal-centered assessment.

Anthropocentrism (*See* Biocentrism; Ecocentrism; Zoocentrism)

The word "anthropocentrism" derives etymologically from the Greek words *anthropos* (human) and *kentron* (center) and is used to classify systems or perspectives centered on the human, on humankind. The Greek suffix—*ismós* (Latin—*ismus*) expresses the general scope of the word to which it is added. Thus anthropocentrism refers to the different doctrines that privilege man as the supreme being—the only being having an intrinsic, absolute, and unconditional value. It considers man as the center (as well as the beginning and end) of all thoughts and actions to whom everything else should be subordinated. Anthropocentrism has been the dominant perspective of Western religions, cultures, and philosophies. Man is regarded as the only rational being in nature (according to Hellenistic philosophy) and the only being created in God's image gifted with reason and free will (according to the Judeo-Christian tradition). These historical backgrounds justify the uniqueness of man among all living beings and his superiority. In contemporary philosophy, especially in utilitarianism (18th century), the superiority of man started to be challenged by valuing the capacity for suffering over other human characteristics and by arguing that this capacity is not exclusive to man. All beings capable of experiencing pain and suffering (i.e., all sentient beings) have moral worth and are part of the moral community thus inaugurating a zoocentric perspective. Later (in the 20th century), anthropocentrism was accused of being at the origin of environmental degradation and climate change through the selfish use of natural resources and biodiversity exclusively articulating their instrumental value in relation to human goals and putting at risk all life on earth. These criticisms introduced an ecocentric perspective. Anthropocentrism today tends to reject a radical position that neglects other forms of life and their habitats beyond man himself and adopts a moderate position (in other words, the weak anthropocentrism developed by Bryan Norton, Eugene Hargrove, and James P. Sterba). On the one hand, anthropocentrism advocates a qualitative difference between man and all other living beings as the only being capable of theoretical thought and free will and therefore the only moral agent and part of the moral community. On the other hand, it recognizes the inherent value of all forms of life and their habitats and underlines man's responsibility for the well-being of other beings and for the sustainability of the environment as his own moral obligation. It is not animals and ecosystems that have rights, but man that has duties. Besides, it is also argued that it is not possible to suppress the anthropological perspective because it is always man who is the author of value and of morality. Anthropocentrism has always been stronger in the Western world, where scientific knowledge and technological civilization is more developed reinforcing man's power and the illusion of self-sufficiency, than in other parts of the world where closeness with and

H. ten Have and M. do C. Patrão Neves, *Dictionary of Global Bioethics*,
https://doi.org/10.1007/978-3-030-54161-3_58

the dependence of man on the environment and biodiversity is traditionally stronger as well as religiously and culturally justified. Although global bioethics is not necessarily associated with anthropocentrism, it cannot reject this perspective either.

Anticommons (*See* Commons)

Globalization has reactivated the interest in commons. Commons are shared domains, materials, products, resources, and services that have played a central role in the history of humankind. The Commission on Global Governance in 1995 discussed global commons as an opportunity—not as a tragedy. They are resources for the survival of humanity; they are the global neighborhood that is the home of future generations. The challenge of global commons is that they require cooperation at various levels: local, regional, national, and international. The commission argues that it is imperative to develop consistent global governance regimes. However, the problem is not only one of management or governance. The real tragedy of commons is their ongoing destruction by enclosures and restrictions. The potential benefits of commons are therefore no longer accessible. This is called the tragedy of the anticommons: privatizing a commons blocks people from using it. The area of drug research and innovation is an example. Patenting of discoveries may lead to multiple owners of, for example, gene fragments. This multiplicity requires negotiating, bargaining, and pooling arrangements to promote research and innovation. However, in biomedical research patent anticommons are difficult to overcome. Groups of owners (pharmaceutical companies, universities, and public agencies) are very diverse and have conflicting agendas. Owners usually overvalue their own patents. The result is that privatizing commons in medical research slows down the development of new medical products and increases the prices making them unaffordable for many in need.

H. ten Have and M. do C. Patrão Neves, *Dictionary of Global Bioethics*,
https://doi.org/10.1007/978-3-030-54161-3_59

Applied Ethics

Applied ethics is the branch of ethics focused on the resolution of concrete moral problems raised by different socioprofessional activities. Applied ethics differs from professional ethics in that it promotes the citizen's perspective (wider while open to all citizens) emphasizing regulation by non-professionals, while professional ethics focuses on the professional's point of view (narrower while close to the professionals in question) emphasizing self-regulation. Applied ethics is not limited to professional practice (i.e., to how professionals should behave in their practice), but involves all citizens who can be affected by the social activity in question and who therefore have the right and the duty to influence social practices. Applied ethics can press for changes in professional ethics (e.g., therapeutic privilege was the physicians' prerogative until the patients' rights movement pressed to abolish it as disrespect for patient autonomy). Applied ethics is grounded on common morality and requires efficacy (i.e., an effective solution for the ethical problems identified). Applied ethics emerged in the 1970s due to a confluence of different factors. The first, more remote, and philosophical factor was the decline of moral universals in the 19th century. People who once would just follow the designated universal moral rules knew how to behave in each situation, but that was no longer the case. In the absence of universal rules it was up to everyone and society as a whole to establish the right way to act. Moreover, after the Second World War the human rights movement strongly intensified and progressively expanded to different countries and different social groups empowering them and encouraging them to make their own decisions on what concerned them. Another important aspect was the groundbreaking scientific and technological progress that also followed the war and that should be subordinated to the interests and welfare of human beings. This entails citizens' involvement and participation in determining the ends and means of scientific research. Applied ethics broadly increases awareness of the power citizens have to intervene in any activity that can affect them bringing all and each of them to the debate and widening the public sphere. Since the 1970s there has been an impressive range of areas covered by applied ethics such as business ethics, engineering ethics, media ethics, military ethics, pharmacy ethics, veterinary ethics, and science ethics. The most prominent since then right up to today are bioethics and environmental ethics. More recently other areas of applied ethics have been established such as nanoethics, neuroethics, and cyberethics. Although they are all transdisciplinary endeavors and require the contribution of several disciplines (at least two: ethics and the field to which it is applied), they go further than the simple deduction of ethical theories to concrete problems inaugurating a new field of thought and of practice. Nevertheless, applied ethics is theoretical and practical thus requiring, first, theoretical reflection at the foundational level on the principles that ground and justify human action to guarantee coherence of acts and objectivity in

H. ten Have and M. do C. Patrão Neves, *Dictionary of Global Bioethics*,
https://doi.org/10.1007/978-3-030-54161-3_60

decision-making and, second, practical intervention at the normative level in the establishment of rules that guide human actions and behavior to guarantee the ability to satisfactorily solve complex moral dilemmas (i.e., the efficacy of applied ethics). Having such requirements makes applied ethics very demanding indeed. However, there is a tendency currently to narrow it down to an acritical and unfounded provisory consensus weakening its theoretical basis and thus its soundness.

Artificial Insemination

Artificial insemination is an assisted reproductive technology (ART) consisting in the artificial retrieval of sperm via masturbation, treatment (sperm washed in a laboratory) and selection (and concentration), and insertion of sperm directly into the woman's cervix (fallopian tubes or uterus). It is also called intrauterine insemination (IUI). Matters then proceed naturally with the sperm meeting the oocyte, fusion of the gametes, and implantation of the embryo to the uterus. The sperm used in artificial insemination can belong to the father-to-be or to a (anonymous) donor or can be used in vivo or cryopreserved (from a living or deceased individual) with each situation raising different ethical problems. Artificial insemination is quite simple and the least invasive ART for a couple. It is also the one that covers fewer infertility factors. It addresses some male infertility problems such as when there is a very low sperm count or when sperms are not strong enough to swim up to the fallopian tubes to meet and fertilize the oocyte. It is also recommended for other clinical situations such as a means of preventing transmission of genetic disorders. Artificial insemination was the first ART to be developed and there is information about its use in the 18th and 19th centuries. At that time medical doctors and couples wanted to keep the procedure secret because of social criticism (although some information leaked out and records were made). It was widely considered immoral and equivalent to adultery with the children considered illegitimate at least until the 1960s. Although the cryobanking of sperm (i.e., collecting, depositing, freezing, and storage at a sperm bank) has been possible since the 1950s, it really started to develop in the 1970s. Since then artificial insemination became widely accessible to infertile couples and started to be used for non-clinical situations raising additional ethical problems. In the wake of the transverse ethical issues raised by ARTs the major problems of the procedure refer to the owner of the sperm and its non-clinical uses. Some highly controversial situations intersecting both have been registered such as the use of sperm from a single man to father hundreds of children who thus have the same lineage, creation of a genius sperm bank hoping to give birth to genius children too, selection of sperm from deaf or dwarf males to increase the possibilities of having a deaf or dwarf child, and procurement of postmortem sperm to give birth to an heir. These cases are crude examples of the commodification of children violating their human dignity. Globally, the sperm market is dominated by two companies—one American and the other Danish. These companies have grown because most European

H. ten Have and M. do C. Patrão Neves, *Dictionary of Global Bioethics*,
https://doi.org/10.1007/978-3-030-54161-3_61

countries no longer allow donor anonymity. Denmark stills allows anonymity and this has allowed Viking sperm to become the largest sperm facility in Europe. The California Cryobank now also accepts non-anonymous donors. Both banks have no limits to how many families a single donor can contribute. It is argued that anonymity today is no longer a useful concept since genetic data are increasingly accessible on the internet.

Artificial Intelligence

Artificial intelligence (AI) refers to the replication of human intelligence by a computer system such as visual perception, speech recognition, and decision-making (i.e., in a digital format). The remote antecedents of AI go back to the 17th century and to Thomas Hobbes' mechanics perspective of human intelligence as a combination of mathematical symbols. Much later (in 1936) the mathematician Alan Turing (pioneer of AI together with Alonzo Church and Kurt Gödel) formulated several important contributions to computational theory. In 1950 he predicted that 50 years hence there would be machines capable of intelligent behavior and language skills that could interact with humans without being recognized as machines (Turing test). He was not wrong. Perhaps the most interesting of Turing's foresights was the suggestion that, instead of designing programs to replicate human intelligence, research should aim to have a program that would enable machines to learn by themselves. Development of such an idea started in the 1960s. The history of AI as an academic field goes back to 1956 and the conference attended by several pioneers of AI at Dartmouth College (New Hampshire) organized by John McCarthy and Marvin Minsky. Some research groups were created with the aim of artificially reproducing those areas of human reasoning such as mathematical problem solving and playing board games. In 1959 Arthur Samuel designed a software program capable of playing chess. The program later beat him. This led to pursuit of a long and complex path resulting in other outstanding achievements in the 21st century such as the IBM computer Watson (cognitive computer system) that won the Jeopardy contest (2011), Apple's initiative to integrate a virtual assistant Siri in the iPhone 4 (2011), Amazon's launch of another virtual assistant Alexa to advise people on purchasing decisions (2014), and the chabot Eugene (a computer program/AI "bot" that conducts a conversation or "chat"). Eugene was mistaken as being a real person by about a third of those to whom it talked! Despite such advances it has been very difficult to reproduce the human brain, especially in its interaction with the surrounding world. This led to another strategy in which AI was developed to produce an artificial brain that can be taught to learn an automatic learning process from experience (just as happens with children). With the right automatic learning algorithm a computer can learn from human behavior and perform the same tasks such as driving a car. The trend today is to combine different technologies in the expectation of showing intelligent, complex, and more autonomous behaviors such as the creation of digital assistants for different jobs in call centers, financial service offices, and daycare facilities for the elderly and children. Indeed the possibilities for AI to substitute humans seem endless. At the same time, the social impact this will have is still highly unpredictable. What is certain is that the nature of human relationships and the kinds of jobs available will change

H. ten Have and M. do C. Patrão Neves, *Dictionary of Global Bioethics*,
https://doi.org/10.1007/978-3-030-54161-3_62

dramatically. All this begs a number of important questions. How far will AI substitute humans? What kind of control will humans keep over their creation? What kind of power will AI have over humans? Although AI is still at an early stage of development, in 2015 Stephen Hawking along with about a dozen experts on AI called for accurate prognostication (and more research) of the societal impacts of AI warning for the dangers of creating something that humans will not be able to control. He later said that AI would "either be the best thing that's ever happened to us, or it will be the worst thing. If we're not careful, it very well may be the last thing."

Artificial Nutrition and Hydration

Nutrition and hydration are vital for all living beings including of course human beings. The lack of adequate nutrition and hydration quantitatively or qualitatively leads inevitably to death. In recent decades there has been a growing number of situations in which patients as a result of their specific pathologies are not able to feed or drink by themselves naturally. This led to different means of artificial nutrition and hydration (ANH) being developed: enteral (nutrients are artificially introduced into the human body and naturally absorbed) by a feeding (nasogastric or gastrostomy) tube; and parenteral (when nutrients are directly introduced into the bloodstream by an intravenous tube). ANH places a lot of demands on professionals, involves a multidisciplinary team (clinical nutritionist, pharmacists, physicians, and nurses), and requires specific tailoring to the individual in need. ANH was originally developed to provide short-term support. It was used as a temporary resource to improve outcomes in patients who were expected to resume eating and drinking. Such a situation does not raise specific ethical issues since it is generally welcomed (e.g., as happens with premature newborns or comatose patients). However, ANH also started to be used to provide long-term or even permanent assistance (e.g., in elderly or terminally ill patients). In such cases it can be quite invasive (causing pain) mostly due to (accidental or self) extubations and to the risk of infections. Sometimes it is the only care provided to keep the patient alive. It is not expected to help the patient to recover, to gain weight, or to improve the patient's health or quality of life; it just extends life or postpones death. This led to ANH being considered a therapeutic resource, part of medical intervention, and subject to evaluation of its efficacy in the particular case in question just like any other treatment (e.g., the use of ventilators). ANH can then be withdrawn when considered an extraordinary medical intervention. Such an interpretation is highly controversial. The major ethical issue raised by ANH today is whether it should be considered an essential good to sustain life, as nutrition and hydration are for all living beings, or an ordinary clinical resource for short-term use and an extraordinary clinical resource for long-term use such as other biotechnological means of intervention. In the first case it has always to be provided regardless of circumstances. In the second case the initiation, withholding, and withdrawing of ANH has to be considered according to the clinical status of the patient. This topic is highly controversial and has required the adoption of guidelines from different institutions providing guidance to physicians, caregivers, and families. A different situation in which ANH can also be used is for hunger-strikers who can indeed refuse it according to the principle of respect for autonomy.

H. ten Have and M. do C. Patrão Neves, *Dictionary of Global Bioethics*,
https://doi.org/10.1007/978-3-030-54161-3_63

Artificial Organs

Artificial organs refer to all engineered devices (originally mechanical organs) or tissues both external (e.g., augmenting senses such as sight or hearing with glasses and hearing aids) and internal (e.g., replacing organs such as the heart and kidney with pacemakers and hemodialysis) to the human body. Such devices are the product of biotechnologies aimed at restoring functions that for different causes are failing, deficient, or missing. Although the idea of a mechanical device to substitute a human organ dates back to the 17th century, the procedure remained highly experimental and unsuccessful until the biotechnological advances made in the second half of the 20th century. Artificial organs today can indeed restore lost human functions. They can be permanent or just temporary (e.g., to gain extra time while waiting for the right donor for a transplant). Strictly speaking the production of artificial organs corresponds to a stage in the development of transplantation medicine. Such a stage includes diversifying the sources of organ procurement to fight the chronic scarcity of biological organs. It has evolved into widening the base used to recruit donors from family members to friends to anonymous donors and failing all else to artificial organs. Bioengineering has currently shown a lot of promise in producing organs from stem cells and creating organs artificially in a lab using 3D printing. Artificial organs have been used to replace biological organs thus overcoming the shortage of organs for transplantation and histocompatibility obstacles. Moreover, they have great potential in the fight against organ trafficking. There is little doubt that artificial organs are beneficial. From an ethical perspective the major problems artificial organs present in therapeutic use are related to cost and limited access (e.g., when prices increase, access decreases). An artificial organ can be more expensive than a corresponding biological organ as shown by hemodialysis becoming ever more expensive than kidney transplantation the longer it is used. Moreover, life constraints imposed by hemodialysis make an organic kidney transplantation highly preferable. Although artificial organs were originally conceived and designed for therapeutic purposes, they have evolved as resources to enhance deficient human functions. An example is bionic limbs that perform better than the biological body parts they replace. It is the use of artificial organs for enhancement purposes that raises the most complex ethical problems.

H. ten Have and M. do C. Patrão Neves, *Dictionary of Global Bioethics*,
https://doi.org/10.1007/978-3-030-54161-3_64

Assisted Reproductive Technology

Assisted reproductive technology (ART) refers to those procedures aiming to achieve pregnancy through manipulating one or both male and female gametes (i.e., sperm and oocyte). The main fertilization techniques are artificial insemination, in vitro fertilization (IVF), intracytoplasmatic sperm injection (ICSI), and gestational surrogacy. Each makes use of different biological materials and organs such as sperm, oocyte, and the uterus of different persons such as parents-to-be or donors. Each also addresses specific infertility causes be they male, female, or idiopathic (unknown). These techniques were gradually developed over a long time span. Artificial insemination was the first to appear and has been known since the late 18th century. ICSI is the most recent with the first baby born as a result of its use in 1992. Such reproductive technologies were originally developed as infertility treatments and later used for other clinical situations under the principle of beneficence. Moreover, they are all currently used for nonclinical situations under the principle of autonomy. There are a number of major ethical issues across the entire range of ARTs. The first focuses on the nature and purpose of the procedure. Are ARTs a subsidiary or an alternative method of reproduction? In the first instance they should remain faithful to their original goal and be limited to helping heterosexual couples overcome infertility. When a couple cannot conceive because of a certain disorder, ARTs should help them. In the second instance ARTs can be used by any competent adult regardless of their reasons or particularities (infertile or fertile, heterosexual or homosexual, single or with partner). ART is then considered a social tool to be used at will. The second focuses on the rights of children referring to the anonymity of both the process of life creation via ART and the biological progenitors or gamete donors. Do children born through ARTs have a right to know how they were conceived and who their biological parents are? Some argue that anonymity is essential to assure the requisite number of gamete donations and that the decision to disclose how the child was conceived belongs to the parents who are protected by professional confidentiality. Others say that children have a right to know their origin and who their biological parents are even if they put aside any claim they might have against such parents. The third focuses on the rights of adults who want to be parents. Do reproductive rights include the right to a child? Some argue that these rights refer to being free to decide whether to reproduce or not and at a time that guarantees reproductive health; others include the right not only to have a child, but also to choose some characteristics of this child. ARTs open up all options. How such questions are answered defines humanity today and will influence humanity tomorrow.

H. ten Have and M. do C. Patrão Neves, *Dictionary of Global Bioethics*,
https://doi.org/10.1007/978-3-030-54161-3_65

Assisted Suicide

Assisted suicide refers to the deliberate act of helping someone to bring about their own death or being clinically helped to do it. In such a case assistance is required to guarantee the person dies effectively (ensuring the suicide attempt does not fail) and efficiently (ensuring it is painless). These goals can only be assured by clinical assistance through the ingestion of prescription drugs thus narrowing the gap in meaning between assisted suicide and euthanasia. Nevertheless, these two actions are quite different: euthanasia is performed by someone at the request or on behalf of someone else, while assisted suicide is self-performed although someone else can facilitate it by providing the means and/or the instructions. Formally, assisted suicide is more restrictive than euthanasia since it is performed only by adults who are fully aware—not physically incapacitated by disease. Therefore, it is also less problematic from an ethical perspective because it relies on someone's autonomous decision exclusively regarded as free will. Still, the person's competence and the genuine free will to die should be psychiatrically assessed—something that does not always happen. The willingness to die can be dictated by such feelings as loneliness, sadness, frustration, or lack of hope for a better future all of which converge in the belief that no one really wants to die and no one chooses death. People just do not want to suffer physically, psychologically, affectively, and socially. Broadly speaking the arguments for and against assisted suicide are similar to those regarding euthanasia and rely on the principles of autonomy and human dignity (differently interpreted by both sides), the ownership or sanctity (inviolability) of life, individual rights, and societal duties. Assisted suicide is forbidden in most countries, although the legal framework can vary considerably. In Switzerland assisted suicide is admissible, but euthanasia is illegal. Moreover, assistance to suicide is performed by volunteers working for non-governmental organizations (NGOs), whereas healthcare professionals are forbidden to collaborate. In Belgium euthanasia has been decriminalized, but assisted suicide is illegal.

H. ten Have and M. do C. Patrão Neves, *Dictionary of Global Bioethics*,
https://doi.org/10.1007/978-3-030-54161-3_66

Authenticity (*See* Altruism)

The concept of authenticity derives from the Latin *authenticus* and the Greek *authentikos* that translate as the quality of what is real, true, genuine, and original. It is a common word in many different fields such as legal affairs and psychology (mainly in existential philosophy where it can assume particular significance). Authenticity was also introduced in the realm of moral thought where it acquired the specific meaning of being oneself, being faithful to oneself, resisting being changed by others or by society (in the wake of Rousseau's thesis on man's natural innocence and of Romantic philosophy), refusing to exhibit changes in personality according to circumstances (lying to oneself, bad faith as denounced by Sartre, and existentialism in general). Therefore authenticity expresses one's true identity, autonomy (being who one decides to be according to one's own free will), and responsibility (the capacity to respond for oneself) toward oneself and others. It is possessing the disposition to present oneself as one truly is. It is therefore regarded as a virtue associated with other virtues such as truthfulness (real in opposition to fake), honesty (integrity in opposition to corruption), and genuineness (originality in opposition to imitation). Such a notion as a result of its positive connotations (authentic is good and recognized as an ideal to be pursued and a virtue to be practiced) has grown stronger in contemporary moral philosophy leading to some advocating an ethics of authenticity. Such an ethics requires acting according to who one is. Authenticity becomes a criterion for morality since being faithful to oneself is morally good. There has been criticism pointing out how a culture of authenticity can lead to narcissistic individualism (proud of being who one is and of one's uniqueness) and thus to antisocial behavior with an emphasis on self-sufficiency. Moreover, accepting who one is can also lead to lack of effort to self-improve or simple recognition of one's faults and to atomistic self-indulgence. As a result of the work of Canadian philosopher Charles Taylor there is currently an effort to recover the moral significance of authenticity by rebutting its common perception as involving a turn inward and self-determining freedom (self-centeredness) and restoring its self-transcendence (differently underlined by existentialism that considers authenticity also as openness to transcendence). Although authenticity does not enclose one in oneself, it does demand the recognition of others and an intersubjective perception of good or shared values as a condition required for its accomplishment. Authenticity continues to be a relevant moral concept, an ideal of self-fulfillment within social relationships, and as a concept that rejects subjectivism and individualism.

H. ten Have and M. do C. Patrão Neves, *Dictionary of Global Bioethics*,
https://doi.org/10.1007/978-3-030-54161-3_67

Autonomy (*See* Respect for Autonomy)

The word "autonomy" derives etymologically from two Greek words *auto* (self) and *nomos* (law, rule). Thus it literally means self-government. Autonomy at the social level refers to the laws people establish to regulate themselves (synonymous with independence) and at the individual level it refers to living by one's own laws (synonymous with liberty). The ideal of liberty (Latin *libertas*) or determination of action beyond the traditional notion of free will (Latin *libero arbitrio*) or freely choosing between two alternatives of action has its roots in 18th century political theory associated first with Rousseau, but mostly with Kant, and later with Stuart Mill's *On Liberty*. Human dignity for Kant requires treating people as ends in themselves—not merely as means for one's own interests. This entails respect for autonomy. For Mill the individual pursuit of happiness, though perceived differently by each person, requires autonomous thought and action. Since then liberty, as a dominant strain in liberalism, has become an essential concept in moral and political contemporary philosophy, although sometimes criticized as leading to a too narrow and autistic individualism. Autonomy as the capacity to freely decide about what concerns one's own interests entered into bioethics in 1979. It did so first with the *Belmont Report: Ethical Principles and Guidelines for the Protection of Human Subjects of Research* by a US Congress National Commission and then with the *Principles of Biomedical Ethics* by Tom Beauchamp and James Childress. The Belmont Report, referring to the principles that ought to guide human experimentation, stated that respect for persons required treating them as "autonomous agents." *Principles of Biomedical Ethics*, identifying the principles of biomedical ethics, introduced the "principle of autonomy" (later referred as respect for autonomy) as the "right to hold views, to make choices, and to take action based on their personal values and beliefs" (6th ed., 2009: 103). Although the authors presented four prima facie principles—autonomy, beneficence, nonmaleficence, and justice—autonomy was commonly understood as being superior to the others, particularly to beneficence. The major argument still used today is that when autonomy and beneficence conflict in a concrete situation, autonomy prevails. The classic example is a Jehovah's Witness complying with his/her religious beliefs by refusing a blood transfusion (respect for autonomy) and thus accepting death as the consequence (overriding the principle of beneficence). Indeed, the principle of autonomy has a deontological nature (i.e., compliance is independent of the consequences); and beneficence is a teleological principle (i.e., compliance requires the accomplishment of good consequences). When the two principles conflict, as in the example above, autonomy can be respected regardless of the consequences, but beneficence cannot. Even saving one's life is not a beneficial outcome when the individual in question is condemned to lose the purity of his/her soul. It is down to the different moral nature of the two

H. ten Have and M. do C. Patrão Neves, *Dictionary of Global Bioethics*,
https://doi.org/10.1007/978-3-030-54161-3_68

principles that autonomy seems to be superior to beneficence. Unfortunately, this still common mistake is responsible for the overvaluation of autonomy and the undervaluation of other principles all of which are needed for the ethical practice of biomedical ethics.

Avian Flu

Emerging infectious diseases are among the best examples showing that biomedical and environmental ethics cannot be separated. Pandemics of viruses such as avian influenza, Ebola, Zika, and COVID-19 are the consequences of human interventions such as deforestation for the sake of economic development. Avian influenza or bird flu is caused by becoming infected with influenza type A viruses. Such viruses circulate among aquatic birds and can infect other birds (both wild and domesticated like chickens) and other animals. The resulting diseases are primarily zoonoses that rarely infect human beings. Human infection is the result of direct or indirect contact with infected live or dead poultry. Since viruses circulate in aquatic birds they represent a vast natural reservoir that is impossible to eradicate. There are several types of viruses. Influenza type A viruses are considered the most dangerous and have the potential to cause pandemics. Such viruses are classified on the basis of two surface proteins: hemagglutinin (HA) and neuraminidase (NA). Avian virus subtype H5N1, for example, is different from human influenza viruses. It causes a severe infection and has a high mortality rate. In December 2003 the first global outbreak of this type of influenza spread from Asia to Europe and Africa. Then there was the horror scenario represented by the Spanish flu. This was a pandemic of the H1N1 influenza virus in 1918–1920 that killed approximately 50 million people. Global governance is focused on early detection. Surveillance and fast diagnosis are therefore important. This will also allow the early development of vaccines against any emerging subtype of virus. Many infections originate in Asia and are related to bioindustry. Hence the need for a broader concept of health that urges health professionals and authorities to include animal health in health policies. Since outbreaks first occur among animals the response mostly entails the culling of all animals within the infected area (sometimes hundreds of millions of chickens). Clinical treatment of human victims is limited. Although some antiviral drugs are available, their effectiveness is questioned. Focus is mostly on prevention, especially personal hygiene and protective measures.

H. ten Have and M. do C. Patrão Neves, *Dictionary of Global Bioethics*,
https://doi.org/10.1007/978-3-030-54161-3_69

Behavioral Economics

Combining psychology with economics has produced the new discipline of behavioral economics. This examines the actual processes of economic decision-making and studies why people make some decisions rather than others. The discipline provides empirical correctives to the prevailing model of *Homo economicus* that dominates economic discourse. Such a model assumes that human choices and decisions are primarily motivated by self-interest and guided by utility maximization. It assumes that the ideal actor is the rational individual who carefully weighs relevant information and makes decisions that will produce the most benefit. Empirical studies, however, show that in practice economic decisions are influenced by psychological biases, limited cognitive resources, and care for fairness. There are also many subliminal and unconscious stimuli that steer decisions toward specific directions. The way food is displayed in a supermarket will stimulate buyers to reach for items displayed at eye level rather than lower to the floor. This information can be used to direct choices in healthcare. If healthy food is placed close to the entrance of the store or higher on the shelves, then consumers can be manipulated into choosing them. Choices for healthy food can therefore be encouraged by "choice architecture" or "nudging." Although the effectiveness of human decision-making can be enhanced in this way, nudging is manipulation. It has been called a form of libertarian paternalism. Although it is not coercion and leaves alternative options open, the values that are nudged are selected and presented by other people (often in commercial business or health promotion) who want them to be chosen. This means there is a positive outcome of individual decisions that is promoted for specific reasons that can be good or bad, but that are often not revealed or explained. In this sense nudging is manipulative. It also questions the assumption that individual choices are based on respect for autonomy.

H. ten Have and M. do C. Patrão Neves, *Dictionary of Global Bioethics*,
https://doi.org/10.1007/978-3-030-54161-3_70

Behavior Modification

Behavior modification comprises those procedures aimed at changing behavior patterns. Another term is behavior therapy. It has been promoted since the 1970s by the school of behaviorism, especially by John Watson and B.F. Skinner. They argued that the usual approaches of psychology with their emphasis on introspection are not working. Psychology should be redefined as a natural science since its goal is the prediction and control of human behavior. Psychology today has developed new procedures based on conditioned Pavlovian reflexes and operant conditioning that can help to improve the behavior of individuals. Such procedures can therefore be used to treat dysfunctional behavior. Behaviorists have argued that a new approach to ethics is required because the traditional view of ethics is wrong since it assumes that human beings have free will. Although they are supposed to be able to freely choose among various options, in fact their behavior is determined by genetic and environmental factors. Human history shows that coercion has often been used to control behavior. Free will is used in these approaches to justify the use of coercion rather than genetic and environmental factors determining behavior. However, if behavior is controlled by genetics and the environment, then a different, more scientific approach is needed based on insights into how physiological variables can be influenced. Behavior can then be controlled more effectively with the use of procedures that are not coercive. An example is positive reinforcement. Systematically rewarding specific behavior stimulates particular responses to stimuli so that desirable behavior will be repeated and reinforced. Such an approach is used, for example, in parenting and education but it can also be applied in treatment of clinical behavior disorders.

H. ten Have and M. do C. Patrão Neves, *Dictionary of Global Bioethics*,
https://doi.org/10.1007/978-3-030-54161-3_71

Benefits and Harms

Article 4 of the Universal Declaration on Bioethics and Human Rights presents the principle of benefits and harms: "In applying and advancing scientific knowledge, medical practice and associated technologies, direct and indirect benefits to patients, research participants and other affected individuals should be maximized and any possible harm to such individuals should be minimized." In fact, this principle advocates two ethical principles that have been constitutive of medical ethics from the beginning of medicine: the principle of beneficence and the principle of non-maleficence. Both principles are grounded in the Hippocratic tradition. Later they were based on the inherent dignity of human beings. This generates the moral obligation to do good for other human beings and avoid harming them. Both principles can be taken together as one principle or they may be regarded as separate principles. Non-maleficence at the most basic level can be regarded as the lowest level of beneficence in that it is imperative not to inflict harm. Usually, beneficence is regarded as more encompassing since it requires not merely the removal of harm but it would be more compelling for it to prevent harm. The highest level of beneficence is to promote good. The "good" here was initially only the clinical good. It was only after the Second World War and introduction of the principle of autonomy that the "good" in beneficence started to refer to the good of the person. The benefits and harms in medical practice and research are not clear-cut and obvious. Although there are always risks and benefits cannot be predicted, there is also the possibility of unintended harm. Although patients cannot be absolutely protected against harm, benefits can never be guaranteed and harm can never be completely avoided. For example, every drug no matter how beneficial it is may have toxic adverse effects. This means benefits and harms need to be balanced in practice. The ethical requirement is of course to maximize benefits and minimize harms.

H. ten Have and M. do C. Patrão Neves, *Dictionary of Global Bioethics*,
https://doi.org/10.1007/978-3-030-54161-3_72

Benefit-Sharing

Article 15 of the Universal Declaration on Bioethics and Human Rights (UDBHR) formulates the principle of sharing benefits. It states that "Benefits resulting from any scientific research and its applications should be shared with society as a whole and within the international community, in particular with developing countries." The principle is based on the notion of the common heritage of humankind. If resources are common property, then their benefits should be shared by everybody. It is directly derived from the human right to enjoy the benefits of scientific progress and its applications included in both the Universal Declaration of Human Rights (UDHR) and the International Covenant on Economic, Social and Cultural Rights (ICESCR). The UDBHR specifies the relevant benefits: (a) special and sustainable assistance to, and acknowledgement of, the persons and groups that have taken part in the research; (b) access to quality health care; (c) provision of new diagnostic and therapeutic modalities or products stemming from research; (d) support for health services; (e) access to scientific and technological knowledge; (f) capacity-building facilities for research purposes. Benefits therefore are not only monetary but can also cover a wide range of possibilities. The principle of sharing benefits is currently used in an increasing number of contexts. The Convention for Biological Diversity (CBD) applies it to genetic resources. The CBD was adopted in 1992 and has three purposes: conservation of biological diversity; sustainable use of its components; and fair and equitable sharing of benefits from the use of genetic resources. Medical research is another area of application. The Declaration of Helsinki introduced the obligation of post-trial access by study participants in 2000. Another context is the migration of healthcare professionals. Health professionals like everybody else have freedom of movement. Although many migrate to richer countries because of attractive salaries or better work conditions, their medical education was provided and paid for by the home countries. This meant host countries benefited by being able to address health shortages without investing in education. The benefits of migration therefore are unequal and unjust. A problem with the principle of benefit sharing is that it is not clear what "benefits" are, especially from the perspective of those receiving them. Although there are few good examples of successful sharing, it is important not to confuse benefits with profits. Monetary benefits (access fees, fees per sample, research funding, and joint ventures) should be distinguished from non-monetary benefits (sharing of research results, participation in teaching and training, and capacity-building). Furthermore, what counts as benefit is not the same in all countries but is dependent on the social and cultural context and varies according to local needs. Sharing benefits is a theoretical

H. ten Have and M. do C. Patrão Neves, *Dictionary of Global Bioethics*,
https://doi.org/10.1007/978-3-030-54161-3_73

principle based on the principles of justice and solidarity. Its aim is to protect vulnerable populations through countering inequality and exploitation. Furthermore, benefit-sharing underlines the importance of protecting the environment because the biosphere and biodiversity are common goods. The practical challenge is how to do this in a fair way.

Big Data

Big data refer to the extremely large datasets that have been produced, to the procedures used to collect and store such datasets, to the way in which they are organized algorithmically, to their computational analysis in significant associations and trends, and to their presentation patterns all of which can be used to reveal or discover new realities and knowledge for better decision-making in an increasing diversity of domains. Big data represent a recent phenomenon born in the digital era to which everybody contributes. This has resulted in the production of an increasing amount of information that is ineradicable and piles up permanently and indefinitely in the virtual world. A quick glance at this phenomenon reveals that human beings are currently producing more data every 2 days than all the data collected since the origin of humanity right up to the year 2000 (about 90% of all data in the world were generated in the last 2 years). This amazing amount of information can only be collected and stored by very powerful computers (quantum computers will make short work of big data) assisted by artificial intelligence that organizes, analyzes, and presents the data in meaningful patterns (creating pattern recognition algorithms). The information this produces should lead to unbiased decision-making. Big data are usually characterized by the 3Vs: volume (the huge amount of data generated and collected); velocity (the speed at which the flow of data arrive and are processed); and variety (the wide range of types of information used). Lately a further 2Vs have been added: veracity (the level of trust in the results produced) and value (the rate of return on the knowledge delivered). The range of applications for which big data can be used is expanding to cover not only an increasing number of human activities, but it is also growing in importance, especially when it comes to business interests. Its application to healthcare and biomedical research has already revolutionized many specific fields such as oncology, neurology, autoimmune diseases, and rare diseases. It improves diagnosis (faster and more accurate), therapies (more efficient thus saving money), and prognosis (maximizing good outcomes). Therefore it is becoming part of clinical practice and changing it through, for example, digital assistants. Although highly beneficial, big data raise challenging ethical questions the major ones being privacy issues related not only to data collection, but also to data use and retention (e.g., data owners such as people from third world countries do not benefit from the revenues produced by their data); sharing issues related to erosion of the distinction between the private and public sphere (e.g., automatic indexation of data and the possibility of reconnecting anonymized data with the individual to whom it belongs); and transparency issues related mainly to secondary uses of datasets (e.g., from which tech or net giants benefit by acquiring unprecedented economic and political power despite

H. ten Have and M. do C. Patrão Neves, *Dictionary of Global Bioethics*,
https://doi.org/10.1007/978-3-030-54161-3_74

not being representative of any population). Big data are still recent phenomena in healthcare where currently only 20% of data are being collected and whose potential has yet to be seen. Global action is needed to keep this borderless phenomenon in check.

Biobanking

A biobank is a repository that stores biological samples for use in research. There is a huge range of human tissue sources that can be used in research. Special banks have been established to collect brain tissue, blood, umbilical cord cells, sperm, and other tissues for current and future use. Many countries today have established biobanks. In 2009 the Organization for Economic Co-operation and Development (OECD) adopted guidelines for human biobanks. The Recommendation on Human Biobanks and Genetic Research Databases addresses the establishment, management, governance, operation, access, use, and discontinuation of biobanks. It emphasizes that fostering scientific research is the fundamental objective of biobanks. It is important that human dignity and human rights are respected. Ethical concerns related to biobanking are informed consent, privacy protection, return of individual results, and ownership of data and samples. A basic issue is consent. Obtaining and using samples for research will require consent. This is important since it shows respect for the autonomy of the donor. It fosters a climate of trust that is necessary for scientific research. However, it is argued that in the context of biobanking solidarity is also an important value. Sharing human tissues is crucial for scientific progress and will contribute to the public good. It is also questionable whether consent can really be informed. Biobanks store materials over a long time. It is not known in advance what uses will be possible and interesting. Does this mean that donors should be recontacted if a different type of research is undertaken or can they provide blanket consent? Moreover, donors have the right to withdraw. Even here it is not clear what the practical implications will be: no further contact, samples no longer accessed, or destruction of the samples? Privacy concerns are morally relevant. Even if information is anonymized and links between samples and individuals are removed, it is possible with the current information available on the internet to identify individuals. It is therefore imperative to have not only a strict privacy protection mechanism, but also policies to prevent discrimination and stigmatization. Another ethical issue is the return of individual results. If research produces information that is relevant to the specific donor, then should personal results be returned? Many biobanks have a policy of not doing so. However, if minors are involved and the information indicates preventable or treatable early onset conditions, then information should be returned. A fourth ethical concern is the ownership of samples and data. Are the donors the owners so that they decide what will happen with the information? Or is ownership with the researchers or with the institutions that house the banks? Such issues are not clear. Since data are increasingly commercialized and sold it is important to clearly determine ownership issues.

H. ten Have and M. do C. Patrão Neves, *Dictionary of Global Bioethics*,
https://doi.org/10.1007/978-3-030-54161-3_75

Biocentrism (*See* Anthropocentrism; Ecocentrism; Environmental Ethics; Zoocentrism)

The word "biocentrism" derives etymologically from two Greek words *bios* (life) and *kentron* (center) and designates a perspective centered on all forms of life—animals, plants, and microorganisms—regardless of their particular characteristics such as sentience or the capacity to experience sensations, particularly pain. The Greek suffix—*ismós* (Latin—*ismus*) expresses the general scope of the word to which it is added. Thus biocentrism refers broadly to the different doctrines that consider life as a good in itself (i.e., the supreme good) attributing an intrinsic value (an inherent worth) to all manifestations of life (i.e., to all organisms). From an individualistic perspective life has a value in itself that does not depend on its utility to human ends (as an instrumental value) and requires to be respected and protected. The appreciation of life in all its manifestations is not a contemporary novelty: ancient Greek philosophy expresses a deep harmony between humans and nature (under the same and unique *logos*) and the Judeo-Christian tradition recognizes that life is God's creation (and reproduction is a form of participation in creation). Several cultures (i.e., indigenous peoples) and religions such as Buddhism mostly in Asia have long venerated life. The rise of modern and experimental science imposed a distance between the subject searching for knowledge and reduced what is to be known to the status of object even if it is a life-form. Biocentrism considers that every living being has a good or well-being of its own that follows a teleological perspective (i.e., advocating that each living being is a unique individual pursuing its own good in its own unique way) that should be protected as a unique and precious manifestation of life. Biocentrism today cherishes all living beings and life itself. Sharing such a conviction of the unconditional value of life that identifies biocentrism different theories have developed ranging from biological egalitarianism (Paul Taylor) to others that consider the possibility of establishing degrees in intrinsic values (L.G. Lombardi). Commonly, biocentrism extends the moral community to the biotic community including all living beings deserving (equal) moral consideration, although biocentrism acknowledges that not all of them are moral agents. The latter have moral obligations toward all manifestations of life. Although some consider biocentrism a theory within environmental ethics, strictly considered it surpasses anthropocentrism and zoocentrism by extending moral value to all forms of life. However, it is more limited than ecocentrism that attributes moral worth to ecosystems as well. Global bioethics is not restricted to human life since it includes the widest and most complete meaning of *bios* thus becoming co-extensive to life itself.

H. ten Have and M. do C. Patrão Neves, *Dictionary of Global Bioethics*,
https://doi.org/10.1007/978-3-030-54161-3_76

Biodiversity

The Convention on Biological Diversity (adopted in 1992) defines biodiversity as variability among living organisms from all sources including terrestrial, marine, and other aquatic ecosystems and the ecological complexes of which they are part. This includes diversity within species, between species, and of ecosystems. In practice, the term biodiversity has become synonymous with all living things (i.e., with the richness and variety of life on Earth). Measuring biodiversity is often limited to counting species as the basic unit of diversity and to making lists of endangered species. Biodiversity here is interpreted as the richness (or loss) of species. It is still disputed whether the focus should be on the number of species or on how wide the range of species is. Regardless of theoretical and practical controversies, it is remarkable how quickly the notion of biodiversity has been disseminated in popular and scientific discourses. This stems from the sense of crisis. For example, in the 1970s one species was lost every day, in the 1980s one species was lost every hour, and in the 1990s 74 species went extinct every day (i.e., three every hour). Such losses are tragic because they are irreversible for once a species is lost it cannot be brought back. Extinction of a life-form is of course much worse than deaths of its individual members. It is also tragic because species unknown to science are also being lost. Although there are approximately 1,800,000 species known to science today, the true number of extant species is unknown. It is estimated that there are between 5 and 30 million species. The focus of science on the extinction of species stems from a deeper concern: the extinction of humanity itself. Human survival is dependent on biological diversity. Loss of biodiversity is the most fundamental global environmental problem because it endangers the survival of humanity. The term "biodiversity" introduces a change in perspective that focuses more than ever on effective policies that advocate taking action before it is too late instead of elaborating theories or initiating more research. The causes of biodiversity loss are known: habitat destruction and fragmentation, invasive species, overexploitation, diseases, and climate change. Although we know the causes and what to do about biodiversity loss, still not much is done. The new concept of biodiversity shifts the perspective in three ways. First, biodiversity can be used as a wake-up call to save species or ecosystems from extinction. Preventing extinction is much more important than theorizing about stability among species or determining the health of the ecosystem. Second, biodiversity changes the perspective from preservation to conservation. Preservation means that nature needs to be preserved for its own sake and should be protected against present and future consumption. Environmentalism traditionally focuses on the preservation of endangered species and the creation of special reserves and national parks. Conservation means that nature should be defended for the sake of humans and should be saved for future consumption. It implies human use should be more prudent, but does not exclude such

H. ten Have and M. do C. Patrão Neves, *Dictionary of Global Bioethics*,
https://doi.org/10.1007/978-3-030-54161-3_77

use absolutely. The emphasis on biodiversity is associated with conservation as a more feasible goal for policies. Third, biodiversity moves the attention from protection to sustainable use. It is a conceptual and practical instrument to solve the long-standing tension between conservation and development. The notion links ecological and economic views. By considering biodiversity as a global resource that generates significant benefits, especially in developing countries, it should if properly managed reconcile conservation with economic development.

Bioengineering

Bioengineering is a relatively young field that combines the knowledge of living systems with engineering principles. It is often applied in medicine and the life sciences when it is then called "biomedical engineering." Forms of biological engineering can also be applied in other areas such as agriculture, food, pharmaceutical development, and the environment. The term "bioengineering" was first coined in the 1950s. Biomedical engineering is specifically applied to the engineering of cells, genetic materials, and genetic tissues and is controversial. Genetic engineering has long been a subject of ethical debate. Engineering is also applied to the development of synthetic materials, prostheses, and implants where the aim is to produce artificial body parts that can replace defective components or miniaturize artifacts that can restore or monitor body functions. The emphasis here is on enhancing or augmenting human functioning. Biomedical imaging technologies that can lead to better diagnoses represent another area of engineering. Neural engineering is a new area of engineering and deep brain stimulation is a current example. The emergence of bioengineering has raised new ethical issues at the intersection of medicine, biology, and engineering such as human enhancement, genetic engineering, biomimetic robots, and artificial intelligence. Such ethical issues fall into three types. The first relates to research and development such as experimentation, clinical use of biomaterials, integrity of research, and conflicts of interest. The second concerns medical applications such as the stage of development it is deemed safe to use new devices in human beings. The third questions the impact on society and culture. The engineering approach assumes there are technical solutions to problems and often implies a reductionist view of human beings as determined by genes or algorithms. Approaches that advocate changes in lifestyle or socioeconomic conditions are therefore not preferred. Engineering is therefore attractive in that it can favor technical priorities over other possible approaches. It is also closely connected with the commercial context of healthcare (with many companies promoting devices, smart drugs, or robots) raising questions about justice and equal treatment.

H. ten Have and M. do C. Patrão Neves, *Dictionary of Global Bioethics*,
https://doi.org/10.1007/978-3-030-54161-3_78

Bioethical Imperialism

One of the controversies in global bioethics is whether ethical principles are universally applicable or limited to a specific cultural context. Some anthropological and sociological studies advance the idea that globalization is in fact a form of Westernization. Moral values are dependent on cultures and traditions and hence can only be applied in specific cultural settings. According to this view global bioethics is the imposition of a specific form of local ethics (i.e., ethical principlism as developed in Western countries). The claim that global bioethics is a universal discourse actually imposes a restricted Western view on other parts of the world. The claim is regarded as a form of bioethical imperialism or colonialism. The accusation of imperialism disregards the fact that global bioethics is a multifaceted and complex phenomenon, that global bioethics is based on common and shared values, and that the alternative would be bioethical relativism. Such an accusation also sets the global against the local in that it is assumed that the global is context-free (i.e., a neutral space that encompasses diverse circumstances). In reality, the global is articulated through local activities. Global bioethics is not simply the rational and deductive application of universal principles but the result of deliberation, communication, and negotiation. There is always a dialectic interaction between universalism and local moral diversity. Global bioethics is an intercultural process—not a finished product that can be applied or imposed across the world.

H. ten Have and M. do C. Patrão Neves, *Dictionary of Global Bioethics*,
https://doi.org/10.1007/978-3-030-54161-3_79

Bioethics and Religion (*See* Religion and Bioethics)

When bioethics first appeared many practitioners of the new discipline were theologians and religious scholars. The original core of the bioethical literature was produced by philosophers and theologians. Both disciplines were associated with broad and critical perspectives on relevant moral challenges. However, such disciplinary characteristics quickly disappeared when bioethics evolved into a separate area of study with its own professional and disciplinary specifics. Theologians (and philosophers) rapidly morphed into bioethicists distancing themselves from exclusively academic analysis and focusing on practical issues in clinical medicine and research. Bioethics became a specific discourse that was secular, rational, and analytic. For example, Albert Jonson who was educated as a Jesuit narrates in his history of bioethics how he transmuted into a bioethicist. The Society for Health and Human Values was the first professional organization and originated from the initiatives of religious ministers. Theologians advocated a broad approach to medical humanities. Advances in science and medicine raised questions about human values that were usually addressed not only by philosophy and theology but also art and literature. Theology also brought with it specific perspectives that went beyond scientific materialism and furthered a particular vocabulary emphasizing destiny, creation, grace, sin, blessing, responsibility, and personhood. Theologians argued that there were larger themes at stake in moral life that related to the ends of scientific progress and technological innovation. However, this initial influence of theologians soon disappeared. The new area of bioethics offered increasing numbers of opportunities and the professional identity of theologians faded. A common thread in the histories of bioethics was that the more bioethics flourished the less the engaged theologians used their disciplinary background. Religious perspectives were increasingly translated into secular language. The histories explain why and how such a marginalization of theological contributions happened. One common argument is that the increased recognition of pluralism gave rise to the development of secular ethics as a way of applying a neutral common language. Although Western countries are marked by having an increasing plurality of religious and non-religious beliefs, this is even more obvious at the global level. Therefore bioethics should transcend this plurality of value systems by developing a secular, neutral approach based on rational criteria. Another reason to transcend this plurality is the growing involvement of governments and the need for public policy. National committees identified principles and drafted guidelines that would govern the conduct of scientists and healthcare professionals. However, such principles were also supposed to be acceptable to everyone (i.e., all citizens) so that trust in science and healthcare could be restored and maintained. Although the focus of such efforts was on common morality, it was unclear what contributions theologians could make. In policy areas religious claims were

H. ten Have and M. do C. Patrão Neves, *Dictionary of Global Bioethics*,
https://doi.org/10.1007/978-3-030-54161-3_80

translated into the language of principlism. The result is that religious discourse and imagery lost influence in bioethics. Founder of the Hastings Center Daniel Callahan pointed out that bioethics was accepted because it pushed religion aside. However, bioethics remains strongly influenced by religious perspectives in non-Western countries.

Bioethics, Clinical

Clinical bioethics is a practical field of applied ethics that assists health professionals in identifying, analyzing, and resolving ethical issues in clinical practice. The focus is on the application of ethical analysis within clinical settings and on specific patient cases. It can be regarded as a subdiscipline of medical bioethics. An ethics consultation begins with an examination of the practicalities of the case and encounters between patients and healthcare providers. Such a consultation has become a clinical service in many hospitals, especially in the United States and Canada. Consultants come from many different backgrounds such as medicine, nursing, social work, law, and chaplaincy. The majority are practicing healthcare professionals and not necessarily experts in ethics. This has raised a number of issues such as qualification and education. Against this background the American Society for Bioethics and the Humanities launched the Code of Ethics and Professional Responsibilities for Health Care Ethics Consultants based on core competencies. It also inaugurated a certification and accreditation program for healthcare ethics consultants by setting examinations and tests. The role of consultants is to assess factual information, analyze ethical questions, and evaluate the outcomes of the consultation. Such a bedside role is primarily pragmatic facilitating clinical decision-making. However, there is no agreement on the various goals of ethics consultation. Although the role of the clinical ethicist has been regarded as that of interpreter, facilitator, Socratic guide, educator, and mediator, he or she is often not supposed to make ethical judgments or justifications. Common issues concern, for example, surrogate decision-making, advance directives, and (mis)-communication between patients and health providers. The general goal of clinical ethics is to enhance the quality of care. It makes use of an interdisciplinary approach in which clinical ethicists work together with health team members, patients, and family members. The methods used can be different. Although basic ethical principles such as individual autonomy, beneficence, and non-maleficence are often used as the theoretical framework, casuistry has also been promoted as the most appropriate such framework.

H. ten Have and M. do C. Patrão Neves, *Dictionary of Global Bioethics*,
https://doi.org/10.1007/978-3-030-54161-3_81

Bioethics, Education

The number of ethics-teaching programs rapidly grew in the early 1970s, primarily in medical schools in the United States. In a relatively short period of time almost all medical schools introduced ethics education. Currently, such schools are required to include bioethics in their curricula to be accredited. Other countries followed this pattern of dissemination. Since then the scope of bioethics education has significantly widened. Ethics-teaching came to be offered not only in undergraduate programs but also in graduate, specialization, and postgraduate education, especially in clinical settings. Bioethics-teaching was furthermore introduced in the professional training programs of other health professions such as nursing and scientific disciplines such as biology, genetics, and life sciences. Finally, bioethics education has become relevant outside the professional training context as a resource for experienced practitioners, members of ethics committees, policy-makers, journalists, and interested parties in public debate. This growth of bioethics education is in line with the wider notion of bioethics as a new discipline that combines scientific knowledge with philosophy and ethics to analyze and comprehend the contemporary challenges facing science and technology for health, life, and care. Bioethics education, particularly at the global level, is confronted by several challenges. Although bioethics-teaching programs mushroomed in the 1970s and 1980s in the United States and European countries, the situation has stabilized since then and in many countries all medical schools now have ethics-teaching programs. However, there is a risk that this situation will deteriorate under economic and political pressures put on universities as a result of which experienced staff are replaced by temporary adjuncts and online courses. Ethics-teaching is also regarded by policy-makers as a curious type of palliative remedy. Every time professionals infringe important ethical norms the need for ethics-teaching is reemphasized as the antidote. In response to a repeated cycle of cases of scientific misconduct and ethical problems concerning financial conflicts of interest, the National Institutes of Health and the National Science Foundation in the United States have required as of January 2010 that researchers funded by their grants must have received ethics education focused on promoting research integrity. Education in ethics is seen as a remedy to deficiencies in professional behavior. However, it is obvious that the impact of bioethics education is limited if the systemic and structural causes of such misconduct are not addressed. Although bioethics-teaching is undertaken, in most countries it is not very impressive in terms of volume, time, and commitment. Studies show that bioethics education in the United States, although required, comprises only 1% of the medical school curriculum. Many educational activities are sporadic and occasional. In Europe most hospitals have only short-term educational initiatives instead of longer courses and programs, while nobody

H. ten Have and M. do C. Patrão Neves, *Dictionary of Global Bioethics*,
https://doi.org/10.1007/978-3-030-54161-3_82

seems to take responsibility for the activities of bioethicists. Moreover, there is a serious lack of qualified teachers. Fewer than half the bioethics instructors in the United States have published an article in bioethics. Research is the primary academic focus of such teachers. Another challenge is related to bioethics education itself as borne out by the enormous heterogeneity of the field. Although different types of programs are offered within the same country, didactic approaches and methods differ, the number of teaching hours vary widely, and ethics courses are not scheduled in the same phases of the curriculum. Major controversies exist concerning the objectives, methods, content, and evaluation of teaching activities. However, such diversity does not imply there is no consensus at all. Over the last few decades scholars have come to agree that certain approaches to teaching are preferable. For example, there is a need for longitudinal and integrated programs making ethics part of daily care routine—not an isolated, one-time event; there is a need for team teaching with close cooperation between ethicists and clinicians; there is a need for a student-centered approach to bioethics education focused on active learning (preferable since it encourages critical thinking and reflections); and there is agreement on the need for comparative studies. Developing teaching programs is often not informed by experiences elsewhere. In many cases the wheel is reinvented as a result of the few descriptive and analytic studies of specific programs that have been published. Finally, many efforts have been made to define a common core for bioethics education (e.g., the core proposal in the United Kingdom and the core curriculum for bioethics launched by UNESCO).

Bioethics, Environmental (*See* Environmental Ethics)

In Potter's distinction between medical and environmental (or ecological) bioethics the last form of bioethics is concerned with the survival of humanity. It concentrates on responsibilities toward future generations and takes a long-term view associated with the preservation of biodiversity and ecosystems. The scope of ethics in Potter's view should be expanded beyond the ethics of individuals. He developed the environmental dimensions of bioethics in which the risk for extinction of humanity was his main concern. Preservation of the biosphere and a healthy environment were necessary to guarantee human survival and future disasters could be avoided only if ethical perspectives changed. Deficient water supplies, toxic waste, pollution, acid rain, and global warming seriously affect human health. A broader notion of bioethics is therefore urgently needed. The problem is that the new discipline of environmental ethics has been developed without connecting it to bioethics. Major textbooks, introductions, and overviews in this new area do not generally include any references to Potter or bioethics. Obviously, the idea that the ethical study of environmental problems should be part of the broader discipline of bioethics has not found advocates among the promoters of environmental ethics. This is curious since the orientation of this new field is close to Potter's notion in advocating an ethics of life in all its manifestations. It is focused on human survival rather than individual well-being, on long-term perspectives, and on the connectedness of human beings with the natural world. Such advocacy was part of a trend at that time with well-known personalities such as Albert Schweitzer and Aldo Leopold. At the same time the orientation of environmental ethics is theoretical. Reflecting on the moral dimensions of relationships between human beings and their surrounding environment has generated two different ethical approaches: one human centered and the other life centered. The first anthropocentric approach proceeds from the fundamental principle of respect for people in that it identifies duties that human beings should have, employs the notion of future generations, and emphasizes the usefulness of the biosphere for human beings. The second non-anthropocentric approach proceeds from the fundamental principle of respect for nature in that it delineates the duties people have to living things and emphasizes they are entities that have intrinsic value or a good of their own. This second approach has two variants: biocentrism and ecocentrism. The first assumes that all life has inherent worth corresponding to the view of Albert Schweitzer who espoused an ethics of reverence for life. The second variant ecocentrism is the normative theory that species and ecosystems such as forests, lakes, deserts, and wetlands have moral standing independent of their component individuals. The land ethic espoused by Aldo Leopold was such a holistic view claiming that preserving the integrity of the biotic community was the most important ethical obligation.

H. ten Have and M. do C. Patrão Neves, *Dictionary of Global Bioethics*,
https://doi.org/10.1007/978-3-030-54161-3_83

Bioethics, Global

After Potter coined the term "bioethics" in 1971 the new discipline rapidly developed. Nevertheless, Potter was disappointed at what he felt was its narrow medical focus. He was dismayed that little attention was given to ecological and social issues. This prompted him in 1988 to introduce the notion of global bioethics in a new effort to broaden the scope of bioethical analysis and discussion. "Global" in this notion means two things: unified or broad, bringing together various concerns, and worldwide or planetary, taking into account planetary problems and solutions—not just local and regional ones. Global bioethics for Potter encompassed not only the ecological and social dimension but also the traditional medical ethical perspective focused on individual patient care. Globalization promoted the emergence of global bioethics, especially since the 2000s. It is not the scale but the type of ethical problems that is different due to particular processes of globalization (mainly, economic policies and neoliberal ideology) that are fundamentally changing the social, cultural, and economic conditions in which people across the world are living. For example, in many countries the benefits of scientific and technological progress are not available to the majority of the population. Ethical problems are the result of social inequality, injustice, violence, and poverty. The changing social context negatively impacts human health and well-being. Globalization generally also leads to more private and commercial healthcare services and less social security and governmental protection. This makes healthcare more accessible to the wealthier members of societies while other groups will be more vulnerable. The economic context of science and healthcare is furthermore associated with corruption, trafficking of organs and body parts, and scientific misconduct. If such globalization is responsible for ethical problems, then global bioethics will be a new kind of bioethics—not a different stage. Global bioethics differs in two principal ways from mainstream bioethics in that it has a broader agenda and a broader ethical framework. A new set of moral problems is associated with globalization an early example of which was the HIV/AIDS epidemic. Moreover, new problems have emerged such as biopiracy; brain and care drain; corruption; dual use; food security; health tourism; humanitarian disaster relief; conflicts of interest; pandemics; poverty; refugees; and the trafficking of organs, tissues, body parts, and humans. Global bioethics furthermore applies an encompassing ethical framework that can be used globally. For example, personal autonomy is highly appreciated in the West where individual patients want to be informed about their care and want to decide about possible treatment and intervention. Such a focus on the individual is less outspoken in other cultures where the extended family or community is important. Furthermore, the concept of property is different in various cultures such that genetic information and material cannot be somebody's property since they are God's

H. ten Have and M. do C. Patrão Neves, *Dictionary of Global Bioethics*,
https://doi.org/10.1007/978-3-030-54161-3_84

creation. Such examples underline that fundamental ethical notions such as self-determination and individual ownership differ according to cultural settings and that it is problematic to apply the current ethical framework of bioethics in these settings. A framework of broader ethical principles is offered in UNESCO's Universal Declaration on Bioethics and Human Rights, which goes beyond the well-known principles of mainstream bioethics such as autonomy, beneficence, non-maleficence, and justice (which are in fact incorporated). The UNESCO declaration is not only the first political statement of a global framework, it also reflects Potter's idea of global bioethics covering concerns for healthcare, the biosphere and future generations, and for social justice. The declaration assumes the existence of a global moral community in which citizens of the world not only increasingly connect and interrelate but also share global values and responsibilities. This global community generates certain common principles such as protecting future generations, benefit-sharing, and social responsibility. Various ethical systems in different cultural settings are converging into a single normative framework for all citizens of the world. Such a process is driven by the moral ideal of cosmopolitanism that concerns itself with common heritage, global solidarity and the future of the planet, and thus humanity.

Bioethics, History

Bioethics emerged as a new discipline in the 1970s and as a result put the traditional concept of medical ethics under pressure because of three factors. One was increasing criticism of paternalism of the medical professional. The second was the growing power of science and technology that placed moral concerns about life-supporting technologies, dialysis, transplantation, human research, and reproductive technologies high on the public agenda. The third factor was social change in which patients emphasized the importance of their rights over the virtues and duties of physicians. The transformation of medical ethics into bioethics was also facilitated by high-profile scandals related to medical experimentation that necessitated legislation and regulation of medical activities. In a relatively short time bioethics rapidly developed and was taken up by institutes, centers, teaching programs, journals, and professional associations first in the United States and Europe, but somewhat later in other parts of the world too. Although Potter was the first to use the term "bioethics" in a publication in winter 1970, his claim was diluted by the argument that the word was in use around the same time by the founders of the Kennedy Institute (founded on July 1, 1971). The story is further complicated by the claim that bioethics was a European innovation and that the term was in fact coined long before the emergence of bioethics as a movement and discipline. The German pastor Fritz Jahr introduced the new German word *Bio-Ethik* in a publication in 1927. His concept of bioethics was broad and based on respect for both human beings and other living organisms much like the respect for life advocated by his contemporary Albert Schweitzer. Historically, the globalization of bioethics took place in four stages. In the first stage the dominant paradigm of bioethics was increasingly criticized as individualistic and minimal with a limited agenda. In the second stage bioethics was confronted by a new set of problems. The new disease HIV/AIDS highlighted the limitations of its ethical framework, especially the neglect of public health and the common good. It demonstrated that bioethics could no longer solely focus on relations between individuals. The third stage was the proliferation of international activities and collaborations. The Human Genome Project in 1990 was a major impetus for international cooperation. There was also an explosion of cross-cultural studies in bioethics. Finally, since globalization confronted humankind worldwide with similar challenges—not just particular countries and cultures—there was a search for common ethical principles and values that could guide shared decision-making and policy development. Adoption of the European Convention on Human Rights and Biomedicine in 1997 by the Council of Europe is an example of a regional framework. The Universal Declaration on Bioethics and Human Rights unanimously adopted by member states of UNESCO in 2005 is an example of extending such a framework into a global one.

H. ten Have and M. do C. Patrão Neves, *Dictionary of Global Bioethics*,
https://doi.org/10.1007/978-3-030-54161-3_85

Bioethics, Medical

Bioethics emerged as a new discipline in the 1970s. The term "bioethics" was coined by Potter. Although the same term was also used by the founders of a new institute at Georgetown University in the United States, Potter intended the new discipline to have a broad scope including not only medical but also social and environmental concerns. However, he was disappointed that the subsequent development of bioethics did not go beyond the individual patient perspective. Without a broader perspective he felt bioethics would not be very different from traditional medical ethics. This prompted him to make a distinction in 1987 between medical and ecological bioethics. Medical bioethics is focused on the moral problems facing individual physicians and patients, especially the moral principle of personal autonomy and individual rights. Moreover, medical ethics also takes a short-term perspective on treatment options. In Potter's view the problem was that contemporary bioethics had become restricted to medical bioethics and did not take into account environmental issues. His conclusion was that mainstream bioethics was not essentially different from the traditional approach of medical ethics, although it incorporated more challenges, especially because of the development of new biotechnologies.

H. ten Have and M. do C. Patrão Neves, *Dictionary of Global Bioethics*,
https://doi.org/10.1007/978-3-030-54161-3_86

Bioinvasion (*See* Invasive Species)

Bioinvasion has been regarded as a major threat to biodiversity especially since the 1990s. Bioinvasion is the introduction and spread (intentional or accidental) of non-native species outside their natural past or present ranges that causes the extinction of native species and changes existing ecosystems such that ecosystem services are affected, with important consequences for food, water, and health. The effects of what are called "invasive alien species" are mostly presented as negative. They can produce health problems such as allergies and skin damage. Many disease vectors have mostly unintentionally invaded new territories and disseminated diseases where they were not known before. Invasive species reduce yields in agriculture, fisheries, and forestry and decrease water availability. The economic damage is significant. Article 8 h of the Convention on Biological Diversity declares that contracting parties shall prevent the introduction of, control, or eradicate those alien species that threaten ecosystems, habitats, or native species. As a result many countries have developed policies to protect native biodiversity against such species aimed at prevention, early detection, and rapid eradication. The phenomenon of bioinvasion has recently been more critically examined. As a result of globalization non-native species are increasingly spreading over the world and will continue to do so unless trade, travel, and tourism are curtailed. Hence containing biological invasions seems an impossible task. There is also a more positive assessment of alien species. Human beings have always introduced species to other countries and continents. Exotic species were often introduced for commercial purposes or simply because people liked them. Introducing non-native species to new habitats apparently manifests a human tendency to enhance the lifeworld (i.e., to increase knowledge, satisfy desires, and grow economic output). Agriculture has been greatly enhanced in this way. Although the terminology of "alien" and "invasive" presupposes an ethical judgment, evidence shows that many non-native species are not harmful to biodiversity and can help to restore damaged ecosystems. Most invasive species can be used to recover biodiversity rather than being attacked with chemicals to protect a pristine nature. They can also be used to develop novel ecosystems with mixtures of native and alien species.

H. ten Have and M. do C. Patrão Neves, *Dictionary of Global Bioethics*,
https://doi.org/10.1007/978-3-030-54161-3_87

Biolaw

Biolaw refers to legislation concerning the development and use of biotechnologies throughout the entire process. This includes everything from project design (e.g., stem cell gene editing or embryo genome editing), to the procedures implemented (e.g., the use of surplus embryos or embryo production for research), through to potential impacts (e.g., research embryo discard or transfer). Biolaw has a broad scope in that it concerns individual and social levels, humans and biodiversity, the environment, the planet, the present, and the future. The amazing advances made by biotechnologies mostly in recent decades have indeed brought about quite new situations, totally unknown realities, and conflicts and dilemmas never before experienced. Societies did not know how to deal with such situations and lacked a clear perception of what was good or bad and what should be allowed or forbidden. A classic example is the case of Karen Quinlan who suffered a cardio-respiratory arrest in 1975 and was resuscitated using innovative and advanced techniques. Although she remained in a persistent vegetative state for 11 years, she was not biologically dead. She remained in a state between life and death until she contracted pneumonia when family and healthcare professionals agreed that any treatment would be a disproportionate therapeutic measure. During this time there were court rulings concerning her case and other similar cases that provided guidelines on how to deal with the ethical issues raised by this and other new realities created by biotechnological advances. Although biolaw has always been associated with bioethics, the relation between both followed two different approaches depending on whether the Anglo-American legal system was employed or the continental European legal system (i.e., civil law). The Anglo-American system is a common law system in which principles and rules are grounded in universal custom or natural law. They are then developed and applied by courts. Court rulings (jurisprudence) establish a precedent that can later be cited and can be used to ground other court decisions. In common law jurisdictions most guidelines to tackle bioethical issues are provided by the courts. The continental European system is based on civil (or Roman) law in which the long legal tradition in each country is applied to new situations. Such new cases are also discussed from a bioethical and social point of view and an assessment is made about the need for new legislation to address unprecedented cases. Both case law in the Anglo-American system and the new laws required by the application of biotechnologies in the continental European system can be said to configure biolaw. However, in the Anglo-American setting the usual designation given to this specific legal field is "health law" or "law and bioethics," while "biolaw" is a common concept in the continental European legal framework where it strictly refers to jurisdictional translation of social agreements and attributes legal strength to ethical consensus. Biolaw follows bioethics either at the national or international level and is currently considered a specific branch of law.

H. ten Have and M. do C. Patrão Neves, *Dictionary of Global Bioethics*,
https://doi.org/10.1007/978-3-030-54161-3_88

Biological Weapons (*See* Biosecurity; Dual Use; Weapons)

Soon after the terrorist attacks in the United States five letters containing weaponized anthrax spores were mailed (late September 2001). They made 22 people ill, 5 of whom died. The FBI investigation took a long time since the assumption was that foreign terrorists or governments were responsible. Although the FBI concluded in 2008 that a biodefence expert working in the US Army Research Institute of Infectious Disease was the perpetrator, this conclusion is still controversial. The anthrax attack raised the issue of biosecurity and placed concern about biological weapons high on the public agenda. Contemporary science makes it possible to develop biological agents, especially disease-producing agents such as bacteria, viruses, and toxins, and use them as weapons against human beings. Biological warfare is not a new phenomenon. There are many examples in history of the deliberate use of disease agents to destroy enemies. Several countries have set up biological weapons programs in the last century. In 1972 countries agreed to prohibit the development, production, and stockpiling of biological weapons. They signed the Convention on the Prohibition of the Development, Production and Stockpiling of Bacteriological (Biological) and Toxin Weapons and on Their Destruction (BWC). This convention came into force in 1975. It replaced the Geneva Protocol of 1925 that banned the use of biological (and chemical) weapons, but not their possession or development. However, the United States and the Russian Federation continued research for defensive purposes. Examples of the use of biological weapons for bioterrorism are scarce. In 1984 people in Oregon were intentionally infected with *Salmonella* by followers of Bhagwan Shree Rajneesh. In 1994 the Aum Shinrikyo sect in Japan released anthrax in the Tokyo subway. In 2002 the British police arrested six suspects who attempted to produce ricin in their apartment in Manchester. Such examples illustrate that the threat of biological warfare is no longer restricted to the state level but now includes that of non-state actors. Ethical debate on biological weapons relates to the issue of dual use. Although modern bioscience is developing many new products and technologies that can be used for the benefit of humankind, at the same time it can be misused for sinister purposes. For example, although instructions about how to modify or create new viruses generate new knowledge, they can also be used for bioterrorism. Terrorists are more likely to develop biological weapons than nuclear weapons. The emphasis therefore should be on vigilance. This raises the question as to how the threat of biological weapons can be minimized without obstructing the advance of science and the free flow of scientific information. The life science community itself has a growing responsibility to prevent the risks of dual use.

H. ten Have and M. do C. Patrão Neves, *Dictionary of Global Bioethics*,
https://doi.org/10.1007/978-3-030-54161-3_89

Check for updates

Biometrics

Biometrics is the measurement of the physical and behavioral characteristics of people. The assumption is that every person can be identified on the basis of his or her physical and behavioral characteristics such that each individual can be authenticated. Biometrics technology is therefore used to identify individuals for various purposes. Physical identification can be based on facial recognition, fingerprints, retina scanning, voice recognition, and DNA typing. Behavioral identification is based on behavioral patterns such as gestures and walking gait that are unique to individuals. Biometric verification is increasingly used since it does not require passwords, passports, or other security measures. Biometric data are usually stored in databases. Biometrics poses two ethical concerns. One is that the growth of biometrics promotes the expansion of surveillance systems and is therefore a component of biopolitics. Huge databases are built that include many data on citizens that could be used for various purposes. India set up the Aadhaar project to provide every citizen with a single unique identifier. Although this has been designed to facilitate government services, it will also be very attractive for private transactions. China is using biometrics to monitor political opponents and mandatorily collects fingerprints, eye scans, and DNA samples from people in Xinjiang to monitor the Muslim population. Security cameras are everywhere in Chinese cities. People are surveilled to build a "social credit system" where citizens are ranked according to their behavior. Bad driving, gambling, buying too many video games, not paying your bills on time, or smoking are considered infractions and can lead to consequences such as restricting travel, barring stays at hotels or taking vacations, not being able to get loans at cheaper rates, or having access to better schools. The other ethical concern regards privacy. Although biometric data are usually collected and stored with the consent of the person involved, it is not clear how secure the data are. Even though data cannot commonly be cancelled or changed (unlike passwords), they can be disclosed and used for various purposes that are no longer under the control of the individuals concerned.

H. ten Have and M. do C. Patrão Neves, *Dictionary of Global Bioethics*,
https://doi.org/10.1007/978-3-030-54161-3_90

Bionics

The word "bionics" derives etymologically from the Greek word *bios* (life) and the ending of the word "electr*onic*." It thus basically refers to an association between biology and electronics. Bionics today designates a new science developed at a time when sciences commonly converge. It combines biological knowledge with expertise in electronics and applies knowledge about how biological systems function to help solve engineering problems (in electromechanical engineering) and to bring about electronic or mechanical advances in electromechanical devices to reproduce and enhance biological systems (in medicine). Although the neologism goes back to the 1960s, it was only in the late 1990s that the first bionic arm was successfully implanted. Currently, most people think of bionics as the production of artificial human body parts that are electronically or mechanically powered. Initially such parts were mostly limbs reproducing as close as possible the original function of the organic body part. Bionic implants today such as legs, arms, and eyes can in fact achieve a higher performance than the organic part they substitute. This is the reason bionic implants are used to replace amputated limbs. Not only are they able to substitute the lost function, more recently they are also used to enhance the performance of some (deteriorated) functions. From the ethical perspective there are two major lines of reasoning. The first stresses bionics as facilitating the replacement of human body parts and the recovery of lost functions. At such a level bionics contributes to the well-being of individuals such as amputees thus overcoming the problem of scarcity of body parts, as happens with transplantation. However, accessibility to such implants is a major concern due to their price and lack of coverage by most national health systems or health insurances (e.g., cochlear implants). Although the availability of this technology will surely increase and the price decrease thus widening access to it, social justice is an important ethical issue. The second ethical line of reasoning concerns the potential hybridization of human beings by the addition and reinforcement of a new electromechanical dimension (artificial organs) to the natural organic dimension of human beings. The ethical issues raised within such a framework are common to those of human enhancement and transhumanism with some arguing that the artificialization of human beings is a human achievement and part of human evolution (to cyborgs) and others arguing that human identity is in danger by creating a new species for which we have no authority.

H. ten Have and M. do C. Patrão Neves, *Dictionary of Global Bioethics*,
https://doi.org/10.1007/978-3-030-54161-3_91

Biopiracy (*See* Bioprospecting)

Bioprospecting has long been commonly carried out without any concern for benefit sharing with source countries and indigenous populations. From the perspective of developing countries and environmental NGOs it is a kind of theft in which resource extraction is unfair much as it had been in earlier times. Notwithstanding what little positive discourse there is regarding bioprospecting the practical reality is that resources in developing countries are appropriated and monopolized by scientists and international companies in developed countries. Such injustice is called "biopiracy." Although it has long existed, it is not by accident that biopiracy became a contested issue. The context of bioprospecting significantly changed in the early 1990s as a result of two legal treatises. First, the Agreement on Trade-Related Aspects of Intellectual Property Rights (TRIPS) and, second, the Convention on Biological Diversity (CBD). TRIPS is a legal agreement established in 1995 between member states of the World Trade Organization (WTO). It introduced Intellectual Property Rights (IPR) to the international trade system. TRIPS has been in force since 1995. Being part of the WTO it presents a stringent governance regime backed up by enforcement and dispute settlement procedures such that countries can be forced to implement its regulations. The primary concern of TRIPS is the protection of property rights and it makes no reference to the protection of traditional knowledge. Its basic assumption is simple: if things are not patented, then they are not owned; if they are not owned, then they belong to the global commons. In other words they are *terra nullius* and belong to everybody and can be taken by anybody. Such a position conflicts with that of the CBD that came into force in 1992. The CBD is the only major international treaty that acknowledges indigenous communities as the owners of biodiversity. Its goal is to find a balance between conservation, sustainable use, and sharing of benefits. One way of doing this is through benefit-sharing arrangements in which bioprospectors are granted access to biodiversity and traditional knowledge. It is clear that the approaches taken by TRIPS and the CBD conflict. The biopiracy discourse in fact criticizes the priority given to intellectual property rights and thus patenting. TRIPS encourages the commercialization of biodiversity and monopolization of natural products without addressing benefit sharing and thus in fact promotes biopiracy. Bioprospecting in reality ends up most often in exclusive monopoly control without any compensation for countries and populations. Vandana Shiva is an activist from India and was the first to raise the issue of biopiracy. She rightly pointed out that the discussion about TRIPS, patents, and IPR should be about the ethics of how we relate to other species and about how our biodiversity is used and controlled—not just about trade. In her view it is an issue of justice and human rights. Since bioprospecting frequently results in claims of exclusive ownership of natural resources and products that have been used for centuries by

H. ten Have and M. do C. Patrão Neves, *Dictionary of Global Bioethics*,
https://doi.org/10.1007/978-3-030-54161-3_92

traditional and indigenous cultures, it can be regarded as a form of exploitation if such cultures are not respected and compensated. This is the reason bioprospecting has frequently been redefined as "biopiracy" or the unauthorized and unfair exploitation of biological resources and/or associated traditional knowledge. This highlights the ethical problem of global injustice. To address this problem the notion of benefit sharing has been introduced as one of the goals of the CBD, specifically in the Nagoya Protocol that established the Access and Benefit-Sharing framework (ABS). Such an approach to the problem is based on the notion of justice as an exchange between two parties. The underlying idea is that countries rich in biodiversity provide access to resources in return for benefits that are derived from their resources. Sharing benefits is thus a compensation mechanism or a form of commutative justice (justice in exchange).

Biopolitics

Biopolitics is a neologism consisting of the Greek words *bios* (life) and *politikos* (that which relates to the citizen or to the state). It refers today to the organization and administration of people and public life, of nations and states, and of common goods. Etymologically, biopolitics refers to the political power applied to life either in its natural manifestation such as preservation of biodiversity or in its artificial manipulation by bioengineering such as the production of genetically modified organisms (GMOs). Historically, the concept of biopolitics is not univocal and has been used with quite different meanings. Thomas Lemke believes these depend mostly on whether life is viewed as determining politics or whether the object of politics is said to be life. The word "biopolitics" was first used in the mid-1960s by North American political scientists. Some of them did not agree with the mainstream idea at the time that social behavior is only socially determined, but considered that biological influence could not be ignored. They developed an intellectual movement advocating recovering the social primate genetic legacy (i.e., the social behavior of primates including humans becoming part of their genetic heritage within an evolutionary perspective) as key to understanding the political organization and dynamics of human societies. Although this is the original neo-Darwinian meaning of biopolitics grounded in political sciences and biology, it has been fading away. A second and perhaps still the most immediate and common meaning attributed to biopolitics is the one proposed by the French philosopher Michel Foucault about a decade later in a series of lectures between 1975 and 1978) (i.e., in *The Birth of Biopolitics*, 1978–1979). Foucault introduces the concept of biopower as a new model or strategy of governing populations and life itself pointing out how political power then recently enabled by technology has extended over all major processes of human life. Biopolitics refers to public policies regarding the application of biotechnologies to life sciences, controlling life and populations, intervening at the level of everyday issues such as reproduction or mortality in the name of—we would add today—health and human rights. Foucauldian biopolitics has its roots in philosophy and sociology and has been widely and critically developed by scholars such as Giorgio Agamben, Antonio Negri, Lemke, Nikolas Rose, and Paul Rabinow. A third meaning was attributed to biopolitics with the creation of the Biopolitics International Organization (BIO) in 1985. Biopolitics refers here to all policies that put life and the environment at the center of decision-making aiming to protect the gift of life and environmental sustainability as essential for the future of humanity and mobilizing all communities and all nations to acknowledge the value of diversity and to develop respectful relationships. This third concept of biopolitics derives from and almost coincides with environmental ethics. Finally, there is a fourth meaning of biopolitics that is becoming more common mostly because of its

H. ten Have and M. do C. Patrão Neves, *Dictionary of Global Bioethics*,
https://doi.org/10.1007/978-3-030-54161-3_93

global dimension along with the widespread impact of bioethics. It refers to national and international political decisions concerning bioethical issues. As bioethics developed from the personal level and intersubjective relationships to the social level and valuing common good through to the international level and global cooperation, its articulation with politics became more important. Such a perspective advocates that the ethical consensus (bioethics) should be legally established (biolaw) and politically implemented (biopolitics) in a process that should unfold at the national and the international level. Although such an articulation between bioethics, biolaw, and biopolitics is minimalist, it should contribute in a very significant way to global bioethics with core values and guiding procedures extending worldwide (opposing such phenomena as medical tourism).

Bioprinting

Bioprinting is a new technology enabling the production of three-dimensional (3D) tissue and organ structures (organoids). Such a technology is increasingly being developed in life sciences and basic research studying cellular mechanisms focused on tissue regeneration (i.e., skeletal, muscular, nervous, lymphatic, endocrine, reproductive, integumentary, respiratory, digestive, urinary, and circulatory systems) and applied to building tissues and organs for implantation (i.e., heart valves, myocardial tissue, trachea, and blood vessels). The ultimate goal is to be able to produce fully functional whole organs and in so doing solve the two major problems of compatibility and scarcity in transplantation. Bioprinting has also been used for in vitro drug-testing models thus contributing to personalized medicine. Although the range of beneficial applications of bioprinting is impressive and significant progress has been made in the field, the clinical translation of this technology is still in the future. Bioprinting offers new hope for transplantation and regenerative medicine in the fight against the lack of donors and the shortage of organs. It does so by fabrication-based research focused on tissue regeneration. However, there are significant challenges that hamper the clinical utility of bioprinting technology.

H. ten Have and M. do C. Patrão Neves, *Dictionary of Global Bioethics*,
https://doi.org/10.1007/978-3-030-54161-3_94

Bioprospecting (*See* Biopiracy)

Bioprospecting is the systematic search for biological and genetic resources in plants, animals, and microorganisms in the wild. The basic idea is that such a search will potentially deliver genes and chemical products that will benefit humanity, especially by delivering pharmaceuticals. Many pharmaceutical companies today have grown into multinational corporations thanks to bioprospecting and ethnobotanical research. Plants, animals, and microbes hold great potential as medication as borne out by human beings using them for thousands of years. Many effective drugs are products of nature or derived from them. More than half the most prescribed drugs in the United States are derived from natural sources (e.g., simvastatin and cyclosporine). In their search for new drugs researchers and pharmaceutical companies have long undertaken bioprospecting of biological and genetic samples, particularly in developing countries that have a rich biodiversity (especially also off their coasts). Biodiversity is regarded as providing a kind of genetic insurance against future unknown diseases. Since most plants and animal species have yet to be described and analyzed, who knows what is out there or what new medications are waiting to be discovered? However, biodiversity is rapidly lost and the arsenal of potential new drugs is shrinking every day. Bioeconomically, this means that commercial opportunities will be lost. New promising drugs will never be discovered unless more resources are invested in countering biodiversity loss. Interest in natural products is currently resurging primarily because natural products have superior novelty and chemical diversity. Natural products are repositories of genetic information that lead to synthesizing new chemical entities. The very act of identifying a medicinal plant is interesting because it involves looking at promising molecules, suggests blueprints for modification, and can be the source of novel structures—not because it is raw material or a substance that can directly be applied as a final drug. Although bioprospecting has the dual purposes of discovering new products (especially food and medication) and conserving endangered ecosystems, the underlying idea is that it is commercially valuable. The collection and assessment of biological samples may lead to useful products thus providing an economic incentive for biodiversity protection and conservation. Such an idea is expressed in a common definition of bioprospecting as the exploration of biodiversity for commercially valuable genetic and biochemical resources. This raises a couple of issues such as collectors taking interesting natural products from

H. ten Have and M. do C. Patrão Neves, *Dictionary of Global Bioethics*,
https://doi.org/10.1007/978-3-030-54161-3_95

biodiverse countries and developing commercial products without compensating the source country. While recognizing that biological and genetic resources can have a medical as well as a spiritual and cultural value (thanks to the traditional knowledge of indigenous populations who have used these resources for decades or even centuries), the question arises as to who owns nature and traditional knowledge. Developing countries, especially in the 1990s, increasingly complained that rich countries exploited their natural resources often without compensation. In this context the term "biopiracy" was coined.

Biosafety (*See* Biosecurity)

Biosafety is the discipline covering the safe handling and containment of infectious microorganisms and hazardous biological materials. Many laboratories today work with pathogenic organisms and their toxins. Military biological research laboratories, particularly in the United States and the Russian Federation, have been working with dangerous pathogens for decades in their biological weapons programs. Infectious diseases such as COVID-19 have recently emerged and little is known about how pathogenetic they are. Since treatments and vaccines are often unavailable more research will be required. The focus of biosafety is on containment such as preventing laboratory workers from being exposed and preventing the escape of pathogens from the laboratory. Several safety layers have been developed such as the need to provide workers with immediate protection from exposure to chemical and biological hazards. Another is to prevent contamination by incorporating special architectural and mechanical designs at the facility. A number of biosafety levels can be distinguished according to the risks (the United States have four) such as all laboratory staff are required to have special biosafety training; the need for continuous risk assessment; and categorizing biological agents in risk groups based on pathogenicity, availability of countermeasures or prophylactic treatment, and ability of the disease to spread. However, it is impossible to eliminate all risks.

H. ten Have and M. do C. Patrão Neves, *Dictionary of Global Bioethics*,
https://doi.org/10.1007/978-3-030-54161-3_96

Biosecurity (*See* Biosafety; Bioterrorism)

The term "biosecurity" was initially used to protect agriculture, livestock, and the environment against invasive species and diseases. Human health was later included when infectious diseases were regarded as security threats. The term has been widely used since the War on Terror to focus on concerns about biological terrorism. Biosecurity has become much broader and now covers the management of biological risks to the environment, agriculture, animals, food, and humans. It focuses on pathogens that can affect populations whether by natural occurrence as in the case of emerging infections, by accidental release, or deliberate use. Emphasis over recent decades has been on human actors such as states, terrorists, criminals, and scientists as sources of biohazards. Biodefence infrastructures have been built, particularly in the United States, such as Bioshield (providing countermeasures like stockpiling vaccines and drugs against the most dangerous biological agents, increased research to develop new drugs, and fast-track approval by the FDA), BioWatch (early warning and detection), BioSense (collecting real-time health data for the whole country), and the establishment in 2005 of the National Science Advisory Board for Biosecurity. The difficulty with biosecurity is that an agreed definition is lacking and it is used to cover heterogeneous practices reflecting different concerns. For example, in the United States the emphasis is on bioterrorism and laboratory safety, in Australia and New Zealand primary concerns are with invasive species, while in Europe agriculture and food safety are priorities. The emphasis given to biosecurity is now increasingly criticized for a number of reasons such as the common benefits expected have not materialized; when the funding for Bioshield multiplied the number of emergency departments in US healthcare diminished and grants for non-biodefence microbial research decreased; and the number of epidemiologists working in environmental health dropped. This led many to conclude that the emphasis on and investment in bioterrorism was counterproductive and negatively impacted public health efforts. More importantly, growth in the security apparatus had the opposite effect and in fact multiplied biothreats as borne out by the rapidly increasing number of biosafety laboratories (now 1,300 with 14,000 laboratory workers) raising the probability of accidental release or escape of pathogens. For example, between 2003 and 2009 nearly 400 accidents were reported in US labs involving dangerous pathogens and toxins. The focus on biosecurity (and bioterrorism) changed the bioethical debate on health and disease in two ways. First, it securitized health and, second, it militarized security. Regarding disease as a security threat is little more than continuing and intensifying the military vocabulary that has always been used to characterize infections and alien species. Connecting disease and terror in such a way is not new since the fear and dread of past pandemics have led to social disruption,

H. ten Have and M. do C. Patrão Neves, *Dictionary of Global Bioethics*,
https://doi.org/10.1007/978-3-030-54161-3_97

economic collapse, and cultural change. However, present day biosecurity is no longer metaphorical since it has created structures, systems, and methods that survey, monitor, contain, and respond to potential threats and has reinforced the role states play in securing their borders. Moreover, biosecurity has developed a risk discourse to deal with uncertainties suggesting that prediction is a matter of scientific data. Framing health as an issue of security introduces a specific normative perspective that emphasizes that infectious diseases not only have an individual impact but also endanger societies and social institutions, economic exchange, and political stability. Although framing health in this way forces states to address infectious disease, it bypasses other ways of framing health such as regarding disease as a medical or scientific challenge, a rights issue, or a humanitarian crisis.

Biosphere

The term "biosphere" was first coined by Austrian geologist Eduard Suess in 1875 to refer to the layer approximately 20 km thick on and around Earth in which life exists. The term was popularized by the Russian scientist Vladimir Vernadsky in his book *The Biosphere* (1926). The biosphere comprises a number of layers including the lithosphere (the solid and rocky surface layer of the Earth); the atmosphere (the layer of air), and the hydrosphere (water-covered areas of the Earth). The biosphere is a complex ecosystem consisting of many specific ecosystems and is essential to the survival of life on Earth. In 1971 UNESCO launched the Man and the Biosphere (MAB) Programme to promote sustainable development by setting up a network of biosphere reserves in an effort to establish a working, balanced relationship between people and the natural world. The biosphere as the zone of life is currently threatened by human activities that are responsible for climate change and pollution. The global bioethical concern about how to preserve the biosphere is expressed in the Universal Declaration on Bioethics and Human Rights. Article 17 formulates the principle of protection of the environment, the biosphere, and biodiversity.

H. ten Have and M. do C. Patrão Neves, *Dictionary of Global Bioethics*,
https://doi.org/10.1007/978-3-030-54161-3_98

Biotechnology

Biotechnology has a long history as borne out by microorganisms being used in the production of beer, yogurt, and cheese and by humans domesticating animals and selectively improving crops. However, modern biotechnology is different in that it uses new scientific methods such as genetic engineering, tissue culture, and embryo transfer; it is capable of crossing the genes of one species with another; and it works much faster than traditional biotechnology. Biotechnology as the application of biology for human purposes can be ethically assessed from two perspectives one of which is the consequentialist view where only the consequences are relevant. If biotechnology can improve human health or eliminate diseases, then the benefits and harms of biotechnological interventions should be determined and calculated. This raises the major question as to whether they are safe for human beings, for biodiversity, and for the environment. Once safety and risks are established, then ethical decision-making can proceed. However, such an ethical framework is rather limited and in many cases consequences cannot be predicted. Even if safety can be ensured, it should be remembered that the word "safe" does not mean the same as "better." Often a limited view prevails that focuses on the harms and benefits for individuals, while longer term consequences for the environment are disregarded. The other ethical perspective emphasizes intrinsic arguments in which biotechnological interventions are regarded as unnatural. For example, many people believe genetically modified food is abhorrent since it is completely artificial. The producers of such "Frankenfood" fail to recognize there are limits to what humans can manipulate. Other people argue that the advocates of biotechnology are arrogant in assuming they can improve creation by "playing God." It is further argued that the advocates of biotechnology do not respect the dignity of living beings. By assuming technologies are there to change and improve life, they overlook the fact that living beings also have intrinsic value and should be respected for what they are—not merely instrumental value.

H. ten Have and M. do C. Patrão Neves, *Dictionary of Global Bioethics*,
https://doi.org/10.1007/978-3-030-54161-3_99

Bioterrorism (*See* Biosecurity)

Bioterrorism involves the intentional release of biological agents such as the anthrax letters sent in late September 2001 in the United States. Five letters with powder containing weaponized anthrax spores were mailed and made 22 people ill, 5 of whom died. Although it was long assumed that foreign terrorists were responsible, in 2008 the FBI concluded that a biodefence expert working for the US Army Research Institute of Infectious Diseases was the perpetrator. Many experts have become skeptical about bioterrorism because threats are rare and often exaggerated. Except for a few, exceptional, and small-scale cases, no biological attacks have happened. The risk of dying from accidents or gun violence is many times higher than bioterrorism. Nevertheless, the fear of bioterrorism has led to the establishment—perhaps more in the United States than other countries—of a specific infrastructure and an ideology that has important ethical implications four of which are morally questionable. First, the focus on security assumes opposing sides in that threats (invasions) are foreign coming from outside and "others." They should be contained and enclosed so that 'safe' can be separated from 'unsafe.' Instead of enhancing relationships and interconnections with other humans, other species, and the nonhuman world the emphasis is on identifying, monitoring, restricting, and controlling them. A consequence of the biosecurity discourse is that it promotes antagonism rather than relationships and may therefore hinder international cooperation and downgrade the moral vocabulary of global solidarity. A second consequence of securitization is that priorities are distorted. It is not simply about securitizing efforts being directed against specific pathogens (those that can be used by terrorists) or particular places from where invasions originate (usually Africa and Asia), the focus of biosecurity is on keeping the dangers out—not on addressing the conditions in which they are produced. Moreover, some dangers and threats cannot be militarized thus explaining why biodiversity loss and damaging neoliberal market practices are not considered security threats. Although there clearly is a conflict between the management of biological risk and the neoliberal ideology of free trade, such a conflict is rarely regarded as an ethical problem. A third implication for the bioethical debate is that the security framework implies a specific approach to biological threats that in its perpetual war scenario proceeds with continuous surveillance and vigilance. The best response to threats is preparedness and preemption (eradication) rather than prevention. In 2005 the WHO launched its Global Influenza Preparedness Plan urging countries to make national biopreparedness plans. The rationale was that the future was full of imminent catastrophes and the assumption was that such threats could not be prevented (i.e., there was no way of predicting when they would emerge). Furthermore, future pandemics were inevitable. Since the risks were always potentially devastating it was better to be prepared and

H. ten Have and M. do C. Patrão Neves, *Dictionary of Global Bioethics*,
https://doi.org/10.1007/978-3-030-54161-3_100

take preemptive action before the pathogens were an actual threat. For public health this meant stockpiling vaccines and drugs as well as early mass vaccination. Finally, the security framework has social and cultural implications in that it works as a mechanism of depoliticization. With its emphasis on risk assessment as an objective and scientific tool it is foreclosing public debate and regarding risk evaluation as an expert matter. However, the aim of biosecurity measures is the evaluation of risks that are acceptable—not zero risks (there is incomplete knowledge and thus fundamental uncertainty and unpredictability). The basic question raised is ethical in ascertaining what is acceptable and to whom. Risk assessment therefore is an ethical and political issue related to the common good—not a technical issue. Presenting it as a technical issue the biosecurity framework usually imposes itself as a top-down approach to regulation and communication—not involving citizens, engaging communities, or encouraging public debate. Although this does not allow space for dialogue, solidarity, trust, and convergence around common interests, it does affect society and culture because of the peculiar nature of the threat. Although biological threats seldom happen, they are always possible and the consequences are very serious. In addition to their potential to emerge at any time biological agents are invisible and difficult to distinguish from natural sources. This context of uncertainty, urgency, and threat produces a perpetual state of emergency in which everything and everyone can be a threat. Against such a background countries have introduced new legislation and practices restricting and violating human rights, reducing and surveying the public sphere, and practically eliminating privacy. Accountability, transparency, trust, participation, and engagement as core principles of democratic societies are no longer regarded of primary relevance in the fight against biothreats since security demands compliance and docility—not critical citizens.

Birth Control (*See* Contraception; Fertility Control)

Birth control is the use of methods or devices to prevent pregnancy. Different options are available such as condoms, contraceptive pills, intrauterine devices, implants, the rhythm method, spermicides, vasectomy, and tubal ligation. Although birth control was applied in historical times, it only became safe and effective in the twentieth century. Birth control methods encourage family planning allowing parents to choose how many children they want to have and over what period of time. Governments today are required by human rights agreements to provide family planning, contraceptive services, and information. Although access to these services should not be limited, birth control raises religious issues. The Roman Catholic Church is opposed to birth control and only accepts natural family planning. Protestant Churches vary from opposing to supporting as is the case with Judaism. Islam allows contraceptives as long as they do not impact health. Ethically, birth control is deemed morally wrong because it is unnatural, antilife, and separates sex from reproduction. It is also argued that it raises a number of negative issues such as posing risks to health, preventing the birth of people who might benefit humanity, bringing about drops in populations, being used as a tool to pursue eugenics policies, and misusing it propagate population control for racist purposes. A final argument is that it may promote immoral behavior (e.g., it makes it easier for people to have sex outside marriage). The arguments in favor of birth control are that it allows women to have control over their bodies and allows them full access to the labor market without the worry of becoming pregnant.

H. ten Have and M. do C. Patrão Neves, *Dictionary of Global Bioethics*,
https://doi.org/10.1007/978-3-030-54161-3_101

Brain Death (*See* Death)

Most countries now require brain death to be determined before organs can be removed from patients for postmortem organ donation. The development of life-supporting technologies and the emergence of transplantation technologies have encouraged debates about what criteria should be used to determine death, especially in patients who have permanently lost consciousness and are receiving artificial life support. Although the notion of brain death was first formulated in the late 1960s, transplantation practices are now framed by the dead donor rule that states organs can only be taken from people who are dead and physicians may not induce death to obtain organs. This means patients need to be declared dead before their organs are taken. There is no consensus on the definition of brain death or of the diagnostic criteria used to determine it. There are three different definitions of brain death. The first is brain stem death as a result of which the brain as a whole can no longer function adequately and consciousness is permanently lost and within a few hours the other parts of the brain stop functioning and die. Such a definition of brain death is used in several countries (e.g., the United Kingdom). The second definition is total (or whole) brain death that involves the irreversible loss of all functions of the brain including the brain stem, the absence of brainstem reflexes, and the presence of apnea. Such a definition is used in most Western countries. The third definition is higher brain death in which the function of the higher brain has irreversibly been lost resulting in the permanent incapacity to return to consciousness. Since such functions as perception, thinking, and volition depend on the cerebral cortex and are determinative of being a person the death of the person should therefore be distinguished from the death of the organism (as in the total brain concept). Such a definition of brain death has not been adopted in any legislature.

H. ten Have and M. do C. Patrão Neves, *Dictionary of Global Bioethics*,
https://doi.org/10.1007/978-3-030-54161-3_102

Brain Drain (*See* Care Drain)

One of the phenomena brought about by globalization is the brain drain, which entails skilled health professionals from developing countries moving to the developed world. Most developing countries lack the requisite number of qualified personnel. Such a global shortage is estimated by the WHO at approximately 4.3 million healthcare workers. Health professionals should be free to emigrate to countries where salaries and workloads are better, especially when violence, war, and corruption are rife in their home countries. However, such global movements raise ethical concerns. One is that the brain drain raises the problem of how to balance individual freedom (i.e., movement) and that the common good (the public interest and healthcare of citizens in the home country). Another is that the brain drain impacts equality and justice. Healthcare organizations in developed countries have recruitment agencies and programs that deliberately publicize migration encouraging health professionals to migrate, while there are shortages in the home countries that have invested in their education. Such practices are increasingly criticized since they are not compatible with the global ethics discourse of solidarity and justice. Moreover, they are contrary to the principle of benefit sharing in which advantages accrue to one actor by harming another. In an attempt to balance the benefits the World Health Assembly adopted the WHO Global Code of Practice on the International Recruitment of Health Personnel in 2010. However, the code is voluntary. A useful strategy to fight the brain drain is to offer scholarships or grants requiring health professionals to work at least the same number of years spent in education in the country that financed their studies.

H. ten Have and M. do C. Patrão Neves, *Dictionary of Global Bioethics*,
https://doi.org/10.1007/978-3-030-54161-3_103

BSE (Bovine Spongiform Encephalopathy)

Food security has become a major concern for global bioethics. Most food is no longer grown by individual people, but mass-produced by large agrobusinesses. Such industrial production is often accompanied by technological innovations to increase productivity. In countries such as Belgium and the United States growth hormones are used in animal farming. More than half of all antibiotics in the world are applied in bioindustry. Using economics to determine the way animals are treated for food production can lead to serious problems as illustrated by the bovine spongiform encephalopathy (BSE) scandal. In 1986 a strange disease associated with degeneration of the brain and the spinal cord was detected in cows in the United Kingdom. This disease (a.k.a. mad cow disease) was caused by prions (abnormal protein particles). The neurodegenerative disease was disseminated because the corpses of dead animals were reused as food for other cows—not destroyed or burned. The motives were purely economic. Although the British government banned the feeding practice in 1988, infected meat had already entered the human food chain. When the outbreak peaked in 1993 almost 1,000 cases of infected cattle were being reported each week. Since then the numbers have dropped significantly. In 1996 it was disclosed that the infection had also affected human beings. Consuming meat from infected animals led to a new type of Creutzfeldt-Jacob disease (a fatal neurodegenerative brain disorder). As of 2019 a total of 231 cases of this disease had been reported globally. It is estimated that a few million infected cows were used as food during the outbreak. The European Commission prohibited the export of meat from the United Kingdom (only lifted in 2006). In the United Kingdom 4.4 million cows were killed in eradication programs. Nevertheless, sporadic cases are still detected. BSE is often used as an example of an inadequate government response to a devastating disease. The UK government acted slowly, did not spend enough money to control the outbreak, and did not base its decisions on science. The risks for human health were initially downplayed with the Agriculture Secretary (unsuccessfully) feeding his daughter a beefburger on television to demonstrate that it was safe. After 14 years the government finally accepted (in 2000) after a long independent inquiry that the failures of successive administrations had contributed to the BSE catastrophe. It was also evident that there was poor oversight of slaughterhouse practices and lax controls on human food.

H. ten Have and M. do C. Patrão Neves, *Dictionary of Global Bioethics*,
https://doi.org/10.1007/978-3-030-54161-3_104

Business Ethics

Business ethics is the branch of applied ethics that studies the ethical principles and problems that emerge in businesses. It applies to the conduct of individuals as well as organizations. Relevant issues are corporate governance, insider trading, bribery, discrimination, and social responsibility. Interest in business ethics started in the 1970s around the same time as the emergence of bioethics. In the United States the Society for Business Ethics was established in 1980. The society publishes the journal *Business Ethics Quarterly*. The European Business Ethics Network was founded in 1987 and the International Society of Business, Economics and Ethics in 2000. Although initially there was little interaction between business ethics and bioethics, in the 1990s the two areas of applied ethics became closer as a result of two developments. The first was the commercialization of healthcare in many countries in which healthcare institutions and organization were encouraged to operate as businesses such that the usual principles of bioethics needed expansion. The second was that the processes of globalization were now driven by economics and raised many new challenges for bioethics. Such developments posited a wider range of relevant ethical issues for business ethics as well as for bioethics. Examples include the need to respect cultural traditions and religious perspectives across the world not only in doing business but also in applying healthcare, the impact of unethical practices such as child labor on health, pharmaceutical industry practices such as biopiracy and bioprospecting, and unfair pricing of medication. Such issues stimulated the search for universal values and global standards to avoid ethical imperialism, discrimination, inequality, and injustices. Although globally the focuses of business ethics and bioethics are often aligned, the debate on intellectual property rights and patenting requires expertise in both areas. Virtues such as honesty and integrity are highlighted in both ethical discourses to eliminate abuse and fraud. The common focus today is on reducing environmental degradation and pollution and seeking "green" business practices that are not only environmentally sustainable but also better for health. A special concern is corporate social responsibility the idea behind which is that businesses should take responsibility for the impact their operations have on human beings, society, and the environment. Most pharmaceutical companies today have strategies for corporate social responsibility and special units devoted to this task. Such companies are usually members of the Global Compact, an initiative of the United Nations to encourage businesses to implement sustainable and socially responsible policies.

H. ten Have and M. do C. Patrão Neves, *Dictionary of Global Bioethics*,
https://doi.org/10.1007/978-3-030-54161-3_105

Capabilities

The notion of capabilities is central to the capabilities approach developed by philosophers Amartya Sen and Martha Nussbaum. It is a framework for the analysis of global problems such as poverty and inequality. Capabilities focus on what people are able to achieve and how they are able to function with the means they have. Instead of focusing on commodities the central question is how people can convert commodities into achievements. The abilities of individuals to do this vary since they do not have the same needs and values. The availability of resources is therefore not determinative of human well-being, but it provides an opportunity to choose to do what people value. According to Sen, a better understanding of development is one that associates human flourishing with freedom and human capabilities. In this connection he makes the well-known distinction between functionings (i.e., the actual achievements of a person or "what he or she manages to do or to be") and capabilities (the ability to achieve a functioning or the freedom or opportunity to achieve it). Capabilities are more important than functionings. Available commodities (e.g., food or bikes) are used to achieve a functioning (e.g., being adequately nourished or riding the bike). Capabilities (e.g., the ability to avoid hunger or the capability to move freely) are opportunities to function and reflect freedom of choice. To illustrate the difference Sen frequently compares a starving and a fasting person. Although both are hungry and have the same functioning, they have different capabilities. The starving person is deprived of the capability to eat. Therefore his functioning is not the result of free choice whereas it is for the fasting person. Sen's framework has been broadened by other scholars, most notably Martha Nussbaum. The new view of development as an expansion of human capabilities is elaborated in her many publications into a general theory of justice and human flourishing beyond the domain of economic development. A good and just society is concerned with the distribution of capabilities. Quality of life is determined more by the opportunity to function than by people's actual functioning. For Nussbaum this implies that there is a minimum level of capabilities that should guarantee a life that is worthy of the dignity of the human being. Without certain functions human life itself will be in danger or human dignity will be violated such that human life will either come to an end or lose its specific human character. Since there is a basic social minimum, each human capability has a threshold beneath which truly human functioning is not available to citizens. She also holds that certain capabilities are universal since they are important for every person regardless of who and where he or she is and whatever the person chooses. As a result of these considerations she proposes a list of central human capabilities.

H. ten Have and M. do C. Patrão Neves, *Dictionary of Global Bioethics*,
https://doi.org/10.1007/978-3-030-54161-3_106

Capacity (*See* Capability; Capacity Building; Competence)

Capacity generally refers to the maximum amount of something such as material goods that can be accommodated in a particular container. It can also be defined as the potential or suitability to hold or store something referring either to objects or to intangible assets such as knowledge or personal experiences. Capacity is often used as a synonym for ability (an actual skill, a natural aptitude, or an acquired proficiency) or for capability (a quality or faculty susceptible to being developed) meaning one's power to perform an action. However, capacity differs from such other concepts in that it expresses the potential to develop a skill or an intrinsic characteristic. The word capacity is used in bioethics in two ways. First, the capacity of someone to carry out a specific task, sometimes also described as competence. However, competence is a legal term that within bioethics often refers to someone's mental condition to be accountable for his or her decisions or actions. Therefore competence is traditionally determined by a judge. Capacity refers more specifically to a general assessment of a person's ability to make reasonable decisions and give informed consent, which can be assessed by a medical doctor (particularly by a psychiatrist). Second, the frequent use of the term capacity in the field of bioethics is in the expression capacity building, referring to an individual or an organization's ability to absorb new knowledge and implement this in practice to bring about change effectively. There are a vast number of ethics capacity building programs worldwide.

H. ten Have and M. do C. Patrão Neves, *Dictionary of Global Bioethics*,
https://doi.org/10.1007/978-3-030-54161-3_107

Capacity Building (*See* Capacity)

The application and implementation of bioethics presupposes that appropriate capacities exist. A certain infrastructure needs to be present to introduce ethics teaching programs, to develop proper legislation, to have a public debate, and to provide ethics consultation in clinical settings. In many developing countries such infrastructure is generally inadequate. International cooperation is now more focused on capacity building. International organizations such as the WHO and UNESCO especially focus on this nowadays. Initiatives have been taken to network national bioethics committees. In 1996 a global summit of national advisory bodies was convened and supported by the secretariat of the WHO. Since then summits have been organized every two years, the 12th of which took place in Dakar (Senegal) in 2018 with participants from 71 countries. Another initiative was the Assisting Bioethics Committees Project of UNESCO that became operational following the adoption of the Universal Declaration on Bioethics and Human Rights. Article 19 of this declaration defines the main features and functions of "independent, multidisciplinary and pluralist ethics committees" to be established, promoted, and supported at the appropriate level to "assess scientific and technological developments, formulate recommendations and contribute to the preparation of guidelines on issues within the scope of this Declaration; foster debate, education and public awareness of, and engagement in, bioethics." Since then national bioethics committees have been established in many countries such as Côte d'Ivoire (2002), Guinea (2007), Madagascar (2007), Togo (2007), Gabon (2008), Colombia (2009), Jamaica (2009), Malawi (2011), and Comoros (2015). However, the establishment of bioethics committees does not guarantee that they work effectively. There will be a continuous need for education, support, and cooperation. Essential conditions for capacity building are collaboration and partnership and crucial factors are personnel development and training. However, most education programs are based in foreign countries. Institutional resources and facilities are required. When they are lacking, trained bioethicists returning to their countries face challenges because no bioethics activities or infrastructure exist. Should ethics committees exist and be functional, then they are mostly concerned with reviewing clinical trials and do not have a broader scope. Moreover, legislation and national codes in the area of bioethics have not been developed in many countries. A further hindrance is the role of the media whose job should be to promulgate debate such that the public can be engaged. However, in many countries the media are heavily controlled by governments and free debate is not encouraged. Although professional organizations do operate in most countries, their role is often limited. Effective capacity building therefore is urgently needed across the globe.

H. ten Have and M. do C. Patrão Neves, *Dictionary of Global Bioethics*,
https://doi.org/10.1007/978-3-030-54161-3_108

Capital Punishment (*See* Death Penalty)

Capital punishment refers to the death penalty. Both expressions are synonymous despite focusing on different aspects. The former stresses the notion of punishment (i.e., as a consequence of one's actions); the latter stresses the nature of the punishment (i.e., death). Both expressions designate the execution of an offender who has been found guilty by a court. In some parts of the world a court of law and the law itself have a religious nature. There are different methods of execution that vary according to a number of factors that are mainly cultural. Such methods include stoning, crucifixion, beheading, hanging, shooting, gas inhalation, lethal injection, and electrocution (the last three are considered more humane). From a legal perspective and in compliance with the law of the land in question some crimes such as murder under aggravated circumstances (such as premeditation) are punishable by death. Doing so complies with the *lex talionis* that calls for "an eye for an eye, a tooth for a tooth, a life for a life." Christianity revised this to "turning the other cheek." From an ethical perspective there is an intrinsic incoherence and an unsoluble contradiction in the state condemning somebody to die because that person killed someone else and arguing that that deserves death. If murder is intrinsically bad, how can it become acceptable by a court decision that makes it legal, but not ethical. If killing a person is not intrinsically bad, then it is the principle of human dignity (the recognition of each and everybody's absolute and unconditional value) that is at stake. Many consider that the death penalty violates human rights and argue there is no evidence that the death penalty reduces violent crime. There is a tendency worldwide to dramatically reduce the number of crimes that according to the law merit capital punishment. It has become rare and exceptional. Nevertheless, over 60% of the world's population live in countries where the death penalty is still legal.

H. ten Have and M. do C. Patrão Neves, *Dictionary of Global Bioethics*,
https://doi.org/10.1007/978-3-030-54161-3_109

Care Drain (*See* Brain Drain)

The care drain is a similar phenomenon to the brain drain. It is primarily focused on care workers and nurses. One reason for the care drain is the rapidly ageing population of many developed countries. This has created shortages of care workers not just for the elderly but also for the disabled, the chronically ill, and children. Many of these countries take measures to facilitate the migration of highly trained care workers and nurses from poorer, less developed countries. The care drain is controversial since it is primarily a female phenomenon and it may promote gender injustice and gender bias. Although many of the immigrants are well educated and trained, they are often regarded as unskilled workers. Since migration removes mothers from their families and nurses from hospitals it can have devastating effects in their home countries creating a care gap in local communities elsewhere. Developing countries that have invested in training healthcare workers are losing out on such investments. Moreover, in developed countries the care drain can cause problems since the work they do can be dirty, difficult, and dangerous. They can easily be exploited and there are numerous reports of abuse and violence.

H. ten Have and M. do C. Patrão Neves, *Dictionary of Global Bioethics*,
https://doi.org/10.1007/978-3-030-54161-3_110

Care Ethics (*See* Chronic Illness and Care)

Care ethics is a moral theory or moral approach that values human relationships and the supportive network they build. It is motivated by the willingness to care for others that is basic to human existence, especially for those in vulnerable situations, and promotes an overall feeling of well-being. Care ethics entails hosting the other, establishing a close relationship, and addressing the other's need. It is centered on the practice of virtue. Care is often considered a disposition or a virtue. Care ethics is a type of relational ethics that approximates and overlaps with virtue ethics and personalist approaches to ethics. It follows the sentimentalist tradition of moral theory thus presenting an alternative to deontological and utilitarian ethics, but without relapsing into paternalism. Although Milton Mayeroff published a short book *On Caring* in 1971, care ethics only became well known in the 1980s with the work of psychologist Carol Gilligan and philosopher Nel Noddings. Gilligan published *In a Different Voice* in 1982 in which she proposes care is a fundamental notion for moral development and argues that care is a universal necessity, a relational activity, and totally indispensable for the coexistence of the self and others. She interprets human relations in terms of a "web" as opposed to a "hierarchy" and of interconnectedness as opposed to inequality in which care is valued more than justice. Gilligan also associates this new way of thinking to a feminine voice that has been ignored and ought to be strengthened. However, the distinction between a female and male voice is thematic and does not refer to gender. In *Caring: A Feminine Approach to Ethics and Moral Education* published in 1984 Nel Noddings stresses the specificity of gender within a theory of care, which she explicitly classifies as a feminine and feminist perspective. Nevertheless, care ethics is not the same as feminist ethics (although both might overlap since both value features associated with motherhood such as nurturance, empathy, and compassion). The latter approach has criticized the views of Gilligan and Noddings (in particular) for attributing care to women. Men also show a natural tendency to care and should not be excluded from the experience of such an important human trace. Care ethics within healthcare has been particularly applied by nursing mostly because of its focus on empathic relations—not only because nursing was traditionally a female profession. Care ethics requires a close and emotional relationship between the healthcare professional and the patient who should be considered from a holistic and contextualized perspective. However, care ethics is a global ethics that can be employed by all healthcare professionals. It can also be applied to animal and environmental ethics and to public social policies that overcome the limits of a more formal and abstract moral theory. It is always guided by the natural needs of others. Nevertheless, care ethics can be combined with other approaches within bioethics.

H. ten Have and M. do C. Patrão Neves, *Dictionary of Global Bioethics*,
https://doi.org/10.1007/978-3-030-54161-3_111

Casuistry

Casuistry is derived from the Latin word *causa* (case) and refers, in general, to a case-based method of reasoning. It is employed in many fields (particularly in specific branches of applied and professional ethics) and in law (where it has been found to be particularly appropriate to the common law system). Casuistry has a long history harking back to being structured in late scholastic philosophy by moral philosophers and theologians (particularly Thomas Aquinas) as a discipline to help solve cases of conscience. Armed with knowledge of the universal moral law (dictated by divine revelation) and the (a priori established) end to be achieved in a concrete given situation, one is in a position to apply the law to that case, deduce how to act adequately, and choose the best means to accomplish the intended goal. Lying outside moral theology and moral universalism, casuistry became a mere case-by-case approach to ethical problems and hence failed the criteria of theoretical objectivity and moral justice. Since the second half of the sixteenth century casuistry has been the subject of intensive debate and triggered strong disagreement, which contributed to its near disappearance after the eighteenth century. Casuistry has recently been revived, especially in Anglo-American countries and particularly in the field of bioethics, due to the work of Albert Jonsen and Stephen Toulmin such as *The Abuse of Casuistry: A History of Moral Reasoning* (1988). Criticizing the abstractionism of meta-ethics that had dominated the panorama of ethical reflection during the twentieth century in the United States, Jonsen and Toulmin reiterated the philosophical importance of casuistry and argued that it was indispensable to solving concrete moral problems of daily life. They also criticized the common view that universal norms were needed for a fair decision. Casuistry in their view is not based on any universal principle applied to a given situation. Jonsen and Toulmin were mainly interested in the concrete circumstances of singular cases and in the maxims that people invoke in the face of moral dilemmas by examining the paradigmatic characteristics of each case, establishing similarities and differences with other cases and with particular types of cases, and thus constructing a so-called moral taxonomy. A case-by-case analysis can then be carried out at the analogical level. This is the reason some commentators consider casuistry to be theory modest rather than theory free. Casuistry today is not considered a good enough methodology per se to be used for the moral appreciation of bioethical problems nor an exclusive alternative to theoretical models of bioethics. On the contrary, revalued casuistry is neither restricted to a purely pragmatic analysis of cases without attention to the level of theoretical foundation nor is it wary of any standard universality. It goes hand in hand with the development of other theoretical models of applied ethics. Case-based methodology is a bottom-up approach that can be combined with the application of ethical theories to concrete situations thus converting it to a top-down approach.

H. ten Have and M. do C. Patrão Neves, *Dictionary of Global Bioethics*,
https://doi.org/10.1007/978-3-030-54161-3_112

Censorship

Bioethics analysis and debate presupposes the free communication of information. Censorship restricts the freedom to communicate ideas and opinions and should only occur if it can be justified. Several recent cases have made the issue relevant. In 2011 researchers in Rotterdam (The Netherlands) genetically transformed the H5N1 virus such that it became more easily transmissible from human to human. They had therefore created a much more dangerous virus than the common avian flu virus. When they submitted an article on their work to *Science*, an American government body for biosecurity prohibited publication because of the risk for misuse by bioterrorists. Another example concerns unsound scientific research disseminating claims that vaccinations are unsafe and ineffective. The argument is that such information is misleading and should not be provided. Censorship has also been a traditional component of the relationship between patients and health professionals. Physicians argue that they are right to withhold information if it is harmful to patients. This so-called therapeutic privilege has been severely criticized because of the increasing importance of the ethical principle of respect for patient autonomy. The ethical controversy of censorship lies in balancing freedom of speech and potential harm. The harm principle of John Stuart Mill justifies holding back information if this prevents moral wrongs being done to third parties. An example is outlawing holocaust denial since it may promote violence against Jewish people. Potential harm that justifies censorship should be clear and evident. Although there was much debate about censoring publication of the paper on the transformed virus, arguments for censorship were finally rejected and the article was published because the potential harm seemed manageable. Dissemination of false and dubious health claims, especially through the internet, is also much debated. When the important ethical principle is autonomy, the assumption is that individuals have the right to correct and assess facts. However, it is argued that this claim is weak since regular medicine itself often uses treatments that have not undergone a lot of research as shown by the rise of evidence-based medicine that implies medicine does in fact lack evidence. Why give this type of information priority above information from other sources? The conclusion such debates reach is that free speech and allowing free communication of information is generally more important than censorship. The burden of proof that harm will result is on the persons or agencies wanting to restrict this freedom. This can only be established through debate and deliberation by the global community.

H. ten Have and M. do C. Patrão Neves, *Dictionary of Global Bioethics*,
https://doi.org/10.1007/978-3-030-54161-3_113

Children and Ethics (*See* Pediatrics)

Since children are vulnerable they are given special attention in bioethical discourse. The Convention on the Rights of the Child was adopted by the United Nations in 1989. It recognized the importance of the human rights and human dignity of children as a special group of human beings. However, the principles of mainstream bioethics have long been applied to children. The more recent awareness of the significance of vulnerability has modified and specified ethical discourse since the principle of respect for autonomy cannot be applied directly in the case of children. Childhood is a phase of human life that is variable in that it is legally defined differently by countries. However, in most countries adulthood starts at 18 years of age. The vulnerability of children varies according to biological functionality, psychological maturity, and social relationships and the legal age is immaterial. Childhood is characterized by the development of biological and psychological processes within a supportive and nurturing environment. Such processes help children grow into autonomous individuals. In the healthcare of children a unique relationship exists between doctor and patient that also involves the parents. Medical decision-making is therefore more complicated than in the usual doctor–patient relationship since the interests of the child are represented by the parents. During the first few decades of its development bioethics was primarily focused on the decisional capacity of adults. It later emerged that pediatric issues should not merely be a special application of adult-centered ethics and in the 1990s pediatric ethics emerged as a new, specialized field of bioethics. Ethical issues in relation to children concern the traditional principles of bioethics. Beneficence refers to the best interests of the child. Although such interests are determined by values held by the parents, they may affect the health and dignity of the child. Although health professionals have a responsibility to protect the child, this should be balanced against input from the parents. Problems are obvious in the case of child abuse. An older problem regards the refusal of treatment such as by Jehovah witnesses when a blood transfusion is needed by the child but rejected by the parents. The principle of nonmaleficence is another important ethical issue in pain medication. Children often do not receive adequate pain treatment since the effects of medication are not well investigated in children and many physicians are reluctant to provide such drugs because of negative effects. The principle of autonomy is compromised in children since they are unable to make such decisions. This is the reason children are represented by parents or other proxies. Informed consent, for example, depends on the capacity to understand and surrogate consent is often the only option available. In Western countries it is assumed that children between 7 and 14 years of age can give assent to intervention. Although this is not the same as full consent, it is a request that as much information as possible should be

H. ten Have and M. do C. Patrão Neves, *Dictionary of Global Bioethics*,
https://doi.org/10.1007/978-3-030-54161-3_114

provided for the child to understand what is going to happen. If the child refuses, then such a refusal is accepted. The underlying idea is that children should be allowed to participate as much as possible in the decision-making process since this will assist their personal autonomy to grow.

Children and Research (*See* Pediatrics)

The first regulations established for medical research involving children were intended to protect children from such research (like the Nuremberg Code). The emphasis on informed consent and the vulnerability of children excluded them from medical experimentation. The 1964 Declaration of Helsinki allowed research involving children as long as the parents consented. This shift from protecting children from research to protecting them within research was justified by the argument that treatments and drugs are usually only tested on adults such that specific benefits and harms for children are not well known. The concept of consent is one of the fundamental ethical issues in research involving children. Although the emphasis was initially on parental consent, such a concept is increasingly disputed since children may disagree with parental decisions. Although it is also assumed that parents will act in the best interests of the child, this is also sometimes questionable. In some cases children do not have parents to fall back on (e.g., when they are institutionalized) and legal guardians then decide on their behalf thus further complicating the issue of best interests. These concerns have put more emphasis on the consent of the child himself/herself. It is generally accepted that the refusal of the child to participate in research will override the consent of the parents. The 1983 revision of the Declaration of Helsinki requires the child's consent in addition to the parents' consent. The management of harm is an ethical concern when such research does not directly benefit children, especially when children are too young to give consent. In such a case the harms and benefits of research have to be limited, especially when there are no direct advantages for the participants. Research risks and benefits should be assessed separately. Such assessment usually gives rise to extensive ethical debates in research ethics committees.

H. ten Have and M. do C. Patrão Neves, *Dictionary of Global Bioethics*,
https://doi.org/10.1007/978-3-030-54161-3_115

Children's Rights

The Convention on the Rights of the Child was adopted by the United Nations in 1989. It was the result of 10 years of negotiations. Entering into force in 1990 it is now ratified by every member state of the United Nations except the United States. The convention defines a child as "every human being below the age of eighteen years unless under the law applicable to the child, majority is attained earlier." The convention acknowledges that every child has fundamental rights. In all actions involving children the best interests of the child should be the primary consideration. Children have the right to life, survival, and development. They should be registered immediately after birth and have the right to a name and to acquire a nationality. They should not be separated from their parents. Children also have the right to freedom of expression including the freedom to seek, receive, and impart information and ideas of all kinds. Furthermore, children should be protected from violence, abuse, and neglect. The child has the right to be protected from economic exploitation and from performing any work that is likely to be hazardous or to interfere with the child's education. State parties should also recognize that children have the right to enjoy the highest attainable standard of health and to facilities for the treatment of illness and rehabilitation of health. They should ensure that no child is deprived of his or her right of access to such health services. The same parties should recognize the right of the child to education and make primary education compulsory and available free to all. The long list of children's rights stated in the convention (54 articles) are a specification of the human rights mentioned in the Universal Declaration of Human Rights. The United Nations Committee on Children's Rights is responsible for reviewing how these rights are safeguarded in individual countries. The focus on how the rights are implemented can differ. In some countries emphasis is on child labor and poverty, whereas in other countries it is on access to health and social services. Four ethical principles guide the implementation of children's rights: non-discrimination; best interests of the child; right to life, survival, and development; and respect for the child's point of view.

H. ten Have and M. do C. Patrão Neves, *Dictionary of Global Bioethics*,
https://doi.org/10.1007/978-3-030-54161-3_116

Chimera (*See* Research Ethics, Interspecies)

The Chimera was a fabulous creature in Greek mythology (*khimaros*) that had a lion's head, a goat's body, a dragon's tail, and breathed fire. Today it generally refers to something imaginary (and absurd) composed of bizarre and incongruous parts. The term chimera is also specifically used in biology (mostly in botany) to designate an artificial organism produced by xenografts. Recently, it has been used to refer to genetically engineered organisms containing at least two different sets of DNA usually produced by the fusion of two or more zygotes. Although chimeras can be naturally generated, this is not often the case. Medicine refers to human chimeras when a person has two genetically distinct types of cells. Blood chimeras are non-identical twins who share a blood supply in the uterus. It is known that a human fetus can absorb its twin (if this embryo twin dies very early in pregnancy) and present two sets of cells: its own original set and the one from its twin. Currently, there are researchers who are working on producing human–animal chimeras. The production of human–pig fetuses has already been announced. The ultimate goal is to better understand the growth of organisms, to study the progression of diseases, to develop new drugs, and to try to grow human cells (blood cells), tissues (liver or heart tissues), and organs in animal bodies—not to create new imaginary creatures. The most immediate ethical issue raised by the creation of chimeras is the denaturation of life itself. Although many argue that humankind has no moral authority, it does have the scientific/technological power to interfere with natural life and alter it. Some may accept genetic modifications for therapeutical reasons or even for the purpose of intensifying agriculture and livestock—but not for the production of new creatures. From an animal ethics perspective the production of chimeras does not respect animals' rights, but treats them as simple objects for human purposes. From a human perspective animal–human chimeras challenge respect for human dignity since human cells are simply used as raw material. What is more worrying is the real possibility to effectively create a creature that is half-human and half-animal. Although human–pig fetuses (with human cells in pig's tissues) were not allowed to develop to term, it might be permitted in the future (slippery slope argument). Such new possibilities call for society to adopt adequate measures to protect and preserve human integrity and dignity.

H. ten Have and M. do C. Patrão Neves, *Dictionary of Global Bioethics*,
https://doi.org/10.1007/978-3-030-54161-3_117

Chronic Illness and Care (*See* Care Ethics)

Although chronic illnesses such as diabetes, hypertension, and asthma have always existed, they have become a major challenge for contemporary healthcare systems. Chronic illness is defined as a long-term health condition that requires ongoing medical management for years and even decades. When the course of a disease is longer than 3 months, it is labelled chronic. Heart disease, cancer, and diabetes are regarded as the most frequent chronic diseases. They are a major cause of death and disability worldwide. They are also called non-communicable diseases. According to the WHO such diseases are responsible for 71% of all deaths globally. Such illnesses disproportionally affect people in low- and middle-income countries where more than three quarters of global deaths (32 million people in 2018) occur. Chronic illness is often associated with old age. However, a substantial number of deaths occur between the ages of 30 and 69 years when they are regarded as premature. Due to globalization, unhealthy lifestyles, and ageing, chronic diseases are now a global problem. The risks for chronic disease increase as a result of unhealthy behaviors such as tobacco use, physical inactivity, unhealthy diets, and alcohol abuse. Since chronic illnesses can be treated though mostly not cured the emphasis of medical management is on prevention. The focus is on reducing the risk factors associated with these illnesses. The affected person needs medical care for a long time and often for the remainder of his or her life. Treatment is usually demanding for patients in that it has the potential to affect quality of life and well-being, as well as posing financial burdens. Chronic illness raises specific ethical questions because the resources for long-term care are often limited, the burden of care is frequently placed on women, people with chronic care are vulnerable, and the difficulty such people have in accessing healthcare. This is related to the usual focus of healthcare on acute care. Chronic illness requires continuous monitoring and interdisciplinary cooperation. Moreover, the emphasis is often on self-care management where patients themselves have to take responsibility for treatment, prevention, and management. Although such an emphasis can reduce dependency, paternalism, and apportioning blame, it also requires sufficient community support. The importance of care for such patients has led to the basic approaches of mainstream bioethics being rethought and promoted the rise of care ethics.

H. ten Have and M. do C. Patrão Neves, *Dictionary of Global Bioethics*,
https://doi.org/10.1007/978-3-030-54161-3_118

Circumcision, Male

Male circumcision like female circumcision has a long history and is a common practice in Jewish and Islamic religion. It is estimated that worldwide 30% of adult males are circumcised. The procedure can be performed for reasons other than religious ones such as therapeutic (e.g., in case of phimosis), preventive (e.g., half of newborn males are circumcised in the United States for hygienic purposes), and cultural (a kind of rite of passage within a cultural tradition). A distinction is usually made between male and female circumcision since male circumcision is not merely a cultural practice and does not have the same harmful impact as female circumcision. There is also no legislation prohibiting male circumcision. Although the procedures are similar in involving intervention in the external genitalia, often any medical need for doing so is lacking. Ethical issues are related to the evidence or lack thereof of medical harms and benefits. Circumcision is regarded as a simple procedure. Nevertheless, complications do occur but are rare. Although the benefits of male circumcision are questionable, it does seem to reduce the risks of urinary tract infections. Another ethical issue relates to respect for the rights of the child and those of parents. Circumcision of newborns is especially problematic. Protection of the rights of the child can only be guaranteed if circumcision, especially when it is non-therapeutic, is postponed until the child can give consent. Bodily integrity is another ethical concern since the human body has a value of its own and its wholeness needs to be preserved. Even though people own their bodies, they are not morally allowed to do whatever they wish with them. Another moral consideration is cultural and begs the question: Should male circumcision as a cultural practice be respected or not?

H. ten Have and M. do C. Patrão Neves, *Dictionary of Global Bioethics*,
https://doi.org/10.1007/978-3-030-54161-3_119

Citizenship, Biological

Biological citizenship (a.k.a. biocitizenship or medical citizenship) defines citizenship from a biological standpoint in that people belong to a community and claim rights because the diseases and injuries they are prone to as well as their genetic status are similar. Such a notion was promoted by the work of French philosopher Michel Foucault in his writings about biopower. Biological citizenship as a concept has been increasingly used since 2000. It is described as an active form of citizenship in which new forms of belonging can be produced and biotechnological resources can be accessed. The term was first introduced in 2002 by anthropologist Adriana Petryna in her study on the Chernobyl disaster. She defines biological citizenship as "a massive demand for but selective access to a form of social welfare based on medical, scientific, and legal criteria that both acknowledge biological injury and compensate for it." Biological citizenship can be more broadly interpreted in that it produces new connections between self-identity and biology. Biotechnology can be used to facilitate specific ideas of what it means to be human in which the body is considered a manipulable mechanism that can be exploited and reshaped by enhancement technologies. Although such views no longer find favor in national politics, they dominate the global marketplace where people are regarded in the first place as biological consumers rather than citizens with rights and duties. The notion is criticized as being a dangerous alternative to social and political citizenship. It is argued that this kind of citizenship overvalues the biological and genetic dimension of human beings while neglecting many other factors that determine human identity. The notion can also easily lead to discrimination and exclusion.

H. ten Have and M. do C. Patrão Neves, *Dictionary of Global Bioethics*,
https://doi.org/10.1007/978-3-030-54161-3_120

Citizenship, Ecological

Ecological (or environmental) citizenship is a recent notion in which cosmopolitan citizenship is demanded. It refers to the idea that our identities are being reshaped on the basis of increasing knowledge of how living conditions on our planet are changing and how all human beings now and in the future will be affected by our current ways of living. It is argued that the current environmental crisis will require a new notion of citizenship. Such a notion was first introduced by political theorist Andrew Dobson and obliges people to reduce their ecological footprint. It is an obligation in which the citizenship status of people is shifted from the public to the private sphere. Such an obligation is non-territorial and asymmetrical in that people who reside in regions where the negative impact on the ecosystem is greater have a greater obligation to reduce such an impact. Therefore being a good citizen means being a minimal impactor on the environment. Although the obligation implied in the notion of ecological citizenship is not merely a question of individual responsibility since citizens always exist in a community, it also requires community and political action in attempting to change social systems and transforming collective practices on a much larger scale than that of the individual.

H. ten Have and M. do C. Patrão Neves, *Dictionary of Global Bioethics*,
https://doi.org/10.1007/978-3-030-54161-3_121

Citizenship, General

Citizenship refers to membership of a political community. All members of such a community have equal status and rights. Although different concepts of citizenship have been advanced in history, the most common view is that citizenship is associated with birth or domicile in a sovereign state. In ancient Greece, for example, citizens were male, free, and native-born adults residing in a particular city-state. Although they had some privileges, they also had political and moral duties. The Roman Empire expanded citizenship to all free inhabitants of the empire. However, slaves and women were still excluded. Contemporary notions of citizenship argue that it should be further expanded to include immigrants. Although most agree that immigrants should be protected by law and be provided benefits by the state, it is controversial whether they should be granted full citizenship or whether they should share the basic features of the ethnic identity of the country in question such as language, religion, tradition, and history. On the other hand, it is argued that getting immigrants to share such features can be done through naturalization. A more recent development is global citizenship or citizenship that knows no borders. Although many people regard themselves as citizens of the world, the difficulty that arises is that there is no world state prepared to guarantee the fair distribution of benefits and burdens. Nevertheless, citizenship has been created at the regional level (e.g., the European Union). However, the same objection applies in that there is no European government and many people question the feasibility of the European Union being a political community with common values and commitments. However, this may change in the future. The emergence of a human rights discourse is promoting the cosmopolitan idea that human beings share similar rights and responsibilities independent of where they live. The worldwide rise of populist governments and political leaders wanting to close borders or break away from communities is having a negative influence on efforts to promote the cosmopolitan trend.

H. ten Have and M. do C. Patrão Neves, *Dictionary of Global Bioethics*,
https://doi.org/10.1007/978-3-030-54161-3_122

Citizenship, Genetic

The rise of genetics and the growing dissemination of genetic information have promoted the concept of genetic citizenship. It is a specification of the notion of biological citizenship that is now focused on the genetic constitution of people. The application of assisted reproduction technologies and genetic testing have identified new global connections in which the idea that people are members of a genetic community has been mooted. Technology can also be used to identify the family or state individuals belong to (e.g., when surrogate children are born but not accepted by the commissioning parents). DNA testing is used in many countries to reunify immigrants with their families in which the genetic ancestry of individuals is ascertained and as a result of which they are attributed rights. The notion of genetic citizenship is criticized because it risks a return to the eugenic projects and racialized politics of the past. It implies a new governmental regime directed at managing risk at the population level and at individual management of genetic risks. The use of DNA analysis for family reunification purposes represents a form of migration control targeted at particular populations such as happens in Germany. This is especially the case with those who originate from blacklisted countries mostly from sub-Saharan Africa and Central and Southeast Asia. The use of DNA tests to determine whether an individual should be allowed to stay further reveals the selective format of the debate on genetic citizenship. Discussion has often stressed that the genetic constitution of people should be the basis of claims about social inclusion, recognition, and democratic deliberation. However, genetic citizenship fails to take into account some important dimensions in the regimes of contemporary migration. The use of DNA testing in immigration policies often serves to reaffirm and rearticulate traditional forms of classification and exclusion. An example is the use of genetic tests in Israel to determine whether potential immigrants are Jewish or not. Another aspect of the notion of genetic citizenship is that it can be used to articulate responsibilities rather than rights. An example is the argument that genetic citizens should participate in genetic databases and biobanks and that individual consent for collecting and using biological samples should not be needed.

H. ten Have and M. do C. Patrão Neves, *Dictionary of Global Bioethics*,
https://doi.org/10.1007/978-3-030-54161-3_123

Civil Disobedience

The active refusal of citizens to obey certain laws or government orders is regarded as civil disobedience. It involves protesting and carrying out unlawful acts deliberately and publicly. It is characterized in three main ways. First, the law is broken knowingly and deliberately (i.e., the aim is to break the law). Second, civil disobedience is often a conscientious protest (i.e., it expresses deep dissatisfaction with some government policies). Third, acts of civil disobedience are public (i.e., designed to be witnessed to get popular support to bring about changes in public life). Although civil disobedience should always be non-violent, there are those who argue otherwise. A famous example of civil disobedience is Mahatma Gandhi's satyagraha movement that adopted deliberate and non-violent disobedience to claim Indian independence from the British. The oldest example can be found in the Book of Exodus where Hebrew midwives did not follow the command of the Egyptian king to kill male newborns. Another well-known example is the Boston Tea Party in 1773. Civil disobedience should be distinguished from revolution, which seeks the overthrow of the government. It generally accepts the established authorities and legitimacy of the legal system. Two kinds of civil disobedience can be distinguished: direct and indirect. Disobedience is direct if it focuses action at the law against which the protest is aimed. Indirect disobedience is the violation of other laws than the one that is the focus of protest such as is the case with sit-ins or blocking traffic. Civil disobedience is most often justified by moral arguments that reside outside the legal system. Justification requires arguments in which the normal obligation to obey the law should be seen as unjust and therefore must be overridden such as people arguing the law ought to be broken because they deem it to be grossly unjust. Justification can appeal to higher laws such as international human rights law. Other justifications are utilitarian. Protesters can argue that disobedience will lead in the longer run to a better and more just society than that promoted by the current law. Although moral justifications are always disputable, civil disobedience is often an effective way to protest against government policies and press for changes.

H. ten Have and M. do C. Patrão Neves, *Dictionary of Global Bioethics*,
https://doi.org/10.1007/978-3-030-54161-3_124

Civil Society

Civil society considers society as a community of citizens connected by common interests and collective activities. It includes the family and the private sphere. It is regarded as an intermediate sphere between state and market. Politics today has increasingly become the domain of civil society. Global activities are undertaken by communities of citizens rather than by traditional political authorities. They participate in direct and horizontal communication, share information, and use publicity to put pressure on such authorities. Instead of using existing institutions, citizens engage with particular causes, develop new collective structures, create social movements, and open up new problem areas using the latest technologies to set up global networks and organize world forums and global summits. The idea of civil society is related to the human rights discourse and cosmopolitan concern for humanity as a whole. Civil society has a moral function in that it can speak on behalf of everyone because it expresses concerns about the common good of humankind—not so much because it represents the rest of the world population. The new meaning given to "world" and the notion of humanity signposts new directions for ethics to follow since it focuses on commonalities rather than differences. The growing importance of the notion of civil society is an effect of the significant change in governance from above to below. It is also related to new ideas about the implementation of bioethical principles. The view of implementation from below implies that local actors play an important role in the diffusion of global principles. They build congruence between these global principles and preexisting normative frameworks through the dynamic process of localization. Although international organizations can declare global bioethical principles, implementation is the work of agents at the local level. The role of individual citizens and groups of citizens is expressed in the concept of civil society. Active citizenship means that individual citizens can unite and take initiatives. Civil society today is no longer restricted to a particular territory. Public debate and political action cross the borders of states. Citizens in one state can link up with others in other countries and organize around a common cause or issue. The growing interconnectedness and the emerging sense of global community have created global civil society. Although this includes social movements and NGOs, it is wider since it applies to every citizen. Global civil society is considered a sphere of interaction between state and market. Armed especially with new media civil society can make room for discourses that are not controlled by governments or commercial forces (even if they try to influence them). Civil society is also a sphere of public conversation and reasoning, of deliberation and participation, and of contestation and conflict. Finally, it is a sphere of public engagement in which self-organizing groups of individuals can undertake collective action. The role civil society plays in global bioethics is demonstrated in the debate about vertical and horizontal health

H. ten Have and M. do C. Patrão Neves, *Dictionary of Global Bioethics*,
https://doi.org/10.1007/978-3-030-54161-3_125

interventions. Vertical programs are preferred by international agencies since they have clear targets and focus on technological solutions. Horizontal programs focus on health systems. They require grassroots participation and programs that are community based. Without the involvement of civil society basic social and economic needs that are responsible for health problems cannot be addressed. Since healthcare services are not discrete interventions they demand a systemic approach guided by local knowledge. The basic features of new governance practice are public debate, involvement of civil society, and the participation and consultation of a wide variety of actors. Improvements in human rights practices are more often the result of efforts of grassroots movements in local settings than the accomplishments of global institutions. Although institutions can create norms, implementation is decentralized and domestic. Actors and networks from civil society take up a cause (e.g., access to healthcare), organize themselves, contest the issue, engage in the struggle, and connect with similar movements and networks in other countries. Although they develop a mode of practice around the right to health, such practice is not simply application of this right. Activities need to be adapted to local circumstances and values and can therefore be more successful in one setting (e.g., South Africa) than in another. Practices looked at this way are constructed through collective labor within a specific context. The implementation or application of global principles and values is therefore a form of domestication in which they need to be transformed and internalized into domestic systems and local contexts. This is usually done by NGOs and individuals—not by governments. Informed consent in medical research is another example. In recent decades it has been incorporated in research cultures and practices and has almost become a bureaucratic routine in ethics review, even though the specific procedures and ways of application can differ depending on the context.

Climate Change

Climate change is the most fundamental global challenge facing the world today. Although deniers still exist in some countries (especially the United States), the overwhelming majority of scientists and policy-makers accept the massive amount of evidence supporting the belief that the planet is undergoing global warming. If emissions of greenhouse gases continue unchecked until 2050, then the consequences for biodiversity will be disastrous. The effects of climate change are no longer concerns for the future since they are currently under way and can readily be seen by today's generation. Since the establishment of the Intergovernmental Panel on Climate Change in 1988 regular assessment reports have been published by the international scientific community. Each subsequent report paints a bleaker picture of the environment. Such reports present observations and predictions in ever more stronger terms such that it is irrefutable that the mean annual temperature at the surface of the Earth has been increasing over the past two centuries. It is now clear that climate change will have a range of effects on natural, social, and human systems such as enforcing human migration, species moving to warmer areas, new types of bioinvasion, diseases emerging in regions where they have hitherto been unknown, and tropical diseases becoming more widespread. Moreover, ancient and traditional cultures will be lost and people will be displaced and forced to move to safer areas. Temperature change will impact precipitation and weather patterns resulting in floods and heatwaves in some parts of the globe and drought, desertification, and extreme cold weather in other parts. Furthermore, global vegetation will change, agriculture will face major difficulties, food production and water availability will be affected, crops will fail, forest and bush fires will increase, extreme weather events will be more common, ecosystems will break down, species extinctions will be more widespread than they already are, and permafrost will melt not only releasing age-old microorganisms but more importantly huge amounts of carbon that will accelerate warming of the planet. Such effects of climate change will lead to significant economic losses in the near future. However, when it comes to the survival of humanity, such effects are perhaps even more important in that they will seriously affect human health but do so in a very unequal way. Although the major contributors to climate change live in the developed world, the existence and health of people in developing countries will be the first and most affected. Many places will become uninhabitable such as the small island states that will disappear as a consequence of rising sea levels. Bangladesh will lose 20% of its habitable land. Such unequal effects can readily be seen today in the phenomenon of climate or environmental refugees. It is estimated that in the next decade tens of millions of people will be driven from their homes due to climate change. Worries about the current migration crisis will be completely overshadowed by what is expected in the near future. The Earth Summit in Rio de Janeiro in 1992 adopted the Convention on

H. ten Have and M. do C. Patrão Neves, *Dictionary of Global Bioethics*,
https://doi.org/10.1007/978-3-030-54161-3_126

Biological Diversity and the UN Framework Convention on Climate Change. This convention acknowledges climate change as a fact and the significance of human conduct in bringing it about. It also accepts that responsibilities for action are not equally divided: developed countries have a larger responsibility than the developing world. Nonetheless, action has been very weak. Targets to reduce greenhouse gas emissions have been voluntary. The Kyoto Protocol signed in 1997 and effective in 2005 placed specific targets on participating countries with the goal to reduce emissions to at least 5% below 1990 levels. The protocol has been lauded as a success because between 1990 and 2012 parties reduced their CO_2 emissions by 12.5%. Parties to the protocol committed to reduce emissions by at least 18% below 1990 levels between 2013 and 2020. However, what success there has been is very limited, developing countries have not committed to emission reductions, worldwide emissions have increased by 50% since 1990, and emissions from China, in particular, have steeply increased. Although 191 countries have now ratified the Kyoto Protocol, it remains unratified by the United States which is the world's largest economy and one of the biggest emitters of greenhouse gases. The impact the protocol has had on climate is therefore very limited. One of the reasons for failure viewed from the ethical perspective is the lack of a comprehensive ethical framework for action. Negotiations among governments have not been based on any ethical principle. The challenge of climate change is fundamental for a couple of reasons: it is intrinsically related to a specific way of life and at the moment there is not a sufficient basis for change. Both reasons refer to the belief that there is no scientific solution to global warming and that the problem is a matter of human values. Numerous studies make clear that drastic cuts in carbon emissions are necessary. This requires enormous changes in the use of energy, particularly fossil fuels, and in people's way of life, especially in high-income countries. The only way to stop global warming is to make societies carbon–neutral worldwide. However, even if the necessary measures are taken immediately, the damaging effects of climate change will carry on for some time. The planet will continue to get warmer for decades to come. Addressing climate change requires something much more ambitious than developing another way in which globalization can be practiced. Climate change poses a number of very basic moral problems: How should we live? What kind of society do we want? What kind of people should we be? Although change is possible, at least theoretically, in practice almost nothing is done. Various excuses for inaction are given by politicians such as uncertainty about the impact of global warming, the prohibitive costs of mitigation and adaptation, and technological innovation will come to our rescue. Denial continues to be a major stumbling block in that people (especially politicians) argue that they do not believe in climate change, they think that innovative and green technologies will provide solutions, they assume that severe impacts will not occur during their lifetime, and they hold that minor lifestyle changes will be sufficient. Moreover, political discourse and negotiations at the government level have reached stalemate, the level of trust between developed and developing countries is low, commitments are not honored, and reciprocity is absent. In such a climate of distrust climate change mitigation is considered an argument deliberately posited to impede the economic development of resource-poor countries and stop the consumption culture of rich countries from being questioned. The fact that climate change is reinforcing patterns of global inequality stands in the way of promoting international cooperation. As long as there is no shared understanding of what solutions will be fair, collective action will always be difficult. The irony is that while climate change is certain, the political will to take effective measures is uncertain. Although such challenges could be paralyzing any action, from the point of view of global bioethics they should not. There are three ethical considerations that will drive approaches to climate change despite global governance being weak and politicians reluctant. The first is the global nature of the problem. It is perfectly clear to all that this problem cannot be solved at the domestic level. The second ethical consideration is that climate change is everybody's problem—not a governmental problem or a specific challenge for politicians and policy-makers. Even though the developing world may well suffer more, in the end everybody will suffer. This means that addressing the problem cannot be left to politicians, particularly when their responses are weak and inconsistent. The third ethical consideration is that addressing climate change is more than just a matter of pragmatism. It is primarily an issue of inequality and injustice in which poor countries suffer the effects despite having contributed almost nothing to the problem, not all nations are equally responsible for climate change, and not all nations suffer equally from the effects of global warming. Although there are significant differences in vulnerability and responsibility, there is also intergenerational harm. This point of view implies that responses to climate change are much more than about the survival of humanity. They are also fundamentally about global inequality despite strategies proposed to address this inequality being controversial.

Clinical Equipoise

A central concept in the ethics of medical research is clinical equipoise. It is regarded as an ethical precondition to involving patients in randomized clinical trials. How can physicians whose first duty is to care for their patients involve them randomly in a trial? The term refers to the professional disagreement among expert clinicians about the preferred treatment. In a randomized clinical trial a novel treatment is compared with a treatment that is routinely used in practice or a placebo if there is no standard treatment. Patients are assigned randomly to different treatment arms in the trial. Randomization is an essential methodological element of the trial and is a guarantee against bias. The term equipoise was introduced by Charles Fried in 1974. He focused on the dual role of physician and researcher. There is an ethical conflict between the obligation of physicians to provide the best possible care and the need to advance medical knowledge. Fried proposed to solve this conflict with the equipoise requirement. If there is no specific treatment preference for a patient, then the dilemma no longer exists. Such a judgment of the physician is based not only on professional norms but also knowledge of the particularities of the patient. If the physician is convinced that one treatment is better than another for the patient, then he or she cannot randomly choose which treatment should be given in the trial. Clinical equipoise has become increasingly criticized. Some argue that the ethical problem of randomization cannot be solved and that randomized clinical trials should not in fact be permitted ethically. Although others agree that clinical equipoise is a proper constraint, they feel it cannot be grounded in the physician's duty of care. The ethics of research should not be based on the ethics of clinical medicine. In fact, research ethics committees review trials before any relationship exists between researchers and patients. Medical practice and research have different goals.

H. ten Have and M. do C. Patrão Neves, *Dictionary of Global Bioethics*,
https://doi.org/10.1007/978-3-030-54161-3_127

Clinical Ethics, Committees

See Committees.

H. ten Have and M. do C. Patrão Neves, *Dictionary of Global Bioethics*,
https://doi.org/10.1007/978-3-030-54161-3_128

Clinical Ethics Consultation

Clinical ethics consultation is strictly speaking a service mostly available in Anglo-Saxon healthcare institutions that assists healthcare providers, patients, families, and other involved parties in complex decision-making processes. Although the institutionalization of this service is less common in continental Europe, it is more often advocated today. Clinical ethics consultation is a service that can also be provided by ethics committees, as is more often the case worldwide, and even in an informal way by the attending physician in specific circumstances. Clinical ethics consultation is a faraway ideal for most developing countries often as a result of inadequate resources and of facilities where healthcare providers are scarce, access to health institutions is limited, and costs of services prohibitive. At the same time there are often cultural and social obstacles to consultation (e.g., due to medical paternalism, gender differences, and discrimination) that prevent the patient and advising ethicists from getting involved in medical decision-making. In the Anglo-Saxon world formal clinical ethics consultation has tended to become part of healthcare services for two major reasons. The first is the scientific and biotechnological development of clinical practice leading to new problems in determining what is humane rather than what is clinical. Although a clinical approach is therefore essential, it is not sufficient to address and help solve such problems. The second is the strengthening of the human rights movement and the growing awareness that each individual has the right to decide on what concerns his or her own life including healthcare decisions. Clinical decisions are no longer the exclusive domain of the attending physician. They involve a team of healthcare professionals, patients, and families who have different goals and values, come from different cultures, and have life stories that are very different. Building up consensus, respecting everyone involved, and providing the best possible healthcare today requires specific ethical expertise. Clinical ethics consultation is usually provided by ethicists acting as consultants and performing roughly the same tasks of clinical ethics committees. Such tasks are educational and normative and involve counseling, case analysis, and review. The advantages of clinical ethics consultation services compared with the regular activity of clinical ethics committees is the level of availability, of expertise (knowledge and experience), and of personal closeness.

H. ten Have and M. do C. Patrão Neves, *Dictionary of Global Bioethics*,
https://doi.org/10.1007/978-3-030-54161-3_129

Clinical Ethics, General

Clinical ethics strictly refers to the ethics of clinical practice. The term clinical ethics can probably be traced back to Joseph Fletcher in the mid-1970s when he drew a distinction between "rule ethics" and "situation ethics." Such a distinction was interpreted by clinicians as clinical ethics and its use has become frequent since the 1980s. It emerged because clinicians needed to discuss unprecedented clinical cases that had been affected by the application of recent biotechnologies to patient care in which crucial ethical dilemmas arose. Although clinical ethics is also an academic discipline and a field of research, it is mostly a practical activity characterized by the application of ethics to clinical practice that is still focused on clinical cases (a bottom-up perspective). The general goal is to objectively identify ethical problems that arise in a clinical case as a result of the clinical encounter between physician and patient, to analyze the moral values and ethical principles that are in conflict, and to try and balance them through a deliberative process that aims at reaching an effective and satisfactory solution for all parties concerned, especially the patient. Clinical ethics is embedded in clinical experience. Case consultation is therefore one of the major roles played by clinical ethics. Although clinical consultation primarily addresses the needs of healthcare professionals, it can also be extended to patients and their families since they are the most important parties of the clinical encounter. Clinical ethics has developed an educational role in which clinical staff and the healthcare community are trained to deal with ethical dilemmas in clinical practice. Research into the best methodology to promote clinical ethics and into outcomes as a result of its implementation is important. Clinical ethics has also focused on health policies and how they influence high-level care while complying with ethical requirements.

H. ten Have and M. do C. Patrão Neves, *Dictionary of Global Bioethics*,
https://doi.org/10.1007/978-3-030-54161-3_130

Clinical Ethics, Methods

Clinical ethics is essentially a practical discipline focused on the resolution of problems that arise in the care of patients. It is a case-based approach to ethical decision-making and as such relies on narratives about the illness experience of patients and families and about the therapeutic challenges experienced by healthcare professionals. Narratives are descriptions of how the reality of a disease was felt, lived, interpreted, and integrated in the life of a patient—not objective and accurate accounts of facts. A clinical ethics case discussion is interdisciplinary in that it embraces all stakeholders, facilitates the sharing of experiences, and contributes to a better understanding of the situation and to a more efficient resolution of the clinical case. The benefits of clinical ethics are widely recognized, especially when it comes to improving healthcare. Clinical ethics is therefore a common practice in healthcare institutions worldwide. Clinical ethics can be promoted through moral case deliberation, ethics rounds, ethics discussion groups, and ethics reflection groups. The cases under ethical analysis can be real examples, current situations that need to be urgently addressed, or former cases that need revisiting in which the outcome is already known and the decision taken at the time can be evaluated to see if it could have been better, thus learning from experience (retrospective case review). They can also be hypothetical cases for training purposes (prospective case review). Such initiatives are often organized by ethics committees or ethics consultants and take place as part of the regular activity of a center or department of clinical ethics hosted in a hospital.

H. ten Have and M. do C. Patrão Neves, *Dictionary of Global Bioethics*,
https://doi.org/10.1007/978-3-030-54161-3_131

Clinical Ethics, Professionalization

Much as is the case with all professions, the professionalization of clinical ethics is a response to a social need demanded by the current predominant healthcare model. Clinical practice today is strongly supported by scientific and technological advances. It has been developed by practitioners with the active participation of patients and families in an attempt to achieve a clinical or medical good (as appropriate to the healthcare profession) that does not always coincide with the personal good (as appropriate to the individual). It is complex and challenging in such a context for physicians to maintain a patient-centered practice. There is a need for professional assistance to address problems of a humane nature such that they can be either prevented whenever possible or solved to the satisfaction of everyone concerned. This is the specific job with which clinical ethicists or ethics consultants are tasked. Such tasks require specific knowledge of the scientific and professional fields involved, skills at the communication level and in consensus building, and experience in dealing with sensitive cases sometimes regarding life or death situations. Clinical ethicists and ethics consultants are fortunate in having available to them an impressive array of very pertinent literature produced internationally. Moreover, there are conventions, declarations, protocols, recommendations, opinions, and landmark court decisions that together with the robust activity of international and national ethics committees and updates of national legislation are all important for clinical ethics. The professionalization of clinical ethics is necessary for all such items to be embraced. Although the professionalization of clinical ethics has yet to become widespread, it is gaining ground in the Anglo-Saxon world. Clinical ethics professionals in other regions are considered either a threat to clinicians (overriding their power) or a management burden (one more person to be consulted). Furthermore, ethics is said to be something everybody is taught—not an academic discipline to be reduced to a technique as a result of its professionalization losing its very identity as a reflective enterprise. Another objection to the professionalization of clinical ethics is the risk of losing independence. A professional who works for a healthcare facility and is paid by that same institution could be tempted or pressurized to defend the institution against the interests of patients and families. Clinical ethics professionals are there to provide clinicians with consultation support—not a litigation service. The importance of clinical ethics cannot be entrusted to volunteers since accountability can only be possible at the professional level.

H. ten Have and M. do C. Patrão Neves, *Dictionary of Global Bioethics*,
https://doi.org/10.1007/978-3-030-54161-3_132

Clinical Ethics, Support

The major goal of clinical ethics is to provide support to healthcare professionals in identifying ethical issues. This can also involve patients and families. Clinical ethics helps healthcare professionals to identify the ethical issues involved in clinical cases and to develop the right approach such that the patient's autonomy is respected while healthcare beneficence is promoted. Although the physician is always the person responsible for the final decision, clinical ethicists can help him or her during the deliberation process in close cooperation with the healthcare team. This makes it easier to take decisions and bolsters the confidence of all stakeholders in such decisions. Clinical ethics can also be used as a powerful resource to fight physician burnout and help improve healthcare quality at the same time. Clinical ethics can help patients and families to cope and better understand situations. Furthermore, it helps the healthcare team to collaborate fully as true partners in the therapeutic process. Some hospitals, mostly in the United States, have centers for clinical ethics that together with ethics consultation and clinical ethics committees frequently provide support to healthcare professionals, offer ethical education, and assist with advice and recommendations in the resolution of concrete problems.

H. ten Have and M. do C. Patrão Neves, *Dictionary of Global Bioethics*,
https://doi.org/10.1007/978-3-030-54161-3_133

Clinical Ethics, Teaching (*See* Bioethics Education)

Teaching ethics in the clinical setting is a specialized form of bioethics education. It is specifically focused on ethics training of residents, physicians, and postgraduates working with patients in clinical institutions. The goal is to improve the quality of patient care. Although such teaching can be offered in formalized programs and courses, it is often also embedded in ethics consultation services. The emphasis is on active learning through doing, reflecting, and experiencing. Ethical quandaries are encountered everyday in hospitals. Although this arguably offers many opportunities for ethics teaching, the advantages are that teaching is at the bedside, different types of health professionals are involved, and teaching happens in real time. Much like other forms of bioethics education, concerns exist because programs in clinical ethics are very heterogeneous. Didactic approaches, teaching methods, number of teaching hours, and scheduling of courses can substantially differ. Major controversies exist concerning the objectives, methods, content, and evaluation of teaching activities. Since the practitioners of clinical ethics are overwhelmingly from healthcare professions there are also worries about the qualifications and competencies of those teaching clinical ethics. Most teachers in practice do not have a substantial background in ethics themselves.

H. ten Have and M. do C. Patrão Neves, *Dictionary of Global Bioethics*,
https://doi.org/10.1007/978-3-030-54161-3_134

Clinical Research (*See* Research; Research Ethics)

Reliable information about medication, treatments, and devices is generated by means of clinical research. This is that part of healthcare science focused on determining the safety and effectiveness of drugs and treatments so that they can be used in clinical practice. New medication such as a new molecule is first identified in the laboratory. It is then subjected to preclinical and animal studies to determine its toxicity and efficacy. If this stage is successful, then Phase 1 clinical trials will involve using small groups of healthy volunteers to test safety and dosage (clinical and biological tolerance) and to study pharmacokinetics and pharmacodynamics. Phase 2 is when the drug or treatment is first administered to patients. It involves a small number of research subjects and only lasts a short period of time. Phase 3 is the final phase and is aimed at studying therapeutic efficacy. It involves many people over a long period of time. If the new drug successfully passes through these phases, then it will be submitted for approval to the national regulatory authority for use in medicine. Clinical trials involving human subjects should be conducted according to national and international legislation and regulations. A coherent ethical framework has been developed over the decades to guide clinical trials. International Conference of Harmonization good clinical practice (ICH-GCP) guidelines describe what clinical trials require to comply with GCP. This also applies to other clinical trials where the research protocol states that it is or has been carried out according to GCP. The Declaration of Helsinki also contains ethical principles that should be used to conduct clinical trials. Different parties are often involved in clinical trials each of whom has different responsibilities. Sponsors are the commissioning party such as the pharmaceutical industry or a non-commercial institution. For investigator-initiated research the party is the board of directors of the institution. The agents involved comprise the investigators such as specialists or general practitioners. Third parties can be hired by the sponsor. They include contract research organizations (CROs), pharmacies, laboratories, and manufacturers of investigational medicinal products. Finally, there are the test subjects such as participants in the trial. Clinical research gives rise to many ethical issues (*see* Research ethics for a discussion).

H. ten Have and M. do C. Patrão Neves, *Dictionary of Global Bioethics*,
https://doi.org/10.1007/978-3-030-54161-3_135

Clinical Trials (*See* Research Ethics; Clinical Research)

Clinical trials are scientific research projects involving human subjects. Such trials are used to evaluate the effects and safety of new drugs or medical devices. They are the most important means of developing biomedical sciences and of discovering or inventing new ways to promote human health and well-being. Clinical trials are preceded by a preclinical stage of research in which laboratory tests and studies on non-humans (animals) are performed with the aim of evaluating the toxicity of the new drug, its potential safety, and its efficacy. Only when the results of this early stage are satisfactory does the clinical stage start and the tests are extended to humans. A clinical trial unfolds in four phases each raising different ethical issues. Phase 1 focuses on the safety of the experimental intervention by testing it on a small group of healthy people to evaluate the correct dosage and possible side effects. Balancing the risks and benefits is essential to guaranteeing the safety of the trial. The major ethical concern here relates to the criteria used to recruit participants and the advantages such intervention offers them. Although enrolling in a clinical trial is done by most people for the right reasons, there are those who enroll to get a free health checkup, free health monitoring, and receive compensation for lost work days, transportation, and inconvenience. There are people who frequently participate in clinical trials. This may damage their health in the long run and affect the results of the clinical trial. Phase 2 focuses on the effectiveness of the new product. It determines whether the product (say, a drug) really works in a larger group of people with the disease or condition the drug aims to cure. The major ethical issue here is the vulnerability of patients some of whom are desperate to find a cure. Phase 3 involves even more people often from different countries who are given different dosages of the drug under trial or combinations of it with other drugs and placebos. The ethical issue here is the use of a placebo for the control group when there is a standard treatment. Patients enrolled in a clinical trial cannot be left without standard treatment (although it is currently recognized that placebos do have a real effect in clinical practice). If the results of these three phases are positive, then the new drug is approved as treatment for a specific disease by the relevant organization. Phase 4 is the last clinical trial and is used to monitor a large and diverse population over a long period of time. There is an ethical requirement at this stage for thorough pharmacosurveillance that is unrestricted. Other ethical issues that need consideration are the ease of access to new drugs as non-prescription medicines, self-medication, and off-label use. Historically, there have been two periods in the development of clinical trials that have raised dramatic ethical issues in which vulnerable populations were abused and enormous suffering and even death was brought about. The first period covered most of the twentieth century from its beginning right up to the 1970s. It was during this time that the experimental method developed and clinical trials became more reliable scientifically. Most human participants were recruited without their knowledge or consent

H. ten Have and M. do C. Patrão Neves, *Dictionary of Global Bioethics*,
https://doi.org/10.1007/978-3-030-54161-3_136

(among them orphans, prisoners, the elderly, and minority ethnic groups). The ethical reaction to this was to require informed consent from all participants and to exclude from clinical trials those groups who were considered vulnerable. The second period endured until recently when clinical trials were delocalized to developing or underdeveloped countries where there was no clinical trial regulation and where research could be pursued in the absence of specific restrictions. The ethical reaction to this was to require the same standard procedure for clinical trials (rejecting double standards) regardless of the geographic region in which they were implemented. Currently, we have entered a new period in the development of clinical trials in which countries in the so-called developing world compete to be the most attractive location for the conduct of clinical trials.

Clone

The word "clone" derives etymologically from the Greek *klon* (vegetable bud) and was first coined in 1903 by the plant physiologist Herbert J. Webber to designate the technique of propagating new plants using cuttings, bulbs, or buds. Its meaning later widened to refer to all forms of life—ranging from a cell to an organism—produced asexually and genetically identical to its ancestor most of which were artificially produced in a lab. Since clones are common in nature among vegetable species, humanity has long taken advantage of this to select and enhance production. Animal clones have been artificially generated since the late nineteenth century and early twentieth century using a number of different techniques. They have been produced for research (as models to study human diseases), for cattle breeding (to enhance production), for sports such as horseracing (as a selective reproduction technology), and as pets (as replicas of dead pets). After the birth of the first mammal clone (the sheep Dolly) from an adult somatic cell in 1996 and subsequent successes with cloning other mammal species (specifically non-human primates), the possibility of producing human clones started to take shape. One of the important ethical arguments against human clones was the deliberate generation of two identical persons despite being fully aware that the uniqueness of each and every one of us is an integral part of human identity. Many rebutted this argument saying that human clones would be born and grow up at a different time and in a different social environment that would make them different human beings despite being genetically identical. Epigenetics today would strengthen this line of reasoning. Nevertheless, other important arguments against human cloning such as the experimental nature of the procedure, the lack of individual or social benefits, and the wide range of possibilities for exploitation dictated the non-legally binding ban on human reproductive cloning (i.e., on the creation of human clones).

H. ten Have and M. do C. Patrão Neves, *Dictionary of Global Bioethics*,
https://doi.org/10.1007/978-3-030-54161-3_137

Cloning, Animal (*See* Animal Cloning)

Animal cloning spontaneously happens in nature via various means. Such cloning commonly refers to the artificial production of animals that are genetically identical to the progenitor. This is brought about by applying one of the many cloning techniques recently developed such as embryo splitting (a.k.a. artificial twinning) and somatic cell nuclear transfer (SCNT). The sheep Dolly was the first mammal to be cloned by SCNT (in 1996). This success triggered the extension of this experimental cloning technique to many other species numbering about 20 at present including dogs, pigs, frogs, mice, cattle, and rabbits. In 1998 scientists cloned the first primate. There are four main fields of interest in animal cloning: livestock breeding, endangered species protection, production of transgenic animals for biomedical research, and pet reproduction. Cloning is more frequently applied to the production of breeding stock (using cloned males for traditional reproduction) than for producing food, although their offspring are used for food production. The main goal is to upgrade the quality of herds by producing more copies of the best animals that will eventually be genetically modified to enhance desirable traits such as being more resistant to disease, producing more milk, or providing higher quality meat. Cloning is likewise used in sports to create clones of champions such as racehorses. There have been a number of issues with animal cloning from the high rate of fetal and neonatal loss to associated abnormalities such as respiratory distress, cardiovascular anomalies, and failures of the immune system. The first mammal to be cloned was a sheep that died (it was euthanized) prematurely at the age of 6 years (sheep can normally live up to 11 or 12 years of age) as a result of having lung infections common in older sheep. Premature ageing (due to shortening of telomeres) is also a problem with cloned animals. Animal health and animal welfare are obviously ethical issues that need consideration. After the successful application of SCNT in farm animals the idea of using the same technique to resuscitate extinct species using recovered DNA caused a lot of interest. However, motion pictures like *Jurassic Park* brought home to many that it might not be a reasonable idea to bring extinct species back to life. Introducing them in quite different ecosystems would be highly prejudicial not only for the animals resuscitated but also for current species. Nevertheless, animal cloning when applied to endangered species and aimed at increasing the number of individuals and boosting their recovery can indeed be a very beneficial use of SCNT. It can in this way protect biodiversity, ecosystems, and the environment in general. This is the only field where cloning is carried out for the benefit of animal life. A further major application of animal cloning is within biomedical research where the objective is mainly to increase the production of transgenic animals. Animals have been genetically engineered to produce human therapeutic proteins, cells, and tissues for xenotransplantation and as

H. ten Have and M. do C. Patrão Neves, *Dictionary of Global Bioethics*,
https://doi.org/10.1007/978-3-030-54161-3_138

models to study human diseases. Animal cloning produces proteins, cells, and tissues faster, simpler, less expensively, and more efficiently. The application of animal cloning has raised transverse ethical issues such as the denaturalization and commodification of animals and concerns about their ontological, legal, and moral status, and their well-being. A more recent field of application for animal cloning that is rapidly growing as a new business is the reproduction of pets (mostly dogs and cats). Although the life expectancy of such pets is limited compared with that of humans, people want to keep their animal companions indefinitely. Although this has allowed pet owners to resort to new agencies that provide cloning services at a very substantial price, the ethical problems raised are many and varied such as the psychological and emotional dependence people have on an animal and the vulnerability they have to aggressive enterprises that offer a kind of continuous and infinite life for pets while knowing that the cloned animal's personality will not be the same as that of the somatic cell donor. Finally, there are concerns about animal welfare such as the anthropomorphization of animal natural life.

Cloning, Concept

The word "cloning" (the production of clones) derives etymologically from the Greek *klon* (vegetable bud). Vegetative cloning generates a new plant that is identical in every way to the original one (buds grow internally and then develop into new shoots the following year). Cloning is the term used to designate asexual reproduction, a form of reproduction without the need for gametic combination (characteristic of sexual reproduction). It thus produces offspring that are exact copies of the original. Not unsurprisingly, the concept was first applied to plants. However, cloning in nature is more common in single-cell organisms such as bacteria that reproduce by a succession of binary divisions. Cloning naturally occurs in single-cell organisms, plants, and animals (even in humans). Currently, cloning broadly refers to the reproduction of genetically identical individuals either spontaneously, naturally, or artificially in a lab. When cloning is natural, it is of course not submitted to ethical scrutiny. When it is induced or produced (i.e., controlled by humans), it should be ethically evaluated according to the nature and status of the living materials used, the procedures implemented, and the goals pursued.

H. ten Have and M. do C. Patrão Neves, *Dictionary of Global Bioethics*,
https://doi.org/10.1007/978-3-030-54161-3_139

Cloning, Food

Food cloning refers to those goods created via cloning and used for nourishment. Although it effectively refers to plants that can also undergo cloning spontaneously and naturally, it also refers to animals. The cloning of plants has been carried out since ancient times (e.g., grafting) as a regular practice in agriculture. The traditional use of cloning in agriculture does not raise ethical problems. Cloning benefits today from lab techniques that enhance agricultural and horticultural food sources. It works via selection in which the best plants are identified and then cloned. In this way the features that make them valuable are perpetuated and the next crop is controlled instead of farmers trusting in it. Moreover, it can increase production. Cloned fruits, vegetables, and flowers are widely available in Western daily markets. In particular, the cloning of mutant fruit such as seedless grapes or watermelons has developed profitable products for fruit farmers. Such fruits are genetically engineered to be seedless and reproduced via cloning. Cloning raises two principal ethical questions in farming. The first refers to the safety of food products. Little stress has been put on this concern because the techniques used replicate traditional ones and no problems have been identified. This may well be the reason the production of such products is not specifically regulated and the reason they are not labeled as such. The second problem relates to the loss of biodiversity due to intensive use of such a highly efficient selection method and the focus on creating elite lines of products. The problem is further exacerbated by the variants created by genetic engineering for cloning are often patented and thus become inaccessible to small traditional farmers who have great difficulty in competing and run the risk of going out of business. Although animal cloning as a breeding technique for livestock uses different and more elaborate cloning techniques such as embryo splitting (a.k.a. artificial twinning) and somatic cell nuclear transfer (SCNT) than traditional farming, the goals are the same: to improve and increase production and consequently profits. Cloning is mostly used for cattle, pigs, and goats. However, the problems with animal cloning are many and diverse. One concern relates to food safety. The consumption of meat and milk from cloned cattle, pigs, and goats has been very controversial (similar to the ongoing debate about genetically modified organisms, although the biotechnology used is different). It is well known that animal clones suffer from abnormalities and unusually large offspring that result in difficult births and neonatal deaths. Nevertheless, all things considered the US Food and Drug Administration (2008) has stated that eating cloned animals is as safe as eating naturally reproduced animals. However, in 2018 the European Commission set out Regulation (EU) 2015/2283 on Novel Food (2018), which requires authorization for "foods originating from plants, animals, microorganisms, cell cultures, minerals, etc., specific categories of foods

H. ten Have and M. do C. Patrão Neves, *Dictionary of Global Bioethics*,
https://doi.org/10.1007/978-3-030-54161-3_140

(insects, vitamins, minerals, food supplements, etc.), foods resulting from production processes and practices, and state of the art technologies (e.g. intentionally modified or new molecular structure, nanomaterials), which were not produced or used before 1997." The goal is to improve the conditions under which new and innovative foods are introduced to the EU market, while maintaining a high level of food safety for consumers. Although the issue relating to the reduction of biodiversity is not major considering the few species that have economic value to be cloned, there is a related and serious problem to be considered: the threat to the genetic diversity of livestock. When compared with agriculture, animal cloning has the additional and very important ethical problem of animal health and welfare.

Cloning, General

Cloning is the term used to designate asexual reproduction whether natural or artificial in which the offspring is an exact copy (genetically identical) of the original, which can be a single-cell organism, a plant, or an animal. Cloning occurs naturally and spontaneously and is the most elementary form of plant and animal reproduction. It happens in nature through fragmentation in which a new organism grows from a fragment of the parent (e.g., starfish), bipartition or binary fission in which the individual divides into two parts each with the potential to grow to the size of the original (e.g., flatworms), gemmulation in which aggregates of cells, mostly archaeocytes, become isolated (e.g., sponges), budding (e.g., rhizomes of grasses, or bulbs), and parthenogenesis (e.g., aphids). Although parthenogenesis does not involve the exchange of genetic material from two different sources, it can also be considered sexual reproduction because it involves germ cells. Nevertheless, most plants and animals reproduce sexually. Cloning has been used by humans throughout history to improve the production of certain plants. This is done by inducing small pieces of plants that have the capacity for vegetative propagation (mostly all types of roots and shoots) to grow independently. It is used in agriculture as a selection procedure and as a way to speed up crop production and therefore productivity. Since the word "cloning" did not enter the public domain until quite recently following the cloning of the first mammal (the sheep Dolly) in 1996, the word today has gained a specific meaning. It generally refers to the biotechnological capacity to produce genetic copies of the living original ranging from DNA (gene cloning or other pieces of DNA) to whole animals. There are different (animal) cloning techniques. The simplest is artificial twinning in which the stem cells (totipotent) of an egg are separated and continue to grow separately into whole individuals. This technique leads to the development of natural identical twins (artificially induced). However, the truly revolutionary technique is somatic cell nuclear transfer (SCNT), which has two variations: the Roslin Technique and the Honolulu Technique. The procedure basically involves the removal of the nucleus from a somatic (body) cell and its injection into an egg cell that has had its nucleus removed. The manipulated egg will then develop into an embryo that can be implanted into a surrogate and then develop into a whole individual that is genetically identical (a clone) to the donor nucleus of the somatic cell. Cloning techniques are being used in a multitude of fields from food production (agriculture and livestock) to clinical research. In every field they have proven to be highly beneficial to the human way of life. This specifically applies to clinical research in which cloning is helpful in transplantation medicine (production of stem cells, and/or genetically modified animal cells for people needing organ or tissue transplants), in regenerative medicine (for degenerative diseases such as Parkinson, Alzheimer, and Huntington),

H. ten Have and M. do C. Patrão Neves, *Dictionary of Global Bioethics*,
https://doi.org/10.1007/978-3-030-54161-3_141

and in the production of sufficient quantities of human therapeutic proteins by creating transgenic animals that are also useful for the study of particular (specifically genetic) diseases. Each of these applications raises specific ethical problems, particularly regarding vegetables and animals. Furthermore, there are some transverse ethical issues (i.e., issues referring to all domains of artificial cloning). The most relevant issue is the manipulation of life and its commodification. Biotechnologies today are generally capable of intrinsically changing natural life. Although this is a recent achievement, technological advances throughout the evolution of humanity have progressively acquired the power to change and shape how life manifests itself, but not to alter them in their very constitution. Although cloning techniques are not unique in being able to control life, they certainly contribute to increasing such control and to reducing life in its different manifestations to a human supply or good. The denaturalization of life and its commodification are the common ethical concerns regarding cloning.

Cloning, Human

Human cloning refers to the artificial production via somatic cell nuclear transfer (SCNT) of a genetically identical copy of a human being. However, current practice limits it to the production of stem cell lines genetically identical to the donor's somatic cell. Human cloning could also refer to the natural process that takes place with monozygotic twins in which a zygote (fertilized egg) is spontaneously split into two or more identical cells that are totipotent and both able to form an embryo. The first successful attempt to artificially clone primates occurred in 1998 with the cloning by embryo splitting (artificial twinning) of a rhesus macaque called Tetra born the following year. This achievement followed the cloning of the sheep Dolly by SCNT and led scientists and the public, in general, to wonder about the feasibility of human cloning. However, SCNT in primates encountered major problems and did not register any success until 2018 when two monkeys called Zhong Zhong and Hua Hua were born out of the 79 embryos produced (i.e., a very low rate of efficiency). Although the birth of the first cloned human was announced in 2002 by the Raelian cult, no proof has been provided. The difficulties for such a possibility are insurmountable and there is little chance of this becoming a reality in the near future. Human cloning can be broken down into two main fields: reproductive and therapeutic. Human reproductive cloning aims at generating a human being who is genetically identical to the donor of the somatic cell. Possible justifications for undertaking such a project beggar belief such as creating an elite lineage of human beings, generating another copy of oneself, and producing a source of compatible organs for transplantation. The mere existence of a human clone would challenge humankind's identity, uniqueness, and freedom. Not surprisingly, the reaction to the possibility of human reproductive cloning was global in its condemnation. The 2005 UN Declaration on Human Cloning prohibits "all forms of human cloning inasmuch as they are incompatible with human dignity and the protection of human life." The support for such a statement was ambivalent and global governance on this issue is still vague. On the other hand, human therapeutic cloning is a well-developed field and holds a lot of promise. It is used in the study of some diseases (enhancing knowledge about cells), in the production of new drugs, in the improvement of genetic therapies, and mostly in cell therapy (growing cells and tissues such as pancreatic tissue and epithelial cells to treat diabetes, leukemia, or burns). Therapeutic cloning produces lines of stem cells that are genetically identical to the donor's somatic cells and are pluripotent (i.e., they can differentiate between all kinds of cells except embryos). Therefore, they can be used to treat diseases in any organ or tissue of the body by replacing damaged and dysfunctional cells (transplant and regenerative medicine) without the risk of rejection. The ethical problem with human therapeutic cloning is the use and subsequent destruction of embryos created for stem cell extraction. This practice uses embryonic

H. ten Have and M. do C. Patrão Neves, *Dictionary of Global Bioethics*,
https://doi.org/10.1007/978-3-030-54161-3_142

human life for supply purposes regardless of the moral status of human life in its early stages and thus violates the principle of human dignity. Advocates of human therapeutic cloning consider that critiques concerning the utilitarian use of human embryos are trumped by the high beneficial potential of this technique. If induced pluripotent stem cells (IPSCs) could be obtained through somatic cell nuclear transfer, then the ethical problem with human therapeutic cloning would be suppressed and all the expected benefits would be maintained.

Codes of Conduct

Codes of conduct are sets of rules, values, and virtues that regulate social and professional activities. They differ according to the profession or to any other social activity they regulate. However, the common ground for all codes of conduct today regardless of the activity or the country they refer to is the Universal Declaration of Human Rights essentially guaranteeing individual rights and their further development. There have been codes of conduct since the days of ancient Greece where they were associated with the practice of liberal professions such as medicine, which was regulated by the Hippocratic Oath. Such a code of ethics governs not only the relationship between physicians and their pupils but also that between physicians and patients. Over time, most codes of conduct were associated with arts and crafts and laid down the relationships between artisans and apprentices. Codes of conduct today not only acknowledge the social responsibility institutions and professions have but also acknowledge the demands of society and the expectations of citizens. Codes of conduct generally start by presenting the major values of the social and professional activity (e.g., respect for life) within its *telos* (end goals) and then present the norms of practice. Such norms can be of two kinds: one is moral in that it considers the good character of the professional to determine his/her appropriateness for professional action (e.g., truth telling), while the other involves administrative law that aims to contribute to the prestige of the profession within society (e.g., preventing conflicts of interest). Some virtues such as empathy are mentioned in a code of conduct since they contribute to good professional performance, whereas others such as integrity are converted into duties and become compulsory. Moral rules imply moral responsibilities, while administrative rules imply disciplinary responsibilities. Codes of conduct today are not as strict as they were traditionally when professionals could be punished for non-compliance. Today's trend is to propose a guide of good practices to which all professionals should be committed. Lack of compliance usually has few consequences since codes of conduct do not prescribe sanctions for violations and work only as guidelines.

H. ten Have and M. do C. Patrão Neves, *Dictionary of Global Bioethics*,
https://doi.org/10.1007/978-3-030-54161-3_143

Coercion

Coercion is the practice of obliging someone else to act in an involuntary manner by the use of threats or force. Its aim is to control the will or behavior of another person. Since coercion restricts the freedom of someone and violates his or her individual autonomy it is usually regarded as morally wrong. Coercion involves using threats to oblige someone to comply with demands on the understanding that his or her situation will worsen if he or she does not comply. For example, certain behaviors may provoke legal sanctions such as when citizens fail to pay taxes. Although in such a case the scope of action of the individual may well be limited, there is no restriction on his or her decision-making. This is different from the use of force where the subject is restrained against his or her will such that there is no room for making choices. Coercion can also be used to describe a situation in which the autonomy of someone is limited, on the assumption the individual is free, competent, and adequately informed beforehand. Coercion can be applied to individuals, groups of individuals, or collectives. Coercion is also conceived as something carried out by agents. Exceptionally, there might be circumstances that are coercive (e.g., natural disasters) when assigning responsibility becomes difficult. Although only nation states should have a monopoly on legitimate coercion, globally various forms of coercion such as intimidation, deception, and exploitation are exercised by other agents. Coercion has been justified with two arguments. One argues that coercion benefits all parties concerned. The other argues that coercion redefines autonomy and freedom such that they are not limited. Both arguments are based on a liberal ethical framework that assumes that limiting freedom is prima facie always wrong. A different way of looking at coercion is based on communitarianism, which is another ethical framework that points out that people are members of a community and that whether coercion is justified or not depends on human relations and connections. Involuntary treatment of people with mental disorders is a paradigm case of coercion in bioethics. It is argued that coercion is in the best interest of patients since they cannot take care of themselves and often harm themselves.

H. ten Have and M. do C. Patrão Neves, *Dictionary of Global Bioethics*,
https://doi.org/10.1007/978-3-030-54161-3_144

Cognitive Sciences

Cognitive sciences study the mind and its processes. They are focused on cognition (i.e., the mental processes of acquiring knowledge and understanding through thought, experience, sensation, and perception). They study intellectual functions such as memory, reasoning, attention, comprehension, and decision-making. Such sciences are interdisciplinary with input from psychology, neuroscience, philosophy of mind, linguistics, anthropology, biology, artificial intelligence, and computer science. Different methodologies are used such as behavioral experiments, brain imaging, computational modeling, formal logic, and deep brain stimulation. Cognitive sciences have been criticized for neglecting the role of emotions. They also ignore the role of consciousness. Human intellectual functions are influenced by and dependent on physical and social environments, as well as the fact that humans are embodied beings. Such challenges are today taken up in many efforts to explain emotions, consciousness, and embodiment as neural mechanisms. However, these efforts assume a naturalistic and reductionist ideology that identifies mental processes as neural or brain events.

H. ten Have and M. do C. Patrão Neves, *Dictionary of Global Bioethics*,
https://doi.org/10.1007/978-3-030-54161-3_145

Commercialism

Healthcare today is regarded by many as a business. Providing care is regarded as a market transaction. Although healthcare professionals have to earn a living, professional standards have always limited commercial interests. However, many hospitals, laboratories, insurance companies, and the pharmaceutical industry today operate for profit. Patients are regarded as autonomous consumers or clients who select the care and treatment they want to receive. They are primarily considered as *Homo economicus* (i.e., concerned with maximizing gains and minimizing harms or costs). Commercialism is defined as the application of the operating and managing principles of business and commerce. This is in tension with the primary demand that medicine should focus on the health of the patient. The main ethical concern raised by commercialism regards the proper role market considerations play in the context of healthcare. The first ethical concern is the connection between professionalism and commercialism. Priorities and goals should not be reversed. Financial gain in healthcare should be a secondary motive. Pay-for-performance programs may provide incentives to increase financial output and therefore generate conflicts of interest. The second ethical concern is the possible impact on quality of care. Avoiding waste and being efficient are not the same as increasing expenses to boost profits. Although such a motive may well stimulate productivity (particularly in those areas that are profitable such as certain diagnostic procedures), it can also lead to overtreatment. The third ethical concern is how market thinking is transforming medicine since it has been shown to encourage thinking in terms of quick solutions, magic bullets, and short-term therapies, while neglecting socioeconomic conditions and health inequalities.

H. ten Have and M. do C. Patrão Neves, *Dictionary of Global Bioethics*,
https://doi.org/10.1007/978-3-030-54161-3_146

Committees, Clinical Ethics Committees

Clinical ethics committees are established at healthcare facilities. They work in the clinical setting assisting staff, patients, and families. They consist of different healthcare professionals, a bioethicist, a lawyer, religious ministers, a patient representative, and/or a member of the local community. The role of such committees is advisory, although they can pursue different functions such as counseling and assisting healthcare staff in decision-making about concrete cases where there is an obvious ethical dimension (specifically moral values and beliefs). Their function can also be normative in that they outline ethical guidelines for specific procedures in concrete situations thus also assisting hospital administration. Moreover, committees can have an educational function such as promoting capacity building of all members of the ethics committee (self-education) and developing staff sensitivity to the ethical dimensions of patient care such that decision-making can be improved and ethical problems can be prevented from arising. Another function of committees is case review in which past cases are revisited to evaluate the advice given at the time in light of the outcome. Clinical committees were the first ethics committees to be informally established in the United States. In 1962 the nephrologist Belding Scribner created the Admissions and Policy Committee in Seattle. This committee was tasked with selecting patients to receive hemodialysis from a long list in need of it. They would do so according to the availability of dialysis machines. This committee became known as God's Committee, so-called by journalist Shanna Alexander in her report on the committee's activities. Later, after the Karen Quinlan case (a young woman in a persistent vegetative state whose parents won the power to make decisions for her), a court in 1975 recommended the creation of what would be known as institutional ethics committees (IECs). They were tasked with working on ethical procedures to cope not only with unprecedented clinical situations produced by the application of new biotechnologies to human beings such as resuscitation techniques, but also problems such as withholding and withdrawing life support from incompetent patients. IECs became mandatory in 1983 after the Baby Doe case (a newborn with Down syndrome who was left to die because her parents did not consent to lifesaving surgery). The President's Commission for the Study of Ethical Issues in Medicine and Biomedical and Behavioral Research considered that healthcare institutions should establish hospital ethics committees to assist staff and the institution in choosing the best course of action as a result of considering the ethical requirements necessary to benefit and respect the persons involved. From then on IECs became common worldwide, although they were not always set up as single bodies. Several European countries along with countries in other continents adopted a mixed model by creating a single ethics committee in hospitals to act simultaneously as an IEC and as an institutional review board (IRB), a research ethics committee. Ethics

H. ten Have and M. do C. Patrão Neves, *Dictionary of Global Bioethics*,
https://doi.org/10.1007/978-3-030-54161-3_147

committees using such a model tend to give priority to regulatory reviews of research projects. Although they act less at the assistance level, they continue to provide a clinical ethics consultation service. In the Anglo-American model IEC competencies can also be combined with the work of an ethics consultant. In such a case ethical assistance tends to be stronger and more effective.

Committees, General

Ethical committees are multidisciplinary, pluralist, and independent bodies that most commonly have an advisory function and work in the biomedical setting. As multidisciplinary bodies they are usually composed of people with different academic and professional backgrounds. They sometimes also include lay people in an effort to give citizens' perspectives. As pluralist bodies they gather people with different cultural, religious, or ideological values that reflect the different sensibilities in society. Committees are also independent and do not serve the interests of any institutions or organizations. Most ethics committees have an advisory role, take decisions on their own initiative or as a result of demand, and assist those who have to make decisions. The birth of ethics committees represents a realization that self-regulation or individual conscience are no longer enough for decision-making in the realm of biomedicine. Such a realization has come about as a result of the growing power biotechnology has over life, the complexity of human problems it can generate, the obligation to respect individual autonomy in a pluralistic realm, and the awareness that collective advice and surveillance are needed. The first ethics committees were local. They were created in the 1970s in hospitals to assist healthcare professionals and researchers in response to pressure from the media and public opinion concerning the best course of action in the face of the unique human situations raised by the new biotechnologies. Moreover, in the 1970s some national ethics committees were created to analyze specific issues arising from the application of biotechnologies to human life and to advise and provide guidance when it came to political and legal powers. Such institutionalization of applied ethics within biomedicine or bioethics continued to grow. In the late 1980s and early 1990s international and worldwide committees were created to establish common principles and procedures regarding how new biotechnological powers should be used. It should be stressed that the impressive development of ethics committees—from local to worldwide and from individual cases to public policies—was first prompted by the need to have a multidisciplinary and pluralistic view of the growing number of unprecedented bioethical issues and their rising complexity. Second, it was supported by recognizing the value such ethical bodies have in taking the advantages of scientific and technological innovation into account, while benefitting and respecting individuals, communities, and humanity itself. Ethics committees today deal with ever increasing demands and pressures from different stakeholders and have to assimilate new and relevant local, national, and international information on a continuous basis. Their members are requested to become more committed by dedicating a growing number of hours to this usually voluntary and gratuitous task. This has led to discussion about professionalizing bioethicists and making them accountable for their recommendations.

H. ten Have and M. do C. Patrão Neves, *Dictionary of Global Bioethics*,
https://doi.org/10.1007/978-3-030-54161-3_148

Committees, International Ethics Committees

The establishment of international ethics committees started in the 1980s in Europe concomitant with the worldwide expansion of bioethics thus acknowledging the importance of having common bioethical standards in different countries and contributing to the validity and credibility of global bioethics. The first international institution of a bioethical nature was the Committee of Experts on Bioethics (CAHBI) set up by the Council of Europe in 1985 to intensify work on the impact that progress in biomedical sciences was having on human beings. It was an ad hoc committee until 1992 when it became permanent under the name of the Steering Committee on Bioethics (CDBI). The name changed again in 2012 to the Committee on Bioethics (DH-BIO). Its main mission has always been to guarantee not only a person's integrity but also respect for human dignity as a result of applying biotechnologies to human beings. It also provides common bioethical guidelines to all member states. Its main achievement was elaboration of the Convention for the Protection of Human Rights and Dignity of the Human Being (1997), which together with its four protocols developed guidelines on different specific issues. In 1991 the European Commission created the Group of Advisors to the European Commission on the Ethical Implications of Biotechnology (GAEIB) to promote ethical discussions about the development of biotechnology and to take into account ethical reflections on the decision-making process when it came to community research and technological development policies. At the beginning it too was an ad hoc advisory body. It became permanent in 1997 under the name European Group on Ethics in Science and New Technologies (EGE). It was only later that worldwide ethics committees were created by UNESCO. The International Bioethics Committee (IBC) was established in 1993 to develop reflections on ethical and legal issues raised by scientific research and technological innovation in the life sciences. UNESCO also set up the Intergovernmental Bioethics Committee (IGBC) in 1998 to examine the IBC's recommendations the best of which the committee proposed to the Director-General for transmission to member states. The IBC is composed of experts who represent the state of science and thinking in their particular fields of expertise. They are selected and appointed by the Director-General and are independent of their governments. Members of the IGBC are nominated and appointed by their governments. Hence they advocate the formal points of view of their countries. These two committees have allowed the process of ethical deliberation to be connected from the start with both political considerations and the scientific state of the art. The main achievement of their joint work was elaboration of the Universal Declaration on Bioethics and Human Rights (2005). UNESCO also established a World Commission on the Ethics of Scientific Knowledge and Technology (COMEST) in 1998 tasked with formulating

H. ten Have and M. do C. Patrão Neves, *Dictionary of Global Bioethics*,
https://doi.org/10.1007/978-3-030-54161-3_149

ethical principles in very different domains of the life sciences for policy decision-makers (e.g., space exploration and climate change). International ethics committees widen bioethical dialogue and consensus building, contribute to the recognition of core values establishing patterns of best practices, and thus promote international cooperation. Although their advisory nature and the soft law they produce can be said to limit their power, their influence on politicians and legislators is significant. Thus the action of international ethics committees becomes more effective when complemented with biolaw (converting ethical consensus into legal rulings) and with biopolitics (converting ethical consensus into public policies).

Committees, National Ethics Committees

The establishment of national ethics committees started in the 1970s as a result of political initiatives and in recognition of the need for an ethical approach to address very complex problems raised by significant biomedical developments, biotechnological innovations, and their application to human beings. The first national ethics committee was created in the United States in 1974 after disclosure of the Tuskegee case (1972). This was at the time an obsolete scientific research project on syphilis conducted on poor black men who were prevented from receiving proper care. The National Commission for the Protection of Human Subjects of Biomedical and Behavioral Research was tasked with establishing ethical principles that should guide all scientific research with human beings. In 1978 this committee presented The Belmont Report: Ethical Principles and Guidelines for the Protection of Human Subjects of Research (commonly called the Belmont Report). The Committee of Inquiry into Human Fertilisation and Embryology was created in the United Kingdom in 1982 after the birth of Louise Brown (1978) who was the first baby conceived through in vitro fertilization. Dubbed the Warnock Commission for its president Baroness Mary Warnock it was tasked with analyzing the ethical and social impacts and legal problems associated with the unregulated field of reproductive and embryological technologies. In 1984 the committee presented its report that proposed extensive and detailed recommendations. The first national ethics committees were ad hoc and established to fulfill a specific task during a given period of time. In the 1980s they became permanent as a result of the need to address the rising number and diversity of bioethical questions and of their growing importance and urgency. In 1983 the French President created a permanent national ethics committee called the National Consultative Ethics Committee for Health and Life Sciences in the wake of the birth of Amandine (1982) who was the first French baby conceived through in vitro fertilization. Its mission was to "give opinions on ethical problems and societal issues raised by progress in the fields of biology, medicine and health." Such a permanent model of an ethics committee has been followed worldwide and international organizations such as the WHO and UNESCO have been engaged in assisting the creation of national ethics committees worldwide. Since 1996 national ethics committees have linked together to create a global network. They meet every two years at the Global Summit of Ethics Committees. Delegates from all over the world gather together with the objective of strengthening dialogue, deepening issues, and discussing bioethical policies and laws.

H. ten Have and M. do C. Patrão Neves, *Dictionary of Global Bioethics*,
https://doi.org/10.1007/978-3-030-54161-3_150

Committees, Research Ethics Committees (*See* Research Ethics; Research Ethics Committees)

Research ethics committees have been established in a wide range of institutions where clinical research is conducted such as hospitals, research centers, and universities. This has been done to ensure that ethical standards are respected in research involving human subjects. Although the composition of such committees depends on the institution at which they are established and on the kind of research they have to evaluate, they should always be multidisciplinary and independent. Most of these committees have the decision-making power to reject projects that do not comply with standard ethical requirements. Research committees were the first ethics committees to be formally established. Following the publication of Henry Beecher's "Ethics and Clinical Research" in the *New England Journal of Medicine* in 1966, which disclosed that human experimentation without informed consent had taken place, the US Public Health Service (USPHS) recommended the creation of institutional review boards (IRBs). Such committees were tasked with reviewing the procedures of all biomedical projects that received federal funding and with assessing their compliance with ethical requirements, notably those established by the Nuremberg Code and other complementary guidelines that started to be put together in the 1970s to address the different problems that arose. Nevertheless, IRBs only became mandatory in 1974 after disclosure of the Tuskegee syphilis case (1972), a scientifically useless and abusive experiment on a vulnerable population. Since that time IRBs disseminated around the world and currently they are established in most countries where clinical trials take place. However, in some countries (particularly in Africa) there are still important constraints to overcome. They can be broken down into legal constraints as a result of the lack of a robust legal framework (soft and hard law) to prevent exploitation of vulnerable populations while making use of the possible (direct and indirect) benefits resulting from research; educational constraints as a result of the insufficient ethical and regulatory review capacity of committee members; and social constraints as a result of the lack of awareness of the general population of the risks and benefits of clinical research and of their rights (e.g., when it comes to informed consent or insurance). Worldwide organizations such as the WHO and UNESCO have been consistently promoting the establishment of research ethics committees in all countries where clinical research is conducted. They have been playing a major part in capacity building for committee members and professionals, in general. The European Union has long been strongly oriented toward centralizing research ethics committees, specifically those dedicated to evaluating clinical trials. The first step in this direction was taken in 2001 by Directive 2001/20/EC of the European Parliament and of the Council. Issued on April 4, 2001 the directive was titled "on the approximation of the laws, regulations and administrative provisions of the Member

H. ten Have and M. do C. Patrão Neves, *Dictionary of Global Bioethics*,
https://doi.org/10.1007/978-3-030-54161-3_151

States relating to the implementation of good clinical practice in the conduct of clinical trials on medicinal products for human use." The directive ruled that each member state should have just one national research committee to evaluate all biomedical research projects involving human subjects. The goals were to avoid different committees and institutions reaching different opinions and to shorten the approval process. The second step reinforcing these goals was taken in 2014 by Regulation (EU) No. 536/2014 of the European Parliament and of the Council. Issued on April 16, 2014 the regulation was titled "on clinical trials on medicinal products for human use." This regulation states that from 2019 all clinical trials should be evaluated by a single body that would cover all European countries. Research ethics committees have also been established in the area of animal experimentation for clinical purposes.

Commodification

Healthcare has long been subject to market thinking where it is associated with commodification. This term was first coined by Karl Marx and refers to the transformation of goods, services, ideas, and people into commodities (i.e., objects of trade that can be exchanged). When applied to human beings commodification means they are reduced to goods with a market price—not with an intrinsic value. The power of the market subjects everything and all manner of activities to economic thinking. This is true not only for human beings as such (e.g., immigrant trafficking) but also for the human body and parts of the body. Almost all body components (organs, tissues, cells, and genes) are now regarded as commodities and commercially traded. Commodification has become a global phenomenon in which ethical assessment is difficult. It can be regarded as a means to promote scientific advances and to disseminate new technologies more widely. On the other hand, it can be regarded as an unjust practice that favors people with money. The notion of commodity is also opposed to the Kantian idea of human dignity that expresses that every human being is worthwhile and has value—not a price. Selling and buying body parts is legally prohibited in many countries. Furthermore, access to healthcare should be a universal right of everyone—not based on someone's financial means. Another ethical concern is that commodification further encourages the privatization of healthcare by placing emphasis on individual consumer choice. If somebody can afford it, then he or she should be allowed to buy it. There should be no restrictions to the market since price and competition will regulate interactions and transactions. This way of thinking undermines the ideas of social responsibility and solidarity that are still significant in many welfare states.

H. ten Have and M. do C. Patrão Neves, *Dictionary of Global Bioethics*,
https://doi.org/10.1007/978-3-030-54161-3_152

Common Good

Common good refers to the idea that there are certain goods that are desirable by society as a whole such as justice and peace. It also refers to the conviction that all members of a community should engage in electing and achieving such desirable goals thus requiring collective action and active citizenship. The notion of common good is rooted in Ancient Greek philosophy and has been an important issue throughout the Western history of thought mostly in moral and sociopolitical philosophy. Although it started with Plato and Aristotle, it continued in medieval times with Thomas Aquinas; in modernity with Machiavelli, Locke, and Rousseau; in later times with Adam Smith, Karl Marx, and Stuart Mill; and today with thinkers such as Keynes and Rawls. The definition of common good certainly differs among the different philosophical approaches proposed over the centuries. Nevertheless, the Aristotelian concept is widely accepted as the basis for the theoretical elaborations that followed: common good is the good of the members of a community (which can only be achieved by the community) and the good of the community that embraces its members (which is shared by them). Although the rise of individualism and the development of relativism mostly since the second half of the twentieth century has undermined the very idea of common good and tried to deny the existence of a shared vision of a substantial (or contentful) good, this idea not only prevails but has grown stronger thanks to communitarian thinkers: communities shape their members' identities and members should engage in shaping their community profile. Bioethically, public health is certainly an agreed common good that as such requires the engagement of citizens and is beneficial to all citizens. To pursue the common good each citizen must put aside his/her own particular interests and give priority to common interests. However, apart from this virtually consensual theoretical point many practical questions remain open: should the common good be left only to individual goodwill, to solidarity, or should it be politically and legally established? How much should be collectively imposed and how much should depend on individual decisions? Libertarians and communitarians take opposite views on the subject. Notwithstanding this, since humans as social beings live together in communities, this, in itself, is an acknowledgment of common goods and of the joint effort to build them.

H. ten Have and M. do C. Patrão Neves, *Dictionary of Global Bioethics*,
https://doi.org/10.1007/978-3-030-54161-3_153

Common Heritage of Humankind

The common heritage of humankind is a notion that was introduced in international law in the late 1960s to regulate common resources such as the ocean bed and outer space. The speech of Maltese ambassador Arvid Pardo to the General Assembly of the United Nations in November 1967 represented a milestone for the notion. He argued that the seabed and ocean floor are the common heritage of humankind and therefore should be used and exploited for peaceful purposes and for the benefit of humankind as a whole. Although it was considered a new concept, it actually went back to the older tradition of Roman law that distinguished a separate category of *res extra commercium* (things outside of commerce; property that cannot be exchanged) such as common property like seas and public property like rivers. Such a category defines objects of law that cannot be economically exchanged (i.e., cannot be sold or owned by anybody). The distinction between private and common property was famously used in 1609 by Hugo Grotius in his book *Mare Liberum* where he argued that the sea is *res omnium communes* (things belonging to everybody; common property) like the air and the Sun (i.e., objects that are used and enjoyed by everyone but cannot be owned by anyone). The idea that there are objects and spaces outside the usual frame of private and public property was initially used for areas with material resources that are outside the borders of nation states such as the ocean floor, the Moon, outer space, and the Antarctic. However, use of the notion gradually expanded to vital resources within the territory of individual nations such as tropical rainforests. The basic elements of the notion of common heritage have been categorized. First, common areas cannot be appropriated and should be accessible to everyone. Second, common areas demand international cooperation to create common management (i.e., across borders). Third, possible benefits should be equitably shared among states irrespective of their geography. Fourth, they should only be used for peaceful purposes. Fifth, they should be preserved for future generations. Even when individual countries manage such resources (often spread over several countries), the very fact of considering them as common heritage constructs a shared interest. They need to be preserved in such a way that humanity as a whole benefits rather than the citizens of one or more countries. Since resources labeled as common heritage are vital for the survival of humankind, global responsibilities have been defined such as proper management and stewardship transcending national interests. In the 1970s the concept of common heritage came to be used for culture and cultural heritage. It is used today in a broad sense to refer to the idea of a global moral community. It is now applied to science and scientific innovations (i.e., the human genome is regarded as the heritage of humanity). The global ethical principles of the Universal Declaration on Bioethics and Human Rights can also be regarded as the common

H. ten Have and M. do C. Patrão Neves, *Dictionary of Global Bioethics*,
https://doi.org/10.1007/978-3-030-54161-3_154

heritage of humankind. The notion has normative implications such as free and open access, peaceful use, benefit sharing, and the protection of future generations. It therefore provides a critique and counterdiscourse to the prevailing neoliberal ideology that dismisses collective property and regards everything as resources that can be appropriated, harvested, used, and regulated. The notion criticizes intellectual property right regimes and the drive to patent genetic resources, traditional knowledge, biological organisms, and all forms of life. It points out the negative consequences of the neoliberal regime when it comes to access to healthcare and to medication. Furthermore, it explains how science and innovation have become restricted. Traditionally, scientific information has always been openly accessible in that permission is not needed to use and elaborate a specific scientific theory or discovery. Any attempt to restrict the free sharing of knowledge and information has a wider effect: it discourages creativity and innovation. Although new ideas do emerge from competition, they also come from cooperation such as sharing ideas in networks, personal communications, open publications, and settings in which critical ideas can flourish. Creative activity requires free exchange and discussion that build upon and transform the ideas of others. It is therefore increasingly argued that the notion of common heritage should be restored. Scientific progress depends on open science, the sharing of data and materials, and refraining from patenting.

Commons (*See* Common Heritage of Humankind)

Commons are shared domains, materials, products, resources, and services. They are oriented toward the future unlike the notion of common heritage that refers to the past and to historical traditions. Both notions go beyond the perspective of the individual, emphasize what all human beings share and what they need to respect and preserve, and assume inclusion rather than exclusion. A classic example of commons is land available for public use that is used and managed by the surrounding community. Irrigation systems managed by local communities are another example. Commons are not owned by anyone. They represent a shared interest and require cooperation and collective action. They are ruled by self-governing institutions set up by stakeholders themselves. In the history of humankind commons were the rule—not the exception. For example, one of the oldest forms of local government in the Netherlands are regional water boards that have been established since the thirteenth century to control dikes, water levels, and water quality. In most countries natural commons have been tribal and communal property. However, the history of commons is one of destruction and dispossession (especially since the eighteenth century). Commons were expropriated based on the legal fiction that land without private owners was empty and unowned in much the same way as newly discovered territories were considered uninhabited before colonization. The first person occupying the territory is considered the owner. Enclosure is the key word here brought about by blocking access: sometimes literally by fencing off land and constructing barriers to common use, other times by legal restrictions and privatization of collective property. Many commons have disappeared because public space has been restructured as private space excluding others from use. At the same time new commons have been created through new technologies and through the interaction between technologies and nature. Examples are digital commons based on the Internet, airspace commons (for international air transport), ether space commons (for radio and television broadcasting), weather forecasting systems, and gene pools of genetic diversity (for crops). Such new commons are cosmopolitan, necessitated by globalization, and facilitated by new technologies. Such cosmopolitan commons transcend national boundaries in that they are shared across technologies and user communities, they require collective action and negotiation, they are valuable for human existence and nature, and they are always vulnerable to deterioration, overuse, and hazards. Therefore they require protection so that they can be sustained. Different types of commons can be distinguished such as natural commons (fishing grounds, lands, forests, and water supplies), social commons (caring arrangements, public spaces, waste removal, and irrigation systems), intellectual and cultural commons (knowledge, information, and cultural products), and digital commons (the Internet and the World Wide Web). Such types have different qualities in that they can be depletable (natural commons) or renewable (social and cultural commons). They can refer to rivalrous goods where use

H. ten Have and M. do C. Patrão Neves, *Dictionary of Global Bioethics*,
https://doi.org/10.1007/978-3-030-54161-3_155

by one person diminishes the amount of goods left for others (as is the case with many natural commons) or non-rivalrous. Global commons are non-excludable (they are open to everybody since access cannot be restricted) and non-rivalrous (use by one individual does not impact use by others). Despite their heterogeneity commons share several basic features. First, they refer to collective property jointly owned by a group of people (as in local commons) or belonging to all persons on earth (as in global commons). Second, they are essential for human subsistence and long-term survival (since they provide water, food, shelter, health, and knowledge). Third, commons need to be protected for future generations such that sustainability is important. Fourth, commons are social practices (some advocate the use of the verb "commoning") —not simply resources. They express a discourse of sociality, reciprocity, sharing, and social harmony. Commons refer to togetherness (commonwealth) not only with people but also with nature, the environment, and land.

Communication, Ethics

Communication ethics refers to the moral standards with which communication should comply in any form and at all levels ranging from the interpersonal to the institutional. Communication was originally a bridge to connect people and should remain so. In attempts to narrow its meaning down there have been a number of definitions that follow two main orientations. The first orientation stresses that communication ethics emerged and was strongly focused on human communication mediated by communication technologies: mechanical and electronic technologies were developed in the first half of the twentieth century and digital technologies started to be developed in the second half of the twentieth century. Such a perspective roots communication ethics in journalism ethics, broadens the scope to media ethics as if they were synonymous, considers communication just at the professional level, and thus excludes social media, a very important form of communication in today's world. The general idea is that the mediation of communication has broadened the possibilities of its manipulation thus requiring stronger ethical safeguards. The second orientation stresses the importance of communication in our contemporary societies where all people are recognized as equals (with the same rights to an opinion), where there is an increasing diversity of communication departments in many different organizations, and where there is an increasing number of professionals working in the field of communication thus also narrowing communication ethics down to its professional dimension. Most social activities are currently assisted by public relations and marketing offices (especially enterprises and businesses). The general idea is that public communication not only reaches many more people today than ever but has a stronger impact too. Although both orientations consider communication ethics at the professional level in interaction with the public, the first reduces it to the ethics of the media whereas the second establishes a specific new domain of reflection and practice focused on the communication between enterprises and the public. Ethical communication in public relations and marketing raises specific ethical problems that need to be correctly addressed when aiming to promote goods or services, while honestly informing citizens (consumers) and thus building a trustful relationship. Such goals require the rejection of propaganda (based on the idea that information should not be partial or biased) and the rejection of manipulation (communication should be persuasive, objective, and accurate). Therefore communication should enhance trustful dialogue and value the public interest above all else. The general goal is to promote authentic dialogue as two-way symmetrical communication (Grunig's model of public relations). Although conflicts of interest can easily arise

H. ten Have and M. do C. Patrão Neves, *Dictionary of Global Bioethics*,
https://doi.org/10.1007/978-3-030-54161-3_156

between employees, employers, clients, and the community, they should be resolved in compliance with the abovementioned ethical principles. Even though losses might be incurred in the short term, having the public's trust will always be rewarding in the long run.

Communication, General

Communication is the transmission of a message (a meaning)—verbal or non-verbal, through body signs or behavior, oral or written—from someone (sender) to someone else (recipient) requiring there to be an effective contact between both (although not necessarily an answer from the recipient). A message can be sent and not received. In this case there is just a transmission (from a sender), but no communication. Communication can refer to thoughts and emotions, be supported by tone of voice and gestures, and be subject to different interpretations according to the cultural context. There is a huge number of variables that influence communication that can be as natural as they are complex. Communication is not specific to humans. Animals can also communicate. However, humans have been able to encode it into language (symbols) thus enabling them to communicate much greater content more accurately and to many more recipients. It was language that, as a result of accumulating information throughout generations, produced culture. Indeed, communication not only builds culture it is also culturally embedded. Over time humans have developed different ways to widen communication starting with prehistoric cave paintings and culminating with written information (i.e., the most successful communicative tool). It was only recently that it became possible to communicate rapidly (even instantaneously) at a distance due to communication technologies such as the telegraph, telephone, radio, television, fax, computers, and the Internet. As communication technologies evolved so have natural barriers been overcome as borne out by the most recent digital communication devices such as computers, smartphones, and tablets. As long as there is access to such devices there are no geographic barriers to global communication. Digital communication has also created a new category of citizens termed the "info-excluded" who do not have access or do not have the skills to use digital communication devices. This is a new kind of social exclusion (i.e., from today's communicative society) that aggravates discrimination and deepens social inequalities. Another form of social exclusion is that of the deaf–mute community who have traditionally suffered from discrimination. However, they have overcome their exclusion by being able to reach out using today's new communicative technologies. The development of communication technologies should not detract from the importance of personal skills of communication often designated as soft (interpersonal) skills. Communication skills are particularly important in a democratic environment where there is horizontal communication (among equals). The real risk here is to mistake good rhetoric for good communication or to mistake the intention to persuade and manipulate for the willingness to inform and engage. The difference between both is dictated by compliance with ethical principles in communication such as truthfulness, transparency, objectivity, and adaptation of the message such that the recipient can understand it. Good communication skills can

H. ten Have and M. do C. Patrão Neves, *Dictionary of Global Bioethics*,
https://doi.org/10.1007/978-3-030-54161-3_157

avoid conflicts and significantly contribute to solving problems (e.g., within the patient–physician relationship). This is the reason high education today (specifically medical education) includes the teaching of social communication skills.

Communication, Media (*See* Media Ethics)

The term communication media refers to the means or systems of receiving, storing, and transmitting information or data. *Media* is the plural of the Latin word *medium* (mediating tool) that designates an intermediate element. The term literally refers to communication that takes place in a mediatized way through specific tools or channels. The media include all mass communication channels such as publishing (print media, newspapers, magazines), broadcasting and narrowcasting (radio, television), and the Internet (information, advertising, entertainment). Although media communication has constantly changed over time, it was only after the invention of the printing press that mass media became a reality. Currently, there are two forms of mass media communication: analogic such as conventional radio, telephone, and television transmissions; and digital such as computer-mediated communication, computer networking, and telegraphy. Such technological advances have led to the range of communication media becoming literally global. All the information produced today benefits from channels that effectively take it to everyone and everywhere. This new reality has a huge impact on individuals' daily lives, societies, and international relations. The major achievement is perhaps bringing the people of the world closer to each other, sharing the same information, eliminating distances, and illustrating the true meaning of a global world (communication media are perhaps the most powerful tools for globalization). The impact of what happens in one part of the world is immediately seen everywhere and depending on the nature of the event can have very different consequences. If it is a natural catastrophe, then communication usually results in people worldwide coming together to help. Media communication can promote education, raise people's awareness, get people to engage across borders, and bring about global justice. Although it has become enormously powerful in the last decades, its social responsibility should be proportionate. Media communication has changed not only the way people communicate, but also the way people think and behave, the way societies organize themselves, and how they relate with one another. In other words a new culture has been created.

H. ten Have and M. do C. Patrão Neves, *Dictionary of Global Bioethics*,
https://doi.org/10.1007/978-3-030-54161-3_158

Communitarian Ethics (*See* Communitarianism)

Communitarian ethics focuses on the importance of the community and emphasizes the influence community has on human beings. The community a person is born in and raised in shapes his/her own personality and morality thus implicitly also obliging him/her to actively engage in the development of society. There is a reciprocal responsibility among citizens and between citizens and society that creates interdependence and solidarity among everybody. Communitarian theories have their roots in Ancient Greek philosophy. Such theories started with Aristotle who subordinated ethics to politics and developed an ethics of virtues that contributes to the common good. The history of ethics has been dominated by Christian morality (defined as personalist). It too is communitarian in the way that Christians belong to a community of the faithful and have the obligation to promote the common good. Such theories have further been developed in modern times by Hegel who criticized Kantian formal and universal morality and its individualistic features, but valued the historical context where human institutions and relationships develop themselves and take shape. Marx whose greatest goal was to create a new community was another to develop such theories. After the Second World War and the triumph of liberalism that followed, the communitarian perspective again gathered momentum by the end of the century in an attempt to restore the value of communities, but mostly as a reaction to liberal individualism. So-called "new communitarian thinking" abandoned the ideal to rebuild the world as idealized by Marx and stressed the need to value and protect existing communities. As a critique of modern liberalism such thinking was developed in different ways by philosophers such as Michael Sandel, Michael Walzer, Alasdair MacIntyre, and Charles Taylor. The main feature of communitarian theories is the value attributed to the community to which each person belongs ranging from the nuclear or extended family to the political state or nation. Thus such theories are strongly critical of individualism as sterile abstraction. Communitarian ethics rejects moral abstractionism and formalism on the basis of cultural traditions, practices, and their history. It rejects the hegemony of individual rights and instead argues for the common good and the responsibility of all members of the community to achieve it. Furthermore, it rejects the deontological formalism of the principle of autonomy while arguing for an ethics of virtues (cooperative virtues). Finally, it rejects the legalism in applying a single ethical principle by pointing out the significance of practical wisdom. Therefore communitarian ethical reasoning in healthcare and

H. ten Have and M. do C. Patrão Neves, *Dictionary of Global Bioethics*,
https://doi.org/10.1007/978-3-030-54161-3_159

in clinical research will tend to acknowledge the interests of the community as well as those of the patient. For this to occur different procedures in clinical ethics will be needed: in some communities informed consent may require community consent, in organ donation for transplantation opting out may be preferable to opting in, and compulsory admission to hospital may be justified in cases where public health is at stake since health is a common good.

Communitarianism

Communitarianism is a theoretical perspective that proposes shifting the liberal focus on individual rights toward communal responsibilities. It was triggered by criticism of John Rawls' *Theory of Justice* (1971) or more broadly by liberalism and the priority it gives to individual rights and liberties over the social nature of the person, the value of the community, and its traditions. The term was first coined by John Goodwyn Barmby in 1841 as a criticism of utopian socialism that was being proposed at that time. However, the concept did not become well known until after 1971 when it was advanced by contemporary philosophers such as Alasdair MacIntyre, Michael Sandel, Charles Taylor, and Michael Walzer. Such a designation (i.e., communitarianism) was never acknowledged by them. Moreover, none of them developed a full communitarian theory. The theory of justice that brought about communitarianism was proposed by Rawls from an imaginary original position. Although Rawls presented his principle of justice as universal, communitarianism argues that standards of justice emerge from social contexts and their conception of good, as well as from the way of life, the history, and the traditions of that community. A particular social context is necessary to elaborate a concept of justice; hence justice varies from community to community. Although communitarianism is important to ground the concept of justice on a concrete reality, the problem is that it easily leads to a relativist position making it impossible to build a global standard of justice worldwide.

H. ten Have and M. do C. Patrão Neves, *Dictionary of Global Bioethics*,
https://doi.org/10.1007/978-3-030-54161-3_160

Community Consent

In medical practice and research the principle of individual prior informed consent has a crucial role to play. However, the emphasis on individuals is not the same across the world. Communities in many cultures and traditions play an important role in determining human well-being and in individuals living their lives to the full. From a global perspective the moral status of a community is recognized in the concept of community consent. When research is initiated in communities, it should be discussed with the community leadership. Getting agreement from the local leadership should be obtained through a process of dialogue prior to the consent of individual members of the community. However, this does not in any way replace individual consent. Even when consultations with community leadership have taken place and community consent is obtained, research participants should still give individual consent. Getting the support of the community leadership is critical to promoting research practices that are culturally sensitive, collaborative, and supportive. The principle of community consent is formulated in UNESCO's Universal Declaration on Bioethics and Human Rights under Article 6: "In appropriate cases of research carried out on a group of persons or a community, additional agreement of the legal representatives of the group or community concerned may be sought. In no case should a collective community agreement or the consent of a community leader or other authority substitute for an individual's informed consent." Ethical problems arise because the concept of community is often unclear. What constitutes community leadership is also unclear. Furthermore, consultation may not always result in agreement and community consent. The fundamental challenge is to balance individual autonomy, community interest, and respect for cultural diversity.

H. ten Have and M. do C. Patrão Neves, *Dictionary of Global Bioethics*,
https://doi.org/10.1007/978-3-030-54161-3_161

Compassion

The word "compassion" derives etymologically from the Latin words *cum* (with) and *patior* (to suffer). The latter is close to the Greek word *pathos* (suffering) that broadly means "suffering with another." The Latin word *compassio* (sympathy) is a synonym of "sympathy" (from the Greek *syn* (with) and *pathos* (suffering)). The prefix *cum* intensifies the feeling such that we end up with "deep sympathy." Compassion can also be considered a synonym of pity. Aristotle uses the word *eleos* (compassion/pity). Of course, these two emotions today have different meanings: pity expresses a degree of condescension toward someone regardless of whether or not that person acknowledges his/her situation; whereas compassion does not involve condescension but requires someone to recognize the unfortunate situation in which another person is living. Compassion, pity, and empathy are altruistic emotions that are either rejected by philosophers who deplore emotions as affecting and impairing good reasoning or valued by philosophers who perceive them as an essential dimension of human identity. The latter attribute high moral value to such emotions considering them an anthropological specificity of humankind. Moreover, altruistic emotions are considered an excellent way to establish genuine and selfless personal relationships. Indeed, it is from this perspective that compassion is considered an important virtue in healthcare (mostly associated with nursing practice). Nonetheless, overly engaging in compassion runs the risk of the professional becoming a friend of the patient—a situation that should never be allowed. Doing so would mean the professional had lost sight of his or her mission of care, assistance, and treatment.

H. ten Have and M. do C. Patrão Neves, *Dictionary of Global Bioethics*,
https://doi.org/10.1007/978-3-030-54161-3_162

Compassionate Use (*See* Pre-approval Access; Right to Try)

Compassionate (or expanded) use is the same as pre-approval access and the right to try both of which refer to access to medication that is not yet approved for use in medical practice. However, the connotation is different. Pre-approval access and the right to try emphasize individual legal rights and the moral principle of personal autonomy. The dominant ethical framework in countries like the United States articulates the rights of individuals and the ethical principle of respect for autonomy. However, the term compassionate use accentuates the moral principle of beneficence. It is compassion that drives the argument to make medication available to patients suffering from a disease at an early stage of development. However, such an appeal to compassion is less recognized as an argument. The FDA in the United States now discourages use of this term. Seriously ill patients argue that the availability of unauthorized investigational drugs may offer treatment options when no other treatments are available irrespective of uncertainties about benefits and harms. Since it takes more than a decade before an investigational drug becomes available in medical practice, patients argue that they have no time to wait for the outcome of trials. The first compassionate use program was established in 1978 in the United States following a lawsuit about the medical use of cannabis for a glaucoma patient. In the late 1980s patients with AIDS started a movement to accelerate the development of drugs. In 1987 the FDA initiated formal programs of expedited development and expanded access.

H. ten Have and M. do C. Patrão Neves, *Dictionary of Global Bioethics*,
https://doi.org/10.1007/978-3-030-54161-3_163

Competence

Competence refers to a set of abilities such as knowledge, skills, values, and virtues necessary to efficiently and successfully carry out a task entrusted to someone or a task for which someone is responsible. Although such a set of abilities differs according to the mission or the situation, competence always designates a high level of professional performance. Competence is the elementary ethical requirement beyond which there is no ethical compliance. Anyone who is deemed incompetent to accomplish a specific task is ethically obliged to step aside and give way to someone competent. Competence is an ethical requirement that is particularly important within healthcare professions where incompetent professionals endanger the lives and physical or psychological integrity of their patients.

H. ten Have and M. do C. Patrão Neves, *Dictionary of Global Bioethics*,
https://doi.org/10.1007/978-3-030-54161-3_164

Compliance

Compliance designates a sequence comprising two movements: the recognition of a duty (an interior movement at the theoretical level) and the willingness to accomplish it (an exterior movement at the practical level). The duties to be acknowledged can be of a legal or moral nature. Lack of compliance (submission) to legal duties entails civil sanctions, whereas lack of compliance (obedience) to moral duties entails remorse and social blame. Therefore the acceptance of duties or compliance can be externally and forcefully imposed by law (an order, a rule, or a contract) or internally and voluntarily imposed by common morality (answering a request or meeting expectations). In the realm of global bioethics—individual or societal healthcare, clinical research or environmental issues, national or international policies—it is not possible to convert all ethical standards or all expectations of common morality into legal rules. Compliance is then particularly relevant since it leads to people fulfilling their obligations both written and lived.

H. ten Have and M. do C. Patrão Neves, *Dictionary of Global Bioethics*,
https://doi.org/10.1007/978-3-030-54161-3_165

Complicity

Complicity refers to sharing the same goals, participating in the same actions, and working together (as a team) to achieve them. The concept is most often used with a negative connotation such as is often the case in law. It designates the secret, furtive involvement of someone in an unlawful project (as an accomplice or partner in a crime). It also takes on a negative meaning when it refers to a passive individual attitude toward a collective decision that is morally wrong. Although there is no active involvement, the person becomes an accomplice if he or she had the power to counteract and did not. Individual complicity is proportionate to collective responsibility. Viewed from this perspective complicity is more relevant within bioethics when every time society (or political leaders) take a position considered wrong according to an individual's religion, culture, philosophical convictions, and experience of life, yet nonetheless he or she passively accepts it without even trying to present other points of view or ways to act. Classic bioethical examples are abortion, euthanasia, the creation of transgenic animals, and the production of genetically modified crops. Complicity can also have a positive connotation (mostly in Latin languages) where it describes the thoughts, desires, and feelings of different people that are perfectly aligned (i.e., perfect understanding and fundamental agreement among them). Viewed from this perspective complicity can be an advantage when living (as a family) and working together (as a healthcare or research team) such as is the case in palliative care, in disability care, and in planning conjoint actions.

H. ten Have and M. do C. Patrão Neves, *Dictionary of Global Bioethics*,
https://doi.org/10.1007/978-3-030-54161-3_166

Confidentiality

Confidentiality refers to the obligation to keep all information obtained within a professional relationship confidential, private, or secret. This is one of the most ancient and continues to be one of the most important duties of healthcare professionals (it also applies to other professionals such as lawyers). Indeed, it is mentioned in the Hippocratic Oath. From the time of Ancient Greece right up until today confidentiality has always been grounded on the principle of beneficence: keeping all information regarding the patient secret has the purpose of protecting him/her thus benefitting and strengthening trust within the patient–physician relationship. Currently, it is broadly recognized that medical confidentiality contributes not only to good medical practice but also protects the patient. On the one hand, it helps the clinician formulate an accurate diagnosis: the greater the amount of information gathered the better the diagnosis. The patient will only provide the physician with all the necessary information if he trusts that it will remain confidential. On the other hand, patients often have to expose themselves physically, psychologically, emotionally, and spiritually to healthcare professionals thus becoming very vulnerable. Professional confidentiality guarantees that no one under any circumstances will take advantage of such vulnerability. In the second half of the twentieth century as part of the Human Rights Movement healthcare confidentiality was recognized as a patient right (as articulated in the Lisbon Declaration of 1981). Such a shift in the nature of confidentiality from a professional duty to a human right entailed another shift concerning its foundation from the principle of beneficence to the principle of autonomy. It is the patients who have the right to privacy and require the information about themselves to remain confidential. This in turn imposes professional secrecy on healthcare providers. Currently confidentiality is not only an obligation with which all healthcare professionals must comply, it is also an ethical and legal patient requirement. Indeed, some would say that it is the professional duty of secrecy or privacy that demands confidentiality, that it is the person's right to confidentiality that imposes privacy, and that privacy and confidentiality are two sides of the same coin. Although medical confidentiality over time has always been absolute (i.e., without exceptions), that was challenged in 1969 with the Tarasoff case. Tatiana Tarasoff's boyfriend told his psychiatrist that he would kill his girlfriend when she returned from vacation. Although the psychiatrist took the threat seriously, he did nothing to prevent the tragedy from happening. The psychiatrist defended himself in court by appealing to the duty of confidentiality he had with his patient. However, the court considered such a duty could and should be breached when there were potential victims at stake. Today ethical issues related to the patient's right to confidentiality and the healthcare professional's duty to keep the patient's information secret refer mainly to exemptions to confidentiality. The consensus is that medical

H. ten Have and M. do C. Patrão Neves, *Dictionary of Global Bioethics*,
https://doi.org/10.1007/978-3-030-54161-3_167

confidentiality can and should be overruled when a third party's life is at stake from someone with a psychiatric disorder, when the physical integrity or health of a third party are at stake from someone with a communicable disease, or when there is an imminent public health threat from someone with an infectious disease. The current multidisciplinary team model of medical practice and the digitalization of medical records present new challenges for confidentiality.

Conflict of Interest

Conflict of interest is a bioethical topic that is receiving rapidly increasing attention. It is an important topic in care, research, and education. It refers to situations where secondary (often financial) interests influence medical/professional judgment and action. Examples include doctors receiving a bonus from companies in the pharmaceutical industry when they prescribe such companies' medication, researchers being encouraged by their sponsors to publish positive rather than negative results of trials, medical students receiving gifts and free lunches when they attend educational programs sponsored by industry, and medical doctors simultaneously acting toward one and the same patient as both physician and researcher. Such examples demonstrate that conflicts of interest can occur at both the individual and the institutional level. An example is a university that receives large donations from wealthy benefactors while silencing researchers and projects that are critical of these donors. Patients trust their healthcare providers in clinical practice and assume that medical decisions and treatment are always focused on their health and welfare. However, many physicians and institutions receive payments from the industry thus biasing the prescriptive behavior of physicians. Cooperation between academia and industry since the 1980s has led to many cases of misconduct, fraud, and falsification of research data. Although this is found in almost all countries, most data are from the United States where 60% of department chairs in medical schools and teaching hospitals have personal relationships with the pharmaceutical industry as consultants, speakers at conferences, scientific advisory board members, and members of the board of directors. Sponsoring medical education has also become common with two thirds of clinical departments in the United States receiving industry support to assist with continuing education. Many educational programs are biased since they have become a form of marketing. Another area of concern is the development of clinical practice guidelines. Although they should be based on scientific evidence, critical assessment of the evidence, and objective clinical judgment, studies show that conflicts of interest are common. For example, an expert panel that proposes monitoring cholesterol at lower levels will greatly expand the number of people qualifying for drug treatment. Bioethics itself is contaminated in that members of institutional review boards are supposed to independently assess research proposals, but commonly have relationships with the industry that they choose not to disclose. There is also a growing number of commercial research review boards. Conflicts of interests are ethically problematic since they undermine the trust the public has in medicine and compromise scientific integrity and objectivity. The ethical approach to conflicts of interest has long been based on transparency and disclosure. Researchers and physicians should publicly announce their financial interests and links to industry. However, such an approach is

H. ten Have and M. do C. Patrão Neves, *Dictionary of Global Bioethics*,
https://doi.org/10.1007/978-3-030-54161-3_168

limited in that it depends on the voluntary cooperation of doctors and researchers. A database created in August 2013 by the Physician Payments Sunshine Act (a section of the Affordable Care Act in the United States) shows that more than 800,000 physicians received payments from industry in 2017. The act requires commercial companies to report payments and gifts to physicians in the United States. Studies show that voluntary reporting does not work and that voluntary disclosures are unreliable. Even full disclosure does not eliminate the basic problem. Against this background there are those who argue that conflicts of interest are inevitable. Although attempts at managing such conflicts wrongly assume they are inevitable, they nevertheless cannot be eliminated. Financial conflicts on the other hand are optional. They are the result of a physician or scientist making a choice—not inherent in their activity. They compromise independence and the conduct of disinterested and impartial research. Although preserving independence as a scientist and doctor has a cost, it does make clear to the public that there is no hidden agenda. It clarifies that interventions and decisions are based on the health of the patient or the public—not on often invisible commercial interests. Viewed from an ethical perspective the best experts are those who are not conflicted.

Conscientious Objection

Conscientious objection refers to the legal claim or moral privilege of an individual to refuse to comply with a professional obligation for conscientious, religious, or moral reasons. When an individual faces a moral dilemma in his/her professional practice such as when he or she acknowledges two opposite obligations—a professional duty and a moral responsibility—that are impossible to meet simultaneously, he or she has to choose one over the other even though both need to be fulfilled. Conscientious objection solves the dilemma in that moral consciousness is favored over professional duty. Moral consciousness defines the person and professional duties describe practices. Conscientious objection is recognized as a human right by Article 18 of the Universal Declaration of Human Rights on freedom of thought, conscience, and religion. It was explicitly pronounced by the Human Rights Commission of the United Nations in 1987. Conscientious objection has a long history that extends back to the sixteenth century, is mostly related to religious groups, and often refers to compulsory military service. Conscientious objectors are individuals whose personal convictions bar them from using weapons to kill a fellow human being. It was mostly in the twentieth century that the scope of conscientious objection broadened to include such factors as acceptable grounds and activities covered. In the second half of the century it entered the clinical realm where healthcare professionals used it to avoid performing abortions. Currently, most countries give healthcare professionals the legal right to opt out of specific procedures that conflict with their personal beliefs. Such procedures often refer to reproductive advice given to gay couples or single women, peimplantation genetic diagnosis, and euthanasia. Conscientious objection is also regarded as a medical ethical principle that gives healthcare professionals the right to conscientiously object to carrying out specific procedures on patients and clients. At the same time the rights of patients and clients to have such procedures cannot be neglected either. Although the conscientious objector cannot be dismissed as a result of not providing what is due to the patient and can refuse to perform or participate in an assigned procedure, he or she has to guarantee that other colleagues and/or the institution will deliver the healthcare services needed or required by the patient. More recently there have been strong reactions against conscientious objection from healthcare providers with some advocating that it should not apply to what is legally permitted and that professionals should be forced to fully comply with their duties despite what their conscience tells them;

H. ten Have and M. do C. Patrão Neves, *Dictionary of Global Bioethics*,
https://doi.org/10.1007/978-3-030-54161-3_169

otherwise they should simply either step down or be fired. They further argue that the personal values of a professional are no longer valid as a result of the end of medical paternalism and should not interfere with the physician–patient relationship. Others consider that to achieve a physician–patient relationship that is symmetrical the values of healthcare professionals and respect for personal autonomy should be considered—not only the patient's values.

Consensus

The word "consensus" derives etymologically from the Latin *consensus* (agreement, accord, unity, and assent—common consent) and has gained unparalleled prominence in contemporary moral philosophy. After the demise of universal ethics the only way today to avoid moral skepticism and relativism that would collapse morality itself is to build a moral consensus as wide as possible. Although the idea of consensus can be traced back to Greek Antiquity, it is mostly with the development of contractualist theories in modernity (seventeenth and eighteenth centuries) and pluralism in recent times (twentieth and twenty-first centuries) that it has become relevant in morality. Contractualist theories broadly consider that human societies are only possible due to a (tacit) contract or agreement among all persons and that social relations and political obligations are grounded on social agreements that are extremely useful for the well-being of individuals and for peaceful social relationships. The legitimacy of moral norms and of political government derives from the idea of contract or mutual agreement. Therefore consensus as agreement is the cornerstone of contractualist theories such as those of Grotius, Hobbes, Pufendorf, Locke, and Rousseau, philosophers who sought to establish consensual conditions that can enhance the implementation of norms (moral and political) in society. In recent times the axiological pluralism characteristic of democratic societies and the lack of a recognized authority to establish what is good or right for all require broad consensus. Currently the only way to avoid conflicts or to solve them is through engaging in a rational debate that is participatory and building social consensus. Indeed, it is the procedure followed that legitimizes the results achieved. Several philosophers have developed theories of consensus (models of consensus building). Two of the most remarkable are the North American John Rawls and the German Jürgen Habermas who represent two different moral approaches: contractualist and procedural, respectively. However, both focused on the conditions necessary to assure the universal acceptance of consensus. To identify an impartial level of analysis that guarantees the universal validity of moral principles Rawls goes back to what he calls the "original position"—the very origin of a social contract. In this hypothetical situation representatives of free and equal citizens gather to establish the principles of social and political justice they consider necessary for the well-being of the majority of citizens. They are unaware of any of their own personal and social characteristics. Therefore the principles they establish will necessarily be the fairest to the majority of citizens. Parties at the original position impartially agree on two principles of justice ranked in lexical order (priority order): the principle of equal basic liberties that ascribes equal rights and liberties to all citizens and the difference principle that establishes that the distribution of wealth and income must be done in such a way that inequalities in the distribution of goods

H. ten Have and M. do C. Patrão Neves, *Dictionary of Global Bioethics*,
https://doi.org/10.1007/978-3-030-54161-3_170

can produce a greater total product. Application of these principles to society produces the widest possible consensus. To identify norms of action (moral and political) that can be universally accepted Habermas proposes a communicative rationality oriented toward achieving, sustaining, and reviewing consensus. Under the mediation of language communication is a basic activity through which two or more subjects are capable of spontaneously agreeing on a shared project of action or on a shared view of reality. Communication becomes problematic when it is broken by disagreement or conflict and when consensus no longer exists. Discussion is needed to argue about what is being advocated, to put forward the reasons it is being defended, and to reestablish consensus. Lack of consensus obliges communication to expand into fundamental discussion and to provide arguments for justification (i.e., to develop into a reflective form of communicative action). The universality of moral norms can be reached by discussion that complies with two principles: the principle of universalization (U) that as a rule of argumentation states that a moral norm is valid when the foreseeable consequences and side effects resulting from its observance can be accepted by all concerned and the principle of moral discourse (D) that stipulates that the only valid moral norms are those that can obtain the approval of everyone affected in a reasonable discourse. Regardless of the particular theory of consensus considered, it is the level of consensus reached—taking into consideration the level of participation in the debate, the rationality of the arguments, and the rightness of the procedures—that leads to universal acceptance. The focus is on the procedures previously established—not on the principles or the ends. It is the rightness of the procedures that morally legitimates the decisions taken. This path is particularly important at the international level when it comes to building global bioethics. Consensus is one of the major challenges in today's global world and raises at least two important problems: what is supported by the majority does not imply the truthfulness of statements and the widest consensus is still nevertheless an ethical minimum.

Consent, Informed Consent

Informed consent refers to patients and research participants genuinely consenting to undergo treatments and tests and to participate in research. Such consent is obtained by health practitioners and researchers in a number of ways such as communication, providing information and making sure that the information is understood, that patients and research participants are competent, and that their decision is voluntary. Informed consent is a relatively recent normative requirement in bioethics. In 1914 in the United States there was a case ruling (by judge Benjamin Cardozo) that is considered the beginning of informed consent theory. In 1931 the German government established the requirement of informed consent for human experimentation. However, at the international level it was the Nuremberg Code that first articulated that informed consent was a basic norm for medical research. It was included in the first version of the Declaration of Helsinki. UNESCO's Universal Declaration on Bioethics and Human Rights states that consent is one of the fundamental principles of bioethics. Article 6 states that: "Any preventive, diagnostic and therapeutic medical intervention is only to be carried out with the prior, free and informed consent of the person concerned, based on adequate information. The consent should, where appropriate, be express and may be withdrawn by the person concerned at any time and for any reason without disadvantage or prejudice." The same statement applies to scientific research where it is specified that the information should be adequate, provided in a comprehensible form, and include modalities for withdrawal of consent. The person may withdraw consent at any time and for any reason. Although informed consent is firmly embedded in bioethical discourse, articulated in legal treatises and agreements, and often specified in national and international legislation, it continues to provide ethical challenges. Informed consent is the outcome of a process of communication that is only valid if four conditions are satisfied. First, adequate information should be provided. Although it is impossible to give all relevant information, the rule is that sufficient information should be provided to enable the patient or research subject to make well-considered decisions. Simply handing out a substantial written document with all existing information such as one referring to rare and exceptional side effects will not be sufficient. Second, the patient or research subject should understand the information and the healthcare professional or researcher should verify that the information provided is comprehended. Third, the patient or research subject should make a choice that is voluntary and based on the ethical principle of respect for autonomy that requires the receipt of a treatment or participation in research is the result of free choice. Coercion, manipulation, and persuasion should therefore be avoided. Fourth, patients and research subjects should be competent in that they have sufficient knowledge, judgment, skill, and experience to

H. ten Have and M. do C. Patrão Neves, *Dictionary of Global Bioethics*,
https://doi.org/10.1007/978-3-030-54161-3_171

make responsible and autonomous decisions. Physicians and researchers must assess the competence of patients and subjects. If they find such competence is lacking, then substitute decision-makers must be asked for consent. If the above conditions are fulfilled, then actual consent will be provided ordinarily by signing a written document.

Consequentialism (*See* Utilitarianism)

Consequentialism refers to a specific orientation in moral philosophy: a result-based perspective that evaluates the morality of an action according to the results it produces. Goodwill and good intentions are not enough for an action to be deemed moral because they do not necessarily produce a good end. By focusing on the results or outcomes of human actions consequentialism involves stressing the importance of responsibility for actions taken and their consequences. From the result-based perspective it is fundamental for agents to consider the consequences of their actions before acting and to foresee the impact of their actions on others (important in biomedical ethics, research ethics, and especially in health and scientific research public policies) and on the world (important in environmental ethics and animal ethics). On the other hand, consequentialism neglects other important factors or constituents of moral life such as intentions, feelings, or even respect for certain duties especially when under the given circumstances they do not help to achieve the end pursued. Compliance with moral norms is not independent of the consequences that compliance produces. Therefore consequentialism is opposed to deontological moral theories. The latter are based on recognized ethical principles that should be unconditionally respected (i.e., regardless of consequences) because they are worthy in themselves. The most prominent consequentialist moral theory is utilitarianism that advocates that a human action is moral insofar as it maximizes what is good.

H. ten Have and M. do C. Patrão Neves, *Dictionary of Global Bioethics*,
https://doi.org/10.1007/978-3-030-54161-3_172

Consultation

Consultation is a traditional and common activity in healthcare. When healthcare providers are uncertain about interpreting patients' symptoms or deciding the best approach for treatment and care, they usually ask the advice of other experts. Ethical questions about consultation concern the process, its function, and what it consists of. There is a lack of clarity about the authority of consultants and which stakeholders should be included in such a consultation. Should the patient and/or his/her family be engaged in ethics consultation? Although this is usually the case in the United States, it is not in other countries. Consultation depends not only on the cultural and social context, but also the moral framework. Although honesty is more appreciated in some cultures than others, the emphasis on patient autonomy differs. Moreover, there are major gender differences (e.g., in some cultures women are barred from any involvement in consultation). Consultation has become a common practice in the area of bioethics. The Quinlan case in the United States in 1976 led to the emergence of ethics consultation services. They became mandatory for hospital accreditation in 1992. Although consultation is commonly carried out by ethics committees, there is an increasing number of ethics consultants. Moreover, ethics consultation services have been introduced to address complicated moral problems. For example, in the Netherlands there is a legal requirement to consult a second physician when a patient makes a request for euthanasia.

H. ten Have and M. do C. Patrão Neves, *Dictionary of Global Bioethics*,
https://doi.org/10.1007/978-3-030-54161-3_173

Contraception

Contraception refers to all natural or artificial processes employed to avoid pregnancy. It is said to be as old as humankind. Although it is pursued by both men and women, it is mostly women who do so mainly for biological reasons. It is the woman who becomes pregnant, carries the fetus for nine months, and risks her life during labor. Women also pursue it for social reasons. It is the woman who takes care of the children and is responsible for their upbringing. Although the most common and ancient natural contraceptive options are coitus interruptus (withdrawing the penis from the vagina) and the ingestion of herbs by the woman, neither can be considered efficacious. The same can be said of a more recent natural contraceptive method: basal body temperature. Another ancient contraceptive method is the condom. There is evidence that condoms have been used for thousands of years. They were supposedly first used by the Ancient Egyptians (1000 BC) who made them out of linen and then animal tissues. The aftermath of the Second World War represented a turning point in contraception. It was a consequence of the Women's Rights Movement and the biotechnological revolution that followed. Women who had worked during the war (replacing men who had left to fight) did not want to return to their homes as housewives. They needed to control their fertility to continue and intensify the active life they had started. This corresponded with an exponential increase in knowledge of the human reproductive system. Contraception only became efficient in the second half of the twentieth century as a result of the creation of pharmacological (synthetic drugs) and technical (devices) methods. The most revolutionary contraceptive was the birth control pill in the 1950s that proved to be totally effective. Since then side effects decreased and the variety of pills grew and were adapted to each woman's metabolism. Other hormonal contraceptives include Depo-Provera injections, birth control patches, vaginal rings, and emergency contraception. Contraceptive devices (barrier methods) have also been created such as diaphragms, sponges, cervical caps, female condoms, intrauterine devices (IUDs) for long-acting reversible contraception (preventing implantation of the embryo but not conception). There are also male and female methods for permanent contraception: vasectomy and tubal ligation, respectively. Currently there are two major kinds of ethical issues concerning contraception. The first and oldest refers to the moral acceptance of contraception. In the past contraception was interpreted as refusal of a woman's mission to be a mother going against God's will who wants humanity to "multiply and fill the earth" (natural law). The "wrongfulness" of contraception is still pointed out at the religious level. The second ethical issue focuses on the different nature of the contraceptive methods practiced and devices used distinguishing between those that prevent fertilization (conception) from those that prevent the implantation of the embryo (abortifacients).

H. ten Have and M. do C. Patrão Neves, *Dictionary of Global Bioethics*,
https://doi.org/10.1007/978-3-030-54161-3_174

Although the former are supposedly more widely accepted than the latter, many women choose contraceptives for efficacy and comfort. Contraception has also been used to leverage international cooperation. Since developed countries consider lower birth rates improve the health and education of women and children and increase the income of women, they encourage contraception when offering financial aid. For developing countries such an orientation violates individual and family freedom as well as the country's sovereignty.

Corruption

Corruption is widespread and comes in many varieties and manifestations making it difficult to define. Formal corruption refers to deviating from established rules and policies. An example is bribery. Informal corruption refers to ideals that are broken rather than legal frameworks. Although corruption primarily occurs at the individual level, it also affects institutions and political systems. Corruption can affect bioethics because its institutions are particularly vulnerable to corruptive influences. A first possible concern is conflict of interest. There are many examples illustrating how clinical trials and their review by ethics committees can be unduly influenced by industry sponsors. Committees may be financed by the same companies that they are supposed to regulate thus making it questionable whether the protection of research subjects is really the main priority in such circumstances. A second concern is capture in which sponsors may shop around to find the most lenient research ethics committee. As a result of increasing competition committees have been known to conduct the review process in a fast and lenient way deviating from their mission and ideals. Capture also shows itself in the phenomenon of the "revolving door" in which individual regulators and policy-makers can become ever more lenient in the hope of future jobs and lobbying assignments in the companies they currently regulate. A third concern is gaming the system in which pharmaceutical companies provide new drugs for lower prices or even donate them, particularly in developing countries. Although this improves access to essential drugs, sometimes these drugs are diverted and resold back in developed countries for profit. Preventing and eradicating corruption is difficult since doing so requires a range of strategies at various levels. One strategy is to explicitly ban corruption. Although bribing health professionals is unlawful in most countries, the practice of paying doctors under the table continues often under the guise of gratitude and gifts. Other strategies require transparency. Medical journals today request authors to reveal conflicts of interests and identify their sponsors. Another strategy is independent monitoring by providing incentive mechanisms to motivate better behavior, enhance company and personal reputations, and reduce the risk of litigation. Finally, more attention should be given to corruption in ethics education since this is something that can easily be changed and eradicated.

H. ten Have and M. do C. Patrão Neves, *Dictionary of Global Bioethics*,
https://doi.org/10.1007/978-3-030-54161-3_175

Cosmetic Surgery (*See* Aesthetic Medicine)

Cosmetic surgery differs from aesthetic medicine in that it is solely focused on surgery for aesthetic purposes, whereas aesthetic medicine involves both surgical and non-surgical procedures. Nevertheless, cosmetic surgery raises similar ethical issues. Cosmetic surgery also differs from plastic surgery. Although both focus on improvement of the human body by surgical intervention, the goal of cosmetic surgery is to enhance the appearance of the patient. A variety of techniques can be used such as breast enhancement (augmentation, lift, or reduction), facial contouring (e.g., rhinoplasty), facial rejuvenation (e.g., facelifts), body contouring (e.g., liposuction), and skin rejuvenation (e.g., botox). The goal of plastic surgery is to repair defects such that normal function is restored and to reconstruct appearance as effectively as possible. When birth defects, trauma, burns, and diseases affect the body, reconstruction may be worthwhile. Examples are breast reconstruction after mammectomy, burn repair, cleft palate repair, hand surgery, and scar revision.

H. ten Have and M. do C. Patrão Neves, *Dictionary of Global Bioethics*,
https://doi.org/10.1007/978-3-030-54161-3_176

Cosmopolitanism

The development of global bioethics has been inspired by the ideals of cosmopolitanism. Such ideals have often been expressed in history (e.g., in Stoic philosophy). They consider each human being as a citizen of his or her own community or state (*polis*) and at the same time as a citizen of the world (*kosmos*). In the former they are born and share a common origin, language, and customs with co-citizens, whereas in the latter they participate because they belong to humanity in which all human beings share the same dignity and equality. Since the cosmos includes the whole of humanity, being a citizen of the world liberates the individual from being held captive in such categories as culture, tradition, community, gender, and race. Boundaries here have no moral significance thus implying that it is possible to overcome the limitations of being born in a particular place within a specific culture. Cosmopolitanism expresses an aspiration to live beyond specific, bounded horizons and allows a broader solidarity between people that extends beyond boundaries. The moral ideal is that human beings belong to a universal community in which human well-being is not defined by a particular location, culture, or religion. Therefore global citizens have responsibilities toward other human beings be they near or distant. Cosmopolitanism is often metaphorically referred to as expanding circles of moral concern. However, the ideals of cosmopolitanism are contested. It is argued that the term "citizen of the world" is little more than a metaphor or an abstraction—not reality. There is no world community or state to which such citizens belong. The nation state is the only basic political community. Cultural identity can only be constructed within the specific territory of such a nation state. Social bonds with distant others are ephemeral and illusionary. Nonetheless, globalization in recent decades has shown that "citizen of the world" has become more than a metaphor. Globalization is associated with cosmopolitization of the world at the objective and political level—not only subjectively. First, there is an emerging global consciousness of living in a common world that has been brought about by a number of realizations such as the finiteness of the world (everywhere in the world is known, there is no *terra incognita*, and distances mean little), the smallness of the world (television and the Internet instantly bring worldwide events closer to us, relationships and interconnections are unavoidable when sharing such a small space), and the unity that is the world (there is only one world). Even if the world is not a community like the traditional *polis*, there is a common humanity in which all will suffer and perish if this one planet is ruined. Global consciousness often reveals itself in a negative way through disasters, horrors of war, violations of human rights, and reactions to crises. The suffering of distant others makes us aware that we all share the same vulnerability and similar basic needs. The point is that global interaction and interdependency expand our moral sensibility. Wherever they are, people

H. ten Have and M. do C. Patrão Neves, *Dictionary of Global Bioethics*,
https://doi.org/10.1007/978-3-030-54161-3_177

recognize that they share the same predicament. Second, there is objective cosmopolitization in which the autonomy of nation states is eroded by economic, political, and legal global developments. States find themselves increasingly subject to international law that has global jurisprudence (e.g., the International Criminal Court established in 2002 can prosecute individuals such as leading politicians for crimes against humanity). Moreover, risks today are often globalized and require collective responses as a result of which the rise of global organizations and actors can be seen, particularly non-governmental organizations. Third, cosmopolitanism at the political level questions the notion of sovereignty. Individual states are either powerless or often bypassed in responding to global challenges. Politics today has increasingly become the domain of civil society. Global activities are often undertaken today by communities of citizens rather than by traditional political authorities. Such communities of citizens participate in direct and horizontal communication, share information, and use publicity to put pressure on authorities. Instead of using existing institutions, citizens engage with particular causes, develop new collective structures, create social movements, and open up new problem areas using the latest technologies to set up global networks and organize world forums and global summits.

Cost–Benefit Analysis

Decisions can be analyzed by summing up the benefits and subtracting the associated costs. Such analysis is usually applied in businesses before decisions are taken. Analysis begins with a comprehensive list of all benefits and costs. The costs may be direct such as labor or raw materials. They may also be indirect such as electricity, utilities, and management. They may also be intangible such as the impact on employees or customers. There are also opportunity costs such as building a new plant or buying a plant. There can also be costly potential risks such as environmental impacts. The benefits might differ as well. They can include revenues and sales from new products or increased production. Some benefits will be intangible such as improved safety for employees and customers and higher customer satisfaction. Finally, there is competitive advantage because a higher market share can result from the decision. All the items in the list of costs and benefits will be measured in terms of money in which benefits are not overestimated and costs are not underestimated. Cost–benefit analysis (CBA) is increasingly used in healthcare. Interventions and their consequences are measured in monetary terms so that treatment alternatives can be compared. CBA is most commonly used in large developments such as the building of new hospital facilities. It is not often used in health technology assessment since it is difficult to assign monetary values to health outcomes. CBA is also used to evaluate regulatory policies such as assessing the impact of clean water legislation and increasingly in public health where the costs of interventions occur in the immediate future and the benefits in the distant future. The three main reasons for using CBA are deciding whether to implement a program or intervention, choosing between competing options, and setting priorities when budgets are limited. Ethical problems frequently emerge as a result of using CBA. How can a financial value be assigned to a particular medical condition or outcome such as disease or death? Although cost–benefit studies are often regarded as impartial, objective, and neutral, they hide normative assumptions about the goals of treatment, the selection of treatments, the role of patients, and the just distribution of resources. Many medical practitioners argue that costs should not influence medical decisions and that CBA is therefore unethical. CBA is primarily founded on utilitarian ethics in which it primarily promotes the value of efficiency.

H. ten Have and M. do C. Patrão Neves, *Dictionary of Global Bioethics*,
https://doi.org/10.1007/978-3-030-54161-3_178

Covid-19

Coronavirus disease 2019 (COVID-19) is official shorthand for the disease caused by the novel coronavirus SARS-CoV-2. The WHO announced this name for the novel strain of the coronavirus family on February 11, 2020. The outbreak of COVID-19 was first identified in Wuhan (China) in December 2019. The WHO declared a Public Health Emergency of International Concern on January 30, 2020 and a pandemic on March 11, 2020. By early May 2020 the pandemic had spread to 215 countries where 3,272,202 confirmed cases of infection and 230,104 deaths had been registered. COVID-19 was one of the first pandemics to affect the globalized world and proved to be the deadliest of all coronavirus infections. About 50% of the deaths by COVID-19 were elderly people. They were more susceptible to becoming infected and were more likely to die once infected. In the Northern Hemisphere most of the elderly who died were residing in nursing or care homes. This raised important ethical concerns about the level of protection society provides to the elderly and provoked questions concerning intergeneration solidarity. The major challenge that the COVID-19 pandemic posed worldwide was control of the infection. Moreover, the control measures implemented presented a very pressing ethical problem. The lack of a safe and effective treatment (and cure) or a vaccine led to the main control measure being social isolation and quarantine. Although it was applied within a temporary legal framework that was voluntary in some countries but compulsory in most parts of the world, this strategy went against individual human rights. Nevertheless, it was adopted in an attempt to balance them against the demands of common good and the principle of social responsibility. COVID-19 brought about an unprecedented health crisis, caused a deep economic crisis as a result of using lockdown as the single effective measure to control the spread of the infection, and shut down almost all economic activities. The crisis degenerated into a serious social crisis, mostly as a result of increased unemployment. Against this background coupled with the inability of biotechnologies to find an effective way to prevent or cure the infection and address the cause of the crisis, digital technologies were advanced in an attempt to control the spread of the virus. A specific example was mobile contact tracing. Although digital contact tracing varied from country to country, the general principle was for individuals to download an app to their smartphones that registered all their movements and contacts. Individuals who tested positive for coronavirus and had a contact tracing system would be obliged to isolate and all the people they had met in the previous two weeks (and who also had the same contact tracing system) would receive an alert that they had been in contact with someone infected. Although widely accepted as necessary, this was a threat to citizen privacy and raised serious ethical concerns about the anonymization of personal data, the possible use of such data for other purposes, and the

H. ten Have and M. do C. Patrão Neves, *Dictionary of Global Bioethics*,
https://doi.org/10.1007/978-3-030-54161-3_179

possibility of it being retained after the end of the pandemic. Such a technological approach was more intrusive than the traditional methods of public health that focus on identifying infected persons (supposing a sufficient number of test kits are available), following up with contract tracing with the help of public health professionals, and then quarantining them. Instead of waiting for technological innovations some countries simply resorted to traditional public health measures such as hiring thousands of contact tracers.

CRISPR (*See* Genomic Editing; Gene Editing)

Clustered Regularly Interspaced Short Palindromic Repeat (CRISPR) is one of the most recent gene-editing technologies. It is a new tool for genome manipulation capable of cutting, copying, and binding DNA strands. Although the first CRISPR technique used the protein Cas9 as molecular scissors, the derived techniques that followed the success of CRISPR-Cas9 such as CRISPR SKIP used other resources such as directly modifying the chemical composition of DNA. The discovery of CRISPR is most often attributed to the French microbiologist and geneticist Emmanuelle Charpentier in collaboration with the North American structural biologist Jennifer Doudna. In 2012 they reported what would turn out to be a major breakthrough: the biological defense system could be turned into a cut-and-paste tool for editing gene sequences. Indeed, CRISPR is the work of many scientists from around the world that extends back to 1993 and the work of the Spanish researcher Francisco Mojica who first characterized what is now called a CRISPR locus. Although CRISPR was the third gene-editing technology to be developed, after Zinc-Finger Nucleases (ZFNs) and Transcription Activator-Like Effector Nucleases (TALENs), it was the first to be widely known by the public as a result of its revolutionary features. It was easier to apply, faster, cheaper, more flexible, tailorable, more precise than former methods, and more accessible. Therefore it was used by many more scientists and had a greater impact than all the others. These are the reasons CRISPR is being explored in many different fields and for a growing number of purposes ranging from increasing agricultural production by designing new grains, root vegetables, and fruits that are adapted to edapho-climatic conditions (depending on the soil and the climate), resistant to plagues, and economically profitable to promoting human health by being able to modify and edit the genome. This opens up the possibility to develop new treatments for rare metabolic disorders, to treat and prevent complex diseases (such as cancer, heart disease, mental illness, and HIV infection), to treat genetic diseases (single-gene disorders such as cystic fibrosis, hemophilia, and sickle cell disease), and to improve gene therapy (from genetic correction to genetic enhancement). Its simplicity, efficiency, and great therapeutic potential triggered a great deal of interest in clinicians. However, its wide accessibility, straightforward simplicity, and great potential for misuse raised serious concerns in the scientific community who shared their hopes and fears with society. They did so to raise awareness about the benefits and risks of using CRISPR and to get people to participate in bringing about global governance of germline gene editing involving sperm, eggs, or embryos because it is humanity as we know and live it that is at stake here. The scientific community is fully aware of the beneficial potential of CRISPR and possible dramatic consequences of its misuse. In 2015 at the 1st International Summit on Human Gene Editing regarding appropriate uses of the technology, the scientific community rejected its

H. ten Have and M. do C. Patrão Neves, *Dictionary of Global Bioethics*,
https://doi.org/10.1007/978-3-030-54161-3_180

clinical use (specifically the making of genetically modified children) at least until safety and efficacy issues were solved and broad social consensus reached. Indeed, although techniques have improved in the last few years, the use of CRISPR still does not meet elementary clinical safety standards. Off-target effects are considerable, particularly the risk that altered cells trigger cancer, and interactions with other genes and the environment remain largely unknown. Nevertheless, in 2018 the Chinese biophysicist He Jiankui announced he had edited human embryos (he was jailed for it) in an attempt to decrease their risk of acquiring AIDS and had transferred them to the womb of a mother who later gave birth to the world's first genetically edited newborns. In view of such developments there is an urgent need to call for a global moratorium on all clinical uses of human germline editing (changing heritable DNA to make genetically modified children). The aim here is to provide time to evaluate the technical, scientific, medical, societal, and ethical issues involved and to establish an international framework for appropriate use.

Contract Research Organizations

Contract research organizations (CROs) are a relatively new phenomenon in the field of medical research. They can be companies that contract with the pharmaceutical industry or can be biotechnology enterprises that want to outsource research. Ever more clinical trials are relocated to developing countries. Although the reasons are many such as costs are lower, execution is faster, registering drugs is easier, and expanding the market is simpler, there is usually less regulation in these countries. Outsourcing (hiring an outside company) and offshoring (relocating a company to another country) the research process or its components is then an attractive option. Currently approximately half of all clinical trials are outsourced to CROs. The number of commercial research organizations have seen spectacular growth over the last few decades. CROs are involved in all stages of the research process: recruitment, testing, data collection, and reporting to the industry. It is estimated that there are now more than 1,000 CROs with collective sales up to USD20 billion. At the same time an increasing number of violations have been reported, especially in developing countries. Such companies are primarily focused on making profits—not science or healthcare. They usually operate below the radar. Despite existing regulations being outdated (most of them written in the 1970s) and not covering the work of CROs, the level of monitoring and audit has not improved and remains very low. Regulatory agencies do not have legal control over the operations of CROs (unless the CRO is the sponsor of the research). Another ethical concern is the issue of responsibility. Who is actually responsible for the reliability and integrity of data when they are collected in different countries by different commercial organizations and third parties? Moreover, the trial population in many developing countries is often vulnerable and easily exploited, and healthcare systems are usually not well developed. In countries such as Brazil and India trial participants are often highly vulnerable because of illness and poverty. They are not in a position to refuse treatment or seek treatment outside the trial because inclusion in such a trial is often the only option they have to gain access to healthcare. Vulnerability has also important consequences for the application of informed consent since many people in such countries are illiterate.

H. ten Have and M. do C. Patrão Neves, *Dictionary of Global Bioethics*,
https://doi.org/10.1007/978-3-030-54161-3_181

Cryogenics

The word "cryogenics" derives etymologically from the Greek words *kryos* (very cold, freezing) and *gene* (give birth, beget, production). It refers to the production (and use) of very low temperatures from −50 °C (−238°F) to absolute zero −73 °C (−460°F). However, today it is a branch of physics defined as the production and behavior of materials at very low temperatures. It can also involve studying the effects low temperatures have on living cells, tissues, and organs thus making it a branch of biology. Cryogenics has been applied to living organisms. After being exposed to extreme temperatures (at which molecular motion ceases almost completely) such organisms are preserved at their current level of development with the aim to revive or resuscitate them later. From an ontological point of view cryogenics creates a new life status in which life is suspended or, more accurately, animation is suspended. Such a status stops all biological functions and suspends the natural course of life without endangering the later restoration of functions and the continuation of life. Cryogenics interrupts the very essence of time (i.e., its continuity). Such a technique is very beneficial in transplantation medicine where it preserves harvested organs for longer and extends their viability until they can be implanted in the right recipient. This allows a better match (histocompatibility) to be found between the organ available and the patient and favors the outcome. The most successful and impactful achievement is the cryopreservation of sperm, embryos, and oocytes, which has opened up new possibilities for assisted reproduction. Ethically, cryogenics dissociates in time (and space) what was formerly inseparable (i.e., sexual intercourse and generation of a new life). When it comes to assisted reproductive technologies, cryogenics has reinforced the autonomy of future parents and facilitated the intervention of different donors in the generation of a child. The most challenging hope is cryopreservation of the human body. Some people have decided to be cryopreserved immediately after death (i.e., frozen and cryostored) to prevent tissue decomposition in the hope of being revived sometime in the future when state-of-the-art medical treatment and technology can bring them back to life. This would not be a case of suspended animation and revival but of death and resuscitation.

H. ten Have and M. do C. Patrão Neves, *Dictionary of Global Bioethics*,
https://doi.org/10.1007/978-3-030-54161-3_182

Cultural Diversity

The growing importance of global bioethics has reactivated the significance of the notion of moral diversity. The development of global bioethics demands a broader framework of normative interpretation and assessment. Is it justified to apply the principle of informed consent in, say, Nigeria where there is a significantly different culture? Should we respect, say, the Chinese practice of harvesting organs from executed persons? Viewed from the global perspective the ethical systems of different cultures need to be examined and moral values analyzed and applied in specific contexts. This is generally recognized as necessary and has opened up new and fascinating fields of research. However, the next step brings us back to the old controversy of universal values and local values. Is there a universal framework of principles and values or are principles and values different depending on local, cultural, and religious normative systems? One of the accomplishments of global bioethics is that respect for cultural diversity and pluralism is regarded as an ethical principle in itself. The Universal Declaration on Bioethics and Human Rights is the first international bioethics document advancing respect for cultural diversity as an ethical principle to be applied in the moral assessment of bioethical issues and problems. Article 12 presents the principle of respect for cultural diversity and pluralism: "The importance of cultural diversity and pluralism should be given due regard. However, such considerations are not to be invoked to infringe upon human dignity, human rights, and fundamental freedoms, nor upon the principles set out in this Declaration, nor to limit their scope." However, the status of this principle when balanced against other ethical principles is controversial. It is also unclear how respect for cultural diversity should be situated in the delicate balance between universalism and particularism. The search for global ethical principles focuses on the values we share as human beings. For some bioethicists this would be a futile endeavor since different and contradictory ethics systems exist. If there is no basis for rational verification of ethical judgments, then all efforts to formulate ethical principles as universal means in practice that the dominant system attempts to impose its principles as universal. However, this is a mistaken view. If the challenge for global bioethics is to recognize moral diversity and emphasize globally shared values and principles, then it has to assume the global community is a moral community. Global bioethics then refers not merely to extent (worldwide scope involving citizens of the world who are increasingly connected and related due to processes of globalization), but also to content (identification of global values and responsibilities and establishment of global traditions and institutions). Global bioethics in the spirit of the notion of common heritage is not simply advocating universal values or acknowledging moral diversity. It is a dialectical effort to bridge universalism and particularism. Although global bioethics is not a finished product, it is one

H. ten Have and M. do C. Patrão Neves, *Dictionary of Global Bioethics*,
https://doi.org/10.1007/978-3-030-54161-3_183

that has not yet resulted in a clear normative approach comparable with principlism in first-generation bioethics. The main challenge is to combine convergence and divergence. How can the recognition of differences in moral views and approaches be reconciled with convergence toward commonly shared values? Criticisms of global bioethics often presuppose simplistic views of globalization. Although worldwide interconnectedness bridges the gap between distance and proximity, some scholars assume a radical contrast between moral strangers and friends, while others fear the growth of a bioethical monoculture. However, it is not correct that globalization produces either uniformity or multiplicity since it does both. People today are part of multiple cultures and it is not clear where their roots exactly are. They consider themselves at one and the same time as, say, Dutch, European, and citizen of the world. The same is true for the notion of culture itself in that no culture today is monolithic and pure. All cultural traditions are dynamic, they have changed and are changeable, and they are necessarily a *mélange* of different components. Differences do not exclude there being a common core. The term "interculturalism" is therefore more appropriate than the term "multiculturalism" since it acknowledges diversity and at the same time insists on universal values. The term "interculturality" emphasizes interaction. The prefix "inter" refers not only to separation but also linkage and communication. The supposition is that we can position ourselves between two cultures and can occupy a place between the universal and the particular. This means we recognize similarities between self and other that can be the basis for dialogues between cultures, while maintaining difference and sustaining boundaries between self and other. In other words, we are conceptually and practically in between (i.e., moving beyond dualistic, binary thinking and simultaneously universalizing and particularizing practices). Although multiculturalism emphasizes respect for diversity, individual freedom, justice, and equal treatment, interculturalism introduces a moral vocabulary of interaction, dialogue, participation, trust, cooperation and solidarity. It is not sufficient to have multiple co-existent value systems and respect them. The challenge is to produce and cultivate practices that can create community. If there is common ground, then it needs to be cultivated through interaction and communication. Convergence is not a given but the result of an ongoing activity of deliberation, consultation, and negotiation. It is exactly this interstitial perspective that motivates the development of global bioethics.

Cyborg

The term "cyborg" was first coined by the Austrian neuroscientist Manfred Clynes in 1960. He connected the initial syllables of the words "cybernetic" and "org-anism" to designate a fictional, hybrid being (i.e., a modified human being) composed of two parts: one organic and the other electromechanical (i.e., a human machine) the latter powering the natural being. Hybrid creatures have been represented in literature, paintings, and sculptures throughout the history of humanity. In Ancient Greece there were half-human–half-beast beings that were always stronger than natural human beings. The advances of science and technology made it possible for human imagination to design half-human–half-machine beings. However, it was science fiction and movies that gave life to these powerful creatures and later called them cyborgs. Science fiction often announces the path science is pursuing. This is evidenced in this case by advances in different technologies toward the production of electromechanical body parts that are adjustable to the organic parts of human beings. First, they aimed to substitute for the loss of original functions and played a therapeutic role; then, they aimed at reinforcing different human functions and assumed an enhancement goal; and, later, they aimed at adding new functions oriented toward transhumanism. Ethically, therapeutic actions are considered beneficial when they achieve an individual good that is consensually recognized as such. Although the goal frequently widens to social interests, enhancement actions can be ethically controversial and need to be individually considered. Alienating humankind in favor of creating a new humanoid species in the wake of the transhumanist movement lacks an entity that has sufficient moral authority—consensually recognized as such—to validate such an endeavor. Another ethical question concerns the degree to which—percentagewise or performancewise—electromechanical components should be allowed to turn a human into a cyborg.

H. ten Have and M. do C. Patrão Neves, *Dictionary of Global Bioethics*,
https://doi.org/10.1007/978-3-030-54161-3_184

Data Sharing (*See* Research Ethics, Data Sharing; Virus Sharing)

The idea of commons has been rehabilitated in today's global bioethics and used to redefine the public domain. This is borne out by the current emphasis on data sharing an example of which is virus sharing. The results of clinical trials are used to apply for marketing approval of new drugs from regulatory agencies. Such results are often but not always published in scientific journals. However, patient-level data are not publicly accessible since they are considered the property of sponsoring companies. It is now argued that such practices should change and that sharing data will improve medicine and benefit patients. Clinical trial data are complex in that they need to be analyzed and evaluated by different stakeholders including critical and independent researchers such that evidence can be corroborated and claims of interested parties about safety, efficacy, and effectiveness can be scrutinized. Too many recent cases have shown that published clinical evidence is selective, biased, or incomplete. The public benefit of having more reliable information is ethically more important than data protection or trade secrets. Although data sharing demonstrates that scientific activity is a cooperative enterprise, it has taken years of struggle to gain access to raw data from clinical trials funded by industry. The European Medicines Agency (EMA) announced that from January 2015 it will publish clinical reports that support applications for authorization of medicines as soon as a decision has been taken. From 2014 a new clinical trial law in the European Union requires that all trials are registered and clinical study reports are publicly available. EMA has been publishing clinical data submitted by pharmaceutical companies to support their applications since 2016, which has led to some major drug companies changing their policies and now providing access to data. The campaign for data sharing brought together scientists, media (e.g., the *British Medical Journal*), and NGOs. Data sharing is also gaining momentum in emerging research fields such as synthetic biology an example of which is creation of the BioBricks Foundation in 2006 that aims at ensuring research results are freely available and will benefit everyone and the planet.

H. ten Have and M. do C. Patrão Neves, *Dictionary of Global Bioethics*,
https://doi.org/10.1007/978-3-030-54161-3_185

Death, Concept

The concept of death has been defined as following one of two ways. The first, simpler, and universal definition of death is by the via *negativa* (negative way) in which life is negated or denied. Such a definition (i.e., by denial) is grounded on the fact that the only reality we know is life, and death is perceived as its opposite. Indeed, no one has direct knowledge or experience of death since death always belongs to the other. Such an antithetical perspective of death was well systematized by the Epicureans (in the fourth century BC) who stated that death and life define themselves in their reciprocal opposition with life excluding death and death excluding life. On the other hand, Stoics such as Seneca (in the third century BC) stressed that we start dying the instant we are born. Such a dialectic perspective of death presents it as an inseparable part of life. Although both perspectives are rationally arguable, each is grounded on a different perception of life and a different attitude toward death. The second definition of death is by the via *positiva* (positive way) in which the total, permanent, and irreversible cessation of all vital functions is formulated by biological sciences.

H. ten Have and M. do C. Patrão Neves, *Dictionary of Global Bioethics*,
https://doi.org/10.1007/978-3-030-54161-3_186

Death, Criteria (*See* Brain Death)

Throughout the ages up to relatively recently the criterion used to determine death has been roughly the same worldwide: the cessation of breathing. Since ancient times death has been considered to occur when the *animus* (soul) left the body. The last breath (of life) was considered the consummation of death. This has been not only the traditional interpretation, but also the empirical experience and the first criterion of death scientifically established. The cessation of circulatory and respiratory functions is termed the cardiorespiratory criterion of death. However, the cardiorespiratory criterion was recognized by physicians as being fallible and has always been a subject of concern for physicians. There is evidence of people considered dead being buried alive. This is the reason there is still a delay today between the declaration of death under the cardiorespiratory criterion and the funeral of the deceased. In the 1960s resuscitation techniques, especially cardiopulmonary resuscitation (CPR), showed that the cessation of circulatory and respiratory functions was not always irreversible. Moreover, the performance of CPR was not always fully successful. Although it could possibly stop the process of organic or biological death without recovering consciousness (i.e., personal or relational life), one of the consequences of CPR when performed for too long was to leave the patient in a persistent vegetative state (PVS). Such a consciousless existence represented a new status of human existence. This resulted in the need to establish good CPR practices and establish a more accurate criterion of death. The former incertitude about declaring deceased someone who was still alive gave way to the incertitude of continuing to take care of people who were in fact dead (vital functions could be maintained artificially for a long period of time after the brain had ceased to function). A new (second) criterion of death was then established: irreversible cessation of all functions of the entire brain including the brainstem. Several countries established a list of signs as medical indicators to evaluate brain activity. Such lists were based on that of the North American Ad Hoc Committee at the Medical School of Harvard, which listed the tests to be performed to evaluate the clinical status of the individual. Termed the Harvard Criteria such lists became available in 1968. Not only that but the first clinical assessment had to be repeated a second time before death could be declared. The whole-brain criterion was later challenged by the concept of brainstem death (or higher brain death). A patient declared brainstem dead still registers some electrical activity without any possibility of recovering. Legally, the possibility of pronouncing such a patient dead could mean stopping all healthcare provision and considering him/her as a potential donor for transplantation. Most countries require the whole-brain criterion of death. Brain death is generally poorly understood by the man and woman

H. ten Have and M. do C. Patrão Neves, *Dictionary of Global Bioethics*,
https://doi.org/10.1007/978-3-030-54161-3_187

in the street because they fail to understand how someone can be pronounced dead when still (artificially) breathing, warm to the touch, and even able to blush. When this is combined with the fact that a brain-dead person is often the ideal donor for organ transplantation, suspicion and mistrust are the result. Although moving from the cardiorespiratory criterion of brain death to the brain-dead criterion represented a more accurate determination of death, it also moved the perspective from a biological assessment of death to a legal one that complies with the indicators of brain death legally established. However, they are not exactly the same in different countries as a result of there being international differences in the nomenclature and criteria for the determination of brain death.

Death, General

Although death is an objective phenomenon and biologically defined, it has been subjectively interpreted throughout time and the world mostly from a spiritual perspective. Indeed, burial rituals have been considered the first evidence and expression of spiritual life in the evolution of humankind during the process of humanization. Only when human beings acknowledged their existence had a spiritual dimension to it did they start to respect, honor, and venerate the dead. The very statement of humans having a spiritual dimension led to belief in the immortal nature of human beings. This was in direct opposition to the mortal body where death not only belonged to the body but could also be observed in the deceased. The spiritual part (i.e., the soul) could not be seen in life, hence its death could not be testified. Death became perceived as passage from a material existence to a spiritual existence free from the mortal body. Death has been religiously and culturally interpreted as a mark of finitude, a punishment for sins, a purification, and a phase in the cycle of life eventually leading to another physical life-form (reincarnation). Whichever perception of death prevails, it is always naturally and universally related to life. Humans have always tried to control mortality. Although they did so initially at the spiritual level through its (re)definition, it is now frequently done at the physical (biological) level as a result of biotechnological advances. It is possible today to postpone death by stopping and suspending the dying process (mostly due to resuscitation techniques and life support systems). New scientific research is being developed in two unprecedented directions: (physical) resuscitation and immortality in attempts at denying the reality of death.

H. ten Have and M. do C. Patrão Neves, *Dictionary of Global Bioethics*,
https://doi.org/10.1007/978-3-030-54161-3_188

Death Penalty

See: Capital punishment

H. ten Have and M. do C. Patrão Neves, *Dictionary of Global Bioethics*,
https://doi.org/10.1007/978-3-030-54161-3_189

Declaration of Istanbul (*See* Trafficking; Organ Transplantation)

Basic social and economic needs that are not met can result in health problems that cannot be addressed without the involvement of civil society. Although healthcare services are not discrete interventions in themselves, they demand a systemic approach guided by local knowledge. The Declaration of Istanbul focuses explicitly on the organ trade. The WHO estimates that each year 5% to 10% of all 70,000 transplanted kidneys are from illegal trafficking. The organs are usually sold by the poor in countries such as Pakistan, Egypt, the Philippines, and Moldova. Many countries have adopted legislation to prohibit commercial transplantation. However, in most cases enforcement is difficult or absent. Increasing concern about the organ trade triggered scientists, policy-makers, health insurance companies, and patient organizations to get together in 2008 and adopt the Declaration of Istanbul. Since kidney transplantation is impossible without the involvement of physicians, so professional medical organizations and prestigious surgeons appealed to public opinion and put pressure on colleagues in countries to stop cooperating with the kidney trade. The Declaration of Istanbul brought together two different ethical positions: prohibition of the sale of organs and regulation of the organ market. Both groups (i.e., medical organizations and surgeons) agreed that unregulated commercial organ exchange at the global level should not be allowed and the proposed solution was localization. Since organ trafficking came about as a result of transplantation medicine being globalized, it can only be eradicated through deglobalization. This means action at the local level in local settings because organ transplantation requires surgery and clinics. The solution proposed in the declaration is that each country should become self-sufficient in organ donations and that any shortage of organs should only be eliminated through donations within that country whatever the system used. This suggestion has been criticized because citizens from small countries would have a greatly reduced chance of receiving the life-saving organ they need. In addition to localization there will need to be better legislation, surveillance, and traceability of organs. The declaration clearly states that organ trafficking and transplant tourism violate the ethical principles of equity, justice, and respect for human dignity. This is the reason they should be prohibited. However, there should also be a ban on all types of advertising (including electronic and print media), soliciting, or brokering for the purpose of transplant commercialism. Practices that induce vulnerable individuals or groups such as the poor, the illiterate, illegal immigrants, refugees, and prisoners to become living donors are incompatible with the effort to combat organ trafficking, transplant tourism, and transplant commercialism.

H. ten Have and M. do C. Patrão Neves, *Dictionary of Global Bioethics*,
https://doi.org/10.1007/978-3-030-54161-3_190

Deep Ecology (*See* Ecocentrism; Environmentalism)

Deep ecology refers to a radical environmental movement that emerged in the mid-1970s. It came about as a result of the increasing perception that the planet is teetering on the edge of an environmental catastrophe. It advocates a break with the traditional instrumental relationship of humankind and the environment. It also advocates deepening the ecological awareness of people of the value of the biosphere and of their dependence on nature as the only solution to preventing environmental disasters and imminent ecological collapse. The expression was first coined by the Norwegian philosopher Arne Naess in the journal *Inquiry* in 1973 in his essay "The shallow and the deep, long-range ecology movement: a summary" where he points out that current ecological policies were only tackling a few problems and even then just their surface—reducing them to little more than a readjustment of the technological and economical model (a shallow approach)—and that much deeper concern was needed to deal with the complexity of ecological problems. Since modern society is responsible for such problems what is required is that society itself is transformed along with its views of the world and life. Deep ecology presents a new vision of the world and claims that life has an intrinsic value (bioegalitarianism) and that each organism is an integral part of the whole (inextricable interdependence). It is a new philosophy of ecological harmony or equilibrium termed an *ecosophia* (ecosophy or ecophilosophy). This movement was strengthened by the work of former NASA scientist James Lovelock who in 1979 published *Gaia: A New Look at Life on Earth* in which he presented the Gaia hypothesis that argued planet Earth should be perceived as a single and self-sustaining organism and the biosphere as a self-regulatory entity capable of preserving the planet's health such that life sustains life. Although deep ecology developed as a movement and as an ideology that inspired academics, it had its greatest effect on environmental activists. This movement or ideology is characterized by considering the theories and programs of mainstream environmentalists as insufficient to tackle the enormity of environmental problems and stressing the importance of taking a global perspective on environmental ethics.

H. ten Have and M. do C. Patrão Neves, *Dictionary of Global Bioethics*,
https://doi.org/10.1007/978-3-030-54161-3_191

Deliberation

The word "deliberation" derives etymologically from the Latin *librare* (to weigh in a scale, to balance two things) and the prefix *de* (from, the origin of, comes from). It thus means weighing all possibilities before making a decision or doing so as a result of reflecting on and evaluating different options. Therefore it is an intellectual phase prior to and propaedeutic of (i.e., introductory to) the decision. It is a process—not an action or decision to act. Deliberation as a concept was first used with a specific moral meaning by Aristotle in his *Nicomachean Ethics*, Book III. This was practical wisdom in which knowledge of or the theoretical use of reason were applied to the challenges of concrete actions (practical reasoning or moral reasoning). Thus deliberation focuses on choosing the best means to achieve the ends of a moral action. In the Aristotelian tradition throughout the ages, deliberation remained central to moral life combining the two highest human faculties of reason and will for decision making. Deliberation in contemporary moral philosophy is most often used as a methodology to address dilemmas such as a situation in which two obligations conflict but cannot both be fulfilled simultaneously and between which the agent has to choose and inexorably failing to achieve one of them. More broadly, deliberation is a method used for decision making as an alternative to principlism or casuistry. Indeed, it avoids using a top-down or bottom-up approach and more easily relates to narrative ethics that focuses on story telling about life. The Spanish bioethicist Diego Gracia thoroughly systematized the successive stages in which the deliberation process should unfold for decision-making in clinical ethics to be optimized (although it is possible to apply such a methodology to other domains of human activity). The first stage focuses on the facts aiming to clarify them. This can be considered technical deliberation in which all facts related to the case should be very clearly presented by the professionals in charge and a discussion of the technical aspects of the case should follow to reduce uncertainties as much as possible. The second stage focuses on values aiming to identify those at stake. This can be considered estimative deliberation in which, after identifying all moral problems raised by the case, the one in greatest need of solution and thus the primordial one to be discussed should be identified and the values involved analyzed. The third stage focuses specifically on the deliberation process (stricto sensu) aiming to identify the best course of action. This can be considered moral deliberation. The individuals involved weigh the alternatives of action (ponderation), which requires the analysis of all possible courses of action and tries to satisfy the majority of values involved. They should identify the most extreme courses of action to eliminate them and to focus on intermediate courses of action. Finally, they make their decision based on the widest consensus and thus propose the most prudent and feasible action within the concrete

H. ten Have and M. do C. Patrão Neves, *Dictionary of Global Bioethics*,
https://doi.org/10.1007/978-3-030-54161-3_192

circumstances. This whole process is completed by a final step aiming to assess and confirm the soundness of the decision taken by testing it with current laws, justifying it to others, and considering whether the decision could be changed were there more time. Deliberation involves values, emotions, and beliefs—not just principles. Although those involved should argue reasonably about them and consensus may not be reached, the decision made will be enriched by the different points of views put forward throughout the process.

Dementia

Dementia is characterized by chronic and progressive decline in mental ability. It is not a specific disease but refers to a group of conditions affecting memory, communication, language, ability to focus and pay attention, reasoning, judgment, and visual perception. Such conditions interfere with daily life and the performance of everyday activities. Common types of dementia include Alzheimer disease, Creutzfeldt–Jakob disease, and vascular dementia. Damage to brain cells is the cause of dementia. Certain cells in particular regions of the brain are prone to being damaged. In Alzheimer disease, for example, the hippocampus (the center of learning and memory) is first affected. Dementia is not easy to diagnose since it is difficult to distinguish it from normal ageing, particularly in the early stages of memory loss. Since there is no test to determine dementia, diagnosis is based on medical history, physical examination, and laboratory tests and on identifying changes in thinking, everyday functioning, and behavior. Although there is no cure or treatment that can slow or stop the progression of Alzheimer disease, some medications may temporarily improve the symptoms. However, more attention has recently been paid to prevention. Cardiovascular risk factors are particularly important in the case of vascular dementia. Regular physical exercise may lower the risks of some types of dementia. Diet may help to protect brain health (especially the Mediterranean diet). It is also argued that cognitive training may improve memory. Specific ethical issues are associated with dementia one set of which is related to diagnosis. Although there has been a push for early diagnosis because the condition is incurable and often progressive, it is not always in the interests of the individual concerned to know what is wrong. Moreover, the individual's family may look at things differently. Another set of issues concerns the provision of care. Although more care will be required as the condition progresses, the patient frequently does not recognize the problems and the need for care. Family members often provide the care. The aim of care is to enable the patient to live independently as long as possible. At a certain point, however, professional and institutional care will be necessary but the exact tipping point is often difficult to determine. Institutional care in many countries may well be underdeveloped, not always available, very expensive, and not affordable for most elderly citizens and their families. The third set of ethical issues relates to end of life. Although interest in palliative care for elderly people with dementia is on the increase and termination of life is offered in some countries enabling patients to make advance directives requesting euthanasia, the problem is how to establish that such directives are still valid and that the patient was competent at the time of making the decision. Another related challenge is how to treat intercurrent illnesses such as infections or conditions such as cardiac arrest. The ethical question here is whether such conditions should be treated at all or whether it is better to let the patient die.

H. ten Have and M. do C. Patrão Neves, *Dictionary of Global Bioethics*,
https://doi.org/10.1007/978-3-030-54161-3_193

Demography

Demography is the statistical study of populations. It studies how human populations change through three demographic processes: birth, migration, and ageing (resulting in death). Changing populations influence how human beings inhabit the planet, how they form nations and societies, and how they develop cultures. Concerns have been expressed in the past about the population explosion. There is today increasing interest in demographic change and this has given rise to political debates about the future of societies, especially in developed countries with ageing populations. Although birth rates in these countries are now often below the replacement level of 2.1 children per woman, life expectancy is rising. A related problem is increase in the world population and the pressure that will put on available resources and biodiversity. However, population growth is likely to decelerate in the longer run. Global population will likely peak in 2075 and then decline in 2100. After 2050 many countries will enter a phase of population decline irrespective of political measures taken to curb the increase in population. It is also contestable whether population growth is leading to the reduced availability of food and therefore food shortages and hunger. The availability of food has actually increased over recent decades despite population growth. Demographic studies are essential to epidemiology and public health since they provide statistical information not only about the size and density of populations but also on fertility, mortality, growth, age distribution, and migration in addition to interactions with social and environmental conditions.

H. ten Have and M. do C. Patrão Neves, *Dictionary of Global Bioethics*,
https://doi.org/10.1007/978-3-030-54161-3_194

Dental Ethics

Dental ethics is a special area of applied ethics that is primarily concerned with ethical issues in dentistry. In addition to dentists many different professionals work in dental practice such as oral technicians, health assistants, dental prosthetists, and hygienists. In some countries oral and maxillofacial surgery is a dental specialty. Ethical issues are related to treatment, advertising, professional secrecy, professional communication, and biosafety. Cases of unnecessary treatment are not uncommon and the reason is mainly commercial thus accentuating the conflict between dentistry as a business and as a profession providing care. In many countries dental interventions are publicly advertised and promise perfect and beautiful smiles as a result of aesthetic treatments such as tooth whitening, porcelain crowns, and composite fillings. However, much as is the case elsewhere in healthcare, dentists need to obtain informed consent for treatment. Such consent means that the risks, costs, advantages, and disadvantages of interventions need to be clarified and that confidentiality, privacy, and professional secrecy need to be respected. Moreover, biosafety considerations are important since cross infection is a serious problem in dentistry and appropriate safety procedures need to be followed. Professional dental organizations have developed codes of ethics such as the one launched by the American Dental Association in 2018. It identifies five relevant ethical principles: patient autonomy, beneficence, non-maleficence, justice, and veracity. The practical implications of each principle are given in the code of conduct. For example, patient autonomy implies patient involvement to the extent that copies of records should be provided on request, whereas the principle of veracity implies that unnecessary dental services should not be recommended and that removal of amalgam from non-allergic patients for the alleged purpose of removing toxic substances from the body is unethical. The Code of Ethics of the Indian Dental Association has a different approach. It articulates two basic assumptions: to treat the welfare of patients as paramount over all other considerations and to be courteous, sympathetic, friendly, helpful, and always ready to respond to the requests of patients. It then goes on to specify the duties and responsibilities of dental professionals.

H. ten Have and M. do C. Patrão Neves, *Dictionary of Global Bioethics*,
https://doi.org/10.1007/978-3-030-54161-3_195

Deontology, Moral Theory

The word "deontology" derives etymologically from the Greek *deon* (obligation, duty) and *logos* (study) and hence refers to the study of duties. When used as a noun deontology frequently designates professional ethics. However, when used as an adjective (i.e., deontological) it designates a specific orientation in moral philosophy, a duty-based perspective, or a moral theory such as deontological ethics. Deontological moral theory was developed by Kant in the eighteenth century and is characterized by the primacy of obligations and duties (duty-based ethics) over substantial or contentful definitions of good. From the time of Aristotle (the founder of ethics as a philosophical discipline) to Kant, ethics has always had a teleological nature (end-oriented ethics) that is structured on the basis of identifying an ultimate end (*telos*) to be achieved by human action and coinciding with the supreme good. Deontological orientations differ from teleological orientations in that they are structured by duties to be respected instead of ends to be accomplished. Although deontological theories evaluate the morality of an action according to its level of compliance with the obligations at stake, they do not consider the consequences (e.g., autonomy should be respected regardless of the outcome). The deontological approach differs from the consequentialist approach in that the latter values the conception of good and defines goodness and badness in relation to the outcomes. Deontological ethics differs from virtue ethics in that the latter focuses on the excellence of a person's character or personality, whereas from the deontological perspective an individual or a professional has to comply with specific duties regardless of the nature or character of the virtues involved. Deontological moral theories are particularly important when there is no consensus about what if anything unconditionally good is to be pursued or when the person's character is not appraised and duties orient the moral action.

H. ten Have and M. do C. Patrão Neves, *Dictionary of Global Bioethics*,
https://doi.org/10.1007/978-3-030-54161-3_196

Deontology, Professional

The word "deontology" derives etymologically from the Greek *deon* (obligation, duty) and *logos* (study) and thus literally means the study of duties. When it comes to duties deontology has a normative nature much like morality. Indeed, both concepts were considered synonymous in Ancient Greece. However, since those times deontology has always been used within a professional context such as professional ethics. Deontology designates the particular ethics of a specific profession. The professional standards of good practice are gathered in a deontological code. Each profession has its own deontological code such as that of journalism, physicians, and nurses (although such expressions are more common in Romance languages). Nevertheless, all professional ethics and all deontological codes in democratic countries are based on human rights as their matrix (i.e., set of conditions) thus sharing the same principles such as respect for human dignity and advocating many similar guidelines such as respect for privacy and confidentiality. Although the same profession in different countries may issue deontological codes that are highly similar in the way the *telos* (goal, mission) of the profession establishes its *ethos* (duties), they can also show relevant discrepancies dictated by compliance with respective national laws. Professional deontology always involves self-regulation of the profession (i.e., written by professionals for professionals). Therefore it is also minimalist in that it only looks for the minimum standard required of the professional. The minimalism of professional deontology is often justified in two ways. First, since professional standards are developed by professionals for themselves, they are not likely to impose standards on professionals that are too demanding. Second, if professional regulations were difficult to comply with, then this would likely deter many from taking up the profession. Although professional deontology tends to have strict ethical requirements, they are not unreachable. However, they are decisive when it comes to professional qualifications and the social prestige of the profession.

H. ten Have and M. do C. Patrão Neves, *Dictionary of Global Bioethics*,
https://doi.org/10.1007/978-3-030-54161-3_197

Designer Babies

The children of parents who wish to design their offspring are called "designer babies." The term was coined by the media to refer to genetic interventions in preimplantation human embryos in which the aim was to select or alter the traits future children would have. The entire procedure is shrouded in a mist of fear and wonder. Although biotechnological developments such as assisted reproductive technology (ART) and preimplantation genetic diagnosis (PGD) led to the concept of designer babies, at the beginning it was deemed little more than remotely possible and fictional. The most elementary procedure in choosing the features of a designer baby is the selection of gametes within assisted reproduction. Future parents choose sperm and oocytes according to the desired characteristics of donors such as health status, color of eyes or hair, IQ, skills, and hobbies. Design at such a level is not really under control and will only eventually be expressed in the phenotype. Indeed, it was PGD that opened up the possibility of selection at the genotype level being more controlled. Adam Nash born in 2000 is considered to be the first designer baby. As a result of ART his parents produced several embryos that were later submitted to a preimplantation genetic test for screening to determine which one would be compatible for the blood transfusion that was needed to save the life of their daughter who was dying from Fanconi anemia. Births of three-parent human babies produced with the genetic material of one man (sperm) and two women (oocyte and mitochondrial DNA) through in vitro fertilization have been reported. Although selection for or against a specific genetic trait is commonly justified for therapeutic reasons, there was no benefit whatsoever for the designer baby in the Nash case. Current developments in biotechnology and bioengineering, specifically cloning and gene editing, have raised the concept of designer babies to a totally new level in which traits can be selected, effectively altered, and controlled to a much higher degree. Although cloning techniques such as reproductive cloning are still not possible, they are not the most innovative prospect for designer babies. The creation of embryo chimeras (i.e., new embryos combining cells from different embryos) holds much greater potential. The same is true of incorporating reprogrammed adult cells in a human embryo. Nevertheless, the most recent and revolutionary possibility for designer babies is gene editing that can manipulate the genome by cutting, copying, and binding DNA strands and thus allowing different combinations. Recently it was announced that gene-edited children have been born in China. The possibilities to create designer babies today are enormous and growing. Although the first justifications put forward were and still are therapeutic, there are clearly many other possible motivations for these highly experimental procedures some of which may have unforeseen consequences such as instant personal fame, institutional prestige, and financial rewards. There are many ethical

H. ten Have and M. do C. Patrão Neves, *Dictionary of Global Bioethics*,
https://doi.org/10.1007/978-3-030-54161-3_198

problems associated with designer babies such as the sanctity and dignity of human life, responsibilities toward future generations, and the safety and predictability of results. More specific concerns deal with social injustice, new patterns of health, and normal functioning. There are other issues such as the pressure on children to be as their parents want them to be, their lives become less predictable, and even their freedom is affected. Nevertheless, the most relevant questions are whether it is justifiable to produce genetically engineered children in the first place and whether it is possible to regulate such production worldwide. Although ethical concerns here are massive, absent an effective global governance system such concerns remain futile.

Development

The notion of development encapsulates the idea of growth, progress, and advancement. Since some countries are considered less developed than others, the UN Development Program (UNDP) created the Human Development Index to measure such development and compare countries. The index looks at life expectancy, education, and per capita income such that when lifespan is longer, education level higher, and gross national income per capita greater a country is considered more developed. Countries such as Norway, Switzerland, and Australia head the 2017 rankings. There is also a least developed country (LDC) category that measures socioeconomic development. In 2018 this category included 47 countries among which were Bangladesh, Haiti, and Zambia. The reason for putting the emphasis on human development was the realization that economic growth and wealth were not the sole parameters of human advancement. Growth is not necessarily related to increase in the quality of life of many populations in the world. What is needed are specific policies designed to enhance the quality of life of various groups of people within populations. Such new policies came about as a result of the capability approach of Amartya Sen and Martha Nussbaum that regarded development as a means to enable people to make good use of their abilities. Others focus more on eliminating obstacles to growth such as poverty. Economic and social conditions within countries should be identified and tackled in such a way as to increase the standard of living and make development more sustainable. A new field of development ethics has emerged that argues that development includes choices about value. It addresses such questions as: Are there living conditions that are unfair but avoidable? What changes can be made? Who is responsible for preserving public goods in the process of development? How far are national boundaries ethically relevant? Have former colonial powers special obligations? It is here that development policies are criticized because they often measure activities instead of the achievement of value such as the cost of medicine—not the length of people's lives and how healthy they were during their lives. By concentrating on costs it also excludes types of value such as justice, peace, friendship, or dignity. Finally, the emphasis on national economic product neglects how costs and benefits are distributed within and across generations.

H. ten Have and M. do C. Patrão Neves, *Dictionary of Global Bioethics*,
https://doi.org/10.1007/978-3-030-54161-3_199

Disability (*See* Ableism)

Disability is a complicated notion. The Americans with Disabilities Act defines a person with a disability as someone who has a physical or mental impairment that substantially limits one or more everyday activities. The WHO regards it as an umbrella term that covers impairments, activity limitations, and participation restrictions. It differs from impairment in that it is a problem in body function and structure. Disability is more than a health problem in that it reflects the interaction a person has with his or her body and with the society in which he or she lives. Often social and environmental barriers need to be removed to overcome disabilities. Examples are a person in a wheelchair who cannot access buildings or a blind person who cannot see traffic lights. Although people with disabilities have the same health needs as non-disabled people, they are more vulnerable. A disability is a continuing condition that restricts everyday activity. It reduces the capacity of a person to carry out everyday functions such as communication, mobility, and social interaction. It is often associated with the need for continuing support services. The concept of disability is not uncontested. One position is that disability is an individual state resulting from impairments, while another emphasizes that it is the result of social determinants such as discrimination and exclusion and has nothing to do with impairments. This last position rejects the idea that disability is related to the body and to functions of the body and argues that it is society itself that is disabling. The implication is that a medical approach is insufficient. Advocates of disability groups often criticize bioethics since it frequently follows a medical approach. A famous case concerns the deaf community. When new technologies became available to (partially) overcome deafness such as cochlear implants, many deaf people rejected this intervention. Since deaf people learn to communicate using sign language the concern was that implants would negatively affect the cultural and linguistic community of deaf people and the best that could be hoped for from implants would be imperfect hearing. The case illustrates that the meaning of ability is ambiguous. For people who are regarded as disabled by outsiders the impairment they have is not a pathological state that needs to be addressed but a different way of life that needs to be respected.

H. ten Have and M. do C. Patrão Neves, *Dictionary of Global Bioethics*,
https://doi.org/10.1007/978-3-030-54161-3_200

Disasters

Globalization has amplified global vulnerability in that it calls for the sympathy, solidarity, and generosity of people worldwide. This has become clear in the increasing occurrence of disasters. Definitions and classifications of disasters differ. Although definitions generally combine a number of elements, a basic distinction is made between natural and man-made disasters. The International Federation of the Red Cross and Red Crescent Societies have developed a standardized international classification of disasters. Its World Disasters Report 2018 distinguishes two generic categories of disasters: natural disasters and technological disasters. The natural disaster category is divided into five subgroups that cover 12 disaster types and more than 32 subtypes such as biological disasters (epidemics) and geophysical disasters (earthquakes and tsunamis). The technological disaster category includes industrial accidents and transport accidents. Both categories of disasters frequently occur. Over the last 10 years more than 3,751 natural disasters have been reported of which 84.2% were weather related and approximately 134 million people needed assistance. Another way of looking at humanitarian disasters or emergencies is to divide them into complex emergencies and natural disasters. Complex emergencies are disasters caused by human violence (as in Palestine and the Democratic Republic of Congo). Although these are primarily political events requiring long-term assistance, public and popular funding will generally be limited. The relief effort will never be sufficient since progress is constantly endangered by renewed violence. Natural disasters are caused by natural events such as the 2004 tsunami in the Indian Ocean, the 2010 earthquake in Haiti, and the 2011 earthquake and tsunami in Japan. Such natural disasters evoke widespread public sympathy and receive substantial public funding. Relief efforts are usually focused on short-term frontline activities such as providing food, water, shelter, and medical attention.

Both classifications identify disasters as having the following defining characteristics: they are unexpected in that they come as a surprise or a shock; they cause great damage, loss, suffering, and destruction; they estrange people by making them homeless; and there is no issue of human responsibility because they are accidental and hence nobody can be blamed for the disaster. This last characteristic demarcates natural disasters from man-made ones (e.g., humanitarian disasters brought about by civil war are quite simply the result of human evil). There is a different moral responsibility here. Identifying a disaster as natural therefore introduces a specific moral discourse. Natural disasters create innocent, pure victims and result in a particular responsiveness being generated, one in which we are moved because fellow human beings are hurt and in need. It is in this way that disasters today have a global impact and call for our sympathy, solidarity, and generosity. We are touched by personal stories of humans trying to assist each

H. ten Have and M. do C. Patrão Neves, *Dictionary of Global Bioethics*,
https://doi.org/10.1007/978-3-030-54161-3_201

other. The usual pattern of human interaction based on making a buck and self-interest is suddenly transformed. Our world is disturbed by images of people suffering elsewhere, which makes us realize we are all in the same boat in being fragile and vulnerable and requiring reciprocity and unconditional help. Natural disasters are therefore paradigmatic of humanitarian aid and highlight the essence of ethics. What is the value of ethics if we do not care about the victims of such unfortunate events? However, the usual distinction between natural and man-made disasters and the moral geography it introduces is questionable. The origins of disasters can be different and some are indeed uninfluenced by human beings. Nevertheless, what turns an event into a disaster is its impact on human beings. An earthquake in a completely uninhabited area that has no negative effect on humans is considered a geophysical event—not a disaster. Moreover, in today's interconnected world it is difficult to see how a large-scale natural disaster would not affect human beings. The human impact is what makes an event disastrous. Negative impacts and vulnerability are often the result of prior human interventions. The interplay between natural processes and human activity implies the ethical issue of human responsibility is always present. There are two levels of ethical issues concerning disasters consisting of microethics and macroethics. Emergency conditions require that those injured in a major catastrophe are classified such that they receive optimum care as quickly as possible and this implies triage. Decisions need to be made regarding the order of treatment based on the urgency of need. Triage has today become an essential component in disaster relief. However, there is no agreement on the ethical criteria to make such decisions. In ordinary triage the focus is on the interests of individual patients. Under extraordinary circumstances the focus is unclear: Is it on survival of the greatest number of people or on those who are most likely to survive? At the macroethical level the basic idea is that rather than selecting disaster victims for help we should select disasters in need of major relief and determine the short-term and long-term goals of international assistance. New strategies have been developed in disaster response planning. Although the focus is on preparedness, the basic idea is that catastrophic events such as pandemics and natural disasters cannot be prevented. However, we know that sooner or later we will have to face them and that we need to be prepared. Over the last decade countries have invested billions in preparedness strategies, plans, departments, and agencies and special legislation (preparedness acts) has been adopted. Planning and preparedness are typically undertaken by states (e.g., the Federal Emergency Management Agency in the United States). However, as illustrated by the COVID-19 pandemic such preparedness can easily be downgraded and neglected when a global disaster has not occurred for a while.

Discourse Ethics

Discourse ethics (a.k.a. communicative ethics or procedural ethics) refers to a contemporary philosophical perspective and specifically to moral theory that advocates the only way to live peacefully and have fair rules is to get all people potentially affected to engage in reasonable argumentation, while acknowledging the moral pluralism of our societies and the lack of a recognized authority to establish what is good and bad and what is right and wrong. Discourse ethics is closer to deontological ethics that values what ought to be done (duties) than to teleological ethics that values the goals to be pursued (ends). However, its true nature is procedural, valuing dialogue and argumentation (communication) to build consensus. It allows skepticism and relativism in morality to be rejected in democratic and pluralist societies. It is indeed procedural in that it abandons all pretention to reach a substantial morality upon which everybody could agree. It proceeds as a formal methodology to build consensus from different perspectives. It is communicative in that it focuses on a transparent and honest dialogue to bring divergences closer together. Although it is labeled "discourse ethics" in the English language, in other languages it is more frequently referred to as "discussion ethics." *Diskursethik* (discussion ethics) was first formulated by the philosophers Karl Otto Apel and Jürgen Habermas in the 1970s. In German the word *Diskurs* (discussion) does not have the secondary meaning of discourse (*Rede* in German). It is used when referring to a reasonable argumentative discussion. Although both Apel and Habermas aimed following Kant at formulating a formal and universal morality grounded in reason (reformulating the Kantian categorical imperative), they pursued a critical theory of society within the Frankfurt School. Habermas proposed two procedural principles: a discourse principle (D) and a principle of universalization (U). The discourse principle states that "only those action norms are valid to which all possibly affected persons could agree as participants in rational discourse" and the dialogical principle of universalization states that "A [moral norm] is valid just in case the foreseeable consequences and side-effects of its general observance for the interests and value-orientations of each individual could be jointly accepted by all concerned without coercion." Complying with these two norms should guarantee the impartiality and fairness of all other social rules thus facilitating universal moral rules in democratic and pluralist societies. Procedural ethics has also been developed differently by philosophers such as John Rawls and within bioethics by Tristram Engelhardt. The most noteworthy Rawlsian philosophical contribution is the procedural concept of justice. Indeed, it is the procedure to use when determining a notion of justice that guarantees it as just (i.e., impartial). Engelhardt considered only procedural ethics could adequately address the dilemmas of our secular and pluralist society. Although it is crucial to respect individual autonomy, it is also important to contribute to a

H. ten Have and M. do C. Patrão Neves, *Dictionary of Global Bioethics*,
https://doi.org/10.1007/978-3-030-54161-3_202

peaceful society. This can be achieved with the procedural principle of permission: the only moral authority is the authority of permission, an agreement to decide and to act collaboratively. Discourse ethics can be particularly useful in the work of ethics committees as pluralists and multidisciplinary bodies and in public policies that need majority consensual agreements.

Discrimination

Discrimination refers to making distinctions such as separating and classifying things as different. Usually, it has negative connotations because it reflects on individuals and groups in an adverse way such that they are treated differently. Discrimination violates the basic principle of human dignity because distinctions between human beings are often made on the basis of gender, race, ethnicity, religion, disability, or wealth. It infringes the first principle of the Universal Declaration of Human Rights that states all human beings are born free and equal in dignity and rights. Discrimination has been a target of action by the United Nations. The Convention on the Elimination of All Forms of Racial Discrimination was adopted in 1965 and was followed by the Convention on the Elimination of All Forms of Discrimination against Women in 1979. In UNESCO's Universal Declaration on Bioethics and Human Rights, Article 11 formulates the principle of non-discrimination and non-stigmatization: "No individual or group should be discriminated against or stigmatized on any grounds in violation of human dignity, human rights and fundamental freedoms." UNESCO's International Bioethics Committee (IBC) clarified this principle in its report in 2014. The IBC argues that Article 11 is rooted in international human rights law and the scope of the article is determined by the scope of the declaration itself. Article 11 applies to ethical issues related to medicine, life sciences, and associated technologies as applied to human beings taking into account their social, political, legal, and environmental dimensions. Although the committee does not provide a list of discriminatory (or stigmatizing) practices, some are clearly discriminatory and need no comment. The only thing needed is action: naming, shaming, preventing, and punishing. There are cases where discrimination is not so clearly discernible and practices may change over time. The wording of the principle shows that only those distinctions that impair human dignity, human rights, or fundamental freedoms are rightfully called discriminatory under Article 11. A decision or a practice that is discriminatory is one that infringes upon these fundamental notions making such decisions or practices objectionable. An example is genetic discrimination that occurs when people are treated differently because they have a gene mutation that causes or increases the risk of a genetic disorder. People considering genetics testing commonly fear that employers, insurance companies, and adoption services may use such information. Countries have adopted specific legislation to prevent discrimination and protect people. Nevertheless, many cases of genetic discrimination have occurred; hence the reason for legislative action. Although genetic discrim-

H. ten Have and M. do C. Patrão Neves, *Dictionary of Global Bioethics*,
https://doi.org/10.1007/978-3-030-54161-3_203

ination today has not disappeared and the number of cases is low, most discrimination is not explicit and has taken more implicit and subtle forms resulting in stigmatization. A recent notion is positive discrimination. This refers to policies that explicitly favor individuals belonging to groups that have been discriminated against in the past (a.k.a. affirmative action). An example is giving preference to hiring people from ethnic minorities. In many countries such as the United Kingdom this is illegal since it is still a form of discrimination.

Disease Mongering

Disease mongering refers to the unnecessary widening of diagnostic boundaries of illness for economic benefit. It has been regarded as "the selling of sickness" in which the concept of disease is stretched such that a larger market for treatments is created. In everyday life there is often no clear boundary between normal and abnormal. People have complaints such as anxiety, depression, heartburn, fatigue, or baldness that can make them feel ill. Although in some severe cases such complaints might be associated with diseases or abnormal conditions, this is not usually the case since complaints are temporary and expressive of the variability of the human condition. Disease mongering exploits unclear boundaries where ordinary life ailments such as hair loss or shyness are turned into medical problems, mild symptoms are turned into manifestations of a serious disease, and risk factors such as hypertension and osteoporosis are conceptualized as diseases. There are those who argue that variabilities in human existence are expressions of diseases that are not recognized, that the prevalence of such diseases is underestimated and better diagnoses should be encouraged, and that doctors should be better educated in determining these hidden diseases. The suggestion is that all cases should be treated primarily with medication that is available. This is exactly the point of disease mongering in that it creates a need for treatment and thus expands the market for drugs or interventions. It is not interested in fatal or short-duration diseases. More profitable is fabricating diseases that are chronic and lifelong. Rather than promoting curative drugs the focus is on lifestyle drugs (e.g., for allergies and acid reflux) that you need to take every day for the rest of your life. An older example of disease mongering was the launch of chronic halitosis as a new disease characterized by bad breath. Named for Joseph Lister, the nineteenth century British pioneer of antiseptic surgery, Listerine was promoted by its manufacturer as an antiseptic mouthwash in an aggressive marketing campaign in the 1920s successfully arguing that it was a cure for this new disease. Disease mongering also refers to a number of undesirable activities perpetrated by disease mongers who advocate or try to propagate diseases. The diseases advocated are ghost diseases in that it is unclear whether they really exist or not and are best regarded as little more than illusions designed to frighten people. Most of the time the purpose of disease mongering is to promote the sales of drugs. The harm of disease mongering goes beyond deception, exploits people's anxieties and fears, and reinforces the sense of vulnerability people feel when they are suffering and ageing. By blindly articulating faith in scientific progress disease

H. ten Have and M. do C. Patrão Neves, *Dictionary of Global Bioethics*,
https://doi.org/10.1007/978-3-030-54161-3_204

mongering promotes the idea that there is a pill for everything. Moreover, drug therapy is advertised as the most adequate treatment option while other approaches are downplayed. Although fabricating diseases reinforces reductionist approaches in healthcare making all problems individual and located in the body, complaints can only be explained by somatic models. Another harm brought about by disease mongering is that it creates pseudodiseases that lead to waste of scarce healthcare resources that should be directed to more serious conditions.

Disease

Although disease is a crucial concept in healthcare, it is not only difficult to define but also varies according to history and cultural and social context. Diseases of course are as old as humanity and human beings have always had to face them. The concept of disease plays an important part in determining the role and task of healthcare providers and the care patients need. Identification of a disease not only implies the patient needs care, treatment, and attention, but also excuses the patient from work and often is a basis for compensation. Furthermore, the concept of disease implies the patient is not accountable and responsible for his or her condition. Despite the concept of disease being important there is no agreement on how to define it. One approach in the philosophy of medicine is descriptivism (a.k.a. naturalism) where disease is defined as a biological phenomenon that occurs in nature and an internal condition that disturbs natural functioning. The biostatistical theory of Christopher Boorse is well known and defines disease as a state that interferes with the normal function of an organ or system of organs of an individual. Normal function can be determined by regarding the species of human beings (i.e., *Homo sapiens*) as a reference class. He argues that every species has a species design described biologically as having typical anatomical, physiological, and chemical characteristics. The species design can be identified with the help of statistical analysis. Having a disease therefore is not the same as being ill. This approach has been criticized in defining disease solely in terms of physiology and biology and neglecting the fact that disease is also a social convention. The so-called normativist approach emphasizes that disease is a judgment that somebody is undergoing harm that can be explained in terms of bodily or mental conditions. Human norms of harm therefore play a major role in determining what disease is. Such an approach has been elaborated by Caroline Whitbeck. In her view what is significant is a person's psychophysiological capacity to act or respond appropriately (in a way that is supportive of the goals, projects, and aspirations of the person) in a wide variety of situations. Diseases exist when this capacity is disturbed or impeded; hence they are not natural kinds. In this debate hybrid definitions of disease have been proposed combining ideas from the descriptivist and normativist approaches. Disease is then defined as harmful dysfunction. Although dysfunction can be identified on the basis of natural phenomena, it is a value judgment whether it is harmful or not.

H. ten Have and M. do C. Patrão Neves, *Dictionary of Global Bioethics*,
https://doi.org/10.1007/978-3-030-54161-3_205

Diversity (*See* Biodiversity; Cultural Diversity)

Diversity refers to the notion that individuals are different and unique. As a consequence of the race, gender, ethnicity, age, sexual orientation, and religious and political beliefs human beings are not the same. Although the processes of globalization have led to the notion of diversity being increasingly articulated, there are those who argue that diversity should be explored as a rich dimension of the contemporary world—not just tolerated. Ethically, it is not enough to acknowledge and tolerate differences since showing mutual respect and understanding them as different ways of knowing the world are just as important. Furthermore, differences should incite efforts to build alliances and cooperation such that, for example, forms of discrimination and stigmatization can be eradicated. In bioethics the notion of diversity has been elaborated in such areas as the environment (biodiversity) and culture (cultural diversity). In 2001 UNESCO adopted the Universal Declaration on Cultural Diversity that regards diversity as the common heritage of humanity. Such diversity is embodied in the uniqueness and plurality of the identities of groups and societies making up humankind. Cultural diversity as a source of exchange, innovation, and creativity is as necessary for humankind as biodiversity is for nature. The declaration also emphasizes that diversity implies a movement from cultural diversity to cultural pluralism. Since societies are becoming ever more diverse it is essential to ensure people and groups whose cultural identities are various and dynamic interact harmoniously and are happy to live together. Policies promoting the inclusion and participation of all citizens are guarantees of social cohesion, the vitality of civil society, and peace. Furthermore, cultural diversity is regarded as an important factor in societal development. It widens the range of options open to everyone and is fundamental to development understood not simply in terms of economic growth, but also as a means to achieve a more satisfactory intellectual, emotional, moral, and spiritual existence.

H. ten Have and M. do C. Patrão Neves, *Dictionary of Global Bioethics*,
https://doi.org/10.1007/978-3-030-54161-3_206

Donation, Blood

Blood (red cells, plasma, platelets, and other components) is the most common biological material that is donated since it is the most abundant and the easiest to procure (almost every adult can donate). Moreover, since it is constantly produced it can be regularly donated. Collecting donated blood is less invasive and more safe than other forms of donation. It is not only the most needed resource, but also the target of most campaigns for biological material donation. Nevertheless, demand still exceeds supply as a result of it being essential to the survival of many patients with very different pathologies. Non-direct blood transfusion in which the donor and receiver no longer needed to be in direct contact (to avoid blood coagulation) began in the first half of the twentieth century. Some countries such as the United States had programs that commercialized blood. Today, blood is not generally treated as a commodity, there is a broad awareness of the social good to donate blood. Indeed, blood donation now forms part of some professional curricula as testimony to such professionals' social engagement and voluntary work. Broadly speaking, there are three main ethical issues related to blood donation. The first concerns the expectations of many donors. Although donation should be a totally voluntary act and free of any compensation, it is not rare for blood donors to expect to be given special treatment such as priority in healthcare facilities if and when they need a blood transfusion or lower prices in certain healthcare services. Such expectations jeopardize the gratuity of donation. The second ethical issue concerns the exclusion of specific groups being accepted as donors such as the gay community. As a result of the risk of HIV infection through blood transfusion in the 1980s high-risk groups were not allowed to donate. This decision raised issues about confidentiality of the health condition of excluded persons, on the one hand, and lack of disclosure of the health status of those who posed serious problems of safety, on the other. The response to these problems involved scientific and financial investment to make blood tests as accurate as possible. Although there was a decrease in the number of exclusions and those remaining are likely to be overcome in the near future, the safety of blood transfusion cannot be put at risk and blood donation cannot be used as a weapon against anti-discrimination. The third ethical issue concerns the refusal of blood transfusions on religious or cultural grounds. Many of these cases can be solved through autologous transfusion (i.e., collection and reinfusion of the patient's own blood) or what has been called self-donated blood. As is usual when families are involved the issue of donation does not apply in such cases.

H. ten Have and M. do C. Patrão Neves, *Dictionary of Global Bioethics*,
https://doi.org/10.1007/978-3-030-54161-3_207

Donation, Body (Corpse)

Demand for human corpses is as old as human curiosity to understand how the human body works. Mainly since the Renaissance (fourteenth century) and the beginning of practice-based knowledge (instead of authority-based knowledge) and experimentalism, schools of medicine and scientists needed corpses that they would get from hospitals (hosting the socially excluded) and the streets (the homeless) where destitute people would die. Later, cadavers were stolen from graveyards for the purposes of research and teaching. Although such practices were not allowed by law, they were nevertheless common. Despite news about the stealing and selling of cadavers for body parts, they are of course totally banned today. The supply of bodies needed for research comes from unidentified cadavers and donations. Corpses should be donated during life by the individual or after death by the family. Although human bodies are currently partially substituted by artificial models and digital reproductions, there is still a high demand for corpses of people who died from rare diseases or in particular situations that are of great interest to researchers. There are programs for body donation that have different specifications. Some of which are private and, in many countries, body donation is no longer restricted to medical schools. Giving a family member's corpse to programs whose specifications are private threatens the gratuity and generosity of all donations. There are those who argue that the cadaver should be considered the only true heritage a person leaves to his or her family. The selling of a corpse for body parts can generate high prices (there is a black market handled by so-called body brokers). This idea is highly controversial, and the dominant perspective is that a human cadaver is testimony of a unique and singular life that deserves respect and as such cannot and should not be reduced to an object such as a good for commercialization. Most programs for body donation offer all the conventional services associated with handling cadavers including cremation and as such may be considered as compensation that does not affect the free character of donation—not as a payment.

H. ten Have and M. do C. Patrão Neves, *Dictionary of Global Bioethics*,
https://doi.org/10.1007/978-3-030-54161-3_208

Donation, Embryo

Embryo donation is the process in which an embryo is freely gifted. Such a process was made possible by assisted reproductive technologies (ARTs) that have since the 1970s been able to produce human embryos outside the womb (i.e., in the lab) and then transfer them to the woman's body or cryopreserve them and later transfer the thawed embryos to the woman's womb. At the beginning ART had low rates of success and a tendency to produce a high number of embryos, to transfer more embryos than the desired number of children, and to cryopreserve the remaining ones. Such so-called surplus embryos could later be transferred to the biological and/or legal owner, destroyed after their expiration date, or donated for scientific research purposes and/or for adoption. Although the number of surplus embryos for each procedure of ART has currently decreased, there are still many cryopreserved embryos in the world. Moreover, human embryo banks have been created for storage (eventual future use by the biological progenitors) and donation. A significant number of couples consider donation of their surplus embryos for different purposes and for different reasons when they finish their fertility treatment. Such purposes and reasons entail a number of moral issues such as whether the human embryo should be considered biological material or a subject of rights. The most frequent and morally controversial decision regarding embryo donation is when they are given for medical research, especially research aiming to improve fertility treatments (for which the couple is grateful) or involving stem cells (as a result of the high social expectations regarding this line of research). Such a decision is usually based on the conviction that it is better to award a social utility to embryos than to discard and destroy them (considered as waste). Donation for (prenatal) adoption despite being consensually moral is not common. Nevertheless, it would have the advantage of erasing any criticism concerning the instrumentalization of human life that is inherent in its use in research. It would also comply with the sanctity of the embryo, as advocated by those who consider the human embryo deserves full respect. However, most couples with surplus embryos do not want their biological offspring to be raised by someone else. In any event couples who want to be parents can apply to have fertility treatments and then have a biological link with their own children. Regardless of the purpose of embryo donation, it should always be anonymous. In this way the generosity and free character of the gift have greater meaning, something that is not always the case.

H. ten Have and M. do C. Patrão Neves, *Dictionary of Global Bioethics*,
https://doi.org/10.1007/978-3-030-54161-3_209

Donation, Gametes

Gametes are germ cells that have a single (haploid) set of chromosomes. They can be male (sperm) or female (oocytes) and have the potential for reproduction. Sperm is abundantly produced throughout life in the male body. It can easily be collected and since the 1970s cryopreserved (freezing) and stored in sperm banks. Females are born with all their oocytes (i.e., no more are produced during their lives) and retrieval is invasive in that a woman must undergo ovarian stimulation (daily hormonal injections that encourage the ovaries to develop multiple mature eggs) and when the oocytes are mature a thin needle (attached to a suction device) is inserted into the ovary under ultrasound control via the vagina to aspirate the oocytes. Women are usually sedated during the procedure. The cryopreservation of oocytes was once a difficult procedure and it is only since the late 1980s that it has become possible. The technical procedures involved in collecting gametes and storing and handling them for reproduction have now been mastered. Such procedures were primarily developed for two reasons both of which were medical: men and women who were undergoing medical treatment that affected their germ cells who wanted to preserve their fertility, as well as donation to couples and single women who for whatever reasons (clinical or not) needed gametes from a third party to have a child. Gamete donation has been viewed from two opposing ethical perspectives: one considered altruistic and the other considered irresponsible. The dominant view is that gamete donation is altruistic in that it allows infertile couples or single women to be able to have a child that they could not conceive otherwise. On the other side, there are people (mainly religious) who consider gamete donation is a denial of parenthood. They believe that refusing to acknowledge the birth of a biological child who will not know its parents and not assuming responsibility for it represents such a denial. Although gamete donors are usually financially compensated, any compensation should be equivalent to the costs involved (such as loss of a day's work and transportation) and to the inconvenience caused as a result of the donation. However, since most so-called donors are today students searching for a way to pay tuition fees it can be argued that this is gamete selling (in which personal characteristics are bought)—not a donation. In fact, gamete selling is highly profitable and competitive. Moreover, the demand for gametes is growing. Another issue is the importance of restricting the number of times an individual can contribute in the same region to guarantee gamete diversity. Although genuine donors mostly come from the inner circle of family and friends when the gift will not be anonymous (opposite to what happens in a gamete bank), psychological and emotional coercion cannot always be excluded.

H. ten Have and M. do C. Patrão Neves, *Dictionary of Global Bioethics*,
https://doi.org/10.1007/978-3-030-54161-3_210

Donation, General

The word "donation" derives etymologically from the Latin *donatio* (giving, presenting). It refers to giving something gratuitously and freely (i.e., at no charge) to help someone else. In addition to it being a generous act, it often has some sort of interest, or feelings attached to it involving someone close to the donor. Sometimes it is simply done to garner social appreciation or to foster self-appreciation. Donation has a totally positive connotation. Traditionally used in the legal domain referring to goods, rights, or titles it also has a moral perspective in that it refers to the donor gifting something he or she owns to help or benefit others. In the second half of the twentieth century as a result of biotechnological progress (especially in the surgical, histocompatibility, and immunogenetics fields), it became possible for individuals to donate bodily material for medical treatment or scientific research such as cells, tissues, organs, and corpses. Different body parts have different potentialities and functions and their procurement follows different procedures and has different impacts on both donors and recipients. This means the ethical problems raised by the donation of bodily material can also differ and must be separately considered. Moreover, donation should be considered within different cultural and religious contexts. Although donation of biological materials is broadly viewed as highly moral, there are important variables to take into account. Although demand for biological materials has been increasing and donation of biological materials has also grown, demand always falls short of supply. It is in this gap between supply and demand that the majority of (and the most complex) ethical problems arise.

H. ten Have and M. do C. Patrão Neves, *Dictionary of Global Bioethics*,
https://doi.org/10.1007/978-3-030-54161-3_211

Donation, Organs

Organ donation refers to individuals gifting their own organs for transplantation to save the life of another person. It is perhaps the most praiseworthy way of donating biological material since it is more than an act of donation in that it provides the gift of life and often entails an invasive procedure that has side effects for the donor. Traditionally, the gift of an organ was only made *post-mortem* (i.e., made during the donor's life but only used after his or her death). Thus donation is legally decided during life as becoming effective only after death. Organ donation during life is today on the increase since it is preferable from the medical point of view. However, *post-mortem* donation can involve all organs, whereas donation during life can only include organs that are double (e.g., kidneys), self-regenerable (e.g., liver), or available in sufficient quantity (e.g., pancreas). Vital organs that fall outside these categories and whose removal would cause the death of the donor (e.g., heart) cannot be donated in life (even from parents to children). Although organ donation is an intrinsically good act, it raises a myriad of ethical issues ranging from protection of the individuals involved to the interests of a fair society. At the individual level, ethical issues can concern the donor and the recipient. When it comes to the donor, it is important to guarantee safety (medical acts should comply with the best clinical practices), autonomy (donation has to be totally free of any kind of reward and any form of coercion), and private (strictly following the donor's wish). When it comes to the recipient, it is important to realistically balance risks and benefits, provide necessary information about the procedure, obtain free consent, and respect privacy during the whole process. Although such principles are of paramount importance, ethical problems still arise such as donation between siblings. This may involve psychological coercion that makes evaluating the autonomy of the decision difficult. Another problem is privacy. This is becoming more difficult to respect in such procedures as paired kidney exchange. When the practice of organ donation expanded from anonymous deceased persons to living family members and friends, the organ pool increased but new ethical problems arose especially concerning real autonomy and true donation. At the societal level, ethical problems can include the criteria for admission to and ranking in the transplant waiting list. In many countries the lack of a national public health service puts the onus for being a good candidate for transplantation heavily on financial considerations thus excluding people in serious need. Moreover, there are ethnic groups who for different reasons are not prepared to donate despite being quite happy to receive transplants. Organ matching becomes more difficult in such cases and these individuals can feel discriminated against. The overall problem with organ donation is the scarcity of organs. Public campaigns for the donation of living organs often appeal to public spiritedness as an important way to increase the number of organs available.

H. ten Have and M. do C. Patrão Neves, *Dictionary of Global Bioethics*,
https://doi.org/10.1007/978-3-030-54161-3_212

Donation, Tissues and Body Parts

Transplantation medicine depends on human beings donating tissues such as corneas, skin, tendons, ligaments, heart valves and other cardiovascular tissues, bone, blood vessels, and connective tissues. Although other body parts such as hands and faces can be transplanted, this is not common. Human tissues and body parts can only be legally transplanted *post-mortem*. Donated materials can only be used for transplantation or for scientific research. When used for the latter it is often as a potential starting material for biotechnological products. It thus raises particular ethical concerns such as the quality of informed consent: biological material should not be used for any purposes other than those clearly stated by the owner thus respecting his or her autonomy and the fact that he or she owns the body. Furthermore, it is important to acknowledge that human tissues are financially valuable in addition to being beneficial (life saving). Moreover, potential sources are abundant (i.e., cadavers) and the removal procedure of tissues is not too complex or demanding. Although this may well explain the existence of the black market that operates between the declaration of death and the funeral, it should be remembered that such criminal acts violate the autonomy of the deceased person, the wishes of the family, and violate respect for the human body (which also applies to corpses) reducing the body to a retail product and jeopardizing the safety of transplantation. Human tissues (especially those presenting some specific characteristics) can also be collected for research and used as raw material for biomanufacturing. Many highly profitable biological medicinal products are derived from human tissues. This raises questions about who is authorizing such commercialization and who is profiting from such distribution. The donation of body parts raises issues regarding consent and trade in much the same way as organ and tissue donation.

H. ten Have and M. do C. Patrão Neves, *Dictionary of Global Bioethics*,
https://doi.org/10.1007/978-3-030-54161-3_213

Doping (*See* Sports)

Doping refers to the illicit use of drugs to enhance both performance in sports and the ability of athletes to win. Such a practice is of course contrary to the spirit of sports and can endanger the health of athletes. In 1967 the International Olympic Committee (IOC) published the first Prohibited List (i.e., an annual list of prohibited substances and methods). In 2005 UNESCO adopted the Convention against Doping in Sports. Governments have accepted that it is their responsibility to apply international law to doping to which end the convention harmonized the efforts of countries around the world. The World Anti-Doping Agency (WADA) was established in 1999 as an international agency to monitor the World Anti-Doping Code that was adopted in 2015. However, in recreational sports (i.e., activities whose primary purpose is participation, improving physical fitness, having fun, and getting involved socially) doping substances are commonly used. They are also used to increase everyday performance of members of the public (e.g., for sleeping and concentration). The Council of Europe has defined doping as the administration or use of substances in any form alien to the body or of physiological substances in abnormal amounts using abnormal methods by healthy persons with the exclusive aim of attaining an artificial and unfair increase in competitive performance. The fight against doping in sports is rooted in ethics. It is argued that doping violates the spirit of sports in which values such as fair play, honesty, excellence, and health are breached. Three criteria are considered when deciding whether a substance should go on the Prohibited List: health risks, performance enhancement, and the spirit of sports. Two cyclists died as a result of taking drugs during the Olympic Games in Rome in 1960. This triggered the IOC to take action and the first anti-doping control tests were introduced. The Prohibited List was expanded to include anabolic steroids, caffeine, beta-blockers, and diuretics, later new drugs were added such as growth hormones and erythropoietin, and this was followed by a ban on blood doping. Since 2004 WADA has been responsible for updating the list which now also includes gene doping. Discussion about the criteria for inclusion on the list is ongoing. An important ethical issue is the question of responsibility since it is unfair to put all the blame on the athlete. He or she is nowadays part of a team comprising medical and scientific advisors, coaches, and technicians. Sports physicians have a duty of care to their athlete patients despite being the ones who usually provide and administer the drugs.

H. ten Have and M. do C. Patrão Neves, *Dictionary of Global Bioethics*,
https://doi.org/10.1007/978-3-030-54161-3_214

Double Effect

Double effect refers to the moral legitimacy of an action that causes serious harm as an involuntary, unpredictable, or unavoidable side effect of another action that deliberately aims at performing a commonly recognized good. The doctrine, principle, or argument of double effect originated in Catholic theology (specifically advanced by Thomas Aquinas in the thirteenth century) and resulted from considering the discrepancy that exists between (good) reasons and (bad) effects in which not only actions but also intentions are taken into account. Biomedical ethics today has recovered the pertinence of the double-effect argument in its consideration of traditionally beneficial medical action and possible prejudicial and unwanted effects on the patient. A classic case of the use of the double-effect argument is one that occurs at the beginning of life: ectopic pregnancy in which the embryo has no chance of surviving and developing and the pregnant woman will inevitably die unless the embryo is removed. In such a case most of those who advocate the sanctity of human life consider abortion morally acceptable since the end is to save the woman's life, which cannot be achieved without the death of the embryo. Another classic case is one that occurs at the end of human life: the administration of morphine to treat critically ill patients who are suffering severe pain while involuntarily causing respiratory failure and death. The argument is that if the action pursued is to alleviate suffering then it is in itself good despite the outcome of the action being death (which is neither wanted nor avoidable). Double effect is a highly relevant argument within biomedical ethics as a result of the value attributed to the intention of the action being an important moral element that should not be overlooked when evaluating the morality of an action. Double effect cannot be used as an argument regardless of circumstances since there are two prior obligations to fulfill. First, agents should try to reduce the uncertainty of the situation and to foresee outcomes to the best of their knowledge. Second, agents should balance risks and benefits while respecting the principle of beneficence. The validity of the double effect argument is grounded on compliance with these two requirements.

H. ten Have and M. do C. Patrão Neves, *Dictionary of Global Bioethics*,
https://doi.org/10.1007/978-3-030-54161-3_215

Double Standards (*See* Standards of Care)

Double standards refer to the application of different principles to situations that are essentially the same. The result is that people, events, and conditions are treated differently when they should of course be treated similarly. A common example of double standards occurs in gender inequality such as when women do the same work as men but are paid a lower salary. There are instances when double standards become problematic bioethically such as when the requirement of informed consent is applied to research in developed countries but not for similar research in developing countries. The main ethical problem is injustice since double standards introduce unfair treatment and produce inequality. Much as is the case with research, double standards today also occur in clinical trials undertaken in both developed and in developing countries. The EU Clinical Trial Regulation (2004) abolished double standards in clinical trials, established the same standards for research within the European Union, and got EU enterprises in other countries to promote it.

H. ten Have and M. do C. Patrão Neves, *Dictionary of Global Bioethics*,
https://doi.org/10.1007/978-3-030-54161-3_216

Drugs

Drugs are chemicals that affect the body and the brain. The term is generally used for medicines such as pharmaceuticals. More specifically, the term refers to illicit substances that lead to abuse and addiction. Although four main types of drugs can be distinguished such as stimulants (cocaine), depressants (alcohol), opium-related painkillers (heroin), and hallucinogens (LSD), synthetic drugs are becoming more popular. Drugs can have a number of effects such as their impact on physical and mental health, they lead to social problems, and negatively influence social relations. In jurisdictions where they are illegal using drugs often leads to legal repercussions. In most societies the use and abuse of drugs have increased in recent decades. Repression, care, and decriminalization are the three policies used to address the drug problem. Repressive policies prohibit drug use and trade. Using a military metaphor a war on drugs has been declared. The argument for such a war is that drugs undermine social morality and heroin and cocaine are internal enemies against which all national forces need to be mobilized. Enormous budgets are made available for the police, legal procedures, and prisons. The second policy emphasizes care. Drug use is a public health challenge—not a legal one. Using a medical metaphor drug use is described as an epidemic that affects people whose resistance has diminished. Addiction is a pathological condition that needs adequate treatment and prevention—not punishment. The third policy is decriminalization in which drug use falls outside the legal sphere. Controlling the distribution of drugs has the aim of freeing society from the negative impact of illicit substances. Prominent world leaders have concluded in 2011 that the war on drugs has been lost. There has been an enormous growth in the number of drug mafias, there have been thousands of drug victims and deaths, and the judicial systems in many countries have been overwhelmed (half the prison population consists of people convicted of drug use or trade). Although most countries apply repressive policies, such policies are increasingly criticized. First, they are not effective in that the trading, disposability, and use of drugs is growing rather than declining despite enormous efforts. Second, such policies have serious negative impacts on the legal system. Moreover, corruption is rife and the methods legal authorities use often intrude into the rights of citizens. Third, the relationship between drug use and criminal behavior is still not clear. Criminality seems to be an effect of scarcity and the price of drugs being very high. Thus controlled distribution has the potential to remove the causes of criminal behavior. The problem can finally be reduced to defining drug use primarily in moral terms. Repression considers drug users as autonomous individuals who intentionally use these substances. They are therefore responsible not only for their use but also for harmful consequences. Identification, prosecution, and punishment are the ways to go. In response to these

H. ten Have and M. do C. Patrão Neves, *Dictionary of Global Bioethics*,
https://doi.org/10.1007/978-3-030-54161-3_217

criticisms several countries such as Portugal and the Netherlands have chosen a different approach. They regard drug use by and large as a medical problem—not a criminal one. Treatment, care, and prevention are better than punishment. Although the drugs themselves have not been legalized, the use and possession of small amounts are no longer prosecuted. The focus is on harm reduction such as making needle exchange programs available to intravenous drug users, providing a system of outpatient consultation services for detoxification, and making methadone substitution programs available.

Dual Use

Dual use refers to biotechnological products, vaccines, and microorganisms that can be used for both beneficial and hostile purposes. Although biotechnology is in principle a peaceful technology, today it is a potential source of risks for human security. A famous case was the creation of a new form of the avian flu virus that was transmissible between mammals. In 2011 a Dutch researcher submitted the results of experiments to *Science*. The National Science Advisory Board for Biosecurity in the United States decided that the manuscript should be redacted to make sure that details of the experiments would not be publicly available and could not be used by bioterrorists. Although misuse may be unintentional (i.e., accidental), it can also be deliberate, the risk for which has led to stricter measures and monitoring of biosecurity. However, the potential for accidental misuse is not often recognized and is one of the reasons greater emphasis is placed on better and increased ethics education of scientists and researchers. The international community adopted the Biological and Toxic Weapons Convention (BTWC) in 1975. This was followed by the Chemical Weapons Convention (CWC) in 1997. Despite both treaties banning entire classes of weapons some countries refused to sign up. Although both treaties control dual-use risks, ethical concerns regarding the dual use of biotechnology are directed at reaching a proper balance between scientific progress and freedom of inquiry, on the one hand, and security and community safety, on the other hand. Such a balance has to take into consideration a number of questions: How would scientists know whether their research will be used for hostile purposes? Should there be a general prohibition on some types of research (such as creating new viruses)? Should the precautionary principle be applied? At the very least scientists should be more aware of the potential misuse of their work.

H. ten Have and M. do C. Patrão Neves, *Dictionary of Global Bioethics*,
https://doi.org/10.1007/978-3-030-54161-3_218

Ebola

Ebola is a deadly disease caused by a virus. Ebola virus disease (a.k.a. Ebola hemorrhagic fever) starts with a fever, sore throat, muscular pain, and headaches and is then followed by vomiting, diarrhea, and rashes. Liver and kidney functions deteriorate and people begin to bleed internally and externally. Ebola kills approximately 50% of its victims some 6 to 16 days after symptoms appear. Direct contact with body fluids such as blood is the main source of contamination. Between 2014 and 2016 the world was hit by a major outbreak of Ebola in Guinea, Liberia, and Sierra Leone. First identified in 1976 in Central Africa, Ebola is a rare disease though not unknown. Before 2014 some 24 outbreaks were reported in such countries as the Democratic Republic of the Congo, Gabon, Sudan, and Uganda. Although such outbreaks were limited to a few hundred cases often in poor rural villages, the disease kills half of those affected. Nevertheless, it is not very contagious since it cannot be transmitted through the air like the flu. Countermeasures are known and involve isolating symptomatic patients, tracing contacts, and observing those contacts for symptoms (for 21 days). As long as contact with the body fluids of infected patients is avoided the spread of the disease can be contained. Although such a strategy was successful in countries such as Senegal and Nigeria, unfortunately there is no treatment or vaccine. In the 2014 outbreak 11,000 patients died and this raised ethical questions concerning treatment and prevention. Although experimental treatment was available (developed by the US army), it was scarce and only used for repatriated Americans. Prevention is limited despite the first vaccine now having been approved. The best way to prevent contamination is to avoid touching patients. However, this is difficult in African countries where burial practices often involve washing and touching the bodies of deceased family members. This raised an ethical issue concerning cultural diversity: Should traditional practices be respected? Management of the Ebola epidemic was an example of the failure of global governance. Although Médecins Sans Frontières (Doctors Without Borders) was quick to raise awareness, global responses were very slow or absent, ministries of health were unprepared, and policies were disorganized. It has only been since August 2014 when the personal tragedies of Western missionaries overshadowed the anonymous statistics of casualties that global responses were formulated. The US government labeled Ebola a security threat and decided to send thousands of troops to help build treatment centers. However, even a month later in September 2014 there was still no major UN response. Criticism was levelled at the lack of leadership of the WHO in being late in recognizing the extent and impact of the epidemic, in failing to take initiatives in the early stages, and in not coordinating relief efforts until a late stage of the epidemic. This raised a number of ethical issues such as the moral solidarity and responsibility of

H. ten Have and M. do C. Patrão Neves, *Dictionary of Global Bioethics*,
https://doi.org/10.1007/978-3-030-54161-3_219

humanity and the obligation to assist. Since August 2018 the second worst Ebola outbreak has been under way in the Democratic Republic of Congo and more than 2,000 cases have been identified. The epidemic is difficult to control since it is occurring in an active war zone. The WHO was initially reluctant to declare a public health emergency and assumed the epidemic could be controlled in the short term. An experimental Ebola vaccine is now available for use and in a major trial it has proved to be highly effective. In July 2019 the WHO declared the Ebola outbreak a Public Health Emergency of International Concern. The disease has now spread to Goma, a city of 2 million inhabitants on the border with Rwanda.

Ecocentrism (*See* Anthropocentrism; Biocentrism; Environmental Ethics; Zoocentrism)

The word "ecocentrism" derives etymologically from the Greek *oikos* (house) and *kentron* (center). It is a perspective that considers planet Earth as the house in which all creatures dwell. The suffix *-ismós* (Latin *-ismus*) expresses the general scope of the word to which it is added. Thus ecocentrism broadly refers to the different doctrines (inspired by Aldo Leopold's *A Sand County Almanac*) that extend moral worth not only to all forms of life but also to their habitats, ecosystems, and to the planet Earth itself. Furthermore, it considers the spontaneous dynamics of nature should be recognized as a universal rule to which human beings should comply. The roots of ecocentrism are commonly said to go back to European Romanticism (especially Rousseau). This was an intellectual and cultural movement of the eighteenth century that developed different theories that argued for the unity of nature and perceived the universe as a single unified and interconnected oneness. It led to the birth of ecology as the science of the interrelationships organisms have with one another and with their physical surroundings in the nineteenth century. Ecology has become the scientific foundation of environmental ethics. Nevertheless, it is possible to establish a connection between contemporary ecocentrism and ancient Greek naturalism according to which the intelligible principle that determines the passage from the passage from chaos to cosmos, the *arché*, the *logos*, is one and the same for the entire world and its inhabitants thus making nature the model for human moral action. This has been a fundamental characteristic of ecocentrism throughout human history even before the neologism was coined. Nature as a whole is considered to be a model for moral human action. Ecocentrism expresses the most holistic view of the moral domain and extends its borders beyond all forms of life to populations and species and from life itself to habitats and ecosystems. It also includes all elements such as soil and water. Although individual life-forms can be sacrificed to preserve the integrity of ecosystems, human interests should be subordinated to the protection of life and the sustainability of ecosystems on behalf of the ecological whole. Since ecocentrism argues that human beings are not superior to other forms of life, it is opposed to anthropocentrism and is broader than zoocentrism and biocentrism. Global bioethics should not be restricted to biomedical ethics; it should also include ecocentric perspectives.

H. ten Have and M. do C. Patrão Neves, *Dictionary of Global Bioethics*,
https://doi.org/10.1007/978-3-030-54161-3_220

Egalitarianism

Egalitarianism is a philosophical doctrine that is particularly important in sociopolitical and moral thought and characterized by the central role it gives to equality. However, it is not always evident what sort of equality people are referring to or is at stake (what it is that should be equalized). Such a lack of clarity lies at the root of the different types of egalitarianism such as economic or material egalitarianism (all citizens should have equal access to wealth) and legal egalitarianism (all citizens should be subject to the same laws). The most consensual statement about equality refers to the assumption that all people are equal in dignity and deserve equal rights and opportunities. This is currently the ground used to claim equal treatment across gender, religion, ethnicity, caste, economic status, and political beliefs. A second major idea related to equality and one that defines egalitarianism is the model of distributive justice (distributive equality) in which the distribution of goods should be equal among all members of society. Such a principle has been subjected to strong criticism because it does not prevent inequalities and can even deepen them. If citizens do not have equal social standing (income or wealth) or capabilities (life prospects), then distributive equality will not produce greater equality and should be balanced with other approaches to address equality as a goal. Both egalitarian principles (i.e., equal treatment and distributive equality) are of a normative nature. Historically, egalitarianism is relatively recent in that it was used by John Locke who argued that all persons should be equal and have the same rights, it was disseminated by the French Revolution with equality as one of its ideals, and it was used by Karl Marx who wanted to eliminate all social inequalities and to create a society of equals. In modern times John Rawls is the best known egalitarian advocating that all people should be treated equally and benefit from equal basic rights and opportunities. However, his theory does not result in equality. Rawls recognized that distributive justice would not always entail an equal distribution of goods. Equality is better achieved following other models of justice that take into consideration that people are not in equal concrete situations. Many other contemporary moral philosophers such as Ronald Dworkin, Thomas Nagel, Amartya Sen, Thomas Scanlon, and Derek Parfit share different aspects of egalitarianism. Moreover, egalitarian theory is an essential part of bioethics in that it advocates that healthcare (i.e., a basic healthcare package) should be accessible to all and some egalitarians even campaign for universal access to some social goods. Internationally and globally, egalitarianism plays an important role in encouraging the establishment of a minimum level of healthcare and of social conditions for people to have a decent life worldwide.

H. ten Have and M. do C. Patrão Neves, *Dictionary of Global Bioethics*,
https://doi.org/10.1007/978-3-030-54161-3_221

Electronic Patient Records

Traditional paper patient records are increasingly being displaced by electronic patient records (EPRs). They contain the usual information such as diagnoses, medications, interventions, medical history, and radiological images. Information stored on such data storage devices can easily be transmitted, received, and searched while allowing easy access for patients. The ethical issues related to EPRs concern security and privacy. Since the technical basis of information systems should be secure there is a need for the transmission technology to be appropriate, storage to be error free, monitoring to be continuous, and backup systems to be available. Moreover, it is imperative that the data stored and transmitted are protected to the greatest extent possible. Hackers have attacked hospital systems as a result of which information breaches have occurred. There is also the question of ownership. Often a distinction is made between the material and informational aspects of EPRs. The material aspect refers to the device on which the record is stored and kept. Although such a device is owned by the physician, institution, or corporation that is in charge of it, the data themselves are owned by the patient. However, such a distinction is not made uniformly. For example, in Europe data contained in EPRs belong to the patient, whereas in India they are owned by the physician or institution.

H. ten Have and M. do C. Patrão Neves, *Dictionary of Global Bioethics*,
https://doi.org/10.1007/978-3-030-54161-3_222

Electronic Surveillance

The word "surveillance" derives etymologically from the French verb *surveiller* (to watch, pay attention, check out, oversee, and control). Electronic technologies have allowed surveillance to be carried out from a distance. Although it is commonly used for national security and crime prevention and detection (e.g., closed-circuit television in public places and shops), it is much more widespread (as Edward Snowden revealed in 2013). Moreover, in countries like China it is commonly used to monitor citizens in everyday life and to award social credits or remove them from individuals. Such credits are used to facilitate entrance into schools, access to public services, and to allow travel. Electronic surveillance is rapidly developing in the field of healthcare (e.g., Google compiles health information). People who experience symptoms usually search for information on the Internet. This allows the Google search engine to determine patterns and analyze search terms such that statistical models for the spread of diseases can be created. Remote electronic surveillance is also used in telecare (i.e., personal alarms and security systems). Important ethical issues are raised by electronic surveillance such as privacy and confidentiality since the technology has an impact on individual autonomy. Online public health surveillance should make sure data are not identifiable. Telecare surveillance can be more problematic in focusing on specific individuals who, aware they are being continuously monitored, may start behaving differently and avoid conduct that may raise alarm. Surveillance may also remove the freedom of patients to take or not to take medication because they fear such monitoring may reveal the contrary.

H. ten Have and M. do C. Patrão Neves, *Dictionary of Global Bioethics*,
https://doi.org/10.1007/978-3-030-54161-3_223

Emergency Medicine (*See* Triage)

Emergency medicine (EM) is a medical specialty defined by the International Federation for Emergency Medicine in 1991 as "a field of practice based on the knowledge and skills required for the prevention, diagnosis and management of acute and urgent aspects of illness and injury affecting patients of all age groups with a full spectrum of undifferentiated physical and behavioral disorders." EM is used to care for patients with serious and life-threatening injuries primarily because often little information (if any) is available about such patients and frequently no diagnosis can be made for prompt and appropriate treatment to be provided without knowledge of preexisting conditions. Moreover, critically injured patients are rarely able to communicate thus making it impossible to identify or contact families or friends. Such circumstances are important challenges for the beneficence and autonomy of patients. The emergency triage system is ethically very important since it is used to prioritize the most serious patients. It is a system that prioritizes the severity of injuries in an attempt to save as many lives as possible and provide relief to all affected. Such a system differs from the standard triage system in healthcare in that the latter provides healthcare as people request it on a first-come, first-served basis. However, neither of these triage systems can be applied at times of scarce resources such as happens during war, natural disasters, or pandemics. Rationing and rationalizing are two other triage systems that are used for the most extreme situations. Rationing refers to the distribution of limited resources by following specific criteria that focus on characteristics of the patient such as age, gender, and nationality. Such criteria are ethically required to be transparent. Rationalizing refers to the most rational, logical, and reasonable use of limited resources using the single criterion of making the most of them. The goal is optimizing the resources available and making them as efficient as possible. Rationing focuses on people and their characteristics, while rationalizing focuses on resources and their performance. This is the reason rationalization is the only ethical criterion used in life-and-death decisions such as providing a mechanical ventilator to someone suffering acute respiratory insufficiency. Rationalization does not value individuals over others, optimizes scarce resources, and does not violate the principles of human dignity and social justice. Although EM is strongly patient centered, emergency healthcare professionals are sometimes called to work together with paramedics, social workers, and the police.

H. ten Have and M. do C. Patrão Neves, *Dictionary of Global Bioethics*,
https://doi.org/10.1007/978-3-030-54161-3_224

Emerging Infectious Diseases

Since 1980 more than 35 "new" infectious diseases have emerged in humans (i.e., 1 every 8 months). "New" refers to pathogens new to science and known agents that have crossed borders as a result of globalization. Infectious diseases are now regarded as a growing global threat. They strongly demonstrate the impact the environment has on the interconnection between the individual, environmental, and social determinants of health. Although infectious diseases originate in a specific location and are often endemic, as a result of globalization they rapidly spread across borders. Local and national responses will not be enough to diminish global risks. Emerging infections are due to either new pathogens or known infectious agents. They are termed "emerging" because their incidence or geographic range is increasing. Known diseases such as cholera, malaria, and tuberculosis are resurging unlike the past when they were less frequent and more controlled. The greatest concern is focused on diseases caused by a new pathogen such as HIV, Zika, Ebola, or COVID-19. Although a range of explanations have been put forward, it is clear that both human activities and behaviors have been instrumental in the emergence and pandemic spread of infectious agents. Emerging infectious diseases demonstrate the importance of not separating biomedical and environmental ethics. Pandemics of avian influenza, Ebola, Zika, and COVID-19 are the price human interventions such as deforestation have to pay for the sake of economic development. Such interventions reduce biodiversity and bring human beings into closer contact with wildlife and its reservoir of pathogens. It is estimated that 60% of emerging infectious diseases are caused by zoonotic pathogens. Of these zoonoses almost 72% are caused by pathogens that have a wildlife origin. Until recently infectious disease has not been a major topic of concern in mainstream bioethics. Whatever ethical debate there has been focused on balancing private and public interests. Since infected patients are both victims and vectors of disease, the challenge becomes how to weigh respect for individual autonomy *versus* the common good (e.g., regarding vaccination and quarantine). The emergence of new diseases coupled with the resurgence of ancient ones have placed new concerns that require bioethics to move beyond the perspective of individual interests on the ethical agenda. It is generally acknowledged that most emerging infections are zoonoses. Although birds, bats, and pigs are natural reservoirs of viruses that can lead to animal diseases and be transmitted to humans, in some cases such viruses acquire the capacity for person-to-person transmission. This raises the question of why animal diseases develop? The wet markets in southern China are often blamed for avian flu. In addition to this being a result of the industrial revolution and growing prosperity in this region, China is the world's largest producer of pork. Globally, 21 billion animals are reared annually to feed the world population. The

H. ten Have and M. do C. Patrão Neves, *Dictionary of Global Bioethics*,
https://doi.org/10.1007/978-3-030-54161-3_225

poultry industry rears and provides 6.5 billion birds annually and this is of especial importance when it comes to avian influenza. The rapidly increasing consumption of meat and milk in developed countries has resulted in a livestock revolution and global demand for meat has encouraged the growth of factory farming on an industrial basis. Meat production is estimated to double by 2050. This demand has primarily benefited large agribusinesses. Although factory farming accounted for 30% of world meat production in 1990, now almost all chickens, turkeys, and pigs are reared intensively by factory farming. Thus animal farming has significantly changed and large numbers of animals are grown today in an industrial manner, mostly indoors, and intensively. Since antibiotics, vitamins, hormones, and vaccines are used to improve production it goes without saying that this way of livestock farming makes it easier for viruses to replicate, reassert, and spread. Most avian flu outbreaks occur on industrialized farms that provide the ideal environment for the evolution of more virulent virus strains that have the potential to become influenza pandemics. Although animal health and human health are connected, mainstream bioethics rarely criticizes the industrialization of animal farming (leaving it to other disciplines of applied ethics such as animal ethics or environmental ethics). Looked at through the lens of global bioethics the connection between animal and human health can no longer be ignored. Now that industrialized farming has become one of the sources for the emergence of infections the time is ripe for bioethics to turn its focus to the role played by agribusinesses in this.

Emerging Technologies

Emerging technologies refer to the most recent and most impactful technologies of our time. The context of emerging technologies is that of development that is increasingly guided by technological innovation (rather than by the production of knowledge) and by development that is at its most fundamental level enhanced by the convergence of different technologies. Although scientific and technological advances are today no longer solely dependent on specialization in ever narrower and deeper fields, they are still mostly built on the ability to combine various technologies by enhancing the ways in which they converge. This is also the reason identifying emerging technologies that converge is not always easy. Although new technologies are constantly emerging, some do not develop from their early stages. There are analysts who consider emerging technologies as those whose usefulness is still dubious and controversial but still include such developments on their lists thus greatly widening them. Adopting a narrower approach by considering the huge potential impact of new technologies such that they can be classified as emergent (excluding those that have not yet proved themselves), there are four domains of innovation that are widely accepted as standing out: nano (nanoscience and nanotechnologies), bio (biotechnologies and biomedicine), info (information technologies), and cogno (cognitive sciences). New technologies raise many different ethical questions according to the field in which they emerge and according to their goals, procedures, and potential impacts. Nevertheless, there are transverse ethical concerns that should be stressed. One refers to the high expectations they create by promising to address people's needs such as healthcare or well-being (in general) or by attracting financing without being able to deliver what seemed certain. Moreover, emerging technologies have helped to spread the conviction that technology can solve all human problems such as those contributing to the conquest of immortality and those that former technologies created themselves (mostly applicable to the environment and climate change). Another transverse ethical problem is the restricted access they grant for a number of reasons such as inequalities between the countries where they are produced and prices that are often unaffordable thus deepening global injustice.

H. ten Have and M. do C. Patrão Neves, *Dictionary of Global Bioethics*,
https://doi.org/10.1007/978-3-030-54161-3_226

Empathy

The word "empathy" derives etymologically from the Greek *en* (in, at) and *pathos* (feeling, suffering). It refers to the psychological capacity to feel what another person is feeling, to sense that person's emotions, and to imagine or understand what others are going through and what and how they are feeling. Empathy is deemed a virtue (i.e., a highly desirable predisposition to be and act in a specific way). It is the capacity and willingness of someone to get close to others and show them that he or she shares their emotions thus helping them to bear negative feelings and celebrate positive ones. Virtues such as empathy are encouraged as highly desirable character traits, promoted as highly important in professional practice (mostly in relational or interpersonal professions), but cannot be imposed as an obligation or a duty to be fulfilled. The capacity to sense how others are feeling brings people together, tightens personal relations, and strengthens social bonds. This is the reason empathy has become so highly valued (mostly since the eighteenth century). Philosophers David Hume and Adam Smith strongly emphasized the importance of such virtues for moral and social life. Although these philosophers used the word sympathy—not empathy, its definition corresponds to what is currently designated as empathy. Sympathy defines a natural affection for or an emotional attraction to someone, while empathy is the capacity to put oneself in the other's place and feel what that person is feeling. Empathy is highly valued in healthcare as a means to bridge the gap with the patient and contribute to care that is more humane and personalized. Empathy is considered a core value in nursing. However, when empathy is too deep and intense it can lead to the professional suffering (*pathos*) with the patient and the family, losing professional objectivity to deal with the concrete situation, and contributing to professional burnout as a result of overburdening themselves emotionally.

H. ten Have and M. do C. Patrão Neves, *Dictionary of Global Bioethics*,
https://doi.org/10.1007/978-3-030-54161-3_227

Engineering Ethics

Engineering ethics refers to that branch of applied ethics that examines the actions of and decisions taken by engineers. It emerged as a special field of ethics in the 1980s and focuses on professional ethics in which the responsibilities and duties of engineers, which are often formulated in codes of ethics, are addressed and on individual ethics in which integrity and honesty are addressed. Codes usually deal with the competence of and obligations engineering professionals have to the public such as health, safety, and welfare. Ethical issues in engineering relate to whistleblowing, loyalty, conflicts of interests, risks and safety, and environmental concerns. Whistleblowing is important in that engineering ethics demands engineers have the necessary technical competence and specialized skills such that there is no conflict between the public interest and the interests of employers. Engineers are also obliged by codes of ethics to act as faithful agents to their employers and clients and such an obligation may conflict with the public interest. Conflicts of interest may arise when personal and professional interests clash (e.g., brought about when gifts or worse bribery are involved). Safety and risks are important issues since engineers have a professional responsibility to ensure safety and human health. Examples include the use of inferior materials to construct a bridge and the use of alternative cheaper technologies that may increase risks. Environmental concerns have more recently found themselves high on the agenda of engineering ethics. Engineering projects have long paid little attention to sustainability.

H. ten Have and M. do C. Patrão Neves, *Dictionary of Global Bioethics*,
https://doi.org/10.1007/978-3-030-54161-3_228

Enhancement

Enhancement (more precisely human enhancement) refers to improving human capacities and traits according to designated goals and criteria. At such a general level it could be argued that all educational projects contribute to human enhancement in which case human enhancement has a positive connotation. However, the expression human enhancement currently refers specifically to improving the biopsychological identity (e.g., appearance or memory) of humans and their performance at all possible levels by advances in biotechnologies or assisted by current biotechnologies. Biotechnologies are no longer used for their original therapeutic goal. They are specifically applied such that humans can have capacities and reach performances that are so high that unenhanced human beings cannot possibly reach and that are significantly superior to human nature. Human enhancement refers then to designing (or redesigning) humanity in an effort to inaugurate a transhuman era. Enhancement has become possible as a result of scientific developments in genetics (especially genetic engineering). This is not limited to curing genetic disorders but aims at improving performance thus transforming therapeutic genetics into enhancement genetics. Although human enhancement is today pursued through a sequence of biotechnological interventions that start with the morphologic selection of gametes, embryos chosen by preimplantation diagnosis, pharmacological products applied during the lifespan, blood transfusions and stem cells injections (to live longer), and culminate with new synthetic drugs (alterations to biochemistry to increase resilience to work or to the environment), in the near future it will also be possible to enhance desirable human characteristics by genome editing (CRISPR/Cas9), by improving performance by artificial transplantation, by nanotechnologies, by computer interfaces (brain modifications), and by getting different fields of emergence sciences in a convergent logic. Although some consider the process of human enhancement started long ago with the simple use of spectacles and others more recently with cochlear implants, today convergent technologies just proceed along the same line developing and extending what has long been done. Although this is a major argument in favor of enhancement in that it advocates individuals and societies should invest in human enhancement using all biotechnological resources available, there are those who stress there is a difference between enhancement, which they reject, and human flourishment, which they advocate. When it comes to altering or perfecting human nature, they argue there is a

H. ten Have and M. do C. Patrão Neves, *Dictionary of Global Bioethics*,
https://doi.org/10.1007/978-3-030-54161-3_229

world of difference between qualitative improvement (adding something foreign to humans) and quantitative development (developing everyone's potential). They argue that it is not up to humanity to destroy humanity by artificially creating a new species that, by its very nature based on the infinite acquisition of more power, would destroy moral values such as social justice and global solidarity on which our contemporary societies are built.

Environmental Ethics (*See* Ecocentrism)

Environmental ethics is a recent branch of moral philosophy that has been systematically developed since the 1970s. It is concerned with the moral relationship between human beings and nature and considers the value and moral status of living beings, ecosystems, and the biosphere. This relationship has changed over time from prehistoric times when humans had a mythical mentality and first gained awareness of themselves as natural beings and an inseparable element of nature (syncretism) to Ancient Greek philosophy when reason predominated and humans beings discovered the existence of a universal law that rules the entirety of nature including humans as natural beings and does so in perfect harmony. Later when monotheistic religions predominated humans started to see themselves as created by God in his image and as spiritual and superior beings. Nevertheless, nature continued to be cherished since it was believed to be God's creation too. It was mostly with the onset of modern science that the distance between humans and nature grew as a result of knowledge requiring a separation between who knows (i.e., the subject) and what there is to be known (i.e., the world). Since that time science has progressed and the separation has continued to grow further deepened by technology and how the latter mediates in its relationship with reality. Human beings became spectators of nature much the same as if they did not belong to it, using it, transforming it at will, and dominating it according to their own ends and regardless of any other considerations. It was only recently that this situation began to change when people became all too aware of environmental degradation, the limitations of natural resources, the impoverishment of biodiversity, and the risk for humankind's very survival. This marked the beginning of environmental ethics in which humans were urged to review their relationship with nature, their identity by recovering their natural dimension, the value of nature, and to recover their perception of its intrinsic value. Aldo Leopold in *A Sand County Almanac* (1949) was one of the first scholars to stress the need to develop what he called "land ethics" (i.e., recognizing human moral responsibility toward the natural world that included not only humans, animals, and plants, but also natural elements such as soil or water). The significant development and expansion of ecology as a science since then has allowed environmental ethics to develop and follow two main streams. The first developed from traditional ethics (centered on humans), extended its scope to the environment, argued that the value of the environment went far beyond simple instrumental values, attributed moral duties to humans regarding nature, and thus promoted environmental sustainability. The second argued for a break with anthropocentric ethics and proposed ecocentrism, which assigned an intrinsic value to nature and considered it an integral part of the moral community. Although both streams cover a multitude of different viewpoints, the ecocentric stream is arguable more important in that it is

H. ten Have and M. do C. Patrão Neves, *Dictionary of Global Bioethics*,
https://doi.org/10.1007/978-3-030-54161-3_230

more centered on life-forms and ecosystems thus allowing a holistic view of the biosphere. From an anthropocentric point of view, different perspectives have also recently arisen such as the technological approach (i.e., investment in new technologies to solve environmental problems). Over the last half century environmental ethics has become part of civic education worldwide and well supported by environmental authorities and legislation (local, national, and international in scope; governmental and individual in nature; and subject to hard and soft law). Environmental authorities pressed for its creation and development. The environment was the first issue to be recognized as a global problem demanding a global solution and relying on a global ethics. The UN Environment Programme (UNEP) created in 1972 was the first worldwide agency for the environment, one of the co-founders of international cooperation, and the funding partner of many initiatives in the field. It is up to global environmental governance to set an example by recognizing new problems such as demographic growth as global.

Environmentalism (*See* Ecocentrism)

Environmentalism refers broadly to a wide variety of environmental doctrines ranging from the more radical deep ecology (fighting for a break with the current civilization model) to a more conventional environmental sustainability (accommodating different interests in society). Although environmentalism has been developed at the social level, it pursues political transformations. Such a social movement and ideology criticize modern Western civilizations, especially the way they organize production, the primacy they give to economic values, and the incentives they give their populations to consume. In addition to affecting the Earth's balance and aggravating old and raising new environmental problems, they also aggravate traditional international injustice, create new social vulnerabilities, and deepen poverty worldwide. Although environmentalists are concerned about the impact modern society and human activity have on the environment such as pollution, depletion of natural resources, and climate change, they are also concerned about the impact of environmental changes on human communities such as health problems (e.g., respiratory problems or the spread of endemic diseases) and more recently environmental refugees. Consequently, environmentalism has triggered political initiatives such as the creation of green parties (mostly in Western countries) and NGOs. Although both advocate preservation of the environment by getting humans to change their behavior toward nature, they are also aware that environmental disasters affect, sometimes dramatically, populations that did not contribute to it. Both argue the major threat to global environmental sustainability and international human justice is the way in which modern societies are organized and their dynamics and that changes at the cultural, economic, and political levels have to be made to alter in a fair and equitable way the relationship between human beings and nature. The image of environmentalism today is a far cry from the one it had of radical protests and strategies of direct action in the 1960s and 1970s. It now engages with governments, organizations, and businesses in common actions to bring about a more beneficial and effective outcome for the environment and for society. It also stresses the importance of environmental sustainability as a factor in social development and economic growth (i.e., investment in clean technologies and creation of green jobs). Thus environmentalism requires interdisciplinary approaches to environmental problems, concerted action by stakeholders, and finding new solutions whose scope is global at the political, economic, and social levels. The very complexity of environmental issues illustrates the growing extent to which efforts have converged.

H. ten Have and M. do C. Patrão Neves, *Dictionary of Global Bioethics*,
https://doi.org/10.1007/978-3-030-54161-3_231

Epidemics (*See* Epidemiology)

Infectious diseases can be endemic, epidemic, or pandemic. An epidemic is the rapid spread of an infectious disease to a large number of people in a given population within a short period of time (e.g., measles in a specific country or region). An endemic disease is always present in a population or region (e.g., malaria). A pandemic is an epidemic that has spread across a large region, many continents, or worldwide (e.g., COVID-19). The difference between an epidemic and a pandemic is one of scale. In 2003 an outbreak of severe acute respiratory syndrome (SARS) occurred in Asia killing 800 people. This was an epidemic since it had no global impact. HIV/AIDS, on the other hand, is a pandemic since it is global in nature. Although epidemics have occurred many times during human history, today some diseases (e.g., smallpox) have been eradicated while others have emerged (e.g., avian flu). Epidemics raise a number of ethical questions such as the threat they pose to public health since they are communicable and have serious health consequences. This means that the traditional focus of bioethics on individual interests will not be enough to control epidemics and that individual freedom may have to be limited (e.g., by quarantine measures and isolation of infected individuals) for the sake of public health.

H. ten Have and M. do C. Patrão Neves, *Dictionary of Global Bioethics*,
https://doi.org/10.1007/978-3-030-54161-3_232

Epidemiology

Epidemiology is the branch of medicine that studies the distribution, patterns, and determinants of health-related conditions and the control of disease. It plays a major part in public health. The WHO defines epidemiology as "the study of the distribution and determinants of health-related states or events (including disease), and the application of this study to the control of diseases and other health problems." Epidemiological research involves using various methods such as surveillance and descriptive studies to study distribution and analytical studies to study determinants. The results can be used to plan and evaluate strategies to prevent illness and to guide the management of patients with particular diseases. Ethical issues in epidemiological research relate to public health in that they raise concerns about privacy, about protecting individual rights while protecting public health, and about how to avoid stigmatization. The potential harms and risks of epidemiological research should be minimized by guaranteeing the confidentiality of information (especially when vulnerable populations are involved).

H. ten Have and M. do C. Patrão Neves, *Dictionary of Global Bioethics*,
https://doi.org/10.1007/978-3-030-54161-3_233

Epigenetics

The word "epigenetics" was first used by the geneticist Conrad Hal Waddington in 1942. Use of the Greek prefix *epi* (beyond, above) allowed epigenetics to present itself as a new science as going beyond and above, coming after, and revising and complementing knowledge garnered from genetics. Since mapping of the human genome did not deliver on its promise of providing a full understanding of all the genes of human beings it became ever more clear that genes did not explain each person's reality. Genomics and proteomics followed and deepened knowledge of the structure of genes, the way in which they interacted, and the variety of proteins produced by cells. Nevertheless, much more needed to be understood about the expression of genes such as the mechanism by which they produce phenotypic effects (How can a genotype produce multiple phenotypes?) and transmit them to the next generation. Epigenetics is the study of the way environmental factors such as nourishment, stress, pollution, and climate influence the expression of genes and how such expression can be reversed (acquired traits can be transmitted to the next generation and inherited traits can be lost). The advent of epigenetics led to a new vision of life and abruptly ended the conviction that all life was written in the genes. Epigenetics research led to the conclusion that the evolution of living organisms and their behavior is not a direct consequence of genes since the latter express themselves differently according to environmental factors. This was absolutely revolutionary in that it freed humanity from genetic dictatorship, denied genetic determinism, and returned power (individual freedom) over the future to each individual. Human beings can control and be responsible for who they become making this arguably the most significant consequence of epigenetics. Moreover, epigenetics increases the responsibility to future generations as a result of the link between their behavior and the impaired health of children (transgenerational inheritance of acquired epigenomic signatures). Concerns about eugenic temptation (epigenetic mechanisms that take place during embryo development) and new ways of discrimination (epigenetic marks) have been raised. Epigenetics is also paving the way to new treatments, alternative therapies for genetic diseases, better understanding of mental disorders and ageing, and interesting approaches to cancer.

H. ten Have and M. do C. Patrão Neves, *Dictionary of Global Bioethics*,
https://doi.org/10.1007/978-3-030-54161-3_234

Equality

The word "equality" derives etymologically from the Latin *aequalis, aequalitas* (equal, uniformity of proportions, symmetry, harmony). It designates a qualitative relationship in which a group of different items share at least one quality or one common feature under which they are equal. Although often used in daily life, equality is a controversial concept within theories of justice. Historically, the concept of equality is already present in Greek philosophy in the word *isonomia* (equality of rights). In modernity (eighteenth century to 1930) Hobbes referred to the concept by saying that all human beings are equal only in their risk for death. However, the concept only became widespread when equality together with liberty and fraternity became the leading social ideals of the French Revolution. Since then equality has played an important role as a strategy to achieve fairness, most particularly in contemporary theories of justice. Equality expresses the conviction that all citizens are equal and should be treated in the same way. Although it is the main principle of egalitarian theories, critics argue that different people are rarely in the same social position in real life. Therefore, even if they were all treated the same social imbalances would still persist. It is important to acknowledge John Rawls' contribution to enlightening the meaning of equality. His first principle of justice states that "each person has the same indefeasible claim to a fully adequate scheme of equal basic liberties, which scheme is compatible with the same scheme of liberties for all." His second principle of justice stated that all persons should have "fair equality of opportunity" in society. The consensus today is that the obligation of equality (i.e., equal rights) should be applied to basic rights and opportunities (as a principle advocating equal basic liberties) and that it should be the same for all (e.g., women and men). It is also argued that equality should be the intended outcome of the implementation of justice (e.g., if women and men have the same rights and opportunities, then they should both do equally well in society). Equality in the field of healthcare requires that a basic healthcare package is available to all citizens.

H. ten Have and M. do C. Patrão Neves, *Dictionary of Global Bioethics*,
https://doi.org/10.1007/978-3-030-54161-3_235

Equity

The word "equity" derives etymologically from the Latin *aequus, aequitas* (equal, uniformity, tranquility of soul, impartiality, evenness, justice) and evokes feelings concerning moral dimensions. It refers to a theory or strategy for the implementation of justice in which application of the law or the distribution of goods takes into account the particular needs of the different persons involved. The presupposition for the claim of equity is that people are not really equal when it comes to their social and economic standing or genetic (physical or intellectual) constitution. Therefore, to benefit from the same opportunities and to reach the same level of achievement (equality) they need differentiated social support (equity). Historically, such a concept is already present in Greek philosophy in the word *epieikeia* (adequacy, equal distribution of power and revenues). Plato and Aristotle refer to equity as a kind of practical wisdom concerning application of the law that attends to the particular circumstances of those to whom it applies. This so-called equity tradition stresses that rigorous application of the law regardless of the circumstances of a particular case can sometimes result in greater injustice. In recent times John Rawls recovered and strengthened the importance of equity, specifically in his second principle of justice. This focused on social and economic inequalities and introduced the difference principle that states that the greatest benefit should be given to the least advantaged members of society. This second principle of justice considers that equality (sameness) is not enough for justice to be realized since equity (fairness) is also crucial. In the field of healthcare the WHO states, "equity is the absence of avoidable, unfair, or remediable differences among groups of people, whether those groups are defined socially, economically, demographically or geographically or by other means of stratification. 'Health equity' or 'equity in health' implies that ideally everyone should have a fair opportunity to attain their full health potential and that no one should be disadvantaged from achieving this potential."

H. ten Have and M. do C. Patrão Neves, *Dictionary of Global Bioethics*,
https://doi.org/10.1007/978-3-030-54161-3_236

Ethicists

Ethicists are experts in ethics such as academic scholars who mainly focus on research and the teaching of ethical theories (philosophers specialized in ethics) and practitioners engaged in the application of ethics to different realms of social and professional activity (advisors or consultants in codes of ethics and professional ethics dilemmas). Neither of these professions should be autonomous in that the ethics professional should have a good knowledge of ethics as a philosophical discipline (concepts, history, and theories) that they can use to ground and develop rigorous reasoning and put forward sound proposals to solve ethical dilemmas. Academics should also apply ethical theories to real problems and in so doing demonstrate the importance of ethics in everyday life. An ethicist should have a good understanding of social sciences (particularly of ethics) and the social professional field to which ethical reasoning is applied, together with good methodological skills of case analysis. Unfortunately, this is not always the case. The term "ethicist" only recently became common (in the 1980s) and was restricted to the field of applied ethics. This led to it being narrowly understood as designating a professional of ethics. The first ethicists started work in clinical settings in the United States as consultants. Depending on the setting in which they work, ethicists are tasked with providing ethical education for staff (focused on principles and theories to ground and guide the decisions to be made), helping them to adopt rational, coherent, and valid reasoning in the deliberation process, and assisting them (without taking over) in decision-making. Two major criticisms are directed at ethicists. The first is that their very existence and the kind of work they develop tend to technicalize ethics (i.e., reducing it to a mere technique) and thus the most relevant contributions of ethical reflection to contemporary societies are lost. A second criticism emphasizes that ethics can never be the domain of a professional activity without losing its specificity. However, good ethicists have proven their worth by preventing ethical problems from arising and helping to understand and solve them when they do occur.

H. ten Have and M. do C. Patrão Neves, *Dictionary of Global Bioethics*,
https://doi.org/10.1007/978-3-030-54161-3_237

Ethics

The word "ethics" derives etymologically from the Greek *ethos*. However, in Ancient Greece *ethos* had two spellings: *êthos* in which the first vowel was long was used by pre-classical authors such as Homer to mean a shelter for animals; and *éthos* in which the first vowel was short was used by Aristotle to mean habit or custom. When the two words were later translated into Latin, both spellings of *ethos* were translated to *mos, mores* (manner, custom, usage, habit; i.e., the etymological root of morals). The archaic meaning of *ethos* eventually disappeared and ethics and morals became synonymous and have been used as such for centuries. Recently Heidegger recovered the meaning of the most ancient spelling of *ethos*, adopted the use given it by the pre-Socratics, and presented it as referring to human interiority or intimacy from where actions originate. Following Heidegger's reasoning, ethics refers to the foundation of human action (the reason behind acting in a specific way or its rationality), whereas morals refer to the set of norms for human action (guidance to act). In other words, morals teach people how they should act (i.e., they should not lie), while ethics justifies the reason they should follow moral rules when they act (i.e., lying destroys trust). According to Heidegger ethics grounds morals. Although ethics and morals can be used etymologically and historically as synonymous terms, several authors make a distinction between both concepts. Attributing different meanings to ethics and morals is desirable because such a distinction helps precision of thought and the accuracy of speech: ethics is a descriptive domain that has a foundational nature, whereas morals is prescriptive and normative. Ethics is a philosophical discipline that was first structured and systematized in Ancient Greece by Aristotle in the third century BC. Aristotle's object was *praxis* (practice, human action) that refers to the kind of action that changes the nature of the agent (e.g., the action of helping others makes the agent a supportive person). Although praxis is adequate when it comes to rational critical analysis in attending to sound reasoning and logical development, in being coherent and unitary, and in duly assuming consequences, the specific domain of ethics is human relationships. Restricting ethics to the interpersonal domain prevailed throughout the centuries until recently. In the second half of the twentieth century the scope of praxis greatly widened at two different levels: the first, from actions towards other person to human action to all other living beings, their habitat, the ecosystems and the planet; the second, from the closeness of space and time (*hic et nunc*) and the consideration of direct and immediately consequences of actions to the concern of their distant effects. Currently, the field of ethics has widened considerably.

H. ten Have and M. do C. Patrão Neves, *Dictionary of Global Bioethics*,
https://doi.org/10.1007/978-3-030-54161-3_238

Eugenics

The word "eugenics" derives etymologically from the Greek *eu* (good) and *gene* (birth) and thus literally means well born. It was used in ancient times (notably by Plato) to value the birth of something positive over the birth of something negative according to criteria then in force. It continued to be used in this way despite the difficulty of identifying desirable and undesirable human traits persisting until today. The idea that some are well born while others are not and that the birth of the latter should be discouraged while the birth of the former should be promoted is ancient and has been differently expressed over time and throughout the world as a philosophical, political, social, and cultural conviction. In modern times eugenics was converted from an ideology to a scientific theory and from a conviction to a certitude. This has been the case since 1883 when Francis Galton first coined the word eugenics to represent a science aiming at improving racial stock (moving from archaic to classic eugenism). Galton combined his heredity theory with the evolution theory of Charles Darwin to establish a program to improve the human race by encouraging marriages between people considered physically and mentally superior. During the first half of the twentieth century eugenics was indeed considered a technoscience aiming to improve the genetic heritage of a race and a supposedly successful strategy to increase competition between nations. Many eugenic scientific societies, symposia, and journals were created in different countries and eugenic policies advocating compulsory sterilization were implemented in some countries as Switzerland, the Scandinavian nations, and the United States. During the Second World War Nazi eugenicists pushed these programs to an unprecedented level when they exterminated millions of people from the disabled to Jewish and gypsy people. This led to classical eugenics gaining a very negative connotation that it has never been able to shake off. To make matters even worse there have been cases of ethnic cleansing and genocide around the world that have also been classified as eugenic. As a result of what has happened since the beginning of the twentieth century there are scholars today who argue there is a distinction between eugenics as a technoscience whose purpose is the reproduction of good genes and eugenism (a term introduced by F. Galton) whose purpose is the reproduction of bad genes. Eugenism is an ideological movement accused of freely interpreting scientific information to justify an agenda of racial discrimination and segregation. Indeed, it was with a eugenic goal (and rejecting eugenism) that the Repository for Germinal Choice was created in 1989. This was a sperm bank to which Nobel Prize laureates supposedly donated between 1980 and 1999 (when it was closed down). Reproductive technologies today have chosen to reject such a distinction by simply discarding both classifications and instead selecting gametes and embryos initially according to therapeutic criteria and later social criteria. Other

H. ten Have and M. do C. Patrão Neves, *Dictionary of Global Bioethics*,
https://doi.org/10.1007/978-3-030-54161-3_239

technologies such as prenatal diagnosis or preimplantation diagnosis (followed by selective abortion) have also been classified as eugenics. Despite its negative connotations eugenics still has its supporters (and critics). Supporters point out the advantage of using biotechnologies to improve humanity by selecting the best traits of individuals in attempts at eradicating some diseases, and in light of the fact that natural selection no longer works at the human level. Voluntary sterilization (for Down syndrome in Ireland) or compulsory sterilization (premarital genetic tests for infectious and blood-transmitted diseases in Dubai) are practiced. In both cases the number of sick children has very significantly decreased. Critics consider any kind of human selection a violation of human dignity.

Euthanasia, Active

Active euthanasia commonly refers stricto sensu to the act of putting someone to death to free that person from suffering. Classifying euthanasia as active indicates that an effective action not only exists but will also cause death. This only makes sense in the context of comparing it with passive euthanasia where no such action is involved. Conceptually, active euthanasia adds nothing new of significance to euthanasia and does not change the nature of this action in any way. It is misleading in that it mistakenly leads to the idea that there are different kinds of euthanasia, which is not the case. Euthanasia and active euthanasia refer exactly to the same action: to kill, at request, to avoid suffering. Although it is commonly acknowledged and consensually accepted ethically that there are different levels of moral responsibility between actions (active) and omissions (passive), euthanasia *always* refers to an effective and deliberate action. Describing euthanasia as active or passive is often justified as a social and political strategy in favor of its decriminalization against an adverse historical background. Once so-called passive euthanasia is accepted, it is one step nearer to acceptance of active euthanasia. Although such an approach no longer seems to be pertinent (at least in Western countries), conceptual accuracy should always prevail in honest and transparent debates.

H. ten Have and M. do C. Patrão Neves, *Dictionary of Global Bioethics*,
https://doi.org/10.1007/978-3-030-54161-3_240

Euthanasia, Concept

The word "euthanasia" derives etymologically from the Greek *eu* (good) and *thanatos* (death) and thus literally means good death. Euthanasia refers to the act of putting someone to a "good death" in the absence of any legal sanctions to do so to end a state of suffering considered by the patient as worse than death. This is the reason it is also called mercy killing. Euthanasia as a concept is very complex in that it covers a wide variety of situations all of which have to be legally established where this practice is allowed. It raises many questions: Should it only apply to terminally ill patients? Should it apply to everyone who has an incurable disease? Should it apply to those who claim to be suffering unbearably? Should euthanasia only be requested by the patient? Should a proxy be allowed to request it on his or her behalf? Should it only be available to competent persons? Should psychiatric patients or children be allowed to request it? Should it only be performed by medical doctors or healthcare professionals? Should family members be allowed to carry it out? Should a special profession be set up to carry it out? Should it only take place in healthcare facilities? Should it be allowed at the patient's home or other designated places? The answer to each of these questions will have to take on board specific and important aspects of the concept of euthanasia each entailing different ethical considerations. Euthanasia can be summarized as putting someone to death effectively (in a way that is certain and does not fail as can happen in a suicide attempt) and efficiently (in a painless way).

H. ten Have and M. do C. Patrão Neves, *Dictionary of Global Bioethics*,
https://doi.org/10.1007/978-3-030-54161-3_241

Euthanasia, General

Euthanasia is one of the most challenging issues in bioethics. Although it has been open to debate both before and since the advent of bioethics, it remains controversial and arguably one of the most important topics of bioethics. Contemporary debate on euthanasia began as a result of the unprecedented development of scientific knowledge, biotechnological innovation, and the way in which they were applied to human beings (mainly to patients). At the level of diagnosis and treatment the benefits of medicine have been and still are very significant in improving outcomes for many different pathologies. However, there are disadvantages that have become ever more clear. Although state-of-the-art medicine is sometimes able to control the progress of disease and postpone imminent death, sometimes it is incapable of restoring health and curing diseases. Invasive therapies often inflict a significant level of pain and suffering and sometimes turn out to be futile. Furthermore, a new logic of action developed between healthcare professionals and families frequently led to using all the technological resources available even knowing that the disadvantages would outnumber the benefits. Moreover, it was often done more for the sake of their consciences than for the well-being of the patient. It should be added that the current tendency toward defensive medicine also encourages healthcare professionals to implement extraordinary treatments (i.e., ones that are disproportionate to the patient's well-being). It was in such a context that debate on decriminalizing euthanasia emerged in the second half of the twentieth century. The main arguments of those defending the right to euthanasia are the principles of autonomy and human dignity together with the moral obligation to do whatever needs to be done to avoid causing pain and suffering. Since it is accepted that life belongs to the individual he or she should be free to decide about terminating his or her own life. Such decisions become problematic when patients lose their capacities and cannot let others know the pain and suffering they are feeling, when they lose their relational life but not their biological life, and when they fall into an artificial condition considered undignified and unworthy. Although medicine is constantly advancing, there will always be cases of uncontrollable pain that need to be addressed and legalizing euthanasia could be an answer. The main arguments of those who want to prohibit euthanasia is the sanctity of life, the principle of human dignity, and the slippery slope argument. Most religious people consider life as a gift by God to everyone to make the most of it. Even at the secular level life can be considered sacred in that it should be inviolable as a proto-value (i.e., the ground value that makes all values possible). This is the reason the value of life cannot be measured in any way possible because human dignity is an absolute and unconditional value. More worryingly, experience shows that in all countries where euthanasia has been decriminalized in specific situations these have been progressively

H. ten Have and M. do C. Patrão Neves, *Dictionary of Global Bioethics*,
https://doi.org/10.1007/978-3-030-54161-3_242

enlarged. Many pertinent arguments have been put forward by activists from both sides. Some argue that euthanasia should be an individual decision, while others retort that health professionals and institutions should be involved as a social commitment to making any such decision. Some claim that physicians should care for the well-being of their patients until the end of their lives and that euthanasia should simply be one more resource available to them, while others stress that euthanasia undermines the trust the patient has in healthcare professionals and that it can never be considered a service to be provided but more as a dereliction of healthcare professionals' duty to provide clinical assistance. Some consider that it is better to die than to suffer, while others point out that palliative care needs to be developed and made universally accessible. Euthanasia remains illegal in most countries of the world.

Euthanasia, History

The word "euthanasia" derives etymologically from the Greek *eu* (good) and *thanatos* (death) and thus literally means good death. Although it was practiced in Ancient Greece, there is evidence of its previous use. One of the most famous examples is the killing of disabled newborns in Sparta who were thrown off a cliff to die. There are several primitive societies such as the Jaluah in Kenya who used to leave old people to be eaten by wild animals. Such lives were not considered useful to society and not worth living. Although euthanasia seems to have been practiced in many cultures throughout history (mostly for eugenic and social utility reasons), the word was first used in the sixteenth century. Francis Bacon is commonly credited as the first to refer to euthanasia as a "good death" physicians could give patients for whom medicine could no longer offer any more assistance. It is widely believed that physicians have since then secretly practiced euthanasia as mercy killing to relieve patients from the suffering that physicians could not otherwise alleviate. Euthanasia only became a well-known public issue in recent times. In 1935 the Voluntary Euthanasia Legalization Society was created in the United Kingdom to stimulate debate on euthanasia. This movement spread to other countries (particularly, the United States). The movement suffered a major setback during the Second World War as a consequence of the euthanasia or extermination Nazi program of *Lebensunwertes Leben* (life unworthy of life) under which the handicapped (mental and physically disabled), the weakest, and the elderly (among others) were eliminated for both eugenic and social utility reasons. This mass murder of more than 250,000 people gave the word and any debate on euthanasia a negative connotation that would last decades. It was only in the 1970s that the debate on euthanasia was reopened within much the same framework as today mainly as a result of advances in biotechnologies (specifically, artificial resuscitation techniques). Although biotechnologies are by and large beneficial for human beings, there are situations where they can prevent death but cannot restore life thus inflicting much pain and suffering on patients and families or prolonging biological life in a persistent vegetative state. In 1994 the Netherlands was the first country to allow euthanasia under specific circumstances. This followed a process that started in 1973 when a physician was not legally convicted after admitting she had injected a lethal dose of morphine into her mother who had an incurable disease and who requested her daughter to do so. Today the debate on euthanasia is ongoing in many countries. There are many pro-euthanasia associations and a number of legislative initiatives have been taken in different countries. Although euthanasia is currently permitted in the Netherlands, Belgium, Luxembourg, Canada, and Spain under different legal frameworks, it remains forbidden in most countries.

H. ten Have and M. do C. Patrão Neves, *Dictionary of Global Bioethics*,
https://doi.org/10.1007/978-3-030-54161-3_243

Euthanasia, Passive

Passive euthanasia commonly refers to letting someone die naturally. Although death may be avoidable in the short term and clinically postponed at the cost of lengthening the time the patient suffers, passive euthanasia refers to doing nothing to prevent imminent death and to allowing nature to take its course. Classifying euthanasia as passive indicates the absence of any action to stop death from taking place. In fact, it is arguable whether the term "letting someone die" can be accurately called "euthanasia." There are those who argue passive euthanasia is when a patient dies as a result of medical professionals not doing what is needed to keep the patient alive or when they stop doing something that is keeping the patient alive such as switching off life support machines, thus justifying calling it euthanasia. However, there are others who dismiss this argument because the death in question is just a natural event and has no cause, thus justifying not calling it euthanasia. Letting someone die is qualitatively different from putting someone to death (euthanasia) since it is not a question of degree or intensity of action but an entirely different attitude. Euthanasia is the act of killing on request to avoid suffering. Conceptually, letting someone die cannot be objectively and accurately categorized as euthanasia. Although the expression "passive euthanasia" has led to the idea that "letting someone die" is also a kind of euthanasia, it clearly is not. Describing euthanasia as active or passive is often justified as a social and political strategy in favor of its decriminalization against an adverse historical background. Advocating the withdrawal of all invasive therapies when state-of-the-art medicine is unable to cure is not an option in countries where euthanasia is forbidden. At the beginning of contemporary debate on euthanasia it was common to consider letting someone die as euthanasia by classifying it as a passive attempt to stress the difference. Ethically, the main intention of passive euthanasia is to prevent the extension of suffering when there is no hope—not to put people to death. Although such an approach no longer seems to be pertinent (at least in Western countries), conceptual accuracy should always prevail in honest and transparent debates.

H. ten Have and M. do C. Patrão Neves, *Dictionary of Global Bioethics*,
https://doi.org/10.1007/978-3-030-54161-3_244

Evaluation Ethics

Over the last few decades evaluation ethics has developed into a growing area of research and practical application. Evaluation is the systematic assessment of policies, practices, and individuals based on criteria and standards. It is used to monitor how well such policies, practices, and individuals are performing and whether there is a need for change. Evaluation is a key instrument in evidence-based medicine to improve the quality of health-care. The UN Development Program (UNDP) has listed norms such as evaluation should be independent, evaluators should be free and not have conflicts of interests, evaluation should be transparent (this implies consultation with stakeholders), evaluation should be impartial, it is important that information resulting from evaluations will be used for decision-making, and evaluation should follow ethical standards. Evaluators should have integrity, respect for the rights of individuals and institutions, and be sensitive to the local, social, and cultural environment. Guiding principles for the ethics of evaluation have been proposed by the American Evaluation Association such as systematic inquiry and competence, integrity and honesty (crucial to warranting the process of evaluation), respect for people, all participants in the evaluation process should be respected, and evaluators should take responsibility for general and public welfare and take into account the diversity of interests and values.

H. ten Have and M. do C. Patrão Neves, *Dictionary of Global Bioethics*,
https://doi.org/10.1007/978-3-030-54161-3_245

Evolutionary Ethics

Evolutionary ethics explores what ethics and morality mean for evolutionary theory. It assumes that ethics and the moral sense of human beings (in particular) is the result of natural selection. Morality emerged as a natural phenomenon during the evolution of human beings—not as a result of rational faculties or divine revelation. Morality is the result of adaptation that has given human beings a selective advantage. Edward O. Wilson, who was one of the promoters of evolutionary ethics, interpreted this as meaning ethics could be biologicized in that it is no longer the remit of philosophers but requires empirical studies of the relation between biological determinants such as genes and social behavior. Evolutionary ethics has been criticized for giving primacy to biological determinants of human behavior at the expense of the influence of social and cultural contexts. Philosopher Herbert Spencer (1820–1903) applied the evolutionary perspective to psychology and sociology—not only ethics. He was the first to coin the term "survival of the fittest" and used it as an explanatory principle for social and ethical progress. Spencer argued that policies should discourage "weaker" populations from reproducing. Such views promoted social Darwinism that has often been used to justify eugenics, racism, and social inequality. Epigenetics today has shown that it is not possible to prove the existence of connections between specific genes and social behavior. Another criticism is that evolutionary ethics does not do justice to typical human behavior in that it has difficulty in explaining culture, for example. Research also shows that cooperation is far more common than self-centered behavior focused on individual survival. Although altruism has evolutionary advantages when interpreted as promoting the interests of individuals or species, such an interpretation denies one major feature of altruism in that altruism only occurs when it is not in the interests of the altruistic person to do so. Another problem of evolutionary ethics is the fallacy that it is naturalistic in origin since normative rules cannot be derived from empirical facts. Spencer's argument that nature is focused on the survival of the fittest does not imply that society should not assist its poorer and weaker members. Nevertheless, evolutionary ethics underlines the importance that biological and neurological (specifically) conditions have for the emergence and expression of moral consciousness.

H. ten Have and M. do C. Patrão Neves, *Dictionary of Global Bioethics*,
https://doi.org/10.1007/978-3-030-54161-3_246

Experimentation (*See* Research Ethics)

Experimentation is the use of experiments to examine, test, and validate theories and hypotheses and is usually contrasted with observation. Experiments involving control groups are compared with other groups who have undergone some form of experimental intervention in which the most often used methodology is the randomized clinical trial. Histories of medicine used to present medical progress as the result of medicine redefining itself as a natural science in which measurements, experiments, and instruments are involved. Claude Bernard (1813–1878) was one of the main advocates of experimental medicine and made a clear distinction between physicians who observe and those who experiment. Although medicine has long been a science of observation, now the second period of scientific medicine has arrived: experimental medicine. Medicine was always destined to be an experimental science since scientific progress is only possible through experimentation. Bernard points out the essential differences between observation and experiment in that observation reveals facts while experimentation teaches and gives experience. It is not simply a case of observing activity and passivity since experiments themselves create and produce phenomena. However, experimentation can be active or passive in much the same way as observation. Although an observer cannot of course vary phenomena and studies them as presented by nature, an experimenter tests ideas about facts by modifying phenomena or disturbing them. However, the distinction is not as clear-cut as generally assumed. Experiments have been used throughout the history of medicine such as in the seventeenth century when Santorio Santorio did experiments with a specially devised weighting chair to study metabolism and Reinier de Graaf brought about a pancreatic fistula in a dog to study secretions with the use of a microscope. Moreover, collecting and observing materials was not a passive enterprise in which anatomical and pathological specimens were simply admired or venerated like relics. They were actively examined. They facilitated comparisons such that relationships, similarities, and patterns could be determined. As a result of the central role experimentation played in the production of medical knowledge, interest in medical collections and museums declined, accumulating specimens and body parts was no longer regarded as encouraging science or leading to new knowledge, and the museum came to be contrasted with the laboratory. Moreover, collections lost their significance because medical research was more focused at the cellular and molecular levels. Dissections have massively declined since the 1980s.

H. ten Have and M. do C. Patrão Neves, *Dictionary of Global Bioethics*,
https://doi.org/10.1007/978-3-030-54161-3_247

Exploitation

Exploitation is a word that crops up in many bioethical debates such as those dealing with commercial surrogate motherhood, selling of organs, non-therapeutic medical experiments on prisoners, experiments on cancer patients hoping for cures, and clinical research in developing countries. Although the literature on exploitation is extensive, it is somewhat recent in that it started almost simultaneously with business ethics (where sweatshop labor is paradigmatic) and bioethics. Despite the concept of exploitation being used historically to analyze economic interactions (especially along the lines of Marxist and neo-Marxist theories), more recently it has been applied to non-economic interactions such as in the professional contexts of research and healthcare and in personal and intimate relationships. No single theoretical account can explain all types of exploitation and all the different exploitative practices. Although there is little agreement on the concept itself, four explanations have been given for exploitation to be considered wrong: injustice, lack of freedom, disrespect, and vulnerability. The first is Alan Wertheimer's description of exploitation as a transaction in which one party takes unfair advantage of another. The focus is on the distributive outcome of a transaction irrespective of whether the outcome is fair as measured by a hypothetical competitive market price. According to such a market approach an impoverished vendor of, say, an organ is not exploited because his or her vulnerability is neutralized by offering a fair price for the body part in question. Such a theory argues that harm or invalid consent are not relevant and that many exploitative practices are consensual and mutually advantageous agreements. All that is relevant here is whether such transactions are fair. Other authors feel the focus on transactional fairness is too narrow. Regarding exploitation as taking advantage of injustice implies that harm is actually done even when there is mutual advantage. One party is gaining at the other's expense because by inflicting a loss on the other party the exploiter fails to benefit the exploitee and such a loss cannot be compensated by paying a higher price. Furthermore, unfair exploitation is not confined to exchange interactions and spills over into the field of justice, which requires people should have equal access to healthcare and labor should at the very least lead to a decent standard of living. The second explanation for exploitation being considered wrong deals with lack of freedom. It is not the distribution of the product of a transaction that matters but the conditions of the transfer. Wrongful advantage is taken of others in social arrangements that are coercive, dominating, and alienating. Exploitation is explained as being part of the process rather than the outcome of a transaction and usually occurs in situations where free exchange between equal parties does not take place or in circumstances that are primarily controlled by one party. Redistribution of benefits will not help as long as freedom is denied. The third explanation for exploitation

H. ten Have and M. do C. Patrão Neves, *Dictionary of Global Bioethics*,
https://doi.org/10.1007/978-3-030-54161-3_248

being considered wrong is disrespect. According to this theory exploitation is wrong irrespective of whether injustice or lack of freedom is involved because it is an affront to human dignity or a failure to respect people. Advantage is taken of other people or their situations and they are merely treated as a means of realizing the purposes and interests of the stronger party. What is necessary for the weaker party to flourish or fulfill their basic needs is ignored by the stronger party. Such treatment is humiliating and degrading. The fourth explanation for exploitation being considered wrong is vulnerability. Critical to exploitation is that there are vulnerable people who can be used to derive some benefit for the exploiter. Such vulnerability stops exploitees from being able to influence the conditions of the interaction and often the situation in which they find themselves enables the exploiter to extract benefits. Although these four theoretical accounts offer different explanations of exploitation, they all more or less assume the relevance of vulnerability. Exploitation can only occur in settings where one person has an advantage over another such as when one party has an initial need, insufficiency, or dependency that provides an opportunity to the exploiting party. Such an unequal balance is often interpreted in terms of power. Differences in power are brought about by unequal standing in society or social exclusion and have the effect of making people dependent. Exploitation can briefly be summarized as self-centered individuals exercising power over others less fortunate than themselves. Thus the prime concern is to eliminate dominance such that people become less vulnerable, are given choices, and have a greater say over the process and outcome of interactions.

Fairness

Fairness is frequently used by philosophers to refer to justice and distributive justice (in particular). Justice is interpreted as fair when it is equitable, when it demands individuals receive the same treatment irrespective of what life throws at them, and when it compensates them when they suffer adversity. Distributive justice requires benefits, burdens, rights, and duties to be distributed as equally as possible to improve the situation of the most disadvantaged, especially when they are not responsible for their situation. The concept of fairness (or fair treatment) first became philosophically significant with John Rawls' work *A Theory of Justice* in 1971 in which he systematized justice as fairness (Rawls' paper "Justice as Fairness" was published in 1958 in *The Philosophical Review* 67, No. 2). This marked a major step forward in utilitarianism's reflection on justice (in which maximizing utility is deemed not necessarily fair), correcting libertarianism (in which inequalities of wealth and power are allowed despite not giving all citizens the necessary means to make the best use of their basic liberties), and trying to develop egalitarianism (in which all citizens have equal basic rights and possibilities to be socially and economically equal are brought about by justice as fairness) as much as possible while adopting a liberal position (using legitimate political power to redress natural disadvantages). Fairness is certainly the most common concept used today to describe the type of justice everybody aspires to, hence the reason for it being widely endorsed ethically. Nevertheless, fairness also raises important ethical questions about how disadvantages can be accurately identified such as those brought on by individuals themselves and consequences brought about by the activities of individuals and the effect they have within healthcare (irrespective of whether it is freely provided or covered by health insurance). Such issues are very controversial.

H. ten Have and M. do C. Patrão Neves, *Dictionary of Global Bioethics*,
https://doi.org/10.1007/978-3-030-54161-3_249

Family Medicine

Family medicine is a branch of medicine focused on providing comprehensive healthcare within the family and community. It is also known as general practice or primary healthcare. The World Organization of Family Doctors (WONCA) defines family medicine as the provision of personal, comprehensive, and continuing care for the individual in the context of family and community. Historically, what is now labeled family medicine used to be general medicine before the advent of medical specialties. It has now become a separate specialty itself. Ethical issues in family medicine can be similar to ethical problems faced in other areas of medicine such as dealing with patients with reduced autonomy, the right to refuse treatment, and resource allocation. Specific ethical issues can arise as a consequence of family doctors knowing the family history, being aware of genetic diseases and susceptibilities, having to decide whether to discuss these with individual patients, and being aware of lifestyles. Since the relationship family doctors have with patients is usually much longer than that patients have with medical specialists, shared decision-making is a primary concern within the setting of family medicine. Moreover, family doctors must take a broader approach than happens with strict medical treatment or specialized medicine; have a comprehensive and holistic perspective that takes into account the social, psychological, and economic context of the patient such as when patients need care because they are lonely or unemployed; be more focused on health promotion and prevention; and be responsible for screening for possible risk factors and for discussing lifestyle and medication with the patient. They also act as gatekeepers in that they are the point of entry to the healthcare system in most countries. Most medical complaints are dealt with by family doctors.

H. ten Have and M. do C. Patrão Neves, *Dictionary of Global Bioethics*,
https://doi.org/10.1007/978-3-030-54161-3_250

Family Planning (*See* Fertility Control)

Family planning refers to taking control over pregnancy involving contraception and reproductive technologies in an attempt to help parents decide how many children to have and when to have them (family planning). Although taking control over fertility has been an ambition mostly of women throughout history, it only became a reality in the twentieth century when new methods of contraception together with different reproductive technologies provided efficient control. Good family planning is aimed at reducing the number of abortions and promoting reproductive health. The WHO state that good family planning should prevent pregnancy-related health risks (getting very young women to delay pregnancy and advising older women against pregnancy, especially when there are health risks); reduce maternal and infant mortality, adolescent pregnancies, sexually transmitted diseases, and HIV; and slow population growth. Although family planning is of paramount importance for women, it is also relevant from a social perspective in that it allows women to plan professional careers. The economic stability of families can also benefit from family planning. The major ethical concern related to family planning is ensuring its universal (wide and easy) accessibility and provision by adequate methods that take into account cultural and religious contexts and respect the autonomy of women or couples. Family planning and recent methods used to bring it about have resulted in new situations that raise ethical questions (e.g., post-menopausal motherhood).

H. ten Have and M. do C. Patrão Neves, *Dictionary of Global Bioethics*,
https://doi.org/10.1007/978-3-030-54161-3_251

FGC (Female Genital Cutting)

Female genital cutting (FGC) is a common traditional practice in many countries in Africa, Asia, and the Middle East. It involves the removal of some or all of the external female genitalia for non-medical reasons (often with the purpose of preventing women to enjoy sex). The four major types of FGC are clitoridectomy (partial or total removal of the clitoris), excision (total or partial removal of the clitoris and labia minora), infibulation (narrowing the vaginal opening and cutting the labia minora and labia majora without removing the clitoris), and other harmful procedures such as piercing, incising, and cauterizing the genital area. Although the procedure is regarded as a traditional cultural practice, it has no health benefits for girls or women and can cause serious harm such as severe bleeding, problems with urinating, cysts, infections, complications in childbirth, and increased risk for newborn deaths. More than 200 million girls between infancy and 15 years of age and women today have undergone the procedure in 30 countries. Ethical debate looks at the practice of FGC from three different viewpoints: cultural diversity, human rights, and health. Although the practice is often labelled-differently as female circumcision, female genital mutilation, or female genital cutting, the procedure involved in all cases is the same. What differs is whether such procedures should be portrayed as a problem in need of normative evaluation. Female circumcision is a practice that has existed for centuries in a limited number of countries. Proponents argue it is a rite of initiation for female members of society, an act of purity, and a source of cultural identity. Although the implication is that it should be respected as a cultural tradition, framing it in this way in the 1970s became increasingly contested. From a human rights perspective the practice was regarded as a violation of women's rights and a form of discrimination against women. Since it is most commonly performed on young girls it is also a violation of the rights of children. Although framing the problem in this way encouraged legal action and ever more countries prohibited female circumcision, legislation often had little effect. Such a background gave rise to another framework in which women's health rather than women's rights was emphasized. As more data became available about the short-term and long-term effects on health, it was argued that removal of anatomical structures without medical necessity was mutilation and that the name of the practice should be changed to reflect this. Perceiving the problem this way called for the eradication of FGC by treating it as an epidemic disease. Fortunately, it now falls under the mandate of the WHO. Such an ethical interpretation provides a means for building alliances. Human rights and the right people have to health united NGOs, activists, medical professionals, teachers, social workers, and religious leaders. Younger generations of women actively campaign against the practice as do the media. The United Nations declared February 6 as the

H. ten Have and M. do C. Patrão Neves, *Dictionary of Global Bioethics*,
https://doi.org/10.1007/978-3-030-54161-3_252

International Day of Zero Tolerance for Female Genital Mutilation/Cutting. The number of local NGOs campaigning against FGC has rapidly increased and most African countries where the practice was prevalent have now outlawed FGC. Looking at the practice from the health perspective helped local activists point out the adverse effects on health. They argued it was an attempt to appeal to commonly shared concerns —not a moral judgment on tradition or culture. Moreover, it has become clear that FGC has become a global problem as a result of immigration. Western countries such as the United States and France have a growing population of women and girls forced to undergo FGC. Framing FGC as a health problem helps assessing the practice in a balanced way. There are always different frameworks or even counterframeworks against which normative evaluations have to be weighed. Although the health framework seems the most powerful now, appeals to cultural diversity and accusations of cultural imperialism are less convincing as evidence and awareness of the harmful effects grow. Health is a universal value. In this frame, there are no benefits—only harms. Although legal approaches based on the rights framework are necessary, they are ineffective in that they do not address the systemic determinants of the practice in culture and society. The UN General Assembly adopted a resolution in December 2012 calling for the worldwide elimination of FGC. It was the first time the organization explicitly supported such a ban. Although the practice is in decline in most countries and actively opposed by younger generations, it will not be easy to change such a long-standing cultural practice. Nevertheless, the practice is likely to die out in much the same way as happened with the practice of foot binding in China, which took 200 years of campaigning to be eradicated successfully in the early twentieth century.

Feminist Ethics

Feminist ethics refers to theories that articulate feminist approaches to ethics and is based on the conviction that traditional ethics does not pay sufficient attention to the moral experiences of women. It is argued that methods, theories, and practices in ethics are usually formulated and interpreted from the perspective of men and that a broader perspective can only be advanced when the function of gender in ethical theory, debate, and practice is critically analyzed and corrected. Feminist ethics as an academic field first emerged in the 1970s. A landmark study was Carol Gilligan's *In a Different Voice* (1982), which showed that moral development had been examined without taking the moral experience of women into account. The study further showed that women were usually regarded as morally immature and that care, family responsibilities, and private life were not considered serious issues for ethics. The study brought a response from feminist ethicists who criticized traditional ethics for its emphasis on rationality, abstractness, universalizability, impartiality, and individualism and led to a number of approaches being developed that articulated the importance of care, ethical emotions such as empathy and affection, the significance of relationships, dependency, vulnerability, responsibility, and the context in which moral decisions were taken. An example of such approaches is the work of legal scholar Martha Fineman who argues that human beings are only relational in the context of relations between equals, while vulnerability is inherent in the human condition and defines what it is to be human. Despite being universal and unavoidable, vulnerability is not the same for everyone. Fineman points out that although vulnerability is a complex human feature that manifests itself in multiple forms, it is at the same time particular in that every individual is positioned differently. Moreover, although the embodiment of individuals differs in the ways in which they are situated within economic and institutional networks, the result is that human beings are not vulnerable to the same degree. Fineman argues that vulnerability inevitably arises from embodiment through which individuals' existence is marked by dependency and that this is not merely a developmental stage that is left behind as individuals grow into autonomous subjects but defines them because they are continuously dependent on relations with other people. Since human beings need other people the autonomous and independent subject is a fiction only to be found in the liberal tradition. The thesis that human life is precarious since it inevitably depends on anonymous others is central to the work of feminist philosopher Judith Butler. The inescapable fact is that humans are necessarily related in that they are always exposed to each other. Since such relatedness is not under human control, relationality is not an option. The problem with vulnerability is that it is not equally distributed as a result of interdependency not being symmetrical. However, Fineman and Butler argue

H. ten Have and M. do C. Patrão Neves, *Dictionary of Global Bioethics*,
https://doi.org/10.1007/978-3-030-54161-3_253

that dependency is not a negative condition and regard it as a limitation that should be considered a resource. Since human bodies are vulnerable to injury, pain, and violence and being human means being vulnerable, then the body is not only the site of vulnerability but also that of humanity. Other issues considered in feminist ethics focus on ownership of the body, identity, reproduction, and pornography—just to mention a few.

Fertility Control (*See* Birth Control; Contraception)

Fertility control refers to the power humankind has acquired but is constantly growing to make decisions concerning reproduction rather than leaving it to nature or human biology. Specifically, it refers to the choice to reproduce or not and the number of children to have and when (family planning). Technically, it refers to contraceptive methods or devices (birth control) and methods of assisted reproduction (such as those in infertility cases and those selected by individuals). Humankind has always tried to control fertility mostly with the aim of increasing the fertility of women and the probability of having male offspring. However, fertility control has also been used to prevent undesirable pregnancies as borne out by the wealth of information from ancient times and primitive cultures about different substances (mostly extracted from plants) used to prevent or enhance female fertility. The use of condoms to control male fertility has been known for thousands of years. Abortion was also practiced as a method of birth control brought about by ingesting specific herbs or using elementary techniques such as abdominal pressure or placing sharp tools in the vagina. However, fertility control (use of contraception) only became effective in the second half of the twentieth century following growth in scientific knowledge of the reproductive anatomy and physiology, along with revolutionary advances in contraceptive hormonal methods and devices for women and men (until then condoms were the most popular method of family planning). In short, the increasing number of technical means to control fertility; their growing efficacy, accessibility, and ease of use; and decreased side effects strongly improved fertility control. Put simply, the better the fertility control available the more responsible parenthood became. Moreover, this is the reason fertility control is widely accepted (some religions forbid it as counteracting human biological nature and thus God's will). Furthermore, better fertility control should prevent more abortions from taking place. Such an argument has readily been taken up by those who are against abortion. Strictly speaking, although abortion cannot be considered a contraceptive method, when the woman in question decides to have one then it works as a form of fertility control. Recent assisted reproduction techniques such as gamete and embryo cryopreservation followed by in vitro fertilization strengthen fertility control. Such techniques make it possible for women to postpone a desired pregnancy (even to the postmenopausal stage of life). Nevertheless, such a situation is highly controversial and forbidden in almost every country.

H. ten Have and M. do C. Patrão Neves, *Dictionary of Global Bioethics*,
https://doi.org/10.1007/978-3-030-54161-3_254

Fertility Preservation

Fertility preservation refers to a new reality (made possible by advances in biotechnologies) in which individuals can retain the ability to reproduce even after losing their natural reproductive function as a result of specific (in many cases lifesaving) clinical treatments. The origin of fertility preservation can be traced back to efforts to help young cancer patients of reproductive age and without children who were undergoing chemotherapy or radiation therapy. Although such aggressive treatments cannot be postponed without seriously endangering patients' lives, they do impair their reproductive system. This resulted in young patients being condemned to becoming infertile. Biotechnologies made it possible to retrieve healthy gametes (sperm or oocytes) before therapy and to cryopreserve and store them for the future until the treatment ceased or until the patient (now cured) wanted to have a biological child via assisted reproductive technologies. A more aggressive fertility preservation method for women is ovarian transposition in which the ovaries are surgically removed from the area likely to be affected by radiation to another part of the abdomen and later replaced. Ovarian transposition and gonadal shielding are specifically designed to preserve fertility in children who do not have mature sperm or eggs. Fertility preservation then became available for those pathologies in which loss of the ability to conceive naturally commonly occurs, specifically in patients with an autoimmune disease such as lupus or a genetic disease that affects future fertility. Such procedures for handling gametes and reproductive tissues for therapeutic reasons are generally considered ethically uncontroversial, which is not always the case for fertility preservation. Fertility preservation procedures are currently available to patients undergoing therapeutics that affect their reproductive capacity and to individuals who for different reasons want to extend their reproductive potential (e.g., women approaching menopause or dying men). Each individual situation raises different ethical controversies mostly related to whether the attitude of the adult is selfish and negligent of the interests of the child. The practice of cryopreserving embryos rather than gametes for possible future use is currently quite common as a result of its high level of success. It raises ethical criticisms that are common to the debate concerning respect for the dignity of the human embryo.

H. ten Have and M. do C. Patrão Neves, *Dictionary of Global Bioethics*,
https://doi.org/10.1007/978-3-030-54161-3_255

Fetal Research

Research using living fetuses inside or outside the uterus is called fetal research. It also includes research using embryos (depending of course on the stage of development). Research can use invasive or non-invasive techniques. Examples are tests of the efficacy of rubella vaccines, development of diagnostic techniques such as amniocentesis, chorion villi sampling and ultrasonography, and new techniques for obstetrical anesthesia and treatment of maternal hypertension. Often the aim of the research is to improve the survival of fetuses and the fertility of people who want children. Fetal research is one of the most controversial topics in research ethics. In the 1970s fetal research became subject to regulation and legislation and many abortion laws banned such research. If the fetus had a beating heart, then no research was allowed on it outside the uterus. Although standards have gradually emerged for fetal research, it is acknowledged that fetuses are vulnerable and only research that has therapeutic purposes and minimal risk is allowed. Fetal research involves conducting preclinical studies (including studies on pregnant animals) and clinical trials with non-pregnant women so that potential risks to pregnant women and fetuses can be assessed. There should be no connection between abortion and fetal research, which means no inducements whatsoever should be offered to terminate the pregnancy. Current controversies are focused on research using human fetal tissue such as when cells from dead fetuses are obtained to establish cell lines for research or for use as transplantation material. Although such fetuses are the result of elective or spontaneous abortions, ethical questions arise as to whether the opportunity to donate fetal tissue influenced the decision to have an abortion. There are also concerns about potential conflicts of interests for researchers involved in the retrieval, storage, testing, preparation, and delivery of fetal tissues. As a result of such concerns the US administration in 2019 restricted fetal tissue research. Federal agencies can no longer conduct research using human fetal tissue obtained from elective abortions and researchers applying for federal grants have to explain why other tissues cannot be used.

H. ten Have and M. do C. Patrão Neves, *Dictionary of Global Bioethics*,
https://doi.org/10.1007/978-3-030-54161-3_256

Fetal Surgery

Surgical intervention on behalf of a fetus takes place, of course, inside a pregnant woman's body, hence the reason it is sometimes called maternal–fetal surgery. Surgical procedures are usually done to correct anatomical abnormalities in the fetus such as repairing various forms of spina bifida. Although efforts in the past were directed at post-natal repair, novel intrauterine interventions have become possible because of advances in imaging and instrumentation technology, surgical techniques, and knowledge of fetal pathophysiology. Fetal surgery is rare since the option of surgery is given to just a small number of pregnant women. Surgery needs to be carried out by highly specialized surgical and medical teams who work in a few centers around the world. Ethical considerations include assessing anticipated benefits for the fetus compared with possible risks and looking at the alternative approaches available. Although risks for the fetus and for the pregnant woman should be reasonable and minimized, it is necessary to balance benefits for the fetus against risks for the pregnant woman. What benefits the fetus may harm the woman and what benefits the woman may negatively influence the viability of the pregnancy. Fetal surgery does not always enhance the quality of life of the developing fetus. The ethical issues of fetal surgery are complicated since any intervention is invasive, often experimental, and involves two patients.

H. ten Have and M. do C. Patrão Neves, *Dictionary of Global Bioethics*,
https://doi.org/10.1007/978-3-030-54161-3_257

Food Ethics

Food is fundamental to the survival of all living beings by fulfilling homeostatic needs. Living beings instinctively choose the kind of food that satisfies their nutritional needs and can be processed by their digestive system. However, when it comes to humankind food is not just a natural, basic biological necessity it also has a symbolic value in that humans have attributed symbolic significances to many kinds of food that they consume as omnivores (eating both plants and animals), but symbolically would favor or refrain from eating. Although food is culturally embedded because of the kind of food (plants, animals) available geographically, it has also become an artifact that creates social bonds and acquires different meanings. Food ethics focuses on the human criteria used to select food and the ways to produce, prepare, and consume it according to religious, moral, philosophical, or ideological values and convictions. Indeed, cultures and religions have long placed restrictions on certain foods considered impure (unclean for the soul), while other foods have been given preference for their special (sometimes wondrous) features such as healing, increasing sexual performance, celebrating initiation into a group, and creating a moral identity. Although the criteria used today by people to choose the food they eat have multiplied and diets have diversified, many concerns relate to the way food is produced such as genetically modified, intensively farmed, and high-carbon-footprint foods. People prefer food that is biologically produced and complies with strict environmental rules. Other concerns relate to questionable foods offered for consumption such as certain animal species (sacred or endangered), animals in general, and animal products. Ethical diet choices can range from macrobiotics (consumption driven by health considerations and the healing qualities of food), vegetarianism (excluding animal flesh), veganism (excluding meat and all animal-derived ingredients), flexitarians (semi-vegetarian), to pescatarians (excluding animal flesh other than fish). Such food regimens raise concerns with healthcare professionals when children are involved since they need a more balanced diet to address their natural needs. Although Paleolithic diets (thought to mirror those eaten during the Paleolithic era) are a recent trend toward eating a more natural diet, this neglects the evolution of humans and the sorts of products available today. Although such diets satisfactorily address health concerns, they need to be balanced against respect for people's values. However, health too is a value and healthy food diets are recognized as ethical in that they guide people toward making responsible choices such as eating low-sugar or fat-free food. At the same time health problems related to food such as eating disorders like anorexia and bulimia are growing. The role food plays today socially and for entertainment purposes has grown. Moreover, food has become an art and gastronomic delights are offered at very high prices. Such a cavalier attitude toward food raises concerns about

H. ten Have and M. do C. Patrão Neves, *Dictionary of Global Bioethics*,
https://doi.org/10.1007/978-3-030-54161-3_258

waste and many NGOs have been set up to fight against the dumping and loss of perfectly good food. A broad approach to food ethics presents it as a wide and very diverse thematic field exemplifying conflicting values. The most fundamental challenges are hunger in underdeveloped countries and obesity in developed ones.

Food Security (*See* Hunger; Food Ethics)

Food security and food safety are two interrelated societal issues. Food safety refers to safe sources of food for human consumption that are free from chemical or microbial contamination and are properly stored (cold chain when needed), transported, labeled, prepared, processed, and cooked. While acknowledging food safety requirements, food security refers to the adequacy of food (especially in societies where food has cultural importance) thus guaranteeing the stable supply of food over time and guaranteeing it is accessible (physically and economically) to all citizens such that they can enjoy a healthy life. The World Food Summit in 1996 defined food security as existing "when all people at all times have access to sufficient, safe, nutritious food to maintain a healthy and active life." The four classic dimensions of food security are availability, access, utilization, and stability each one of which includes a set of indicators to measure the levels of food security/insecurity in each country. The level of severe food insecurity in the world population has been rising. According to the report of the Food and Agriculture Organization of the United Nations (FAO 2018) it currently stands at 9.2%. Although it is the responsibility of each state or nation worldwide to assure food security, today countries are not self-sufficient in all the items they need. Moreover, despite countries being major producers and exporters of some products, they are often deficient in others that they have to import. Nevertheless, it is the government's obligation to provide the infrastructure and the logistics to assure the quantity and quality of the food supplied to its population are satisfactory and to be prepared for emergency situations such as natural disasters. Small islands or sparsely populated regions face particular challenges when it comes to food security because geographic constraints can make transportation difficult, affect the availability (mostly of fresh foods), and increase the price. Food security is principally a severe problem in underdeveloped and poor countries that have insufficient food resources, are remote from large distribution infrastructures, and often depend on international help that seldom arrives despite the fact that people need to eat every day. Access to clean drinking water is another serious issue in many regions of the world since water reserves are falling in many countries. Furthermore, land degradation and climate change compound these issues in that they are already affecting food security in different regions around the world. Such issues are likely to intensify as a consequence of the world's population growing.

H. ten Have and M. do C. Patrão Neves, *Dictionary of Global Bioethics*,
https://doi.org/10.1007/978-3-030-54161-3_259

Forensic Medicine

Forensic medicine (a.k.a. forensic pathology) is a medical specialty that makes use of medical knowledge to examine civil or criminal legal cases and determine, say, the cause and time of a suspicious death. Investigations are not limited to victims and suspects of crime, they also include people who committed suicide or died accidentally. Forensic pathologists are engaged at the request of the police, the prosecutor, the defense, or the court and assist in the investigation as medical experts. They are tasked with providing impartial assessment and objective testimony in court proceedings at the request of one of the parties. National regulatory organizations belonging to the International Forensic Medicine Association allocate rights and responsibilities to forensic scientists. Six categories of ethical issues have been identified. The first concerns professional credentials since education, professional licensures, and certifications have been misrepresented in court proceedings. The second concerns problems with laboratory procedures; although most laboratories have established protocols, they are not always followed by forensic scientists. The third concerns the interpretation of data and presentation of testimony since there is evidence of these being deliberately biased to solve crimes. The fourth concerns the increasing number of forensic scientists who are privately employed; since such individuals are subject to less supervision and less peer review and given financial incentives there is a greater risk of malpractice. The fifth concerns the independence of publicly employed scientists; in addition to being independent they must not be part of local government or local law enforcement agencies. The sixth concerns the responsibilities and obligations forensic scientists have to their profession to uphold ethical values and standards such as taking care their work is not misused, maintaining their competence, and updating their knowledge. There have been numerous cases of misconduct in forensic medicine in the United States and elsewhere. In 2015 a lab technician working at a New Jersey State Police drug-testing unit was accused of fabricating test results leading to a review of almost 8,000 legal verdicts. In 2017 the highest court in Massachusetts dismissed more than 21,000 criminal drug cases because of falsifications of results by a forensic laboratory. Evidence tampering and faulty analysis (e.g., inappropriate and inaccurate DNA analysis) can have serious consequences, especially in countries where capital punishment is applied. Although DNA tests can be used to exonerate people, improperly analyzed DNA evidence can lead to wrongful executions. It should be kept in mind that hundreds of death row inmates have been exonerated as a result of DNA evidence being reviewed.

H. ten Have and M. do C. Patrão Neves, *Dictionary of Global Bioethics*,
https://doi.org/10.1007/978-3-030-54161-3_260

Freedom, General

Freedom is a fundamental concept at the moral (individual) level and at the legal and political (collective) levels. It is more frequently defined via negativa (i.e., the absence of restrictions or constraints on how an individual thinks and acts) than via positiva (i.e., an exercise of the human will or self-determination). The idea of human freedom has always been important for identifying humans as beings endowed with reason and will and forming relationships within a community in which different levels of freedom are legally recognized. However, freedom has been understood throughout history as free will (*liberum arbitrium*) (i.e., free choice between given alternatives). According to the general conviction of moral universalism, humans can only choose between acknowledging and complying with universal moral law or refuting it. Freedom defined as the total indeterminacy of action (*libertas*) is a contemporary human reality. In the eighteenth century German philosopher Immanuel Kant decisively established freedom as the foundation (*ratio essendi*) of morality. Rational beings can only be moral if they are free (i.e., if they can decide for themselves to act correctly). Kant wrote, "I ought never to act except in such a way that I could also will that my maxim should become a universal law." In the nineteenth century Stuart Mill presented freedom as essential to the pursuit of individual happiness with individuals following their own path and to the development of a society built by fulfilled citizens. Freedom is also one of the three ideals of the French Revolution, the heart of the human rights movement, and formally established in the Declaration of Human Rights (1948). Freedom is expressed in bioethics by the concept of autonomy or self-determination. Although freedom is a crucial notion in democratic societies and creates the conviction in some people that it is absolute, such absolute freedom is a utopia or an (empty) abstraction. What some consider as absolute freedom is nothing more than libertinage (freely indulging in sensual pleasures). Freedom is always situated (i.e., it always unfolds within a space, time, and specific circumstances). Therefore freedom consists in individuals realizing their full potential while adhering to its true moral context.

H. ten Have and M. do C. Patrão Neves, *Dictionary of Global Bioethics*,
https://doi.org/10.1007/978-3-030-54161-3_261

Freedom (of the Press)

Freedom of the press refers to the media being allowed to publish whatever contents it considers pertinent to the public interest without suffering any kind of censorship or fearing any kind of coercion or reprisal by the state or individuals. It is the core principle of the press (i.e., identifying it as free—otherwise it is just propaganda) and the cornerstone of democracies throughout the world. Any political regime that restrains the press is always a form of dictatorship. Freedom of the press and freedom of expression were first legally established by the Swedish parliament in 1766 and today are core values in all democracies worldwide. Freedom of the press is also an established ethical principle in all international documents on ethics relating to the press, the media, or journalism. The international organization Reporters Without Borders keeps a list ranking countries according to how free the press is inside them. Currently, state censorship and regulation of the press are still realities in most countries. The most recent phenomenon is growing hostility toward freedom of the press in formally democratic countries ruled by populists. Such governments tend to discredit the press (journalists and media) and cast suspicion on the media arguing that they promulgate fake news to affect public opinion thus paving the way to regulating the press and restricting its freedom.

H. ten Have and M. do C. Patrão Neves, *Dictionary of Global Bioethics*,
https://doi.org/10.1007/978-3-030-54161-3_262

Freedom (of Speech)

Freedom of speech (more broadly, freedom of expression) is the right individuals have to express their opinions publicly without fearing consequences. Such a right is established in the Declaration of Human Rights: "… everyone has the right to freedom of opinion and expression; this right includes freedom to hold opinions without interference and to seek, receive and impart information and ideas through any media and regardless of frontiers" (Article 19). Although freedom of speech is legally recognized in all democracies, in dictatorships it is robustly repressed and legally persecuted. However, the right to freedom of expression is not absolute. For example, offensive language (discriminatory, promoting hate, or obscene) is not morally, socially, or (sometimes) legally acceptable. Since it is not always easy to draw a clear line between freedom of speech and offensive language, applying legal penalties here can also be controversial. In recent decades there has been increasing pressure to eliminate particular words in common language because of their negative connotations and to substitute them with axiologically neutral words thus making language politically correct. Nevertheless, the major ethical problem today concerning freedom of speech is social media where moral boundaries seem to have vanished and freedom of speech is taken as absolute—which it is not. For freedom of speech to be a human right it must be articulated with respect for other human rights.

H. ten Have and M. do C. Patrão Neves, *Dictionary of Global Bioethics*,
https://doi.org/10.1007/978-3-030-54161-3_526

Freedom (of Treatment)

Although freedom of treatment essentially refers to the accessibility patients have to experimental treatments, it can also refer to the prerogative patients have to choose a treatment officially approved in some countries but not in the patients' own country. It can be summarized as referring to standards of care and freedom from regulation. Patients claiming freedom of treatment are almost always suffering from life-threatening illnesses and their claim is grounded on the principle of autonomy. Freedom of treatment raises many difficult ethical issues that can be summarized by the dilemma brought about by the autonomy patients have to choose what they consider or trust to be the best for themselves according to the particular clinical situation and by the state's duty to guarantee the safety of products and procedures legally available to citizens. Patients and families in life-threatening situations are extremely vulnerable and sometimes experience intense feelings such as profound despair or unrealistic hopes none of which help them decide the best option. What is worse, experimental treatments are seldom presented in a neutral way leading to patients and families frequently nurturing high expectations. Researchers and pharmaceutical companies have their own interests such as speeding up the process of approval and getting a return on their investment in selectively releasing products at an experimental stage. Although the state has the obligation to protect citizens from possible abuses without restricting their freedom, striking a balance between these two obligations is not easy. In addition to using the general rule of law to approve new treatments and their accessibility by patients, a higher entity needs to be established to make decisions on a case-by-case basis.

H. ten Have and M. do C. Patrão Neves, *Dictionary of Global Bioethics*,
https://doi.org/10.1007/978-3-030-54161-3_527

Futility

The word "futility" derives etymologically from the Latin *futilis* (deprived of value, importance, interest, or usefulness). Medical futility was first used in medical practice when it became more scientifically oriented, technologically assisted (since the 1960s), and more invasive or aggressive. The three elements of medical futility that need to be considered are "futile therapy or treatment" (for particular patients), "according to the clinical case" (in concrete situations), and "following the diagnosis and prognosis." Such elements must always be taken into account to reach a decision about what should be considered futile. One and the same treatment can be said to be futile for one patient but highly recommendable for another. Moreover, it can be highly recommendable at one time and futile in another for one and the same patient. Since there is no list of futile therapies, medical futility has to be determined by balancing benefits and harms (both clinical and personal) in each individual case. When the harms outweigh the benefits, medical futility can be used as the reason to withhold or withdraw intensive and disproportionate care that aggravates the suffering of patients who have no hope of recovering. It is thus very important for end-of-life ethical decision-making. Sometimes medical futility conflicts with patient autonomy or with the surrogate autonomy of family members. Some healthcare professionals in such cases tend to comply with the patient's or family's wishes (mostly in the Anglo-American context), while some others will not provide medical treatment or invest clinical resources when there is no chance of any real benefit (in the continental European context). Physicians will not provide futile treatment under any circumstances (even if it goes against the patient's or the family's wishes). Although medical decisions should be reached in a transparent and realistic dialogue between healthcare professionals, patients, and their families, physicians should always provide comfort care.

H. ten Have and M. do C. Patrão Neves, *Dictionary of Global Bioethics*,
https://doi.org/10.1007/978-3-030-54161-3_263

Future Generations

Future generations is the term used to refer to the next and subsequent generations of humans. Whatever context it is used in it acknowledges that those who do not yet exist matter today and that present generations who are now living have an obligation to think and act not only for themselves, but also for others including those who will follow them in the future. Such a perspective allows the human community not only to reach beyond former borders and to expand to the entire planet as a consequence of globalization, but now also in time to unborn generations —making the worldwide community even more global. The two key areas where concerns with future generations have emerged are the environment and genetics (specifically genetic engineering). Both were studied by the philosopher Hans Jonas who argues in *The Imperative Responsibility: In Search of an Ethics for the Technological Age* (1979) that present generations have a moral responsibility toward future generations, that current generations should not deplete natural resources as if they own them, that planet Earth is not our heritage but our legacy and is not only for us to use but also for us to protect, that present generations do not have the right or moral legitimacy to deliver living conditions to future generations that are poorer than those it received from former generations. Jonas applies the same reasoning to the potential impact of genetic engineering on human biological identity as a natural species and argues that genetic engineering should not interfere in any way with the human genome (the complete set of DNA including all genes) shared by all human beings defining our common identity as humanity and that genome modification should be confined to individuals and carried out at the somatic level (effective only for the individual)—not at the germline level (passed to future generations) since present generations do not have the right or moral legitimacy to choose how future generations should be. In 1997 UNESCO adopted the Universal Declaration on the Human Genome and Human Rights. Article 1 states that "the human genome underlies the fundamental unity of all members of the human family, as well as the recognition of their inherent dignity and diversity. In a symbolic sense, it is the heritage of humanity." In 2005 UNESCO's Universal Declaration on Bioethics and Human Rights explicitly refers to future generations in Article 16 titled "Protecting future generations," which states that "The impact of life sciences on future generations, including on their genetic constitution, should be given due regard."

H. ten Have and M. do C. Patrão Neves, *Dictionary of Global Bioethics*,
https://doi.org/10.1007/978-3-030-54161-3_264

Gender

The WHO describes gender as the socially constructed characteristics of women and men such as norms, roles, and relationships between women and men. While a person's sex is a biological fact, the characteristics of masculinity and femininity depend on the context in which specific roles and identities are articulated. The distinction between biological sex and the role gender plays was first made in 1955. In the 1970s the concept was adopted by feminist scholars and since the 1990s gender studies have emerged as a new interdisciplinary field of study. Gender not only varies from society to society, it can also change because people learn how to interact with others. From the perspective of healthcare, gender is an important concept because it can lead to discrimination, stereotyping, and social exclusion and such consequences affect health. This is the reason the WHO pays special attention to gender. It does so by emphasizing that gender norms, roles, and relations influence people's susceptibility to different health conditions and diseases and affect their enjoyment of good mental health, physical health, and wellbeing. They also have a bearing on people's access to and uptake of health services and on the health outcomes they experience throughout the course of their lives. The concept of gender has been criticized from two angles. One is that gender studies distort the agenda of feminist studies since insufficient attention is paid to the structures of violence and oppression. There is also the fear that studies of masculinity, for example in gay studies, will deradicalize feminist approaches. Another criticism is that gender studies represent an ideology that jeopardizes the traditional family and heterosexuality because they deny that differences between men and women have natural and biological foundations.

H. ten Have and M. do C. Patrão Neves, *Dictionary of Global Bioethics*,
https://doi.org/10.1007/978-3-030-54161-3_265

Gene Therapy

Gene therapy refers to a set of technologies aimed at correcting genes that have been injured (mutations) and cause new diseases as a consequence. Gene therapy involves introducing DNA containing a functioning gene (created by genetically modified viruses) into a patient to correct the effects of a disease-causing mutation. There are two basic types of gene therapy depending on the cells targeted or treated: somatic and germline therapy. Most cells of the human body are somatic cells and transferring DNA to such (differentiated, non-reproductive) cells will just produce effects in this patient without being transmitted to the patient's offspring. Cells of the germline are gamete cells (undifferentiated, reproductive), eggs, and sperm and any change in them will be transmitted to the patient's children. The possibility of gene therapy was first raised in the 1960s and triggered high hopes of finding miraculous cures to serious diseases without invasive procedures (surgery) or aggressive drugs (side effects). However, not only did many experimental trials fail some were severely criticized by the scientific community. The Jesse Gelsinger case was one. It involved an 18-year-old with a mild and well-controlled ornithine transcarbamylase deficiency who died after a gene therapy clinical trial. This led to the provisional shutdown of the North American gene therapy program. The case brought into question the intervention of the private sector in this kind of research and made the public aware of the side effects of clinical trials. Gene therapy failed to deliver what it had promised in the twentieth century. Indeed, it was not until the twenty-first century and genetic-editing techniques that genetic therapy realistically recovered the expectations raised almost half a century ago. Gene editing enables gene therapy at the somatic level and the germline level. Somatic gene therapy has been used to treat severe combined immunodeficiency and Leber congenital amaurosis. It has also showed great promise to treat muscular dystrophy, cystic fibrosis, sickle cell anemia, Parkinson disease, cancer, diabetes, and heart disease. Since germline gene therapy in humans has the potential to produce genetically modified babies (designer babies) it is very controversial and remains forbidden worldwide. There are two major ethical problems with gene therapy. The first is its experimental nature, requiring very detailed information, provided in a realistic (and not overoptimistic) way, and following a careful selection of subjects for clinical trials, taking into account their different vulnerabilities. Beneficence, quality informed consent, and respect for and protection of vulnerability are of paramount importance. The second concerns germline gene therapy which, in addition to sharing the same ethical problems as somatic gene therapy, raises questions about its right or authority to change the genome of future generations. Responsibility toward future generations has to be acknowledged since it is simply not good enough to argue that germline gene therapy would only be used to treat or eradicate diseases.

H. ten Have and M. do C. Patrão Neves, *Dictionary of Global Bioethics*,
https://doi.org/10.1007/978-3-030-54161-3_266

The slippery slope from therapeutic needs to social desires is at risk. Moreover, although new findings in epigenetics can contribute to the treatment of some genetic disorders, altering genes (even in somatic cells) can produce important imbalances and cause new diseases. The importance of applying the precautionary principle in this area cannot be overemphasized.

Gene Editing (*See* Genome Editing; CRISPR)

This entry is discussed under Genome editing.

H. ten Have and M. do C. Patrão Neves, *Dictionary of Global Bioethics*,
https://doi.org/10.1007/978-3-030-54161-3_267

Generic Medication

Medicines that have the same chemical composition as pharmaceutical drugs protected by patents are called generic medication. Generic drugs can only enter the market when patents on original drugs have expired. Since they have the same chemical profile they are considered bioequivalent to the original drugs. They are simply copies of brand name drugs in which the dosage, use, side effects, administration, risks, and safety are the same as the original. Generic drugs work in the same way and provide the same clinical benefits as the brand name version. The advantage is that they are usually cheaper than brand name drugs. For example, metoprolol is a generic drug used for hypertension, while Lopressor is the brand name of the same drug marketed by pharmaceutical companies. Generic drugs are regulated. The FDA has a generic drug program that rigorously reviews the manufacturing of generic medicines, inspects manufacturing plants, and monitors drug safety after they have been approved. Generic drugs are cheaper because applicants do not have to repeat the animal and clinical studies required of brand name medicines. Healthcare systems not only benefit from the savings but substantial resources are also freed up for other purposes. In the United States approximately 80% of all medication is now generic medication. By making drugs more accessible the demand for generic medication has grown rapidly across the world. This has led to developing countries producing generic drugs. Although many companies are based in India, the biggest is Teva Pharmaceutical Industries in Israel. However, regulatory and quality control systems in poor countries are often inadequate such that low-quality and fraudulent drugs can enter the market. Moreover, it does not always follow that prices are lower since competition is limited in some countries.

H. ten Have and M. do C. Patrão Neves, *Dictionary of Global Bioethics*,
https://doi.org/10.1007/978-3-030-54161-3_268

Genetic Counseling

Genetic counseling is the process in which a healthcare professional such as a genetic counselor talks to patients, family members, or a couple about a genetic or hereditary condition that runs in their family. The counselor provides the necessary (detailed) information for the patient to make an autonomous and informed decision. Genetic counselors are healthcare professionals with medical genetics expertise and counseling training who not only provide genetic information to those affected and discuss it with them, but also help to interpret such information in light of personal issues such as expectations, needs, values, and beliefs. Genetic counseling is deemed a preventive intervention within medical genetics. Information about genetic risk is most often requested by individuals who are aware of a genetic condition in their family history. Such individuals include patients, family members, and especially couples who want to have a baby (preconception counseling). Although genetic counseling is frequently rejected by patients who simply would rather not know, it has currently become so important that it is considered necessary for most (if not all) genetic tests and is sometimes put forward as an argument to stop patients accessing genetic tests without a medical prescription such as online tests. Historically, genetic counseling started in the first half of the twentieth century when it was associated with eugenic movements and given the unfortunate name "genetic hygiene." After the Second World War there was a major shift in genetic counseling evidenced by it rejecting former genetic ideologies about the extermination of some people and the sterilization of others considered not worthy to live or to reproduce. Genetics moved on to become a medical expertise in the 1950s and genetic counseling was given a scientific grounding not connected to any ideology. Scientifically based genetic counseling adopted a position of absolute neutrality in which it presented factual information without orientation to any ideology, did not interfere with patient autonomy, and did not pressurize the patient when it came to decision-making. However, such a model was criticized for a number of reasons because it was widely believed that delivering neutral information was not possible, that the emotional status of patients did not always allow them to make autonomous decisions, and that many patients really needed advice. The new paradigm of genetic counseling started to take shape when it fully acknowledged what gene expression meant. It was tasked with delivering information and providing counseling in a way that was never directive or influenced by the counselor's values. Genetic counseling deals with many different ethical issues such as the emotional reaction of most people to their diagnosis and their consequent incapacity to make a rational decision. It also raises issues concerning feelings of guilt, fear of discrimination, abortion decisions, and insurance coverage.

H. ten Have and M. do C. Patrão Neves, *Dictionary of Global Bioethics*,
https://doi.org/10.1007/978-3-030-54161-3_269

Genetic Determinism

Genetic determinism is a doctrine that states that all features of living organisms are dictated not only by their genetic constitution including physical and psychological characteristics but also by their behavior. This doctrine is broadly accepted when it refers to all other beings (except humans). Individuals within the same species are identical and react in exactly the same way when facing the same external stimuli. In contrast, humans have over time developed free will, which allows them to choose to pursue a different path even when they start out in the same situation as others. Their actions determine the person they become. Therefore genetic determinism challenges the collective and individual identity of humans. Genetic determinism started to take shape in the nineteenth century mostly with Gregor Mendel's work on heredity (Mendel is today considered the father of genetics). Mendel's laws of heredity explained how the characteristics of living organisms (i.e., the traits of progenitors) are inherited by their offspring and transmitted from one generation to the next via "particulate matter" (later identified as DNA). Offspring are thus conditioned by inheriting characteristics transmitted by progenitors. Heredity theories have not only developed scientifically they have also evolved ideologically. Socially, philosophically, and politically, the idea that families considered bad would perpetuate themselves took hold mostly between the two world wars and led to the conviction that such families should be stopped from procreating by means of eugenic programs. The eugenic programs of the Nazis led to millions of people losing their lives as a result of which eugenism and the theory of genetic determinism weakened immediately after the Second World War. Scientifically, the discovery of the double helix structure of DNA by James Watson and Francis Crick in 1953 marked a milestone in the development of genetic determinism. This paved the way for genetics (the study of genes, heredity, and genetic variation in living organisms) to make amazing progress. The more genetics advanced, the greater the number of genes identified, and the better the understanding of how they functioned—the stronger the ideology of genetic determinism became and more links were established between behaviors and specific genes. This led to establishment of the Human Genome Project (HGP), which started in 1990 and ended in 2003 after having identified and mapped all human genes. The genome was considered the "book of life" and it was believed that all of human reality was written in the genes and that the complete mapping of the human genome would enlighten everything there was to know. At the time nobody doubted genetic determinism and its absolute sovereignty lasted 50 years (1953–2003). However, in 2003 geneticists became aware that DNA did not explain the whole sequence of all cellular mechanisms in humans. Boosted by the achievements of genetics, initially genomics and subsequently proteomics (new branches of

H. ten Have and M. do C. Patrão Neves, *Dictionary of Global Bioethics*,
https://doi.org/10.1007/978-3-030-54161-3_270

molecular biology earlier envisaged) attempted to go beyond the study of single genes and get an understanding of the interactions genes have with each other and of the proteins expressed by the genome, respectively. More recently, epigenetics (another new science earlier envisaged) definitively developed and showed that environmental factors do influence gene expression. The evolution of humankind is not entirely and inexorably linked to DNA dictated by our genes, gene activity can change because of external factors. Therefore human beings do have some control over the way they evolve, the way even some genetic diseases express themselves, and the way they develop their own singular identity. Although genetics is our given circumstance, it does not determine who we are.

Genetic Modification (GMOs), Animals

Animals have long been genetically modified via traditional breeding techniques or via selection by humans where the aim is to increase the production (and reproduction) of naturally occurring variations of specimens that show desirable traits. Although bioengineering (specifically, gene transfer techniques) has mostly been used in agriculture and in microorganisms to produce particular enzymes, it has also been applied to animals with considerable success. The goal has been to enhance certain features such as size and faster and greater growth or to introduce new traits such as disease resistance. The first complex organisms to be genetically modified were mice (in the 1980s) achieved by introducing new DNA (oncogenes) by microinjection into a mouse embryo. Such a transgenic mouse would be predisposed to develop cancer. The production of transgenic animals to study different human diseases is still one of the main reasons for their creation. More recently, transgenic animals have also been produced for xenotransplantation research and for research into developing human tissues and organs for transplantation. Since that time many other genetically modified (GM) animals have been produced (i.e., pigs and cattle) with the aim of developing livestock that are transgenic the first of which was produced in 1985. Cloned transgenic cattle produce milk supplemented with higher levels of certain proteins (beta-casein and kappa-casein). The main goal in this field is to create animals that can produce different substances that are of interest to humans and at the same time increase their economic value. In addition, the production of transgenic livestock aims at getting animals to grow faster and have greater resistance to their natural vulnerabilities. The first GM animal to be commercialized was the GloFish in 2003. This is a colorful, brilliant, and fluorescent fish created to live in small aquariums as a decorative item. Transgenic pets (such as hairless cats) have become an important field of scientific and business activity. Such cats are considered highly practical to keep indoors, easy to keep clean, and diminish the risk for allergies in humans. Although biomedical research, livestock production, and pet creation are currently the three major fields for GM animals, another field is population control and conservation (biological control) for reducing or mitigating pests and pest effects in nature (e.g., invasive species in a natural environment). Transgenic animals have also been used to control infectious diseases such as malaria by spreading GM mosquitos despite the first experiences not being positive. The ethical problems associated with GM animals are many. In addition to transverse ethical issues such as the denaturalization of life, the uncontrolled

H. ten Have and M. do C. Patrão Neves, *Dictionary of Global Bioethics*,
https://doi.org/10.1007/978-3-030-54161-3_271

dissemination of GM animals, and questions regarding intellectual property, there are specific issues such as the utilitarian use of animals for human benefit, protection of animal well-being, and growing disquiet about the use of animals in scientific research. Although opposing interests often clash in these areas, public opinion is playing an ever more important role.

Genetic Modification (GMOs), Food

Food production is the most relevant reason to genetically modify organisms. Genetic modification first started with vegetables in 1994. Although the Flavr Savr (pronounced "flavor saver") tomato was the first commercialized genetically modified (GM) food, it was not long before such modification extended to animals, livestock, and fish such as salmon. The first animal approved for human consumption was the AquAdvantage salmon in 2015. GM food has been viewed as the solution to hunger in underdeveloped countries and the lack of food security worldwide. However, some critics consider the problem does not lie with the production of food, but downstream of it with food distribution channels and systems. Such a view is enhanced by GM crop development not focusing on food security, but on commercial interests, on the economic value of food products, and on satisfying consumers in developed countries who expect more nutritious products that taste better. Matters are made worse by many GM crops providing raw material for different lines of business such as the textile industry (cotton). However, intensive production to increase the volume and lower the prices is widely accepted as not being environmentally sustainable. Another important problem raised by GM food is safety. GM food is a concern since little is known about it and even traditional food products have been known to contain foreign substances to which the consumer is allergic and can be life threatening because of the risk for anaphylactic shock. The WHO discourages the use of allergenic DNA (unless it can be proved that the gene does not cause the problem), thus the labeling of all substances becomes crucial. Moreover, many people consider there is still not enough information about long-term effects. International organizations such as the FAO and WHO have published guidelines for the safety assessment of GM food (specifically, GM animals and their derived products). In short, the most important ethical issue concerning GM food is the protection of human health, although the environmental impact of its production runs it a close second.

H. ten Have and M. do C. Patrão Neves, *Dictionary of Global Bioethics*,
https://doi.org/10.1007/978-3-030-54161-3_272

Genetic Modification (GMOs), General

Genetically modified organisms (GMOs) are those whose genetic material has been altered via genetic engineering such as molecular cloning, recombinant DNA technology, gene delivery in which the three Ts (transformation, transfection, and transduction) are used to introduce foreign DNA, and gene editing. Organisms (both plants and animals) have been genetically altered throughout history by selection by humans (in which desirable traits in some species are enhanced). However, the genetic modification or manipulation of an organism today is brought about by using different bioengineering techniques that alter the genome of an organism by introducing or transferring fragments of DNA from other species (transgene) to create a new organism (a GMO) that expresses the desired features. Transgenic techniques have greatly improved both since the 1970s when the first experiments were performed and since the 1980s when the first transgenic animal (a mouse) was produced. Such techniques are today not only more widely applied, they are also more efficient. The production of GMOs (both plants and animals) has huge potential for food production, biomedical research, and healthcare. Nevertheless, it has also been highly controversial from the very beginning. Indeed, there are ethical issues that are common to the production of GMOs regardless of the field of application or the goal. The first is the denaturalization of species by crossing species boundaries to produce new organisms. Humankind has the capacity to change nature (which has taken billions of years to evolve) as we know it today. The second is the danger of GMOs disseminating uncontrollably in nature, interacting with natural species, reducing biodiversity, and having unpredictable consequences. With this in mind and in accordance with the precautionary principle the Cartagena Protocol on Biosafety to the Convention on Biological Diversity (CBD, 1992) was approved in 2003. Its purpose is to "contribute to ensuring an adequate level of protection in the field of the safe transfer, handling and use of living modified organisms resulting from modern biotechnology that may have adverse effects on the conservation and sustainable use of biological diversity, taking also into account risks to human health, and specifically focusing on transboundary movements." Environmentalists are strongly opposed to GMOs as borne out by some environmental associations using violence, destroying crops, and threatening farmers. They also lobby politicians to stop them authorizing the production and commercialization of GMOs (mostly those aimed at human consumption). Countries have developed their own regulatory systems regarding GMOs and even the European Union leaves this matter to each member state. Different national laws concerning GMOs make international trading difficult, especially when blocs of countries such as the European Union mandates the labeling of GMOs. Another issue is the genetic modification of microorganisms such as bacteria and viruses.

H. ten Have and M. do C. Patrão Neves, *Dictionary of Global Bioethics*,
https://doi.org/10.1007/978-3-030-54161-3_273

The first medicinal use of genetically modified bacteria was to produce the protein insulin to treat diabetes. Although a new and still experimental generation of vaccines are being prepared using GMOs, genetic manipulation has another downside in that it can also be used to produce lethal biological weapons (bioterrorism).

Genetic Modification (GMOs), Human Beings

Advances in genetic modification techniques (mainly genetic editing) have opened up the possibility of applying them to human beings. Modification of the human genome can be done at the somatic level (altering the individual) and the germline level (altering the individual and his or her descendants). Cloning and gene editing are the principal techniques used and they are applied for therapeutic (gene therapy) or enhancement purposes. Genetic modification in somatic cells for therapeutic purposes raises major ethical issues. Ethical questions here relate to the safety of the procedure (especially since many interventions still take place at the experimental level) and to the quality of informed consent. Genetic modification in the germline is highly controversial irrespective of whether the purposes are therapeutic or enhancement. The recent announcement of the birth of the first two genetically modified babies (twins, designer babies) was by and large condemned by the public and by the scientific community. The overall argument is that no generation has the right to impose decisions that are irreversible on the next generation.

H. ten Have and M. do C. Patrão Neves, *Dictionary of Global Bioethics*,
https://doi.org/10.1007/978-3-030-54161-3_274

Genetic Modification (GMOs), Plants

Plants have long been genetically modified via traditional crop techniques or via selection by humans with the aim of increasing the production of naturally occurring variations of plants that show desirable traits (e.g., sweet corn). This resulted in the wide variety of crops available today. Bioengineering techniques have allowed the genetic modification of plants to become much faster, more controlled regarding specific traits to enhance, and more efficient concerning the outcome. It has also broadened the scope of species targeted. Since plants are by and large genetically modified for agricultural purposes the introduction of new genes from one species into another (not necessarily related) has significantly improved agricultural performance. This has specifically been done by making plants herbicide tolerant and insect resistant, by enriching their vitamin levels, by altering their fatty acid composition, by making them mature faster, and by improving their color. Some examples of genetically modified (GM) plants are tomatoes, cassava, corn, maize, rice, cotton, oilseed rape, and soybeans. The major ethical issues raised by genetically modifying plants are the risk for GM crops escaping into the wild and the impact of their consumption on human health. Escape to the wild has led to a number of legal initiatives determining which plants can be genetically modified and under what conditions and stipulating a safety perimeter between GM crops and others. Environmentalists continue to fight against GM crops and persist in using the arguments they first employed (notably, the denaturalization of life and the decrease of biodiversity). In contrast, farmers point out the significant economic benefit of genetically modified organisms (GMOs) in agriculture (greater productivity) and stress the environmental benefits such as reducing the chemical treatment of plants (resistance to pests or diseases) and soil contamination. However, the patenting of new GM plants that produce sterile seeds (a.k.a. terminator or suicide seeds) obliges farmers to buy new seeds every year for the following season and makes them dependent on the expensive products of big biotech enterprises. Farmers in developing countries still depend heavily on saved seed. This coinciding with traditional crops tending to disappear has resulted in the decline of small farmers and family agriculture. The impact the consumption of GM food has on human health has not provided any evidence of negative effects and scientists generally consider that GM food does not pose any risk to human health and that it is as safe as any other food traditionally grown.

H. ten Have and M. do C. Patrão Neves, *Dictionary of Global Bioethics*,
https://doi.org/10.1007/978-3-030-54161-3_275

Genetic Screening

Genetic screening refers to the study of someone's DNA to identify susceptibilities or predispositions to a specific genetic disorder. The aim is prevention and early diagnosis for better control of that particular disease or abnormality (predictive and presymptomatic testing). Screening does not determine diagnosis. It just distinguishes those who might have a specific condition (those with higher risk for the disease or disorder) from those who likely do not by making an assessment of family history. The main goal of genetic screening is to provide early treatment (newborn screening), to assist reproductive decision-making (reproductive genetic screening), and to identify individuals who could benefit from preventive therapy. Although the goal of genetic screening is preventive, it is becoming increasingly comprehensive in its capability to identify genetic conditions and susceptibilities (i.e., highly sensitive information not only relevant to making reproductive decisions but also to employers and health insurers). This raises a variety of important transverse ethical issues such as privacy, confidentiality, secondary use of information, non-discrimination, and social justice. It is important here to point out that genetic screening should only be recommended when there is family history to justify it (avoiding overdiagnosis). Moreover, genetic screening becomes problematic when it comes to disorders for which there is no cure or effective preventive measures. Screening in such cases is considered unethical because it creates anxiety (even depression) and simply makes matters worse. Furthermore, some predispositions might not express themselves and incidental findings might occur. Although it can be argued that this information might be beneficial at the preventive level, it can be devastating for the individual who did not benefit from his right not to know. In contrast, it can be argued that pertinent information about someone's life should be provided him or her as a right. Another related ethical issue is the screening of minors, which should only be justifiable when it is beneficial to the child, to couples before getting married, or to couples wanting children and even then only acceptable if the autonomy of couples to make their own decisions with the information received is respected. It must also be taken into account that genetic screening is not always accurate and that false positives (results that show an increased chance of something being abnormal when it is not) and false negatives (results that show there is nothing abnormal when it actually is) can occur. Therefore screening should always involve a geneticist tasked with recommending further genetic testing whenever there is a need to accurately confirm findings.

H. ten Have and M. do C. Patrão Neves, *Dictionary of Global Bioethics*,
https://doi.org/10.1007/978-3-030-54161-3_276

Geneticization

Geneticization refers to the sociocultural process of interpreting and explaining human behavior using the terminology and concepts of genetics such that all social interactions relating to such behavior are viewed through the prism of biomolecular technology—not just health and disease. First introduced in the 1990s the concept has developed into a heuristic tool to study and critically analyze the impact new genetics has not only at the individual, societal, and cultural levels, but also within healthcare by influencing and transforming notions of health and disease, diagnosis, and prevention. The concept of geneticization was first introduced by Abby Lippman in her publication on prenatal genetic testing and screening in 1991. Her primary focus of critique is the current tendency to view genetics as the solution to all human problems. The concept not only articulates that genetics in current society is little more than hype leading to exaggerated expectations, it also points to a more fundamental question: How are science, in general, and genetics, in particular, influencing modern society and culture? Western culture is currently deeply immersed in the processes of geneticization. Such processes imply redefining individuals in terms of deoxyribonucleic acid (DNA) codes, a new language to describe and interpret human life and behavior using the genomic vocabulary of codes, blueprints, traits, dispositions, genetic mapping, and the approach of gene technology to disease, health, and the body. Analysis of film, television, news reports, comic books, advertisements, and cartoons shows that the gene is a very powerful image in popular culture. It is considered a cultural icon or an entity crucial to understanding human identity, everyday behavior, interpersonal relationships, and social problems—not just the unit of heredity. The growing impact genetic imagery has on popular culture has been related to genetic essentialism (a.k.a. genetic determinism), the belief that human beings in all their complexity are products of a molecular text. Geneticization allows focusing the analysis of genetic essentialism at two levels. The first is at the surface level of individuals where genetic knowledge not only has the potential to create new possibilities for intervention, treatment, and prevention but also to provide individuals with choices. The second is the deeper, more hidden level of social and cultural expectations that guides particular decisions in one or more specific directions depending on the social discourse and cultural images that have been created by genetic knowledge and technology. Geneticization draws particular attention to socioethical issues that tend to be neglected or disregarded because the moral principle of respect for individual autonomy currently prevails. It also allows bioethics to criticize oversimplifications in current approaches to genetics and to rethink common concepts of disease, health, and body. Genetic discourse today permeates modern society and culture and articulates major values of neoliberal society and culture, while shaping them into an objective,

H. ten Have and M. do C. Patrão Neves, *Dictionary of Global Bioethics*,
https://doi.org/10.1007/978-3-030-54161-3_277

scientific form that presents neutral options such that the preferences and desires of informed and rational consumers can be satisfied. The focus of bioethics should be on the neoliberal ideology that argues that genetics should be used to meet the expectations, hopes, and desires of individuals to overcome ageing, suffering, disability, and finiteness. The concept of geneticization therefore informs bioethics that biomedicine and bioscience should be associated with biocriticism, i.e. critical analysis of the ways life in current societies is conceptualized and operationalized.

Genome Editing (*See* Gene Editing; CRISPR)

Genome editing (a.k.a. gene editing) refers to deliberately modifying DNA by insertion, deletion, or replacement of a gene at a specific site in the genome of an organism or cell. Rapid innovation in genome-editing technology has led to the advent of a number of methods. The first involved artificial restriction enzymes named "zinc finger nucleases" (ZFNs) that have been used since the 1990s. The structures of ZFNs are engineered from naturally occurring proteins to bind to specific DNA sequences in the genome and cut by deleting or replacing the genome at a specified locus. However, ZFNs are difficult to apply, require a long time to prepare, and can only be used once for each target DNA sequence. In 2009 a new genome-editing technology was developed named "transcription activator-like effector nucleases" (TALENs) for a new class of proteins (found in nature). TALENs are restriction enzymes that can be engineered to cut specific sequences of DNA. Although ZFNs and TALENs can both edit the genome with comparable efficiency, the advantage of the latter is its greater simplicity. The success of letting the evolution of genome editing follow such a path was confirmed in 2012 with a third editing technology named "clustered regularly interspaced short palindromic repeats" (CRISPR)—a family of DNA sequences found in the genomes of prokaryotic organisms like bacteria and archaea—in which the protein Cas9 is used as molecular scissors to cut, copy, and bind DNA strands. CRISPR-Cas9 needs only a guide sequence 20 nucleotides long to bind them to the selected site. Therefore it is simpler than former methods that require substantial protein engineering. Moreover, it allows the introduction or removal of more than one gene at a time and can be used on organisms previously resistant to genetic engineering. CRISPR-Cas9 has greatly improved the speed, cost, simplicity, accuracy, and efficiency of genome editing and is a customizable alternative to other existing genome-editing tools. There have been significant advances in CRISPR-Cas9 technology as a result of applying the principles and methods of chemical biology to genome-editing proteins. It has become more precise and can be applied to more complex genomes thus making it more useful (especially for therapeutic purposes). A good example is CRISPR-SKIP, programmable gene splicing with single-base editors, that works as a molecular pen (instead of molecular scissors) directly modifying the chemical composition of the bases (adenine, guanine, cytosine, and thymine). Gene editing can be performed in all life-forms from bacteria, plants, animals (almost all cells in all living organisms contain DNA) to humans. Since it can literally alter all living organisms and be performed at the somatic cell level (at which mutations only affect the individual) and the germline cell level (at which all mutations are passed on to offspring) it has the potential to modify single individuals or alter entire future generations. The variety and complexity of the

H. ten Have and M. do C. Patrão Neves, *Dictionary of Global Bioethics*,
https://doi.org/10.1007/978-3-030-54161-3_278

ethical issues raised by genome editing cover a wide range of applications (agriculture, livestock, medicine), direct impacts (recovery, healing, enhancement), indirect impacts (biodiversity and ecosystems), and deliberate misuse (biological weapons). Hence the growing accessibility of this technology to ever larger groups of scientists requires its application to be subject to global governance and tighter safety measures.

Genomics

The word "genome" derives etymologically from the Greek words *genesis* (birth) and *soma* (body) and literally means a body of genes. The word "genomics" takes the "-ics" ending of the word "technics" from the Greek *techne* (skill) and refers to the application of molecular techniques to the study of genes. It was first coined by the geneticist Tom Roderick in 1986. Genetics naturally evolved into genomics. Genomics is a branch of molecular biology and an interdisciplinary field that studies the entirety of a person's genes (the genome), their structure, function, evolution, mapping, and editing including their interactions with one another and with the person's environment. While genetics focuses on a single gene, studies how it functions, and looks at its composition, genomics focuses on all genes and their multiple interrelationships and identifies the way in which they influence the growth and development of an organism. Genomics only became possible after the human genome was mapped in 2003 (and while it was being mapped) as a result of technical advances in DNA sequencing and computational biology. Contemporary genomics research shows great promise for medicine (in particular) and provides scientists and clinicians with the information and tools they need to understand the role multiple genetic factors acting together can play in complex diseases thus resulting in more accurate diagnosis and treatment. The major ethical concerns of genomics refer to intellectual property and patenting regimes. The patenting of DNA has been an important issue since the launch of the Human Genome Project and the prospect of associated economic profits. It has been robustly argued that discovery and innovation are entirely different achievements and that only the latter should be patented. Since the purpose of patent registration is to encourage more and better innovation, natural DNA cannot be patented. In addition to such ethical concerns, other issues such as biobanks, data use, commercialization, and benefit sharing are at stake because of their effect on social justice when underdeveloped countries are involved. There are other ethical issues implicated in genomics but they are common to many types of biomedical research such as medical engineering, tissue banking, and transplantation. Such issues call for due contextualization (i.e., privacy and confidentiality), informed consent, and right to withdraw from clinical research.

H. ten Have and M. do C. Patrão Neves, *Dictionary of Global Bioethics*,
https://doi.org/10.1007/978-3-030-54161-3_279

Ghostwriting

Ghostwriting refers to authors writing work on behalf of others. It is not uncommon and many famous people produce books that they have not written themselves in this way. However, ghostwriting is relatively new in scientific publications. The revelation that a substantial number of publications had not been written by the named author(s) was a major embarrassment for the scientific profession. Scientists are supposed to be honest, objective, and take responsibility for the results of research and analysis presented and communicated in publications. Since the primary reason they publish their results is to share such knowledge with colleagues and to improve the treatment of patients, readers feel deceived and manipulated when it turns out that the work has been produced by a ghostwriter. Although it is not clear how common ghostwriting is in the world of medical publishing since it is only rarely reported, it is estimated that approximately 10% of publications in biomedical journals are ghostwritten. However, detailed studies of promotional campaigns for drugs (as revealed in legal proceedings) gave much higher figures (over 50% in some cases). It is widely believed that approximately 40% of journal papers on clinical trials of new drugs (and the percentage is even higher for conference presentations) are ghostwritten. Although ghostwriters are identifiable in most cases as professional writers working for medical communication agencies, their names are withheld. Attributing the publication to a prestigious academic who has not contributed to the text is dishonest and makes a mockery of the independence and reliability of the opinion being provided. Such publications cannot be construed as reporting scientific findings honestly even though such findings may be the work of a prestigious academic. Ghostwriters are often part of marketing strategies in which they (or their employers) are usually paid by the pharmaceutical company sponsoring the research. Since it is unclear who is behind the report and what interests are at stake, readers are therefore manipulated and deceived. Although this is often done in a subtle way in which specific drugs are not directly promoted, the problem with ghostwriting is that it can be harmful to public health. Since such publications are misleading in that they do not give an honest presentation of actual risks and benefits, they undermine scientific objectivity since conflicts of interest are not disclosed. Getting a ghostwriter to write not only on behalf of a well-known scholar but to attribute it to the scholar falsely suggests disinterestedness and objectivity. Publications in prestigious journals are sources of authority that should be reliable and objective. It is generally recognized that clinical trials sponsored and published by pharmaceutical companies are not only always uniformly positive about their product they also underreport side effects. By masquerading as interested parties ghostwritten publications cover up the potential biases of medical knowledge. Deception and concealment

H. ten Have and M. do C. Patrão Neves, *Dictionary of Global Bioethics*,
https://doi.org/10.1007/978-3-030-54161-3_280

threaten the trust people have in scientific knowledge. This is the main reason many scholars believe ghostwriting is wrong. It erodes the basic idea that science is reliable and trustworthy not just among scientists but also among the public.

Global Compact

The UN Global Compact is a voluntary initiative launched in 2000 to bring businesses and civil society together on the basis of 10 principles relating to human rights, labor, the environment, and anticorruption. At the moment more than 9,900 companies in 161 countries are associated with the UN Global Compact. Its purpose is to turn business into a force for good and its mission is to promote corporate responsibility by aligning strategies and operations with the 10 principles and taking strategic action to advance broader societal goals such as the UN's Sustainable Development Goals. The 10 principles relate to human rights, labor, environment, and anticorruption. Principle 1 states that businesses should support and respect the protection of internationally proclaimed human rights. Principle 2 states that businesses should not be complicit in human rights abuses. Principle 3 states that businesses should uphold the freedom of association and the effective recognition of the right to collective bargaining. Principle 4 states that all forms of forced and compulsory labor should be eliminated. Principle 5 states that child labor should be effectively eliminated. Principle 6 states that discrimination regarding employment and occupation should be eliminated. Principles 7–9 relate to the environment by stating that businesses should support a precautionary approach to environmental challenges (Principle 7), undertake initiatives to promote greater environmental responsibility (Principle 8), and encourage the development and diffusion of environmentally friendly technologies (Principle 9). Finally, Principle 10 states that businesses should work against corruption in all its forms including extortion and bribery. Most pharmaceutical companies today have special units and programs tasked with bringing about corporate social responsibility. Increasing investment has been given to research into disease in developing countries in an attempt to diminish the 10/90 gap. This relates to a report by the Commission on Health Research for Development in 1990 that states that less than 10% of worldwide resources given to health research went to health in developing countries where more than 90% of all preventable deaths worldwide happen. Global efforts to increase access to medicines are more often publicly monitored and independently assessed (e.g., the Access to Medicine Index) as a result of the UN Global Compact being based on voluntary cooperation and reporting and on the basic idea that corporate responsibility is best realized on a voluntary basis and through self-regulation rather than through external or legal regulation. The problem with this argument is that voluntariness is not a strong enough incentive to bring about real change. Although codes of conduct have been adopted for various types of business, examples of effective changes in business practices are rare. The UN Global Compact is presented as a learning network whose critique is that it seems more interested in sharing experiences than in complying with its principles. It is

H. ten Have and M. do C. Patrão Neves, *Dictionary of Global Bioethics*,
https://doi.org/10.1007/978-3-030-54161-3_281

also criticized as being more of an opportunistic strategy than a moral one. Many companies regard social responsibility as a matter of charity in which donations and responsible conduct are motivated ultimately by self-interest—not as a moral obligation. Although the UN Global Compact is important for companies because it makes them more competitive, research shows that big pharmaceutical companies only get engaged in social responsibility when reputational benefits, employee satisfaction, or new-market creation are at stake. Moral obligation is rarely mentioned. An even more critical argument is that social responsibility is often used as an excuse for neoliberal globalization being a strategy to pacify the increasing critique of the disastrous impacts of corporate power. For example, the collapse of a garment factory in Bangladesh killing many people (in April 2013) pointed to the behavior of Western clothing brands like Gap (which has been a member of the UN Global Compact since 2003) that were mainly interested in cheap production and cared little about workplace safety and labor rights. Consumers in developed countries put pressure on the companies and global activism obliged them to respond and to make clear their social responsibility. Another example of corporate power is the outsourcing of clinical trials usually justified by lower costs, faster ethical review, speedy research data generation, and easier and quicker recruitment of research subjects. However, few companies give priority to researching diseases that affect often vulnerable populations from which subjects are recruited. Social responsibility is often used rhetorically to conceal the fact that outsourcing reinforces existing inequalities. The health needs of countries are not addressed since the benefits go to companies and people in Western countries. Although appeals to social responsibility can express sincere moral convictions, they can also be little more than an exercise in public relations. Since social responsibility as a theoretical component of global governance is playing an increasingly important role, the effectiveness of the UN Global Compact will depend on whether its role is conceptualized as problem solving or as critically transforming global challenges.

Global Fund

The Global Fund to Fight AIDS, Tuberculosis and Malaria was established in 2002 as an independent organization in the form of a partnership made up of governments, civil society, and technical organizations such as the WHO. It is a new funding initiative in which states commit themselves to making contributions and allows them to circumvent existing UN mechanisms. The fund raises and invests USD4 billion per year to assist local programs in more than 100 countries. Donor governments provide 95% of the resources and the rest comes from foundations and the private sector. The Global Fund plays a major part in global health as borne out by it disbursing approximately 10% of all development assistance for health in 2015.

H. ten Have and M. do C. Patrão Neves, *Dictionary of Global Bioethics*,
https://doi.org/10.1007/978-3-030-54161-3_282

Global Justice (*See* Justice)

Global justice refers to applying the principle of justice to global bioethics. There are a number of reasons global justice has become a normative tool in global bioethics: globalization is associated with increasing inequality and inequity; globalization has not led to a world that has a level playing field for everybody; and globalization benefits a minority of countries that control economic and political decision-making. However, most people live in a world pervaded by hunger, austerity, unemployment, growing insecurity, and vulnerability. There are other reasons for emphasizing the importance of global justice: the consumption of natural resources is not equal; total consumption of resources results in an ecological deficit; more is consumed than the Earth can sustain (the main culprits live in just a few wealthy countries); people in most countries do not have a just share of the planet's resources (they suffer disparities in health, disease, food and water); biodiversity is not equally distributed; countries rich in biodiversity are generally poor and have little power within the global system (as a consequence of their resources being exploited by developed countries); and policies designed to protect biodiversity can have unjust consequences. Another reason for emphasizing the importance of global justice is concern about future generations and about the bioeconomic paradigm that has prevailed in environmentalism since the adoption of the Convention on Biological Diversity. Although practical policies regard biodiversity as a global resource that generates benefits and nature as an economic asset, such a paradigm results in appropriation, lack of access, and exclusion thus raising questions of justice and injustice. The principle of global justice transcends the autonomy of individuals and focuses critical attention on the underlying structures and power constellations that determine the social, economic, and environmental conditions under which people live. It not only criticizes the inequalities brought about by neoliberal policies and practices, but robustly argues that the benefits of globalization should accrue to everybody. When it comes to the environment, justice is an abstract principle translated into social movements and activism across the world and draws attention to justice in such areas as health, the environment, food, and water. Global justice is a contested notion. Some argue that the principle of justice is only applicable within the state and cannot be extended at the global level since there are no institutions to implement it. However, from the cosmopolitan perspective we have responsibilities to other people—not only to people in our own community or state; and global justice is based on human rights (there are global governance mechanisms and institutions to implement the principle). However, these two interpretations (the state perspective and the cosmopolitan perspective) are not necessarily incompatible in that states have domestic and global responsibilities. Irrespective of whether or not the principle of

H. ten Have and M. do C. Patrão Neves, *Dictionary of Global Bioethics*,
https://doi.org/10.1007/978-3-030-54161-3_283

justice applies at the global level, it still has implications for the range of duties states have. The first interpretation (state) argues we only have duties to our own community—not to outsiders and that duties are not impartial because we always have special relationships. The second interpretation (cosmopolitan) argues we have duties to all human beings. Duties can be negative (to avoid deprivation and not to cause suffering) or positive (to provide assistance). Since harming others is considered worse than not preventing harm, negative duties are morally stronger than positive ones. The philosopher Thomas Pogge used this to argue that people in high-income countries have a duty not to impose a harmful global order on people in low-income countries. The current global order is a major cause of poverty and a violation of human rights when it should be to the benefit of all. People in rich countries contribute to and participate in an unjustly structured world by claiming intellectual property rights and in so doing violate the negative duty not to harm other people. The cosmopolitan perspective criticizes the emphasis on charity and aid. Although the state perspective argues nations have no duties to outsiders and that providing assistance is morally recommendable, this is not a moral duty but charity. Such a view is criticized because it does not address the injustice of social structures because many inequalities are created by root causes that cannot be removed or remediated by individual sacrifices regardless of how much individuals donate to relieve the needs of others. Placing the emphasis on aid and humanitarian activity fails to acknowledge structural injustice, rights, and obligations. Once the principle of global justice is accepted, debates about what it should include and the consequences of not adhering to it can take place. Such debates should focus on the notion that basic human needs such as food, water, shelter, education, and healthcare are rights that demand protection. Basic needs should transcend cultures such that all human beings can meet their basic needs and fulfill their potential. Therefore policies and practices should focus on providing for needs and states should accept they have duties (both negative and positive) irrespective of how near or distant people are.

Globalization

In the past few decades globalization has transformed human existence. Globalization refers to a multidimensional set of social processes that have created and intensified social interdependencies connecting the local to the distant. There are a number of processes at work here such as globalization being associated with the emergence of a new economic order that regards the world as a single market, new forms of trading and financial infrastructures emerging, and transnational corporations becoming more powerful than many nation states. Globalization is also a political project in which nation states, politics, and governance are increasingly deterritorialized (severing social, political, or cultural practices from their native places and populations). The flow of people, money, and technology has made the nation state less powerful and allowed intergovernmental organizations to play an increasing role. Although globalization is driven by consumerism in which material possessions accumulate and economic welfare increases, at the same time continuous growth endangers the planet's ecosystems. Increasing global interdependence implies environmental degradation will have impacts beyond borders leading to pollution, destruction of biodiversity, and global warming becoming worldwide phenomena. Furthermore, globalization is a process in which cultures themselves are exchanged and the products of one culture are almost instantaneously available in other cultures. Although this may lead to better appreciation of the diversity of cultures, it runs the risk for one culture imposing itself such as the predominance of the English language in scientific publications and internet communication and the fear many current languages will disappear. Finally, globalization is characterized by having an ideological framework guided by what are called "globalisms" (i.e., specific ideas, norms, and values that are taken for granted). The dominant globalism (at least in the earlier stages of globalization) was and still is neoliberalism, which assumes that the market is a self-regulating mechanism whose goal is to remove any constraints on free competition. All of this demonstrates that the processes of globalization have positive and negative aspects. Although they create enormous possibilities for communication, they also run the risk for homogenizing language and culture. Despite making more goods available, they do so at the expense of social inequalities and environmental degradation. Increasing worldwide interconnectedness in healthcare and medical science will likely present new opportunities for and challenges to bioethics. The processes of globalization have transformed bioethics into a global endeavor and have led to the emergence of global bioethics.

H. ten Have and M. do C. Patrão Neves, *Dictionary of Global Bioethics*,
https://doi.org/10.1007/978-3-030-54161-3_284

Good Death (*See* Death, Concept)

The expression "good death" derives etymologically from the Greek *euthanatos* (*eu* = good and *thanatos* = death) and is commonly used synonymously with and euphemistically for the current concept of euthanasia. Since "good" has a positive connotation anything classified in this way leads to its general acceptance and referring to euthanasia as a good death generally entails its approval. However, used more broadly it can lead in the opposite direction. Instead of referring to the possibility of being put to death at one's own request in a situation of unbearable pain and suffering (euthanasia), a good death can also designate death that is clinically assisted (not hastened) in which everything is done to guarantee painlessness and comfort during the natural process of dying (particularly in palliative care). Good death is then one that benefits from professional expertise to address the specific needs (physical, psychological, social, and spiritual) of the patient. An even broader use of the term employed by some refers to dying without being aware of imminent death such as dying in one's sleep. The common feature here is the absence of pain. In short, the expression is never self-explanatory since it has many interpretations and whenever used it needs to be attributed a meaning.

H. ten Have and M. do C. Patrão Neves, *Dictionary of Global Bioethics*,
https://doi.org/10.1007/978-3-030-54161-3_285

Governance

Governance is the process of governing and is often contrasted with government as an institution. The term derives etymologically from the Greek $\kappa\upsilon\beta\varepsilon\rho\nu\alpha\omega'$(to guide or steer). Governance became popular in the 1990s mainly as a result of the work of Michel Foucault. The term refers to the increasing influence non-state actors such as corporations, organizations, media, markets, and civil society have (rather than governments and institutions) on human social life. Governance is a term often applied in global healthcare. Although the WHO and the World Bank play important roles in global health governance, there are other significant roleplayers at the regional and global level such as NGOs, international foundations, and social movements. The traditional mechanisms and procedures of government were only adequate when they could be applied within states. However, as a result of globalization and inter-state cooperation at the global level, they were no longer sufficient. The notion of governance is often applied in global bioethics. The Universal Declaration on the Human Genome and Human Rights (1997) is credited as the first attempt to bring about global bioethics governance and did so by initiating a global debate on moral issues concerning the human body and human life in terms of economic considerations. UNESCO's initiative of setting up a global platform makes it clear that there are two sides to bioethics governance at the global level. The first is the focus on science as global commons that require shared principles. The second is the emphasis on human rights. Both sides recognize the core problem in global bioethics governance as inequity. At the global level the major concern is how global justice can be done rather than how controversies can be dealt with (at the national level) or how different approaches can be harmonized (at the international level). Although other components of governance are similar at a number of levels such as the need for cooperation, variety of actors, various levels of activity, and diverging objectives, addressing them is a much broader task at the global level. Bioethics has become a major component of global health governance and contributes in a major way to global regulation in which different normative instruments such as guidelines, recommendations, declarations, and conventions are developed; to oversight in which a network of ethics committees has developed in many countries providing oversight mechanisms particularly in the area of health research (such review systems are increasingly globally coordinated); to deliberation in which governance requires the involvement of the public in discussions about developments in science, technology, and healthcare; and to interaction in which global bioethics requires the involvement of the public in discussing values and ethical principles. However, the objectives of bioethics governance are unclear as to whether they should focus on solutions or problems. The lack of clarity about these two objectives has given rise to two forms of bioethics governance:

H. ten Have and M. do C. Patrão Neves, *Dictionary of Global Bioethics*,
https://doi.org/10.1007/978-3-030-54161-3_286

administrative and political. Administrative governance with its focus on solutions requires a specific methodology. First, the approach should be pragmatic in which the specific problem that needs to be solved is identified, general debate invoking abstract moral concepts should be avoided, and proceedings should concentrate on specific issues, cases, or a defined area. Second, such a demarcated approach requires a specific rationality in which fact finding and scientific evidence must be the first stage of work, procedures for decision-making should only be applied when the subject of debate is clear, and detailed analysis, argumentation, justification, and explanation can then be used for recommendations. Third, resolving the problem requires consensus and using the language of consensus building. It is the specific expertise of bioethics that not only allows such language to be used and controversial issues to be deliberated, but also provides the conceptual framework in which to analyze and justify particular problems and practices. In contrast to administrative governance, political governance requires another approach. New policies are necessary because of public concerns about science, technology, and commercialization of the human body and body parts. Such concerns need to be explored and articulated. Rather than taking a regulatory approach, engaging in moral discourse not only with policy-makers and scientists but with the general public is a better option. However, such an engagement calls for another methodology. First, an open and democratic approach is needed to clarify the problem(s) since ethical concerns can only be dealt with if they are taken seriously and if deliberation takes into account the concerns of civil society and its various interests. Second, facts and values cannot be separated since controversies about scientific and technological advances already imply moral disputes and diverging ethical views and since scientific evidence only presents information that is factual—not neutral. Instead of using an instrumental, rational approach focused on how to solve the problem, political rationality requires that the problem is interpreted and explored in terms of responsibilities, obligations, relationships, and human rights. The role of scientists in open deliberation is therefore limited since science is not the only source of information and knowledge. Furthermore, public input should be treated differently. Rather than regarding citizens as needing information, as assuming they have an information deficit, and as needing education, public consultation should mean engaging them in dialogue and allowing them to articulate their visions on what the problem is. Consultation should mean participation. Third, the emphasis on consensus often silences moral debate, sets up a particular framework for discussions, and leaves no place for critical reflection in which priority is given to technical and practical matters. On the other hand, political governance emphasizes that searching for common values is more important than consensus. The public process of deliberation in which all relevant views are taken into account is what counts. This implies that voice should be given to dissent, that controversies should not be avoided, and that pluralistic viewpoints should be appreciated.

Grassroots Activism

Grassroots activism is an example of globalization from below. It is commonly assumed in global bioethics that the best way to influence and change practices is by developing and implementing global standards. Such a view takes for granted that the processes of globalization materialize in local contexts from above and that states are the main actors. However, many non-state actors today are involved in globalization. There is also intensive interaction between global and local values. Therefore globalization is not one-sided but operates within many diverse local contexts. For change to occur localization of implementation is what is needed. Grassroots organizations and movements play an important role here. Grassroots is a metaphor referring to local settings in which people or communities organize themselves for political purposes. They take collective action at the local level in an effort to bring about changes at both the local and the broader level. Activities involve self-organization, participation, bottom-up decision-making, and taking responsibility for their communities. Improvements in human rights practices are more often the result of efforts of grassroots movements in local settings than the accomplishments of global institutions. Although institutions can create norms, implementation is decentralized and domestic. Actors and networks from civil society take up a cause (e.g., access to healthcare), organize themselves, take up the fight, and connect with similar movements and networks in other countries. They develop a mode of practice around the right to health, one that goes beyond simply applying such a right. Activities need to be adapted to local circumstances and values and can therefore be more successful in one setting (e.g., the activities of the Treatment Action Campaign in South Africa that were enhanced by the right to health being included in the new constitution of 1996) than in another. Looked at this way, practices are constructed through collective labor within a specific context.

H. ten Have and M. do C. Patrão Neves, *Dictionary of Global Bioethics*,
https://doi.org/10.1007/978-3-030-54161-3_287

Harm (*See* Benefits and Harms)

In bioethical discourse the term "harm" is often used as the opposite of benefit. Since Hippocratic times a basic ethical principle of medical ethics has been *primum non nocere* (first do no harm). The principle of non-maleficence often needs to be balanced against that of beneficence since it is not always easy to make a clear distinction between harm and benefit. Although directly harming a patient clearly cannot be justified, beneficial medical interventions frequently involve some harm (e.g., surgery). At its most basic non-maleficence can be regarded as the lowest level of beneficence (i.e., the primary imperative of not inflicting harm). Nevertheless, the ethical principle of non-maleficence as a result of its consequentialist nature requires that potential harm be effectively avoided. Beneficence is usually regarded as more encompassing in that it requires removing harm and more compelling in that it requires preventing harm. The highest level of beneficence is the promotion of what is good. Since the ethical principle of beneficence is also consequentialist it requires that what is good is effectively accomplished. Benefits and harms in medical practice and research are not clear-cut and obvious since there are always risks, benefits cannot be predicted, and the possibility of unintended harm is ever there. Patients cannot be absolutely protected against harm while benefits can never be guaranteed. Harm cannot be completely avoided (e.g., every drug no matter how beneficial it is may have toxic adverse effects). This means benefits and harms need to be balanced in practice. The ethical requirement is to maximize benefits and minimize harms. However, a difficulty arises here in that four types of harm have been distinguished: physical or psychological harm such as injuries, pain, mutilation, anxiety, depression, and death (in the worst case); harm to interests (harm can vary because not everybody has the same interests); harm as unfairness (somebody is harmed because he or she is treated unfairly); and harm as an infringement of moral integrity (people are harmed because they are intentionally encouraged to break moral rules). The notion of harm in the context of healthcare is usually interpreted narrowly as physical or psychological harm ranging in seriousness from death, violation of the integrity of the body, and pain through to disturbance of mental well-being.

H. ten Have and M. do C. Patrão Neves, *Dictionary of Global Bioethics*,
https://doi.org/10.1007/978-3-030-54161-3_288

Health Education and Promotion

Health education and promotion is part and parcel of public health. Its importance was emphasized in the Declaration of Alma Ata sponsored by the WHO in 1978. The declaration takes a broad view of health as a state of complete physical, mental, and social well-being—not merely the absence of disease or infirmity. Moreover, it reaffirms health as a fundamental human right that all governments and healthcare workers have a duty to promote. In 1986 such ideas were further elaborated in the Ottawa Charter for Health Promotion where health promotion was defined as the process of enabling people to increase their control over health and to improve it. The charter points out that health promotion can only really occur when peace, shelter, education, food, income, a stable ecosystem, sustainable resources, social justice, and equity are in place. In 2006 the Bangkok Charter for Health Promotion in a Globalized World focused on the challenges of globalization and recommended a number of actions. It advocated that health should be based on human rights and solidarity; that investment in sustainable policies should be done to address the determinants of health; and that capacity should be built for policy development, research and legislation, and building alliances and partnerships with civil society. Much like public health, health promotion aims at influencing the social and environmental determinants of health rather than individual behavior. It aims to empower people to promote their health by addressing the social and economic factors that impact their health (e.g., by changing their lifestyles) thus making education an important component of health promotion. Ethical issues relating to health promotion and education are similar to those in public health. The assumption is that health should be a commonly shared value despite it not being the highest value in life for some people. Although the emphasis is usually on persuasion and providing information to encourage individuals to change their lifestyles (e.g., ceasing to smoke and taking more physical exercise), the effect is often so limited that further pressure needs to be applied. Although health promotion and education campaigns often articulate individual responsibility for health, this may have unintended consequences such as stigmatization, blame, lower self-esteem, and psychological harm. Such an emphasis on individual responsibility is of course questionable. In many cases unhealthy conditions are not the result of individual choices but the consequence of poverty, pollution, and marginalization.

H. ten Have and M. do C. Patrão Neves, *Dictionary of Global Bioethics*,
https://doi.org/10.1007/978-3-030-54161-3_289

Health Insurance

Medical needs are often financed by health insurance. It is argued that most people could not afford healthcare without having some kind of health insurance. The WHO has set universal health coverage as one of its health policy goals in which all people should be able to obtain the health services they need without the risk for financial hardship. Such services should be universal (i.e., available to everyone). However, such an ideal has not been reached in most countries and the goal has been limited to universal access to medically necessary care. Although health insurance is an important means of reaching this goal, health insurance schemes vary across the globe. National or social health insurance systems exist in Europe in which employers and employees contribute to sickness funds that provide a comprehensive package of health services (mostly defined by governments). Contributions are sometimes paid through taxation and supplemented by the government. In some countries such as the United States these insurance systems only cover specific populations (e.g., Medicare provides medical insurance for people older than 65), whereas in others voluntary and private insurance systems are increasingly encouraged in addition to public schemes. As technologies advance and populations age, insurance systems will be under ever increasing pressure to prioritize their services to maintain and improve population health. The central ethical issues continue to be access to healthcare and justice.

H. ten Have and M. do C. Patrão Neves, *Dictionary of Global Bioethics*,
https://doi.org/10.1007/978-3-030-54161-3_290

Health Policy

Concerns about health have existed throughout human history. Healthcare was planned and organized long before the term "health policy" came to be used. The WHO defines health policy as "decisions, plans, and actions that are undertaken to achieve specific health care goals within a society." The aims of health policies can range widely from defining a vision for the future that establishes targets and points of reference for the short and medium term, outlining priorities and expected roles of different actors, to building consensus and informing people. Health policies involve much more than healthcare since they are often reflected in regulations, guidelines, and administrative measures issued by governments and NGOs. Health policies are guided by values in which ethical principles such as justice, autonomy, and beneficence are relevant. Although justice is essential in that it determines the distribution of benefits, rights, and responsibilities within a population, policies today also respect the autonomy of individuals. Since they are aimed at populations a crucial ethical dilemma is often the balance between individual freedom and the responsibility of governments. Moreover, policies should be beneficial, promote health and well-being, prevent harm, protect vulnerable groups (e.g., by putting in place special policies for persons with disabilities) and respect human rights.

H. ten Have and M. do C. Patrão Neves, *Dictionary of Global Bioethics*,
https://doi.org/10.1007/978-3-030-54161-3_291

Health Tourism (*See* Medical Tourism)

Although health tourism is basically the same as medical tourism, it has a broader meaning. It not only applies to medical interventions but also to a variety of interventions and applications aimed at maintaining and restoring health. Wellness tourism (a.k.a. health tourism) is a proactive form of medical tourism that promotes health and well-being by getting tourists to engage in physical, psychological, or spiritual activities; is rapidly increasing; and provides attractive packages promoting health and quality of life combined with tourism in luxury resorts. Health screening, relaxation programs, spa services, yoga, detox dieting, and fitness and massage are among the facilities available. Such programs are more correctly regarded as a form of tourism (rather than medical tourism, which is reactive) since they have a significant recreational component. Health tourism differs from pure tourism because travel is undertaken for the purpose of promoting health and well-being through physical, psychological, or spiritual activities. Wellness tourism grew 14% between 2013 and 2015 (totaling USD563 billion) while tourism itself only grew 6.9%. Such growth reflects the increasing concerns people have with health and well-being as evidenced by growing medicalization. One of the benefits of health tourism is that people can combine healthcare with holidays.

H. ten Have and M. do C. Patrão Neves, *Dictionary of Global Bioethics*,
https://doi.org/10.1007/978-3-030-54161-3_292

Health, Concept

Although health (and its restoration) is the primary goal of healthcare, it is a complicated concept not easy to define. Etymologically, health is related to wholeness and integrity. If someone's health is affected, then he or she will be healed (i.e., restored) and made whole again. Although health can mean harmony, balance, and order in regular language, in scientific discourse it often is not described. Medical textbooks usually discuss diseases and make the assumption that elimination of disease constitutes health. Such an approach implies that health is defined in a via negativa way (i.e., defined by what it is not). However, a positive notion of health has recently been articulated in health sciences in which health is more than the absence of disease. An example is the definition of health used by the WHO: it is a state of complete physical, mental, and social well-being—not merely the absence of disease or infirmity. Positive notions of health have been developed from various scientific perspectives such as statistics (health is the most frequent condition of the human body in a specific population), biology (health is adaptation to the environment), psychology (health is optimal self-realization), sociology (health is the capacity to perform social roles and tasks), and anthropology (health is a harmonious relation between human beings and their situation). Negative and positive notions of health have been criticized as either too narrow or too broad and have led to medical philosophers elaborating theories of health. Naturalistic theories articulate that health is objective, empirical, and value-free and along with disease is observable in the natural world. An example is the biostatistical theory of Christopher Boorse in which health is the normal functioning of an organism and its organs that contribute to its survival and reproduction making what is normal statistically determinable. In contrast, normativist theories regard health as a value-laden concept in that what is healthy does not depend on the natural state of someone but on the goals, projects, and aspirations he or she has in a wide variety of situations. For example, Caroline Whitbeck defines health as "A person's psycho-physiological capacity to act or respond appropriately (in a way that is supportive of the person's goals, projects, and aspirations) in a wide variety of situations." Another theory expounded by Daniel Callahan adopts a communitarian perspective in which health is regarded as a common benefit preconditional to communal existence and health is an individual feature that can only be understood from a social perspective. The question that arises is: What level of health is sufficient for society to function well?

H. ten Have and M. do C. Patrão Neves, *Dictionary of Global Bioethics*,
https://doi.org/10.1007/978-3-030-54161-3_293

Health, Global

Global health is defined by the WHO as "the area of study, research and practice that places a priority on improving health and achieving equity in health for all people worldwide." The definition shows just how concerned the WHO is with global health and is the reason for the WHO strategy of health for all. The Global Health Observatory of the WHO provides data on all member states and in this way allows trends in countries, regions, and worldwide to be monitored. Health-related statistics include mortality rates, data on communicable and non-communicable diseases, substance abuse, traffic accidents, health coverage, and essential medicines and vaccinations. Although global health is a notion increasingly used as a policy instrument, it has different interpretations. Sometimes it is used to emphasize health determinants that are outside the national scope such as infectious diseases and immigration. Other times it is used to emphasize inequities and the normative demand to eliminate health disparities. Since contemporary economic globalization has led to benefits and risks being inequitably distributed, monitoring global health is a way to minimize inequities through global health policies. The notion is also sometimes used to relate health to socioeconomic conditions such as poverty and to appeal to collective action. Rather than advocating personal responsibility it argues for structural interventions to be put in place to address health challenges such that global health requires global governance. A variety on the theme is planetary health in which the focus falls on the interconnectedness of human beings, societies, and biodiversity and in which human health cannot be separated from the health of the Earth. Although the concept of global health stresses that the health of all human beings is interconnected and that human health depends on healthy social conditions, the concept of planetary health is wider in that it connects human health and social conditions to environmental concerns.

H. ten Have and M. do C. Patrão Neves, *Dictionary of Global Bioethics*,
https://doi.org/10.1007/978-3-030-54161-3_294

Health, Social Determinants Of

The report of the WHO Commission on Social Determinants of Health (2008) states that human health is determined more by the conditions under which daily life is lived than by medical treatment and healthcare services. The determinants are very broad and range from unemployment, unsafe workplaces, urban slums, lack of education, gender discrimination, food insecurity, and air quality through to degraded natural environments. Viewed in such a broad way health can be seen to be comprehensive and social rather than biomedical. Reports on social determinants convincingly show that major differences in health exist within and among countries. The work of Michael Marmot showed that the conditions under which people are born, grow, live, work, and age significantly impact health. He noted the 20-year difference in life expectancy between the richest (83 years) and poorest (63 years) sectors of society in Baltimore. Similar gaps in life expectancy are found in many other cities (e.g., London). Despite inequalities in health resulting from living conditions diminishing, life expectancy across the globe is increasing. Nevertheless, the gap between populations living longest and those shortest remains the same. Austerity policies in many countries have undermined health and welfare policies and have made matters much worse. The notion that health is primarily determined by social conditions is often associated with the notion that health is a fundamental human right. The International Covenant on Economic, Social and Cultural Rights (1976) affirms this right and articulates that states have the responsibility to protect and improve health in every way possible. The Sustainable Development Goals adopted by the United Nations in 2015 brought the global problems of health, society, and environment together in a practical framework of international cooperation and action.

H. ten Have and M. do C. Patrão Neves, *Dictionary of Global Bioethics*,
https://doi.org/10.1007/978-3-030-54161-3_295

HIV (*See* AIDS)

Human immunodeficiency virus (HIV) alters the immune system by specifically targeting CD4 immune cells (white-blood T-helper cells that detect infections and cell anomalies). When these cells are disabled by HIV, the ability of the body to combat infections is reduced. Therefore HIV infection increases the risk of other infections and cancer and without treatment the infection will lead to an advanced stage of disease called "acquired immune deficiency syndrome" (AIDS). People can be infected with HIV and show no symptoms for a time while others develop flu-like symptoms 2 to 6 weeks after the virus has entered the body. HIV is transmitted through body fluids such as blood, semen, vaginal secretions, rectal discharge, and breast milk. Someone's HIV status can be determined by a specific blood test and early diagnosis is critical to chances of successful treatment. Although no cure is available, antiretroviral treatment (ART) can stop progression of the infection. HIV-infected people usually take a combination of medications such as highly active antiretroviral therapy (HAART) or combination antiretroviral therapy (cART). Early treatment with ART greatly improves quality of life, extends life expectancy, and reduces the risk of transmission such that people today with HIV have the chance to live a long and relatively healthy life. Moreover, the viral load in the blood can be so greatly reduced that HIV is no longer detectable in blood tests. Although treatment is available in developed countries, people in many other countries still die from HIV.

H. ten Have and M. do C. Patrão Neves, *Dictionary of Global Bioethics*,
https://doi.org/10.1007/978-3-030-54161-3_296

Homelessness

The Universal Declaration of Human Rights states that "Everyone has the right to a standard of living adequate for the health and well-being of himself and of his family, including food, clothing, housing and medical care and necessary social services …" (Article 25). Housing is regarded not only as a human right but also as essential to health. Although having a place called "home" is the typical situation of contemporary human beings, the United Nations estimates that approximately 150 million people worldwide are homeless. The percentage of homeless people differs greatly per country: in Portugal 0.03% of the population is homeless, in the United States 0.17%, and in Haiti 23.24%. Estimates of the number of homeless people can also differ because of the way homelessness is defined. Some argue that homelessness is not simply lack of a home or adequate residence but refers to a broader condition of lack of security and privacy. The UN Centre for Human Settlements (Habitat) has defined homelessness as a "condition of detachment from society characterized by the lack of the affiliative bonds that link people to their social structures." Since homelessness stigmatizes people as belonging nowhere rather than simply having no place to sleep, it can thus encompass those living in insecure circumstances, those in hostels, and those forced to sleep without shelter. The causes of homelessness in developing countries are often related to poverty, whereas unemployment, mental illness, substance abuse, broken families, and relationships also play a role in developed countries. Another cause of homelessness is war and violence. Some argue that homelessness often reflects value judgments in which people choose to opt out and deliberately decide not to have jobs or families. However, many others argue that in the overwhelming majority of cases homelessness is not a choice and that social, economic, and political circumstances are responsible for people being homeless. Debate about homelessness is focused on moral rights; meeting basic needs such as food, water, and shelter; and respecting human dignity. To their shame most countries do not acknowledge the rights of homeless people to shelter and security. Countries like Brazil and India do little to address the large numbers of street children let alone give them any attention. The WHO emphasizes the link between housing and health and has recently issued guidelines on what constitutes healthy housing. Another ethical concern is moral responsibility for improving the lot of homeless people. Although individuals often provide assistance and many private organizations and charities provide shelter and help, it is the responsibility of local, regional, and national authorities to address homelessness (implementing human rights that most governments have signed up to). Policies are required to ameliorate

H. ten Have and M. do C. Patrão Neves, *Dictionary of Global Bioethics*,
https://doi.org/10.1007/978-3-030-54161-3_297

structural factors such as economic and societal issues (e.g., poverty, discrimination, and lack of affordable housing), system failures (such as barriers to gaining access to public systems), and individual and relational factors (such as mental health, addiction, and interpersonal violence). There should also be policies to prevent homelessness.

Honor Codes

Honor codes are used to tell people how to conduct themselves appropriately in specific settings. For example, universities and scientific organizations have established rules against cheating, plagiarism, and fabrication of data. Some schools and universities require students to swear an oath that they will follow such codes. The purpose of honor codes is to create and maintain a culture of academic and scientific honesty and integrity. Similar codes of conduct exist in military academies. Honor codes are based on what is regarded as honorable behavior among members of a particular group or community. Some argue that honor codes are obsolete in ethics, that a sense of honor has disappeared in contemporary societies (replaced by the notion of dignity), and that positions in society no longer depend on honor. In contrast, dignity emphasizes the importance of recognizing everybody and thus introduces a more egalitarian point of view. Philosopher Kwame Appiah (*The Honor Code*, 2010) argues that a sense of honor is important to promoting conduct that is good and right and to reflecting the need for recognition and the desire for self-respect. Honor is compatible with the notion of equal dignity and is an important ethical resource since it guides action, constitutes what it means for a human life to be successful, and requires individuals to live up to codes of honor (thereby bringing about integrity). Moreover, by adhering closely to the rules of honesty people aspire to be seen as honest by their colleagues and other members of their community.

H. ten Have and M. do C. Patrão Neves, *Dictionary of Global Bioethics*,
https://doi.org/10.1007/978-3-030-54161-3_298

Hospice (*See* Palliative Care)

Hospices are places usually outside the hospital context where palliative care is provided. In 1967 Cecily Saunders established St Christopher's Hospice in the United Kingdom as the first specialized facility for care of the terminally ill. She started an international hospice movement that resulted in many hospices being set up in many other countries. Hospice care is a special form of palliative care that focuses on the quality of life of patients (and their families) suffering advanced and incurable disease and making what is left of their lives as comfortable as possible. Such care does not aim to hasten or prolong death and is usually provided when life expectancy is less than 6 months. Although the modern hospice movement started in 1967, hospices of some sort or another have long existed. For example, monasteries (especially those along pilgrim routes) had facilities for ill travelers and sick and disabled patients. The word "hospice" was first used by Jeanne Garnier in Lyon (France) in 1842. The word derives etymologically from the Latin *hospes* and *hospitium* (host and hospitality). There are four principal types of hospice care: routine home care in which care is delivered in the patient's home and involves an interdisciplinary team of caregivers (nursing aides, social workers, and spiritual care specialists); continuous home care in which care is more intensive and around-the-clock care services are provided; general inpatient care in which symptoms can no longer be managed through home care (this is usually short term and provided in a hospice unit of a hospital or in a freestanding hospice facility); and respite care in which short-term inpatient care is provided in the day unit of hospices (usually limited to a few days to relieve symptoms or provide a break for family caregivers). Hospices raise similar ethical problems to those of palliative care. Since the focus is on the patient and the family, important ethical notions such as relational autonomy, partnership, and shared decision-making arise. Other ethical issues relate to caregivers frequently facing heart-rending questions about withholding and withdrawing treatment and palliative sedation and issues of resource allocation since hospice care in many countries is limited to those who can afford it.

H. ten Have and M. do C. Patrão Neves, *Dictionary of Global Bioethics*,
https://doi.org/10.1007/978-3-030-54161-3_299

Human Dignity

The word "dignity" derives etymologically from the Latin *dignitas* (to be worthy, to be meritorious of honor) and originally referred to someone holding a prominent social or political position. It was the Roman orator and philosopher Cicero who first used the expression dignity of man in *De Officiis* (44 AD) to indicate high social status (social dimension) and the intrinsic quality of being human (anthropological dimension). Christianity (with Augustine and Aquinas) led to human dignity being enhanced as a common or universal trait of human beings as a result of being created in God's image (ontological dimension). Later, Immanuel Kant in his *Groundwork of the Metaphysics of Morals* (1785) defined human dignity as an intrinsic, unconditional, and absolute value of human beings who should be treated as ends in themselves (ethical dimension). Since all humans are reasoning beings endowed with free will they have value in themselves and cannot and should not be reduced to being used as means for other causes and functions or as objects. Morally conceiving human dignity as value without equivalent, without price, and not interchangeable prevailed right up the present day and was decisively reinforced in the Universal Declaration of Human Rights (1948), Article 1 of which reads: "All human beings are born free and equal in dignity and rights. They are endowed with reason and conscience and should act towards one another in a spirit of brotherhood." Human dignity was understood in the declaration as a moral principle or obligation and the ground on which human rights could be justified. Since then the principle of human dignity has been included in most Western constitutions, international statements (conventions, declarations, opinions, and recommendations), and good practice standards such as codes of conduct. It is considered a matrix principle (i.e., the cultural or sociopolitical environment in which something develops) and grounds the right to healthcare and the rights of patients. The principle of human dignity is part of the Universal Declaration on Bioethics and Human Rights, Article 3 of which reads: "Human dignity, human rights and fundamental freedoms are to be fully respected." Although the idea of human dignity has a long history and is currently recognized worldwide as a foundational principle of morality, its meaning is not unequivocal and its application to practice is not simple. The principle of human dignity has been used to defend conflicting ideas about euthanasia and abortion.

H. ten Have and M. do C. Patrão Neves, *Dictionary of Global Bioethics*,
https://doi.org/10.1007/978-3-030-54161-3_300

Human Rights

Human rights refers to a collection of basic privileges (rights) and freedoms recognized not only as belonging to all human beings but also as entitlements by virtue of being human. Such rights have been argued since antiquity as being natural and based on human faculties and as such are universal, inalienable, permanent, and unchangeable. Although human rights today are legally established, politically implemented, based on human needs, attributed to society, and constantly evolving, they are also a collection of moral principles and norms aimed at establishing a respectful relationship among human beings and protecting individuals and their dignity. The demise of ethical universals such as nature, God, and reason and the rejection of heteronomous (extrinsic) moralities has led to human rights being considered the new universal morality (a conventional resource built by human beings). The historical roots of the modern perspective on human rights go back to the 13th century and the first document asserting individual rights (The Magna Carta, 1215). The contemporary idea of human rights since the 16th century has been developed by prominent philosophers such as Francisco Suarez (1548–1617), Hugo Grotius (1583–1645), Samuel Pufendorf (1632–1694), John Locke (1632–1704), and Immanuel Kant (1724–1804). However, it was the 17th century when this movement started to grow as evidenced by the English Bill of Rights (1689), the French Declaration of the Rights of Man and the Citizen (1789), and the Bill of Rights in the US Constitution (1791). A new era was marked by the Universal Declaration of Human Rights (UDHR) by the UN General Assembly (meeting in Paris on December 10, 1948 and adopting General Assembly Resolution 217A) in the aftermath of Nazi atrocities toward different groups of people during the Second World War. Since 1948 human rights have continued to develop significantly and culminated in 1979 in Karel Vasak's "three generations" (focusing successively on the ideals of the French Revolution: liberty, equality, fraternity). The first generation refers to civil and political rights in which the rights of individuals to non-interference (negative rights) guaranteeing their freedom are provided (rights concerning life, liberty, security; non-discrimination of any kind such as race, color, sex, language, religion, political or other opinion, national or social origin, birth or other status; property ownership; privacy; freedom of movement and residence within the borders of the state they live in; nationality and asylum from persecution; freedom of thought, conscience, and religion; and peaceful assembly and association). In an attempt to reinforce the UDHR the United Nations adopted the International Covenant on Economic, Social and Cultural Rights (ICESCR) in 1966, which became known as the second generation of human rights. This refers to economic, social, and cultural rights and focuses on material conditions to be delivered (positive rights) to guarantee equal conditions and treatment among

H. ten Have and M. do C. Patrão Neves, *Dictionary of Global Bioethics*,
https://doi.org/10.1007/978-3-030-54161-3_301

all peoples (rights concerning social security, work, rest, leisure, reasonable working hours, periodic holidays with pay, health conditions, education). Since 1990 several documents have been adopted defending and protecting rights of a communal nature such as healthy environment, collective ownership (a.k.a. solidarity rights); political, economic, social, and cultural self-determination; economic and social development; humanitarian assistance; sexual, ethnic, religious, and linguistic minorities all of which are considered the third generation of human rights. In 1997 the Convention for the Protection of Human Rights and Dignity of the Human Being with regard to the Application of Biology and Medicine along with the Convention on Human Rights and Biomedicine developed by the Council of Europe (now undergoing ratification) inaugurated what became known as the fourth generation of human rights. The fourth generation refers to the right for genetic heritage to be preserved in the face of advances in biotechnologies and genetic manipulation (a.k.a. rights of future generations). The four generations of human rights are cumulative and have led to the UDHR being the foundation of many treaties and documents such as the Convention on the Rights of the Child (1989) and the World Medical Association Declaration of Lisbon on the Rights of the Patient. It was and still is the foundational document for many international organizations such as the Council of Europe.

Humanitarian Intervention

Humanitarian intervention refers to organizations or states (or a coalition of organizations and states) taking action to alleviate human suffering within the borders of a sovereign state. If human rights are systematically violated (e.g., genocide and ethnic cleansing) by states or because civil order has collapsed, then the international community not only has the right but also the obligation to intervene. Such a doctrine has been widely criticized since it does not respect the sovereignty of states and gives the impression of liberal imperialism. However, humanitarian aid is a much less controversial form of intervention. Confronted with serious human rights violations the international community has reframed the debate about humanitarian intervention as the responsibility to protect (R2P)—not as an argument about the right to intervene. It is the responsibility of every state to protect its citizens from genocide, war crimes, ethnic cleansing, and crimes against humanity. If states fail to do so, then it is the responsibility of the international community to protect that state's population. This is in accordance with the Charter of the United Nations. If peaceful measures prove to be inadequate, then military force may be used. Since the right to provide humanitarian assistance is recognized by international law, impartial humanitarian aid cannot be condemned as interference or infringement of a state's national sovereignty. Humanitarian intervention has a broader scope in that it is often used to provide assistance during natural and environmental disasters.

H. ten Have and M. do C. Patrão Neves, *Dictionary of Global Bioethics*,
https://doi.org/10.1007/978-3-030-54161-3_302

Hunger (*See* Food Security)

According to the Food and Agriculture Organization (FAO) of the United Nations in 2019 some 842 million people were suffering chronic hunger, did not have enough food to conduct an active life, and lived by and large in developing regions. The FAO further pointed out that the prevalence of undernourishment in developing countries is estimated at 14.3% of the population, that almost 5 million children under the age of 5 die of malnutrition every year, and that malnutrition is the single largest contributor to disease in the world. Heads of states gathering at the World Food Summit in Rome in 1996 concluded that the world can produce sufficient food for every person in the world. They observed that even though the world population had doubled in the preceding 30 years, food production had increased even faster. As of 2016 agriculture is capable of making 2,770 calories available to each person on the planet per day (a daily intake of 2,200 calories is recommended). Since FAO studies estimate that 3,070 calories will be available by 2050 it is reasonable to conclude that global resources are sufficient and that food security is not the problem when it comes to explaining food shortages and hunger. Although population growth has long been assumed as the major cause of hunger and starvation as a result of people multiplying faster than food can be produced, such an assumption does not hold up. Despite population growth the availability of food has increased. Countries such as the United States and Australia produce much more than their populations need. Australia exports more than half its agricultural products. Even as early as the 1990s it was observed that the majority of malnourished children in the developing world lived in countries with food surpluses. The common view today is that hunger and undernourishment are the result of poverty and unequal global distribution—not a shortage of food. Therefore ethical debate on hunger means moving the focus from food production to better access. The poor are hungry because they cannot afford to buy food and what food there is ends up being exported to developed countries. An example is India where large-scale agriculture produces surpluses while 250 million rural farmers go hungry. The argument that contemporary agriculture should produce ever more food because the world population is growing is therefore questionable. The moral imperative to produce sufficient food for humanity is not only currently being met but can also be met for future generations. The major concerns regarding food are ethical—not technical or scientific.

H. ten Have and M. do C. Patrão Neves, *Dictionary of Global Bioethics*,
https://doi.org/10.1007/978-3-030-54161-3_303

ICSI

Intracytoplasmatic sperm injection (ICSI) is an assisted reproductive technology (ART) procedure consisting in artificial retrieval of sperm and oocytes (via masturbation and suction, respectively), treatment and selection of gametes, injection of a single sperm directly into a mature egg, and transfer of the embryo into a woman's body in the expectation that implantation will occur naturally. ICSI is the most recent ART and specially designed to help men who present very poor sperm parameters (quantity and quality). In an attempt to counter the continuous decline of sperm concentration in Western men over recent decades, ICSI makes reproduction possible even for men with low numbers of viable sperm. The first ICSI baby was born in 1992 in Belgium. Since then its success rate has been growing and is now higher than conventional in vitro fertilization (IVF) because the encounter and fusion of sperm and oocyte are not left to chance but artificially achieved. Thus ICSI started to be used as a substitute for IVF for couples with failed IVF attempts, for older couples, and for instances where the cause of infertility was unknown and as an alternative to using gamete donors. In the wake of transverse ethical issues raised by ARTs and those raised by IVF, ICSI presents the particular problem that boys conceived using this method when they become young adults present significant lower sperm quantity and quality than would be considered normal thus aggravating the existing problem of declining Western male fertility rates.

H. ten Have and M. do C. Patrão Neves, *Dictionary of Global Bioethics*,
https://doi.org/10.1007/978-3-030-54161-3_304

In Vitro Fertilization (*See* Assisted Reproductive Technology)

In vitro fertilization (IVF) is an assisted reproductive technology (ART) procedure consisting in artificial retrieval of sperm and oocytes (via masturbation and suction, respectively), treatment and selection of gametes, placing sperm and oocytes together in a petri dish, waiting for fusion to happen naturally (i.e., the sperm fertilizes the egg on its own), and transfer of the mature embryo into a woman's body in the expectation that implantation will occur naturally. Gametes (sperm and oocytes) may come from the parents-to-be or originate from donors, may be used in vivo or cryopreserved, and may raise different ethical issues in different situations. IVF is a technically complex procedure and physically and psychologically stressful for the couple. Moreover, as a result of the low rates of success generally attributed to ARTs couples usually undergo several IVF procedures thus increasing their anxiety. Nevertheless, IVF represents a very significant improvement over artificial insemination (AI) in that it addresses the factors involved in female and male infertility and solves difficulties not only at the sperm transfer level, but also at the fertilization level when the gametes meet and fuse. Despite IVF unnaturally creating an embryo outside the woman's body, the first IVF baby (Louise Brown) was nevertheless born in 1978 in England. The procedure was regarded with great suspicion, the effects of artificial production of a human life were unknown, and multiple fears were raised (even in the medical community) surrounding her birth, the various stages of her physical and psychological development, and her capacity to learn and pursue normal activities. Louise was clinically followed up from birth, throughout her early life, and right up until she naturally conceived and gave birth herself. Observation showed that Louise was no different from any other girl of the same age. In addition to IVF becoming standard treatment for infertility, it was also used for other clinical situations such as preventing the transmission of genetic disorders, of AIDS, and preserving the fertility of cancer patients going through chemotherapy. The cryobanking (i.e., collecting, depositing, freezing, and storing at a gamete bank) of sperm and later of oocytes contributed to wider use of IVF (specifically in non-clinical situations) and raised additional ethical problems. Among the many transverse ethical issues raised by ARTs, the major ones raised by IVF refer to a series of disruptions of the reproductive process together with the increasing number of possible persons implicated in the birth of a new human life. Other issues are temporal disruption of a natural continuous process such as gamete retrieval, fertilization, and embryo implantation. Moreover, there is fragmentation, atomization, and lack of accountability of human reproduction in which legal or social mothers and fathers do not coincide with biological progenitors. Arguably the most characteristic IVF ethical problem and probably the most controversial is the

H. ten Have and M. do C. Patrão Neves, *Dictionary of Global Bioethics*,
https://doi.org/10.1007/978-3-030-54161-3_305

production of surplus embryos. During the fertilization process a high number of embryos are produced to improve the procedure's chances of success, but only a few are transferred. Those that remain may be saved by parents for implantation in the near future or they may be discarded. Manipulating human life in such an extreme way can lead to human life being objectified and handled as a means to satisfy individual desires and in so doing threaten human dignity.

Indigenous Ethical Perspectives

Indigenous ethical perspectives look at the way indigenous knowledge is dealt with. Indigenous knowledge is local knowledge unique to a given society and culture who are characterized by their geography (i.e., indigenous populations who have long lived within a particular territorial setting and have their own specific language, culture, and political and economic institutions) and their genealogy (i.e., linking them to past generations by assuming their ancestors are still there and by believing they cannot survive if their ancestors are no longer around). Indigenous knowledge has certain features that distinguish it from dominant and scientific worldviews. It expresses different ethical perspectives characterized by specific cosmologies emphasizing the interconnectedness of all beings and the holistic nature of human beings in that they are a union of body, mind, and spirit, are embedded in nature, and thus are part of nature. In such a worldview knowledge is collectively owned since it is the legacy of ancestors that has been developed over a long period of time. Resources in developing countries are appropriated and monopolized by scientists and international companies in developed countries (a phenomenon of injustice called "biopiracy"). Although intellectual property rights (IPR) do not acknowledge indigenous rights, significant changes in critical thinking have occurred in recent decades. Scholars have shown that various concepts of ownership exist in traditional communities, that the focus on individual rights (of a specific inventor or author) is therefore biased, and that the accumulation of knowledge is a collective activity. Since no one owns traditional knowledge there is no way it can be viewed as someone's property other than that of the people or the community. The ethos of indigenous people is based on knowledge being inherited from their ancestors, shared by the people, and belonging to today's people such that natural resources and biodiversity can be preserved for the sake of future generations.

H. ten Have and M. do C. Patrão Neves, *Dictionary of Global Bioethics*,
https://doi.org/10.1007/978-3-030-54161-3_306

Indigenous Knowledge

Indigenous knowledge refers to knowledge accrued by people indigenous to a particular region or environment. Since indigenous knowledge is unique knowledge endemic to people from a particular culture or society it is also known as traditional knowledge or local knowledge. Approximately 370 million indigenous people live in 70 different countries. Indigenous knowledge has certain features that distinguish it from dominant and scientific worldviews: it is holistic in that it regards human beings as somatic and spiritual entities; it is contextual in that it regards human beings as embedded in their environment; it is comprehensive in that it considers biological diversity as inherently linked to cultural diversity; and it regards knowledge and practices based on nature as collective wisdom (i.e., the common heritage of humankind that cannot be considered as exclusive individual property or used for profit). Such characteristics imply that indigenous knowledge is non-systematic, information is not verifiable, knowledge systems and practices are variable, and the concepts of health and illness used are incompatible with those of modern medicine. The importance of indigenous knowledge and the experience of traditional healers in pointing out plants they employ medically for healthcare led to physicians and ethnobotanists discovering the therapeutic benefits of using such plants, opened up new areas of medical research, and generated in a number of cases enormous profits for pharmaceutical companies. Moreover, traditional medicine is an important source of care as borne out by the WHO estimating that 80% of people in developing countries rely on traditional healthcare. Thus traditional healthcare is normal care for most people in developing countries. Although the percentages may differ per country and continent, in many African countries there are more traditional practitioners available in primary healthcare than modern physicians. The main reasons for attempts in recent years to rehabilitate indigenous knowledge and traditional medicine have to do with biodiversity and changing ideas about wilderness and protected areas. Adoption of the Convention on Biological Diversity focused attention on the unjust effects of policies designed to protect nature from humans rather than engage humans with nature and on the sustainable management of biodiversity and its dependence on the continuous stewardship of people who rely on it for their survival. Indigenous and local communities have built up experience over centuries in dealing with the environment, knowing how to take care of the ecosystem, and conserving biodiversity on which they depend for food, water, shelter, and medicine. Another reason for renewed interest in indigenous knowledge is its importance for healthcare. In 1978 the Declaration of Alma Ata became the first normative instrument acknowledging the role traditional practitioners play as providers of primary healthcare. In the following year the WHO established a program that initially focused on

H. ten Have and M. do C. Patrão Neves, *Dictionary of Global Bioethics*,
https://doi.org/10.1007/978-3-030-54161-3_307

investigating traditional medicine for research and drug discovery purposes, but later the emphasis broadened. An important step was the Beijing Declaration adopted by the first WHO Congress on Traditional Medicine in 2008. The declaration advocates integrating traditional medicine into national health systems and urges governments to formulate national policies such that traditional medicine can be used appropriately, safely, and effectively; systems for quality control, standardization, and regulation can be set up; and stringent education and training can be given to practitioners. A few months later the declaration was endorsed by the World Health Assembly. The major ethical concern with indigenous knowledge is exploitation. Although indigenous knowledge is usually accumulated in biodiversity-rich countries and then used to produce drugs in developed countries, indigenous people often do not benefit from such discoveries and so-called "new" products. Therefore they find themselves confronted by biopiracy in which the indigenous people who long cultivated traditional plants for use in medicine are often ignored, marginalized, displaced, and discriminated against. Although commercial promises are made and honored, the way in which natural products are exchanged is often exploitative and unjust and the reason indigenous rights have been formulated.

Indigenous Rights

As a result of the importance of indigenous knowledge for biodiversity and healthcare and the many examples of exploitation such as biopiracy and unjust patenting, the United Nations in 2007 adopted the Declaration on the Rights of Indigenous Peoples. The document states that such peoples have "the right to their traditional medicines and to maintain their health practices." Moreover, the declaration makes it clear that full ownership, control, and protection of their cultural heritage and intellectual property is theirs and only theirs and that traditional knowledge is not just important because it has instrumental value for conservation, food, or new drugs, it is also important because it has an intrinsic value of its own. Despite the declaration expressing respect for cultural diversity and human rights, there are those who argue the current framework of global governance (especially the intellectual property rights regime) does not protect indigenous rights. It is often very difficult for indigenous communities to show that indigenous knowledge belongs to them since there is no written evidence and without evidence such knowledge is often not recognized by courts. Matters are made worse by patents privatizing knowledge and products as a consequence of which collective rights of local and indigenous communities are not recognized. One remedy is to document traditional knowledge as was done by India when it established the Traditional Knowledge Digital Library in 2001, which now houses 290,000 medicinal formulations and has resulted in patent claims filed on Indian traditional knowledge declining by 44%. Matters are further muddied by traditional knowledge not always being common knowledge since it is often practiced secretly, only known to a few healers, and thus not documentable. In many cases the rights of indigenous peoples are still not recognized despite the declaration. This may have something to do with the declaration taking a very long time to be adopted and developed countries making sure it was non-binding.

H. ten Have and M. do C. Patrão Neves, *Dictionary of Global Bioethics*,
https://doi.org/10.1007/978-3-030-54161-3_308

Infertility

Infertility is defined as the inability of a woman or a couple to achieve pregnancy in one year of regular unprotected sexual intercourse. Historically, infertility has been traditionally associated with women. It was considered a curse, a sign of sin, or a punishment and no matter how it was interpreted it always stigmatized women, made them secondary in relationships, and excluded them from groups or communities. Reproductive medicine has sufficiently evolved today for everyone to be aware there are also male causes of infertility. Although the percentage of infertile women is slightly higher than that of infertile men, it is increasing among men (especially among men from developed countries as a result of declining sperm counts). Infertility is today considered a disease by the WHO, which is concerned not only by the fact that infertility is on the increase but that it mostly affects men in developed countries. Although the causes are multifactorial such as postponing the birth of the first child, prolonged use of contraception, pollution, stress, and even tight jeans and trousers, there are still idiopathic (i.e., unknown or unexplained) causes of infertility. There are two particularities of infertility as a disease the first of which is that it is almost always diagnosed when a couple (or an individual) wish to have a child, have been trying to conceive for at least a year, and the fear of infertility has made matters worse. The second is that infertility is never cured and only circumvented by assisted reproductive technologies (ARTs). Although such technologies have specifically been developed to help couples have a child, the baby will be biologically related to just one partner of the couple (or to neither) irrespective of whether the woman can bear the child or not. In short, fertility is not cured but hidden from view. There are those who argue that classifying infertility as a disease hastened the search for treatment, imposed physical and psychosocial stress and suffering, and dismissed other alternatives such as adoption and the acceptance of infertility as a personal condition. Feminism stresses that a woman's social role extends beyond motherhood and that what is more important is to empower women to make their own decisions free from the medical model of infertility (diagnosis, treatment, and so-called cure).

H. ten Have and M. do C. Patrão Neves, *Dictionary of Global Bioethics*,
https://doi.org/10.1007/978-3-030-54161-3_309

Information Ethics

Information ethics is a branch of applied ethics strictly focusing on the creation, control, use, and impact of information technology at the personal and social level. More broadly, it can also cover libraries, journalism, media, and computers. Although issues related to information ethics arose in 1980 with Barbara J. Kostrewski and Charles Oppenheim's paper "Ethics in information science" in the *Journal of Information Science*, the expression "information ethics" was first coined in 1988 by Rafael Capurro in his paper "Informationethos und Informationsethik" in the journal *Nachrichten für Dokumentation* and Robert Hauptman in his book *Ethical Challenges in Librarianship* (in 1992 Hauptman started the *Journal of Information Ethics*). In any event it came in the wake of the development of applied ethics, which emerged and multiplied as new scientific-technological domains advanced and raised new ethical problems (or old ones altered by the new context). UNESCO defines information ethics as being "concerned with ethical, legal and societal aspects of using information, and information and communication technologies." It is a field of ethics concerned with the best or right way to act in particular situations triggered by information technologies, its current and predictable impacts in society, and the regulation of this new field. UNESCO also points out the essential principles on which information ethics should be based: "the right to freedom of expression, universal access to information, the right to education, the right to privacy and the right to participate in cultural life." However, such principles are insufficient to properly address the major ethical issues that are constantly arising such as information ownership and access to intellectual property (e.g., downloading someone else's intellectual property), type and quality of information disseminated (e.g., how to build a bomb, the genome of a deadly virus, fake information), and restricted access to data and privacy. There are also specific ethical problems that have recently attracted growing attention such as internet censorship and surveillance by political dictatorships, termination of the principle of net neutrality in some countries (i.e., the United States) that have determined that such a public service is no longer open and equal to all users, the commoditization of information and information services, and unfair competition from tech giants (the European Union fined Google several times). The most challenging ethical problem concerning information as it currently flows online is to balance the need to regulate its use while rejecting censorship.

H. ten Have and M. do C. Patrão Neves, *Dictionary of Global Bioethics*,
https://doi.org/10.1007/978-3-030-54161-3_310

Information Technology

Information technology (IT) refers to the design, development, use, and implementation of computer-based information systems (hardware and software). It includes creating, storing, processing, securing, and distributing all forms of electronic data (mostly digital), networking, and systems that facilitate communication (physical devices, infrastructures, and processes). The expression "information technology" was first coined by Thomas L. Whisler and Harold J. Leavitt in a paper published in 1958 in the *Harvard Business Review*. At the time it just referred to the process of storing information in large organizations that needed to store data. Since then (mostly from the late 1970s on) the advent of computers and the internet, the rise of artificial intelligence, and the exponential capacity to gather information (big data) have led to the field of IT hugely expanding to all kinds of social activities and to technologies intensely used and developed by big enterprises for business purposes. We live in an information age where information is power. IT has grown to be a powerful industry not only assisting all economic activities but also becoming a big business sector in itself training and providing high-level IT experts certificated as computer scientists, computer engineers, system analysts, and computer programmers needed in the majority of business sectors that depend on IT. This has resulted in IT becoming part of daily life and to IT experts becoming indispensable not only to companies but also to regular computer users. Moreover, this has led to the rise of hackers who use computers to gain unauthorized access to all kinds of data and for a variety of purposes. Notwithstanding the overwhelming advances of IT, it is still a vulnerable technology and as such makes enterprises vulnerable too.

H. ten Have and M. do C. Patrão Neves, *Dictionary of Global Bioethics*,
https://doi.org/10.1007/978-3-030-54161-3_311

Informed Consent (*See* Consent)

See under the entry Consent.

H. ten Have and M. do C. Patrão Neves, *Dictionary of Global Bioethics*,
https://doi.org/10.1007/978-3-030-54161-3_312

Institutional Ethics (*See* Organizational Ethics)

Institutional ethics (a.k.a. organizational ethics) refers to the application of ethics in such institutions as hospitals, professional organizations, and corporations. It regards institutions as moral agents with responsibilities and accountability. Although ethics is traditionally related to individuals, their behaviors, and practices, institutional ethics emphasizes the importance of institutions also being the object of moral evaluation—something that is not evident since it is often argued that institutions are run by individuals. Institutional ethics in the context of healthcare emerged because of the increasing role institutions or organizations play in the delivery of healthcare such as health maintenance organizations and managed care organizations that determine how patients are treated. Ethical assessment will require a multidisciplinary approach combining bioethics with business ethics, professional ethics, law, and social policy. Many healthcare institutions today have institutional ethics policies that usually emphasize such values as respect for people, collegiality and individual responsibility, and honesty and openness. Such policies also have procedures for conflict resolution.

H. ten Have and M. do C. Patrão Neves, *Dictionary of Global Bioethics*,
https://doi.org/10.1007/978-3-030-54161-3_313

Integrity Concept

The word "integrity" derives etymologically from the Latin *integritas* (totality). The adjectival form of integrity is *integer* (integral, complete, perfect, intact, untouched) and thus expresses a specifically moral sense of purity, innocence, honesty, and probity. The common root of both the noun and the adjective is the Latin verb *tangere* (to touch). When this is preceded by the negative prefix "in" the original etymological meaning of *integritas* (integrity, totality, the condition of being untouched) is restored. Therefore integrity refers to a physical reality in which something is presented as being complete, in no way touched, affected, altered, or corrupted, and remains completely inviolable. It also relates to a moral reality in which something is presented as untouched, uncorrupted, impartial, right, and even pure. In short, integrity refers to a totality that cannot be corrupted and to the factual and evaluative levels to which it is applied. Since integrity is generally formulated as referring to a whole considered in its unaltered completeness without gaps affecting or corrupting it, such an intact totality can be of a physical, psychologial, or moral nature; can be structured at the personal or professional level; and can refer to human beings or the totality of beings. The reason integrity has so many meanings is a consequence of the different realms to which it applies and the different adjectivations it assumes. Indeed, the adjectivation of a term always restricts its range of application and its very meaning when specifying the concept. Although the word "integrity" is common in everyday communication, only recently has its usage been applied to moral issues regarding personal and professional behavior and particularly scientific research.

H. ten Have and M. do C. Patrão Neves, *Dictionary of Global Bioethics*,
https://doi.org/10.1007/978-3-030-54161-3_314

Integrity, Personal

Personal integrity can refer to someone's physical, psychological, or moral status. Physical integrity refers to conserving someone's natural state in which his or her body is not invaded without permission. Medical procedures (especially the most invasive) often affect personal physical integrity and hence can only be performed after informed consent. Psychological integrity means the same but at the mental level. It involves safeguarding someone's convictions and personal identity and ensuring they cannot be adulterated or destroyed inconsequentially. Personal psychological integrity can be breached by a wide range of different procedures from emotional manipulation to pharmacologically induced behaviors (with drugs). Moral integrity refers to someone's honesty and impartiality in his or her deeds and judgments and the transparency with which they are presented and rationally justified (i.e., unaffected by private interests or external influences but changing according to subjective circumstances instead of objective ones). By preserving their moral integrity people can avoid becoming self-centered, individualistic, and selfish, and instead acknowledge the legitimacy of other viewpoints and work in the best interests of common good and social justice. Personal integrity is frequently used synonymously for moral integrity, which is considered a virtue (i.e., a character trait in which people are predisposed to act according to their values even when doing so is disadvantageous to their own interests). Such a virtue is deemed defensive in that it is not oriented at developing something concrete but at defending or protecting something against potential external damaging acts.

H. ten Have and M. do C. Patrão Neves, *Dictionary of Global Bioethics*,
https://doi.org/10.1007/978-3-030-54161-3_315

Integrity, Professional

Professional integrity stricto sensu refers to integrity being practiced in the professional realm (i.e., behaving honestly, being impartial, and being above reproach in the professional setting). Although codes of professional ethics frequently attribute this very meaning to integrity, they have changed (knowingly or unknowingly) the moral statute of professional integrity from a virtue (a personal predisposition to voluntarily act in a certain way or a character trait to be encouraged) to a duty (an obligation that has to be complied with). Integrity is then recognized as essential to the credibility and social prestige of professions and to trust in professionals. Thus it cannot be left to the whims or willingness of professionals and has to be established as compulsory as a professional obligation. Professional integrity is mostly used today in its broadest meaning in which all duties associated with the profession (those explicitly established and those implicitly expected by beneficiaries or by society) are fulfilled—not just single key duties. Given such a broad meaning it has been used synonymously for ethics such that it testifies to and strengthens the growing importance of integrity in professional practice.

H. ten Have and M. do C. Patrão Neves, *Dictionary of Global Bioethics*,
https://doi.org/10.1007/978-3-030-54161-3_316

Integrity, Research (*See* Research Ethics; Integrity)

Research integrity refers to sets of ethical principles that must be respected such that good scientific practice and sound and trustworthy science can be assured. Historically, awareness of the importance of research integrity only developed as a consequence of scandals in the scientific community. Unethical behavior and procedures in science do great harm to research, the team, the institution, the country, and today's world of borderless science. Scientifically, economically, financially, and socially, countries and institutions worldwide realized that they could only prevent such behavior by establishing research guidelines. Such a collective effort culminated in sets of ethical duties that differ in the various international documents focusing on research integrity. The Singapore Statement on Research Integrity (2010) refers to the principles of honesty, accountability, professional courtesy and fairness, and good stewardship. The European Science Foundation and All European Academies (ESF/ALLEA) Code of Conduct for Research Integrity (2010) mentions the values of honesty, reliability, objectivity, impartiality and independence, openness and accessibility, duty of care, fairness in providing references and giving credit, and responsibility for the future. The revised edition of the ESF/ALLEA document (2017) stresses four main principles: reliability, honesty, respect, and accountability. The Montreal Statement on Research Integrity (2013) establishes different levels of responsibilities (general, collaboration management, collaborative relationships for research outcomes) for individual and institutional partners in cross-boundary research collaborations. The Global Research Council Statement on Principles on Research Integrity (2013) points to honesty, responsibility, fairness, and accountability. Despite the wide-ranging nature of such sets of ethical principles it is nevertheless possible to identify some common principles of scientific research within a minimalist approach such as honesty in the commitment to truth, independence in the preservation of freedom of action in relation to pressures outside the profession, and impartiality in the neutrality of professional practice in relation to private interests beyond research. There is currently a tendency to regard research integrity or scientific integrity as the ethics of science or as a separate field from applied ethics (a philosophical discipline), rather than as a particular ethical issue. In the minimalist approach research integrity is a self-regulated set of principles that are institutionally checked as part of administrative procedures. In the ethics of science approach research integrity has to be coherent with common morality in which inputs from stakeholders (civil society, in general) are accepted and welcomed since scientific research ultimately responds to them.

H. ten Have and M. do C. Patrão Neves, *Dictionary of Global Bioethics*,
https://doi.org/10.1007/978-3-030-54161-3_317

Intensive Care

The intensive treatment and close monitoring of seriously ill patients is called intensive care. It requires highly trained professionals, sophisticated technology and monitoring equipment, and specialist hospital wards called intensive care units (ICUs) or critical care units (CCUs). Patients on such wards are usually suffering some form of organ failure and often are unable to breathe on their own. The reasons for being admitted to an ICU range from being involved in a serious accident, having a short-term condition like a stroke, suffering a serious infection like sepsis, to undergoing major surgery such as a heart transplant. ICU equipment includes ventilators, intravenous lines and pumps, feeding tubes, and drains and catheters. ICU patients are usually given sedatives. Some patients only need intensive care for a few days, while others need to stay in the unit for longer. The complexity of care and the serious condition of ICU patients means intensive care is associated with many ethical problems such as the duration and effectiveness of treatment, the withholding or withdrawing of treatment (especially when the treatment is life sustaining), the granting of consent and refusal of treatment (informed consent should always be obtained from patients whenever possible otherwise surrogate decision-making procedures need to be followed), previously expressed wishes (such as when competent patients have expressed the wish not to be transferred to an ICU or have an advance directive stating such a preference), futile treatment (in which the principle of proportionality plays a role in balancing the benefits and burdens of intensive treatment), and the role triage plays in admitting or discharging patients from such units (since ICUs are usually small with a limited number of beds).

H. ten Have and M. do C. Patrão Neves, *Dictionary of Global Bioethics*,
https://doi.org/10.1007/978-3-030-54161-3_318

Interculturality

Interculturality differs from multiculturalism and acculturation in its focus on interaction. Although the prefix "inter" hints at separation, linkage, and communication, it acknowledges diversity, global values, and common perspectives. In contrast, multiculturalism has become problematic in that it does not really examine how people can live together and construct new syntheses and social relationships despite acknowledging the existence of multiple cultures and value systems. Although worldwide interconnectedness bridges the gap between distance and proximity, multiculturalism assumes there is a radical moral contrast between strangers and friends. Multiculturalism is often associated with policies of acculturation (in which the other is integrated or assimilated into the dominant culture), indifference (in which the other is tolerated as peacefully coexisting in different cultures and producing parallel and separated communities that do not interact), or articulating and even retreating into traditional identities (in which the differences between "us" and "them" are reinforced since the other produces fear of losing one's self and one's relativism). Multiculturalism emphasizes respect for diversity, individual freedom, justice, and equal treatment, whereas interculturality introduces a moral vocabulary of interaction, dialogue, participation, trust, cooperation, and solidarity. Interculturality is more interested in what unites people than what divides them, has a more positive thrust than multiculturalism, argues that much more than coexistence is needed to bring about practices that can create community and that whatever common ground there is needs to be cultivated through interchange and communication. Intercultural dialogue is driven by the quest for unity or commonality—not uniformity. Convergence will only come about as a result of translating the various moral languages since no single moral language is a given that underlies or is more fundamental than other moral languages. Hence the communication process cannot be bypassed, a universally acceptable transcultural reference point does not exist, and there are only interstitial spaces where cultures interact and overlap and where people communicate. The first step in reaching understanding is to recognize the existence of radically different moral languages since consensus only becomes a possibility once differences have been expressed and acknowledged. We have little choice but to allow our various moral languages to interact and search for a common understanding since convergence is not a given but the result of ongoing deliberation, consultation, and negotiation, and global bioethics is the result of dialectics between global aspirations and local operations (i.e., involved in the making of the finished product—not the product itself). It is the provisional result of exchange, learning, deliberation, and negotiation in which not only disagreement is articulated but also the possibility of convergence. The notion of interculturality can be used to help explain such a dialectic interchange. It clarifies how recognition of divergence in moral views can be reconciled with convergence toward

H. ten Have and M. do C. Patrão Neves, *Dictionary of Global Bioethics*,
https://doi.org/10.1007/978-3-030-54161-3_319

commonly shared values. Globalization brings about uniformity and multiplicity and is not an "either or" situation since people today participate in multiple cultures irrespective of where their roots lie or the specific heritage they acknowledge. The same is true for the notion of culture since no culture today is monolithic and pure. Although cultural traditions are hybrid and dynamic, they do change as a consequence of being a blend of different components. Similar arguments can be made for bioethics. Although various ethical approaches and systems do of course exist, differences do not preclude there being a common core. Therefore interculturalism is more appropriate than multiculturalism. Any new global world order will be characterized by the continuous interaction and reciprocal learning of multiple value systems—not merely by their existence.

International Law

International law consists of laws that govern the relations between signatory countries, lies outside the legal jurisdiction of individual states, and is closely related to human rights. One of the primary goals of the United Nations was the development of international law. Such law defines the legal responsibilities of states by regulating their conduct with each other and the treatment of individuals within their boundaries, global commons such as the environment and outer space, and global challenges such as migration, refugees, and conduct of war. The four sources of international law mentioned in the Statute of the International Court of Justice are treaties between states, customary international law derived from the practice of states, general principles of law recognized by civilized nations, and judicial decisions about the intellectual property rights of authors. One of the greatest achievements of the United Nations has been the development of a body of international law made up of conventions, treaties, and standards (more than 500 international treaties have been adopted). The International Court of Justice was founded in 1946 to settle disputes; has considered more than 160 cases; and has addressed international disputes ranging from economic rights, rights of passage, non-use of force, non-interference in the internal affairs of states, diplomatic relations, hostage taking, to the right of asylum and nationality. States bring such disputes before the court to find impartial solutions to their differences on the basis of law. By peacefully settling such questions as land frontiers, maritime boundaries, and territorial sovereignty, the court has often helped to prevent the escalation of disputes. When it comes to international criminal justice the United Nations has the power to set up tribunals along the lines of the trials at Nuremberg and Tokyo after the Second World War. In the 1990s ad hoc tribunals were established for the former Yugoslavia and for Rwanda. Massacres in Cambodia, the former Yugoslavia, and Rwanda led to the establishment in 2002 of a permanent international court: the International Criminal Court (ICC). Its jurisdiction is worldwide allowing it to prosecute individuals who commit genocide, war crimes, and crimes against humanity. The ICC is legally and functionally independent of the United Nations. Although international law is closely connected to human rights with international human rights law representing a significant part of it, the two are not equivalent. A series of international human rights treaties and other instruments adopted since 1945 have made inherent human rights a legal requirement and developed the body of international human rights. Declarations connecting global bioethics to human rights such as the Universal Declaration on Bioethics and Human Rights have resulted in international law becoming increasingly relevant in bioethical discourse.

H. ten Have and M. do C. Patrão Neves, *Dictionary of Global Bioethics*,
https://doi.org/10.1007/978-3-030-54161-3_320

Internet

The internet is a global computer network providing all manner of information and facilitating various forms of communication. Although it is often used synonymously for the World Wide Web, they are different. The internet is the global communication system in its entirety including the hardware and infrastructure needed to access it, whereas the web is one of the services available over it. The internet provides various instantly available communication services such as email, video and audio, movies, games, instant messaging, social networking, internet forums, financial services, and online shopping. As a consequence of global bioethics being associated with and promoted by the internet, bioethics online and e-bioethics are growing areas. Bioethical communities and networks have been established and have brought bioethicists from across the world together. As a result of online platforms making information available and facilitating communication and exchange the new interconnectedness provided by the internet has created many opportunities for teaching, research, healthcare, and communication and in so doing greatly enhanced the opportunities for global bioethics as borne out by the number of online ethics teaching courses rapidly growing, virtual communities being created, and many e-publications appearing. The internet generally promotes bioethics in three principal ways: by making the dissemination of relevant information much easier, faster, and less expensive; by supporting deliberation, analysis, and exchange of arguments such that the pros and cons of particular decisions can be weighed and balanced by more input from experts across the globe; and by facilitating interaction in which new developments can be explored and potentially promising therapies can be identified. These are the reasons it is widely argued that access to internet services should be unrestricted. In contrast, a lot of the information is unreliable, fragmented, and biased and cyberspace is an ideal place for harmful and fraudulent practices to take place. Therefore greater regulation has been advocated, which culminated in UNESCO launching a Code of Ethics for the Information Society in 2011 that stipulates that every person irrespective of where they live, their gender, education, religion, or social status shall be able to benefit from the internet and use of information and communication technologies; everyone shall be able to connect, access, choose, produce, communicate, innovate, and share information and knowledge on the internet; and everyone has a right to freedom of expression, participation, and interaction on the internet that should not be restricted except in narrowly defined circumstances based on internationally recognized laws and universal human rights standards. Furthermore, it articulates the importance of privacy in which individuals have a right to their personal data and private life on the internet being protected against unlawful storage, abuse, or unauthorized disclosure of personal data and against any intrusion into their privacy.

H. ten Have and M. do C. Patrão Neves, *Dictionary of Global Bioethics*,
https://doi.org/10.1007/978-3-030-54161-3_321

Invasive Species (*See* Bioinvasion)

An invasive species is any organism that is not native to an ecosystem and is believed to harm it. Bioinvasion (a.k.a. biological exchange, species transfer, relocating or transplanting life) was first regarded as a major threat to biodiversity in the 1990s. Article 8 h of the Convention on Biological Diversity (1992) declares that "contracting parties shall prevent the introduction of, and control or eradicate those alien species which threaten ecosystems, habitats, or species." The idea is that by spreading outside their natural past or present ranges the non-native species introduced cause the extinction of native species and change existing ecosystems such that ecosystem services are affected with important consequences for food, water, and health. The effects brought about by invasive alien species are mostly presented in a negative light. They have been known to cause allergies and skin damage; many disease vectors have unintentionally invaded new territories and disseminated diseases where they were not known before; invasive species reduce yields in agriculture, fisheries, and forestry and decrease water availability; and the economic damage is significant (e.g., EUR12.5 billion per year in Europe). Consequently, many countries have developed policies to protect native biodiversity against such species aimed at their prevention, early detection, and rapid eradication. Although it took a while, the European Union issued Regulation 1143/2014 on Invasive Alien Species in 2014 (it entered into force on January 1, 2015). Recently the phenomenon of bioinvasion has been subject to greater critical examination. This concluded that containing biological invasions is impossible, that their introduction will continue and likely increase, and raised an interesting question as to why this phenomenon is presented as bad and threatening. Human beings have always brought species to other countries and continents as they migrated. The earliest known cultivation of grapes to make wine took place in what is now Georgia and Armenia 8,000 years ago and it was from there that viticulture was introduced in many countries. Exotic species were often introduced for commercial purposes or simply because people liked them. Although rhododendrons and domestic cats are non-native in Europe, they are highly appreciated despite the damage they both do in gardens. Bringing non-native species to new habitats is apparently a manifestation of the human tendency to enhance the lifeworld (i.e., to increase knowledge, satisfy desires, and grow economic output). The most common motive is the desire to make new food items widely available by moving domesticated animals and crops long distances. Although no one doubts that farming and consumers have greatly benefited, the disadvantages of introducing new species are more clear today. Overall biodiversity has been reduced as borne out by endemic species and ecosystems being affected and (more worryingly) the consequences of invasive species being unpredictable. The very terminology immediately generates a negative

H. ten Have and M. do C. Patrão Neves, *Dictionary of Global Bioethics*,
https://doi.org/10.1007/978-3-030-54161-3_322

response in which invasion is seen as wrong and bringing disease and thus should be resisted and attacked. Objectively, the phenomenon could be called "biological exchange," "species transfer", "relocating" or "transplanting" life. It is true that species introduced as biological agents to manage pests have become pests themselves in some cases. Evidence makes clear that native species can also be damaging. Nevertheless, the science of invasion biology shows that most invasions fail and that most invaders have minor consequences. In fact, the only fundamental difference between exotics and natives is their origin. If this is the case, then why are non-native species regarded as the enemy, aliens, killer weeds, or biological pollution? Why is the distinction between native and alien used let alone the basis for global and national policies? Bioinvasion illustrates just how connected diseases are to both environmental and social conditions. They do not just happen, emerge, or spread as natural events but are associated with and often are the result of human activities whether intentional or accidental. Although bioinvasion can be examined, described, analyzed, and managed scientifically or biologically, it should always be interpreted within the context of benefits and harms. Does it increase biodiversity or does it harm it? Do imported species bring about new ecosystems or do they disturb the balance and order of nature? Defining species as alien presupposes a view of nature as characterized by stability, permanence, and predictability or a pristine wilderness that needs protection rather than as dynamic, resilient, and constantly changing. Defining species as invasive further suggests that what is bad and what should be done has already been established. With this background in mind, people find it difficult to accept that nature has always been a mixture of native and alien species that has given rise to novel ecosystems (new wilds) and in many cases assisted in the recovery of biodiversity—not decline. Furthermore, bioinvasion complicates the notion of globalization because invasive species show that natural barriers have broken down, that borders are irrelevant, that differences between species are disappearing, and that diversity is diminishing. Such environmental cosmopolitanism is down to the mobility of life and analogous to cosmopolitanism in many other areas. Although bioinvasion is a common feature of the processes of globalization rather than an exceptional one, current approaches to bioinvasion are localized. The Convention on Biological Diversity regards states as guardians of biodiversity and does not see the need for global governance because it does not regard biodiversity as the common heritage of humankind. Threats to biodiversity such as bioinvasion are national concerns and any sense of global solidarity will not therefore easily develop. Other difficulties for global governance are the intrinsic uncertainty about invasive species and the unpredictability of what will happen with new invasions.

Institutional Review Boards (*See* Research Ethics; Research Ethics Committees)

Institutional review boards (IRBs) are biomedical or clinical research ethics committees whose job is to decide whether clinical research should be approved or not. They were first established as legal entities in 1974 by the US Congress and tasked with deciding whether biomedical research involving human subjects and federally financed could be performed. The first reference to what would become an IRB was made in 1966 by the North American National Institutes of Health (NIH), which recommended the establishment of centers dedicated to biomedical research. Such a recommendation came about as a result of the strong reaction by society (particularly) and the scientific community to the publication of Henry Beecher's paper "Ethics and Clinical Research" in the *New England Journal of Medicine* in 1966 disclosing the fact that of the approximately 50 clinical trials being undertaken in the United States 48 did not comply with the requirement for informed consent. The Nuremberg Code in 1948 had already made it clear that informed consent was an ethical requirement for scientific research involving human subjects. However, the recommendation of the NIH was not taken. This was all to change in 1972 with the Tuskegee case, which revealed that a group of 399 black men with syphilis had been since 1932 in Tuskegee denied the chance of treatment and cure with penicillin (accepted as treatment for syphilis since 1945) all done to enable scientists to carry out a long-term a study of syphilis. The media and public opinion were outraged and condemned this study and all its key players. In the aftermath the US Congress created the National Commission for the Protection of Human Subjects of Biomedical and Behavioral Research in 1974 and tasked it with coming up with "Ethical Principles and Guidelines for the Protection of Human Subjects of Research." It also established mandatory international review boards. In short, the creation of IRBs was a response to public outrage and the need to get scientists to comply with basic ethical requirements in their pursuit of biomedical research, especially when this involves human subjects. The requirement to get previous approval from an ethics committee before any biomedical research project involving human subjects could be carried out was adopted in all developed countries and is now spreading worldwide. Although such biomedical or clinical research ethics committees have acquired various names in different countries, the mission and the ethical principles are the same everywhere: informed consent, positive balance between risks and benefits, and sound science. The current trend is for IRBs to lose their institutional basis in some countries, become external, and even be commercialized. The argument behind this change relates to the increasing number of multisite studies and the need to rely on a single IRB opinion. However, IRBs with commercial purposes, although they can invest in ethical training of researchers, do have an interest in a high level of approval what can conflict with international research good practices and society's interests.

H. ten Have and M. do C. Patrão Neves, *Dictionary of Global Bioethics*,
https://doi.org/10.1007/978-3-030-54161-3_323

Journalism Ethics

Journalism ethics refers to ethical standards or behavior considered right, appropriate, and desirable by codes of ethics for journalism to be practiced. Ethical requirements for journalism were seen as necessary as the profession gained power in democratic societies. True journalism does not exist in dictatorships since the information disseminated is propaganda. It is only in democratic regimes that journalism really exists where voice is given to all sectors of society and to all citizens. A free press is the barometer of democracy and journalistic freedom is a sign of a healthy democracy. Consequently, the power inherent in true journalism calls for responsibility and adherence to the ethical standards of the profession presented in codes of ethics or codes of conduct (self-regulated by independent bodies set up by newspapers). Such codes establish the rights and duties of journalists to publishers, proprietors, colleagues, their sources, and to affiliations (political or economic), interested parties (private or public institutions and associations), and above all to the public and society. Historically, journalism ethics started when journalism became a profession, when democracies became more abundant, and when the first codes of ethics were written in the early 20th century. In addition to all democratic countries today having a journalistic code of ethics, there are also several codes that are international the first of which was issued in 1971 by European federations of journalists and other international organizations interested in bringing this about. In 1978 UNESCO published the Declaration on Fundamental Principles concerning the Contribution of the Mass Media to Strengthening Peace and International Understanding, to the Promotion of Human Rights and to Countering Racialism, Apartheid and Incitement to War. Then in 1993 the Council of Europe adopted a Resolution on the Ethics of Journalism. Although there are differences between these international codes (mostly down to the social and political environment in which they were formulated) and between national codes (mostly down to the cultural and legal framework in each country), there are core values worldwide for the profession. Although the major core value is freedom (independence, exemption from prosecution, fighting all kinds of censorship), the mission given journalism is to serve the public interest (different from the interests of the public), provide the information needed (objectively and accurately, without false accusations, and avoiding all forms of sensationalism), promote active citizenship, and acknowledge its social responsibility. Other essential and transnational ethical principles in journalism are respecting privacy, rejecting all forms of discrimination, protecting sources of information, distinguishing between facts and opinions or interpretations, and correcting wrongful information. Although codes of ethics are enforced by journalistic associations and national governments, journalism is today subject to added pressure such as the development of social media and the imperative

H. ten Have and M. do C. Patrão Neves, *Dictionary of Global Bioethics*,
https://doi.org/10.1007/978-3-030-54161-3_324

to achieve economic sustainability. Journalists cannot ignore or lag behind flows of news in social media despite not having the time to confirm them all and not being allowed of course to plagiarize them. The consequences of not being the first is lost publicity contracts and, more importantly, lost financial revenue. Furthermore, journalists have to contend with citizens acting like them but not having to comply with ethical standards. This runs the risk not only of journalism ethics weakening, but also journalism itself (and possibly even democracy).

Justice, Global

See Global justice.

H. ten Have and M. do C. Patrão Neves, *Dictionary of Global Bioethics*,
https://doi.org/10.1007/978-3-030-54161-3_325

Justice, Intergenerational-Intragenerational

A new feature of global bioethics is the extension of ethical concerns to future generations. Since people are becoming increasingly aware that their existence depends on the survival of the planet and the preservation of a common heritage the conviction that global justice should not only be intragenerational (current generations) but also intergenerational (future generations) has been reinforced. The major problems of today require international cooperation for humanity and the planet to survive. Protection of the environment, preservation of natural resources, safeguarding the biological, genetic, and cultural diversity of humankind all call for intergenerational justice. Although present generations are using the common heritage of humankind and enjoying the benefits of the achievements of previous generations that have preserved and sustained basic recourses for the continuation of human life, they also have a responsibility to pass on such a heritage to future generations precisely because it is a shared responsibility and because it concerns basic resources that are common property. The concept of intergenerational justice is challenging and raises a number of questions: How can we have responsibilities and duties toward people who do not yet exist? Do we mean by "future" generations children or grandchildren who have just been born, human beings not yet born, or distant generations? One answer is that we do not have obligations to future people since responsibilities can only exist between real actors in reciprocal relationships, that such a reciprocity with future generations is fictional, and that only intragenerational justice is realistic. Other answers argue that generations have moral relationships with each other and that the implications of these relations get stronger as the impact of human activity becomes clearer and more predictable. Therefore one position argues we have moral responsibility primarily to one or two future generations. The other position argues that even distant generations can make a claim to justice because concepts such as common heritage and commons apply to all generations. The last two positions raise another query: What kind of obligations do we have to people who do not yet exist or might never exist? The problem is that we do not know what the needs of future generations will be since humankind is constantly changing as borne out by the needs of people a century ago being very different from those of today and there is no way they could have imagined the needs of our generation. Future generations by definition cannot tell us what their needs are likely to be. Many efforts have been made to give voice to future generations. A longstanding practice in health research is the creation of special institutions and mechanisms to protect vulnerable people or populations who cannot protect their own interests and to speak on their behalf. Offices of public guardians have similarly been established to represent posterity at the national, regional, and international levels. Such a role could be played by the UN Trusteeship Council formed in 1945 to help

H. ten Have and M. do C. Patrão Neves, *Dictionary of Global Bioethics*,
https://doi.org/10.1007/978-3-030-54161-3_326

colonial territories transform into independent states. Now that colonial territories by and large no longer exist, it has been proposed to transform the council into an international body to protect global commons for future generations. Similar arrangements for representing posterity have been proposed and introduced by a few countries such as France establishing the Council for the Rights of Future Generations in 1993 (it has not been active though), Wales passing the Well-Being of Future Generations Act in April 2015 (and establishing a Future Generations Commissioner to act as an advocate), Hungary creating an Ombudsman for Future Generations in 2007, and Finland creating a Committee for the Future in 1993.

Justice, Theories

Justice is one of the three basic principles of bioethics as formulated in the Belmont Report (1976) next to beneficence and respect for persons. Beauchamp and Childress's *Principles of Biomedical Ethics* (1979) promulgated justice as one of four principles on which the emerging field of bioethics (next to respect for autonomy, non-maleficence, and beneficence) should be based. UNESCO's Universal Declaration on Bioethics and Human Rights (2005) adopted justice (together with equality and equity) as one of the 15 principles of global bioethics. Article 10 states: "The fundamental equality of all human beings in dignity and rights is to be respected so that they are treated justly and equitably." The notion of justice as a moral and political concept has a long history in philosophy and ethics. Aristotle defined justice as the first moral virtue (a personal disposition whose objective is to enhance the well-being of the community or city-state). He distinguished between commutative, distributive, and punitive justice by arguing that the first concerns the equitable exchange of goods between individuals; the second concerns the distribution of goods, services, and rights among individuals being balanced; and the third concerns striking a balance between crime and punishment. Aristotle further argued that justice can be defined formally and materially. Although a formal definition emphasizes that equal cases should receive equal treatment (and unequal cases unequal treatment), such a definition does not allow what ought to be done in specific situations to be determined. This is where the material definition comes in since it is used to decide what criteria (such as merit and need) should be considered as equal. Contemporary theories of justice can be utilitarian, libertarian, or equalitarian. Utilitarian theories (Bentham and Mill) are focused on the consequences of actions such as maximizing happiness and avoiding pain (e.g., resources in healthcare should be distributed such that they maximize the outcomes of interventions). Libertarian theories (Nozick) emphasize that all individuals are entitled to the product of their work. Although social equality is not an aim of justice, equalitarian theories (Rawls) emphasize that equality is fair distribution (i.e., justice is fairness). Rawls came up with the difference principle in which social and economic inequalities should be arranged in such a way that they deliver the greatest benefit to the least advantaged.

H. ten Have and M. do C. Patrão Neves, *Dictionary of Global Bioethics*,
https://doi.org/10.1007/978-3-030-54161-3_327

Law and Bioethics

Although law and bioethics have been related since the birth of the latter, the way in which they correlate does not always follow the same direction in terms of the nationally adopted legal systems. Common Law uses case law based on past legal precedents to rule on present cases, whereas Civil Law (Roman Law) uses law based on statutes in which already written law (approved by parliament) is applied to provide a ruling in each case. North America uses the Common Law system in which the most complex or unprecedented cases caused by the development of biotechnologies are brought to courts and decided by judges. The rulings of judges establish a strong precedent at the legal level and a pattern of practice at the bioethical level. A number of medical cases can be presented as examples of such rulings. The Baby Jane Doe case (1983) in which the ruling imposed limits on decisions taken by parents on behalf of their children. In this case the parents refused to authorize surgery to allow the newborn who had Down syndrome to be fed with the result that the baby died. Ever since decisions taken by parents that threaten the lives of children are stopped in their tracks and parental rights are suspended. This is the standard way today of dealing with Jehovah Witnesses when their children need a blood transfusion. Another example is the Tarasoff case (1976) in which the ruling imposed limits on professional confidentiality. In this case a psychiatrist could have prevented the death of a young woman had he released information that her boyfriend intended to kill her. The psychiatrist kept this information secret and the girlfriend was murdered. Ever since (despite professional confidentiality still being the rule) exemptions apply when a third party is at risk. This is the standard procedure today for infected patients whose behavior can put other people at risk (especially in the case of HIV/AIDS). Continental Europe uses the Civil Law system in which the relationship between law and bioethics proceeds exactly in the opposite direction (i.e., ethical reflection and consensus are built regarding a controversial issue that are then used to guide legislators to develop the law). This is also the procedure adopted by international ethics committees (the vast majority have no legislative power) to develop what is today considered soft law (i.e., guidelines that do not have formal legal power but are highly influential in shaping national and international legislation). Some major examples can be found in declarations, recommendations, and opinions such as UNESCO's Universal Declaration on the Human Genome and Human Rights (1997) stating that the human genome is inviolable because it unites "all members of the human family" and expresses "the heritage of humanity" and the Protocol on Cloning (1998) of the European Convention (1997) banning all human cloning for reproductive purposes.

H. ten Have and M. do C. Patrão Neves, *Dictionary of Global Bioethics*,
https://doi.org/10.1007/978-3-030-54161-3_328

Law and Morality

Law and morals are two normative discourses (i.e., discourses on how things ought to be) of human action well established since ancient times. The Greek tragedy *Antigone* by Sophocles in the 5th century BC narrates Antigone's dilemma (or conflict) between obeying her king and a royal edict (the law) or following her own conscience and complying with what was commonly considered good in her community (morality). Antigone chose to follow her conscience and was sentenced to death. Ever since law and morals have been recognized as different. In recent times Guy Durant (1989) summarized the main differences between law and morals. The first difference is that morals address human innerness and involve personal convictions, whereas the law only cares about the exterior or visible submission of the individual and institutions to current regulations. The second difference is that morals just make sense universally, whereas law always refers to a particular community (in time and space). The third difference is that morals proceed over time and take into consideration the future of humanity, whereas the law focuses on the short term and is concerned with the current management of individual and institutional rights and duties. The fourth difference harks back to Antigone by pointing out that morals mostly unfold at the level of ideals and sometimes call for heroism, whereas the law just requires a few rules to comply with that do not demand extraordinary efforts. The fifth and final difference is that moral offense can raise social disapproval and personal remorse, whereas a legal infraction entails a penalty. Viewed in this way morals are prior to the law, more extensive, and more demanding and it is morality that grounds and justifies the law. There is also an opposing view that argues that the law alone constitutes common morality and that the individual is only obliged to adhere to it and nothing else. Such a narrow view does not take into consideration the difference between legal obligations with which we have to comply and moral duties we ought to acknowledge and strive for. In short, law contributes to people living together peacefully, whereas morality encourages people to be better.

H. ten Have and M. do C. Patrão Neves, *Dictionary of Global Bioethics*,
https://doi.org/10.1007/978-3-030-54161-3_329

Leadership

A leader is someone who leads or commands a group, organization, or country. There are many ways to define leadership but the principal one is the capacity to translate vision into reality. Such a definition refers to certain qualities that are characteristic of leadership such as establishing a clear vision and sharing it with others such that they will follow, providing information and knowledge to realize the vision, and coordinating and balancing conflicting interests of everybody involved. Unlike management, leadership cannot be taught. Good leaders have certain qualities that make them successful such as clarity (they know what they want and how to get it done), decisiveness (consistently pursuing their visions), courage, passion, and humility (a characteristic of good leaders). They admit when they are wrong and regard criticism as an opportunity for improvement. However, although humility is a quality needed by leaders, it is unfortunately rare as borne out by recent questions raised about the ethics of leadership. Citizens expect leaders to be ethical but feel there is a lack of ethical integrity among them. Since leadership is aimed at intentionally influencing others to change behavior and beliefs, it is important to clarify its underlying values and principles. The leader must be perceived as an ethical person demonstrating such values as honesty, integrity, care, and openness and making use of such leadership tools as communication and discipline to impress on others the importance of ethics. Leadership involves applying fairness in the treatment of other people and is strongly linked to the ethical culture of an organization or community. Although leaders may be competent, intelligent, and industrious, without ethical qualities they are ineffective. In contrast, there are those who argue along the lines of the Machiavellian tradition that leaders never reach power in an ethical way. However, precisely the opposite applies in that the more ethical someone is the more vulnerable he or she is too. Power strategies unavailable to ethical people are used without restrictions by unethical people as demonstrated by dictatorships.

H. ten Have and M. do C. Patrão Neves, *Dictionary of Global Bioethics*,
https://doi.org/10.1007/978-3-030-54161-3_330

Legal Ethics

Legal ethics refers to ethical standards or behavior considered right, appropriate, and desirable for judges and lawyers in the legal profession to practice their trade. Such ethical standards are presented in a code of ethics or a code of conduct (self-regulated by the profession) that enumerates the rights of professionals and their duties to clients, colleagues, and other professionals to whom the legal profession is more closely related such as legal educators, private investigators, courts, and society. Codes of legal ethics are grounded in common morality (specifically in human rights) and aim to provide guidance to professionals concerning the most complex and frequent ethical dilemmas they face in daily practice. Such codes establish the minimum standard necessary for consensual approval of professionals thus making such a standard reachable for all of them. This allows professionals to apply legal ethics principles to all situations regardless of whether they are in the code or not. This discourages practitioners from abusing flaws in the system or employing obstructive tactics. The codes of legal ethics are enforced by bar associations (in Anglo-American types of jurisprudence) and by national governments. Although they differ in different countries according to national constitutions and national legal systems, they are identical when it comes to core principles. The codes are essential to the credibility of practitioners, the profession, and the legal system and crucial to trust between professionals and clients. Some legal ethics principles are controversial in practice such as confidentiality of information (perceived by some as essential for good practice and by others as failing social responsibility when by protecting criminals the lives of third parties are put at risk) and when the interests of clients and society do not coincide. Indeed, incompatibilities and conflicts of interest are two fields where ethical issues easily arise. Other essential and legal ethics principles recognized worldwide are integrity, loyalty, and good faith.

H. ten Have and M. do C. Patrão Neves, *Dictionary of Global Bioethics*,
https://doi.org/10.1007/978-3-030-54161-3_331

Life Sciences

Life sciences (a.k.a. biological sciences) are used for the scientific study of life. The word "biological" derives etymologically from the Greek *bios* (life) and *logos* (study or science). The systematic study of life started in Ancient Greece by the philosopher Aristotle in the third century BC. Based on the observation of nature his study was teleological and essentialist in nature and had a major influence on how life was understood right up to the seventeenth century. A new perspective on the study of life emerged in the Renaissance as a result of experimentalism and the conviction that all knowledge should be based on experience. It was at this time that instruments capable of direct observation started to be created. Later (in the seventeenth century) Descartes developed a different vision in which life was organized and could be viewed by taking a mechanical rather than an essentialist approach. However, it was not until the nineteenth century when Claude Bernard proposed his scientific method in *An Introduction to the Study of Experimental Medicine* (1865) that life became the study objective of a new experimental science called biology. As Bernard and his pupils were studying the human body as a free and independent life-form and developing physiology in its wake, other scientists committed themselves to understanding the history of life, the interactions brought about by biodiversity, and the processes involved in the evolution of life. Such scientists included Jean-Baptiste Lamarck and Charles Darwin (*On the Origin of Species*, 1859) who helped develop natural history. The nineteenth century is known as the century of biology when the scientific study of life started to grow and gave rise to different biological disciplines such as cytology, bacteriology, morphology, and embryology and to the development of many different instruments and techniques. In the twentieth century the discovery of the structure and function of DNA by Crick and Watson (in 1953) triggered a biotechnological revolution that is still unfolding at an ever growing and breathtaking speed. Life sciences today refer to an ever wider range of disciplines and techniques (covering all living beings and ecosystems) some of which have growing power to control, manipulate, and alter life (e.g., genetic engineering, gene editing, and cloning). Indeed, life sciences have gained the power not only to artificialize life, but also to artificially create new life such as so-called synthetic life (giving rise to the new discipline of synthetic biology). The progress made in life sciences has been invaluable to the knowledge of life systems and human physiology thus facilitating beneficial (and precautionary) interventions and interactions between humans and other beings and the environment. The recent possibility to alter the intrinsic and universal nature of life and of living beings raises ethical questions about what criteria should be used and who should have the authority to sanction it.

H. ten Have and M. do C. Patrão Neves, *Dictionary of Global Bioethics*,
https://doi.org/10.1007/978-3-030-54161-3_332

Life, Definitions

Although life is of course an intimate and common reality for all of us, the first attempt at defining it was relatively recent and remains controversial. The concept of life did not arise until the early 1800s with the systematization of the new science of biology or the science of life (although the nature of living beings had been studied since Ancient Greece). Life was said to be the same for all living beings despite their different appearances and lay essentially in their organization. In the nineteenth century the major goal was to identify properties common to all life-forms and to distinguish them from inanimate objects by taking a vitalist view rather than the former mechanist view that explains life by means of physical and chemical laws. Vitalist views (differently espoused by Lamarck, Cuvier, or Claude Bernard) go beyond the description of physical and chemical phenomena, whereas mechanist views (differently espoused by Buffon, Bichat, and T.H. Huxley) tend to identify the locus for life (sought since ancient times) and determine how its animation affects living beings. Nevertheless, by the early twentieth century most biologists were no longer interested in formulating a definition of life. There is general agreement that not only is there no single consensual definition of life, but also that any possible definition of life would be even more difficult today. Applying modern technology to life (biotechnology), creating life (assisted reproductive technology), suspending life (artificial life support systems), and altering life (genetic engineering) have made defining life all the more complex. Augustine in his *Confessions* (fourth century) pondered the sentence "If you don't ask me, I know; if you ask me, I don't know" and argued that such is life.

H. ten Have and M. do C. Patrão Neves, *Dictionary of Global Bioethics*,
https://doi.org/10.1007/978-3-030-54161-3_333

Life, Extension

Life extension refers to increasing the longevity and life expectancy (average total years a human is expected to live) of humans and to the processes used to prolong the lifespan (maximum years a human can live). Human life expectancy has significantly increased (by about 27 years) in the last century in Western countries (life expectancy is increasing at a rate of 5 h a day) mostly as a result of the discovery of antibiotics and vaccines. According to the WHO global life expectancy of a newborn in 2016 was 72.0 years (74.2 years for females and 69.8 years for males), whereas in Europe life expectancy of a newborn in 2017 was 80.9 years. Some scientists believe someone born today could live 150 years. Nevertheless, typical diseases in contemporary societies related to lifestyle such as obesity, diabetes, cardiovascular diseases, and diseases caused by alcohol and tobacco together with resistance to antibiotics can significantly and negatively affect such projections. Although the human lifespan remained much the same throughout human prehistory (about 35 years), today many people reach an age of 100 (the WHO estimates there are about half a million people over 100 at the moment and likely to be about 3.7 million centenarians by 2050). The longest confirmed lifespan was 122 years (Jeanne Louise Calment, 1875–1997). Life extension (be it increasing life expectancy or the desire to extend the human lifespan) raises a number of ethical issues. When it comes to increasing life expectancy, it is important to be aware that the world population (mostly in Western countries) is constantly and rapidly ageing and that urgent social measures are needed to adequately respond to this new reality. Such measures relate to the retirement age, transition from professional life, health and social services, age-friendly communities, and intergenerational coexistence. Assuring quality of life (physical, psychological, social, and emotional) for an ageing population within an inclusive and integrative environment is perhaps the most demanding challenge. When it comes to the desire to extend the lifespan, there are many research projects following different approaches aimed at extending human life as much as possible with immortality as the ultimate goal. Such approaches include pharmacological intervention, regenerative medicine (stem cell injection and organ replacement), and genetic engineering. Although studies on animal models show that life expectancy can indeed be modified, there is a tendency mostly among biogerontologists to consider ageing as a disease rather than a natural phenomenon. This can adversely affect the way we relate to our own ageing and that of others.

H. ten Have and M. do C. Patrão Neves, *Dictionary of Global Bioethics*,
https://doi.org/10.1007/978-3-030-54161-3_334

Life, General

The concept of life is as common in daily usage as it is difficult to explain and the plurality of the contexts in which it is used are as ambiguous as the meanings it expresses. This is the reason the via negativa is the easiest approach to explaining the concept of life (i.e., by saying that life is not death or that life is the opposite of death). A via positiva approach requires a statement about what life actually is. The general definition of the concept of life is that it is dynamic, evolves over time, changes its appearance and its very being, and ends with death. Life has been differently perceived over time according to different geo-cultural contexts and to different religions, philosophies, or ideologies. Historically, life has mostly been understood as a natural fact, a divine gift, and as raw material for human exploitation. People's perception of life influences the way they relate to their own or someone else's life and even the way they relate to death and the kind of funeral (burial, cremation, or cryo-preservation) they want. Thus the multiple perceptions people have about life that shape the way they live become ever more complex when life is subject to adjectivations such as physical, psychological, social, cultural, and spiritual each of which changes its meaning and the reality to which it refers. Therefore life is not a homogeneous reality but one that has many sides each one of which is capable of assuming a specific value and of conflicting with another when it is not possible to maintain both. Regardless of the kind of life referred to, life has always been considered a value or a proto-value (i.e., the foundation of all values) that is currently undergoing two opposite movements. On the one hand, human life has been losing its traditional absolute value as a result of situations in which it can be evaluated (quality of life) and eliminated (death penalty, euthanasia) widening and complying with the ethical principle of autonomy. On the other hand, animal life has been gaining increasing value as a result of situations in which it is required to be protected (scientific experimentation) widening and complying with the ethical principle of responsibility toward all life.

H. ten Have and M. do C. Patrão Neves, *Dictionary of Global Bioethics*,
https://doi.org/10.1007/978-3-030-54161-3_335

Life, Quality of (*See* Quality of Life; QALY)

Quality of life is a highly controversial concept as a result of being confused with other closely related concepts such as health and well-being and by being perceived differently throughout history, during someone's own lifetime, or according to culture. Although the consensual indicators used to consider quality of life are biological, psychological, social, and cultural, there are many others that can be employed to value someone's life. Therefore surprise should not be expressed at the many definitions of quality of life issued by a number of entities and institutions. The WHO defines it as "an individual's perception of their position in life in the context of the culture and value systems in which they live and in relation to their goals, expectations, standards and concerns. It is a broad ranging concept affected in a complex way by the person's physical health, psychological state, personal beliefs, social relationships and their relationship to salient features of their environment." Although the WHO definition is quite general and very broad, nevertheless it does not overcome the controversy the concept has always raised: Who should evaluate quality of life and what should such an evaluation entail? Historically, the concept of quality of life (at least in the biomedical field) has increasingly been used in the ever more extensive and invasive application of biotechnological resources to patients (specifically, the use of extraordinary or disproportionate therapeutic means aiming solely at prolonging life and sometimes imposing serious suffering in situations where there is little hope of recovery). Quality of life can then be introduced as a norm (i.e., an ethical principle) to help professionals decide whether to forgo or withdraw medical treatment by, first, considering the clinical situation of the patient and, second, the health resources available and the need for their optimization or rationalization. Later it was used as a criterion to request euthanasia or assisted suicide. Quality of life is subjective in that it can only be invoked by the self—not by someone else such as the health professional. However, if subjectivity can be reduced or eliminated and presented as fundamentally objective (although critics point out that evaluation of a life always threatens its dignity or absolute value), then someone else can evoke it. Several attempts to quantify or measure quality of life have been made applying different metrics at a number of levels such as quality of life in general (WHOQOL 100), quality of health (sickness impact profile/SIP 136; SIP 68), and quality of health within a specific chronic disease (diabetes DQOL). The very first quality of life scale (QOLS) was defined in the 1970s by American psychologist John Flanagan. Many other instruments have been created to measure quality of

H. ten Have and M. do C. Patrão Neves, *Dictionary of Global Bioethics*,
https://doi.org/10.1007/978-3-030-54161-3_336

life such as quality-adjusted life years (QALYs) that measure the effect of a specific clinical intervention on a patient's quality of life as a result of such life being extended. Although quality of life is a useful concept in healthcare as an indicator for clinical decisions (pain and suffering) or for healthcare allocation (futility of treatment) contributing to prudent decision-making, it cannot be taken as a (single) criterion for decision-making about people's lives.

Life, Sanctity of

Sanctity of life refers to the inviolable character of life. The word "sanctity" has an inescapable religious connotation (sacredness) that conveys the historical/philosophical context of its use in bioethics. The protection of life when viewed from the perspective of biotechnologies being increasingly used to manipulate it is pursued by religious (Christian) arguments such as life being created by God, belonging to Him, and no one being free to alter or end it. This applies to ethical discussions both at the beginning and end of human life not only morally forbidding abortion, embryo research, cloning, genetic engineering, assisted suicide, and euthanasia, but also warranting protection throughout the human lifecycle. The inviolability of life can also be argued rationally from a secular point of view that grants it wider acceptance. From this perspective life can be said to be a proto-value (i.e., the most basic and elementary condition for the manifestation of all other values). The deliberate elimination of life entails suppressing all values. The bioethical principle of the sanctity of life requires full respect for human life, total protection from its beginning to its end (i.e., throughout the entire lifecycle), and the rejection of all forms of relativism (e.g., quality of life). From both the religious and the secular perspective sanctity of life commonly refers to the intrinsic, inherent, absolute, and unconditional value of human life thus grounding the principle of human dignity. Although sanctity of life rejects all intentional attempts against human life, it still acknowledges the natural course of life and the withdrawal of artificial and extraordinary therapeutic means whose sole intention is to keep someone alive without any hope of recovery. The principle behind sanctity of life is strongly criticized for imposing limits on individual autonomy and for compromising the ownership of someone's own body. It is also criticized by animalists for overvaluing human life in comparison with animal life.

H. ten Have and M. do C. Patrão Neves, *Dictionary of Global Bioethics*,
https://doi.org/10.1007/978-3-030-54161-3_337

Lifestyles

Lifestyles refer to the way people choose to live and how their behavior and what they do affect individual and public health. Lifestyles consist of a wide range of individual activities from jobs and hobbies, diets and sports, to relaxation and fun. The relationship between lifestyle and health is increasingly acknowledged and has led to lifestyle being currently recognized as an important determinant of health. At the beginning of the twenty-first century the WHO considered that about 60% of the determinants of individual health are correlated with lifestyle that has been shown to significantly affect quality of life and lifespan. Such a reality underlines the obligation states have not only to promote healthy living conditions for all citizens, but also to invest in health promotion and disease prevention. It is becoming ever more urgent to develop preventive medicine together with curative medicine. The impact lifestyle has on health makes it all the more important for individuals to behave responsibly, make healthy choices, and take care of their own health. Ethical issues concerning lifestyles relate to the obligations of states and individuals to preventing illness and improving health. The major problem is respecting individual autonomy while promoting the common good (e.g., individuals choosing not to be vaccinated affect the well-being of communities). Countries ruled by governments with a liberal or libertarian ideology tend to give primacy to individual freedoms and privacy and hold back from imposing restrictions on personal choices such as obliging citizens to use seat belts. More communitarian-oriented countries try to bring about healthy behaviors by using strategies like campaigning and taxation. Although governments have taken initiatives to advertise health risks such as those associated with smoking, drug and alcohol abuse, or sedentary life, they more commonly impose higher taxes on unhealthy drinks and foods such as sugary drinks or fatty foods and on tobacco. Some countries allow taxpayers to use expenses related to healthy activities as personal allowances such as gym subscriptions or buying a bicycle (some private insurance companies offer lower premiums). In contrast, there is increasing pressure to make people who live an unhealthy lifestyle accountable for their own health costs. Some European countries (Australia too) providing universal coverage through national health services are considering (official or societal) proposals to make citizens accountable for diseases or injuries deemed self-inflicted. Such proposals call for citizens to contribute to healthcare costs and/or to be given lower priority in provision of healthcare. Self-inflicted diseases are often difficult to identify objectively and accurately as are finding direct causal relations between health problems and lifestyle. Such difficulties have led to the realization that moral responsibility cannot be converted into legal or financial accountability and that health education continues to be the best way to promote healthy lifestyles.

H. ten Have and M. do C. Patrão Neves, *Dictionary of Global Bioethics*,
https://doi.org/10.1007/978-3-030-54161-3_338

Literature

Literature is regarded as an important resource for bioethics at two levels. The first of these levels considers world literature as a repository of the fears and desires of humankind throughout history as exemplified in mythology, utopias, and science fiction. Such universal narratives indicate the direction society expects scientific knowledge and technological innovation to follow. Indeed, scientific and technological developments have accomplished things that were impossible at the time authors conceived them such as the fantastic machines and feats imagined by Jules Verne and H.G. Wells. The second of these levels considers literature relevant to teaching bioethics because of the many narratives (often given in the first person) about the need to understand major medical issues from many perspectives—not just the professional one. The different ways illness, pain, and suffering lead to despair or become part of someone's life, the impact of disability on individuals and their families, the experience of ageing, and how the process of dying can be endured are just a few of the many topics extensively described in literature. The narrative model is not the only bioethical approach to have developed from literature since other approaches to medical ethics often use storytelling and classic literature to explain different issues and perspectives. There are those who criticize the value of literature in bioethics denouncing novels as fictional and thus inaccurate.

H. ten Have and M. do C. Patrão Neves, *Dictionary of Global Bioethics*,
https://doi.org/10.1007/978-3-030-54161-3_339

Living Will (*See* Advance Directive)

A living will is a formal document in which someone can freely state his or her own wishes concerning healthcare regarding a future situation in which that person is no longer competent or able to give consent. A living will is an advance directive that extends the principle of autonomy to those who are no longer able to make decisions but allows them to express their wishes or preferences. Most developed countries have legislation enabling citizens to express their preferences for end-of-life situations. Such legislation establishes the procedure that has to be followed to formalize a valid living will, defines limits to what someone can request, and clarifies what cannot be chosen (against state law) such as euthanasia, assisted suicide, and withdrawing of clinically assisted hydration and nutrition. It also establishes what is considered contrary to good medical practice such as someone demanding futile medical examinations or treatments. The advantages of having a living will are important to a number of people such as patients whose formerly expressed will can be respected when they are no longer competent, the patient's family freeing them from the emotional burden of making end-of-life decisions for a loved one, and healthcare professionals releasing them from making difficult choices and at the same time making them feel much more comfortable about the outcome. There are also disadvantages associated with living wills such as the difference between what the living will foresees as a hypothetical situation and a future reality, what the living will anticipates as possible treatments and outcomes for a particular clinical condition, what medical advances will then be able to be provided or accepted had the patient had the opportunity to consider them. Such shortcomings can be reduced when living wills are written with medical assistance. Living wills are only valid for a certain time after which they have to be renewed or updated, they can be updated or discarded at any time, and in an emergency situation a competent patient can orally update the living will. Another ethical issue that needs consideration is whether information about the possibility to write a living will should be disseminated. Should the state publicize it? Should the information be given to healthcare facilities? In the latter case should it be disseminated by administrative staff, physicians, or others? In what specific situation or at what specific ages should it be disseminated? The answer to each of these questions will depend on different ethical interpretations.

H. ten Have and M. do C. Patrão Neves, *Dictionary of Global Bioethics*,
https://doi.org/10.1007/978-3-030-54161-3_340

Malaria

Malaria is caused by parasites transmitted to human beings through the bites of infected female *Anopheles* mosquitoes. There are more than 400 different species of mosquito. Malaria is a life-threatening disease that mainly affects less developed countries. Unfortunately, this is probably the reason discovering an efficient treatment for this deadliest disease in the world has taken so long. Symptoms such as fever, headache, and chills usually appear 10–15 days after someone is bitten. Although the first symptoms may be mild, should no treatment be applied within 24 h then malaria can develop into severe illness and often lead to death despite the disease being preventable and curable. The two ways to prevent and reduce malaria transmission are insecticide-treated mosquito nets and indoor residual spraying with insecticides. CRISPR Cas9 genome editing has specifically been applied to mosquitoes to fight malaria at the experimental level and antimalaria drugs can be used to prevent the disease. Although insecticide resistance is a growing problem, effective treatment of malaria patients is available the best being artemisinin-based combination therapy. Although vaccines against malaria have been developed, they are still in the throes of clinical testing and evaluation. Malaria is frequent and global as borne out by the WHO estimating in 2017 that 219 million people were suffering from malaria in 90 countries and that more than 435,000 of them had died in that year. Although the disease burden is highest in Africa (92% of all malaria cases), malaria and other diseases typical of equatorial regions are now appearing in other countries further north mostly as a consequence of climate change. For example, approximately 1,700 cases of malaria are diagnosed each year in the United States (mainly as a result of immigration and travel).

H. ten Have and M. do C. Patrão Neves, *Dictionary of Global Bioethics*,
https://doi.org/10.1007/978-3-030-54161-3_341

Malpractice

Malpractice is professional negligence (a.k.a. professional incompetence) and applies to professional practices such as medicine, law, and finance. In such practices it is assumed that standards of care, competence, and skills are met by every professional. When a patient is injured because of a negligent act or failure to act by a hospital or healthcare provider it is called medical malpractice. In its 1992 Statement on Medical Malpractice the World Medical Association (WMA) described medical malpractice as "the physician's failure to conform to the standard of care for treatment of the patient's condition, or a lack of skill, or negligence in providing care to the patient, which is the direct cause of an injury to the patient." Examples are failure to diagnose, misdiagnosis, unnecessary surgery, surgical errors, improper medication, premature discharge, and poor follow-up. Malpractice claims include violation of the standard of care (in which patients expect healthcare professionals to provide care meeting such standards), injury caused by negligence where it is down to the patient to prove that the injury would not have occurred in the absence of negligence (an injury without negligence or negligence without injury do not qualify as malpractice), and injury resulting in significant damages (where it is down to the patient to show that the injury led to disability, suffering, loss of income, or significant medical expenditure). The recent increase in medical malpractice claims in many countries may be related to growing awareness of legal firms and the public of medical errors. Although medical malpractice is a legal concept, the underlying notion is ethical in that non-maleficence is a basic principle of medicine. The need to improve patient safety and reduce medical errors has encouraged efforts to report and discuss errors to avoid litigation and at the same time adopt policies of honesty, transparency, and openness. Although harmful errors should be disclosed to patients whether or not there is a risk for litigation, this practice of full disclosure is difficult and often not executed because of embarrassment, fear of litigation, or an inadequate organizational culture in dealing with errors. One of the responses to the growing number of malpractice claims is that it can result in defensive medicine in which many tests and procedures are performed unnecessarily in an attempt to reduce legal liability. Defensive medicine is ethically unjustified not only because it increases healthcare costs and deviates from professional standards, but also because it exposes the patient to unnecessary procedures and anxiety.

H. ten Have and M. do C. Patrão Neves, *Dictionary of Global Bioethics*,
https://doi.org/10.1007/978-3-030-54161-3_342

Managed Care

Managed care refers to organizing healthcare provision in such a way that costs are reduced and quality of care is improved. It is the dominant system of delivering and receiving healthcare in the United States (especially in for-profit healthcare). The primary concern of managed care programs is cutting costs as a result of health insurance companies making contracts with healthcare providers and medical facilities to provide care for members at reduced costs. It is not patients nor healthcare providers who determine and control the provision of healthcare services but the organizations that pay for such services. Programs and strategies to reduce costs may conflict with the goals of medicine and restrict the autonomy of patients and physicians such as limiting the freedom of a patient to choose a physician or medical facility and restricting the freedom of choice of doctors (since they are employees of managed care organizations). A number of ethical challenges arise here such as the impact on patient autonomy, possible conflicts of interests of employees working in managed care organizations may arise (they often receive financial incentives to limit their use of resources despite having a primary obligation to their patients), and the negative impact on solidarity with patients as a result of private, profit-based healthcare systems converting provision of care into a commercial interest with patients as customers. They will receive care as long as they pay to be included in such managed care systems.

H. ten Have and M. do C. Patrão Neves, *Dictionary of Global Bioethics*,
https://doi.org/10.1007/978-3-030-54161-3_343

Marginalization

Treating a person or group as insignificant, powerless, or unimportant is regarded as marginalization. It is a social phenomenon in which individuals and groups are excluded, their needs are ignored, and they are not allowed a voice. Marginalization can occur as a result of someone's nationality, race, gender, sexual orientation, ability, socioeconomic status, age, and religion (minority groups are often targeted). Marginalization often manifests itself by the use of derogatory language that expects targeted individuals to act in stereotypical ways and disregards the importance of their cultural and religious traditions and values, and by the use of patronizing behavior. Marginalization can have a negative impact on health by causing anxiety, anger, depression, self-blame, frustration, and isolation. In addition to its impact on health it is also associated with significant social impacts such as causing reluctance to interact with other people, making it difficult to participate in clubs, sports, and classroom activities, and blocking opportunities normally available to other members of society. It is widely believed by bioethicists that mental illness is often marginalized thus explaining why little attention is paid to the ethical problems of mental illness. Marginalization is related to the concept of vulnerability as borne out by marginalized groups in society very often being the most vulnerable populations.

H. ten Have and M. do C. Patrão Neves, *Dictionary of Global Bioethics*,
https://doi.org/10.1007/978-3-030-54161-3_344

Maximin Principle

The maximin (a.k.a. maximize the minimum) principle was proposed by John Rawls as a central concept in his work *A Theory of Justice* (1991) for use as a reasonable criterion to help someone choose between several alternatives in a situation of uncertainty (or risk). Such a theory of rational choice and decision considers the right decision would be to maximize the minimum outcome (i.e., to choose the alternative whose worst consequences would still be better than the worst of all other alternatives). Although such a principle is both risk and uncertainty aversive, it implies not taking unnecessary risks and playing safe. The philosopher also presents the maximax (a.k.a. maximize the maximum) principle (i.e., choosing the alternative whose best consequence would be better than the best of all other alternatives). Both strategies are reasonable and the choice between them depends on the circumstances. The general aim is to maximize the position of those who are in a more disadvantaged situation. The maximin principle is a strategy that can be used to address the problem of social justice by contributing to a more equitable society and assuring basic needs such as healthcare are met for everyone. Although Rawls did not refer to the right to healthcare, his maximin principle can indeed be applied in defense of just such a right (as did some of his commentators).

H. ten Have and M. do C. Patrão Neves, *Dictionary of Global Bioethics*,
https://doi.org/10.1007/978-3-030-54161-3_345

Media Ethics (*See* Communication, Media)

Media ethics refers to the role media play in society and to the procedural standards they follow. Since it is a branch of applied ethics it reflects the citizen's perspective—not just that of the professional (as is the case with professional ethics). It is the general public (i.e., those potentially affected by media) who express what they expect from the media and what moral values and principles should be followed as borne out by public opinion deciding which if any restrictions should be applied to the dissemination of pornographic material or of violent scenes. The media today are synonymous with mass communication channels since they both have the same broad power to influence people. Therefore the social responsibility of the media (i.e., the capacity to address societal needs and serve the public interest) should be equally broad. It is the respect and compliance of ethical principles such as transparency, justice, responsibility, integrity, and privacy that makes all the difference between public manipulation (serving someone's hidden interests) and public information (empowering citizens and facilitating their active engagement in social life). Currently the media tend to be organized in big enterprises that have significant financial and economic interests (and often political interests). They have a lot of influence over the public agenda choosing what is important and what is not and over the perspective from which information is presented. The more powerful media enterprises become, the fewer private media initiatives there will be thus reducing not only the plurality of media but also their independence and uniqueness. The new digital technologies and the digitalization of media are not only changing the way communication proceeds but doing so in a disruptive way. A significant part of worldwide communication is now done by non-professionals, by citizens (social media and not mass media) who are not and do not pretend to be independent when disseminating content that is not necessarily factual, frequently unconfirmed, biased, and ungrounded. Although the media closely follow what is happening in social media, the possibility of plagiarism and fake news is perhaps the most evident deleterious consequence they have in cyberspace. The media should have enough human and technological resources to investigate all the information they disseminate and to make sure it is accurate. However, this takes time and money that the media increasingly cannot afford thus affecting their quality. Mention should be made of the fact that cybermedia is much more interactive than ever and that the consumer is no longer passive but a participant in the production and dissemination of information.

H. ten Have and M. do C. Patrão Neves, *Dictionary of Global Bioethics*,
https://doi.org/10.1007/978-3-030-54161-3_346

Mediation

Mediation is a process in which arbitration and intervention are used in a dispute to resolve it. Since it is a form of dispute resolution it involves a third party who has the necessary communication skills and negotiation techniques to bring about a settlement and assist the disputing parties in finding the best possible solution. The mediation process can be characterized as voluntary (participants are free to withdraw from the process), collaborative (nothing can be imposed on the participants), controlled (every participant is free to make decisions and has a veto), confidential, informed (expert advice can be invited), impartial, and balanced. A mediator is tasked with assisting each party in a balanced way and cannot favor the interests of one party or dictate any outcomes of the process. Mediation is increasingly used in healthcare as borne out by clinical ethics committees and clinical ethics consultants making use of it to resolve disputes and disagreements. Although adequate training programs often do not exist and practical experiences are frequently not evaluated or published, better training and credentialing are likely to be available in the near future. Bioethics mediation proceeds on the assumption that clinical disputes have specific sources. Since most of such disputes are brought about by problems in communication such as insufficient information, cultural differences, misunderstandings, and different interpretations, there often is no ethical conflict and mediation can clarify matters. When it comes to managing disputes mediators should pay special attention to being as neutral as they possibly can, steer clear of accepting any possible outcome, and be fully aware of the limits imposed on them by law, as well as generally accepted ethical principles and institutional policies. Furthermore, they should avoid being perceived as representatives of one side or the other or biased in some way. Although mediation is tasked with reaching consensus, there may be moral uncertainty and ambiguity and multiple ethical principles at stake when one of many outcomes may be ethically justified. Rather than aiming at consensus regarding principles and values, mediation strives to reach agreement by finding the best possible resolution of the disagreement for the patient concerned.

H. ten Have and M. do C. Patrão Neves, *Dictionary of Global Bioethics*,
https://doi.org/10.1007/978-3-030-54161-3_347

Medical Humanities

Medical humanities is a field of study and teaching in medicine that includes the humanities (comparative literature, ethics, history, philosophy, and religion), social science (anthropology, cultural studies, health geography, psychology, sociology), and the arts (film, literature, theater, and visual arts) and their application to medical education and practice. Although it is an interdisciplinary field that is particularly important for medical education, it is related to the idea that medicine is an art—not merely a science. Physicians need communication skills, empathy, careful judgment, professionalism, sensitivity to the suffering of patients and to the subjective needs of those for whom they care. Learning to be a good health professional therefore requires more than scientific training and thinking; it also demands character formation and sensitivity. Medical humanities can contribute to the development of such skills and attitudes in health professionals a good example of which would be studying Tolstoy's *The Death of Ivan Ilyich* and thus gaining a better understanding of the travails befalling seriously ill patients. Medical humanities is a field that is supposed to humanize healthcare; encourage reflection, speculation, and imagination; help us know ourselves; and make us more aware of our values. When bioethics emerged as a new discipline, it was argued that medicine as a science and technology should be connected to the humanities to address moral values. Bioethicist Edmund Pellegrino argued that medicine is the most humane of sciences and the most scientific of humanities. He was involved in the establishment of the Society for Health and Human Values in 1969 in the United States. The society produced reports covering a number of areas such as medicine and literature, medicine and history, medicine and religion, medicine and visual arts, and medicine and social sciences. As a result of all such reports pointing to the humanities as providing rich resources to address the problems of contemporary healthcare, ever since the 1970s medical humanities has become institutionalized. The Institute of Medical Humanities at the University of Texas Medical Branch at Galveston in the United States was one of the first centers established (in 1973). Specialized journals started to appear such as the *Journal of Medical Humanities and Bioethics* (1979), *Literature and Medicine* (1982), and *Medical Humanities* (2000).

H. ten Have and M. do C. Patrão Neves, *Dictionary of Global Bioethics*,
https://doi.org/10.1007/978-3-030-54161-3_348

Medical Tourism (*See* Health Tourism)

One of the consequences of globalization is mobility as clearly demonstrated in the field of healthcare with health professionals migrating from developing countries to the developed world. Furthermore, patients from developed countries are increasingly seeking medical treatment in developing countries. Such medical tourism is actively promoted in some developing countries such as Thailand and India by offering wealthy patients from developed regions a package of medical interventions. What is more worrying, private and public hospitals in many developed countries are promoting medical tourism as a new and profitable business. Referring to medical travel or health-related travel as tourism is completely inaccurate since medical need rather than leisure is concerned. Medical tourism is the result of industrial policy adopted by commercial five-star hospitals that aggressively markets services at wealthy clients across borders. Patients travel for a number of reasons such as treatments not being available at home, costs being 5 to 10 times higher, and waiting lists for the treatment they need growing ever longer in their home countries. Medical tourism is on the increase as shown by estimates that more than 1.4 million Americans traveled abroad for healthcare in 2017 and that the number of medical tourists is expected to increase by 25% each year. Although the 5 private hospitals in Thailand treated over 104,000 medical tourists in 2010 generating USD180 million in medical revenue, the high-tech care offered to tourists is often unavailable to most of the local population. The most frequent conditions presented by medical tourists are in the areas of dentistry, cosmetic surgery, cardiac conditions, in vitro fertilization, weight loss, dermatology, transplantation, and spine surgery. There are also specialized forms of medical tourism such as transplantation tourism to Pakistan, reproductive tourism to India, stem cell tourism to China, and assisted suicide tourism to Switzerland. In many cases the motivation behind such tourism is that certain practices are prohibited or unavailable at home. A basic concern is whether it is ethical for wealthy nations to use the healthcare resources of developing countries while access to healthcare for the domestic population of such countries is limited. Another concern is quality of care and the questions it raises: What guarantees are there that patients receive quality healthcare and that the appropriate standard of care is applied? What will happen if complications arise after patients have returned to their home country? Returning patients often do not have any records of the procedures and medications they have received abroad despite often needing follow-up. It was against this background that the American Medical Association adopted new guidelines on medical tourism in 2018. The aim of such guidelines is to help physicians understand their responsibilities when interacting with patients who seek or have received cross-border medical care.

H. ten Have and M. do C. Patrão Neves, *Dictionary of Global Bioethics*,
https://doi.org/10.1007/978-3-030-54161-3_349

Medicalization

Defining problems in medical terms (such as illnesses or disorders) and using medical interventions to treat them is known as medicalization. Life is increasingly viewed through the medical lens in which all abnormal feelings are classified as diseases and addressed by prescribing medical drugs. Although the best example is the excess use of antibiotics, there are many others such as hyperactivity, shyness, anxiety, child abuse, and menopause all of which have been transformed into new medical categories and diseases. Medicalization is driven today by commercial and market interests such that fabricating new diseases will create new markets for existing drugs (*see* Disease mongering). Pharmaceutical companies often seek the cooperation of patient advocacy groups to promote their drugs as was the case with the drug Ritalin approved for use in children with ADHD since the 1970s. Industry started to promote its use in adults with the help of patient groups and it was this that first gave rise to the concept of medicalization in the 1970s. In 1972 sociologist Irving Zola argued that medicine had become a major institution of social control overtaking the traditional roles of religion and law. He defined medicalization as a process whereby much of everyday life would increasingly fall under the dominion, influence, and supervision of medicine. In 1976 philosopher Ivan Illich argued that medicine itself had become a source of harm (through iatrogenesis) by expropriating health from the grasps of the individual; creating dependencies on professional power and expertise; and eroding human dignity, autonomy, and solidarity. Although such studies focused on criticizing professional medical power, they also examined discursive practices showing how subjectivity, human bodies, and identities are shaped through medicalization. During the 1990s the focus of medicalization changed from blaming medical domination and power to being understood as a complex social practice backed by multiple driving forces such as the pharmaceutical industry and patient activism. Although neoliberal policies and consumerism are perfectly aligned with medicalization, when it comes to bioethics it raises a number of ethical questions: How should it be assessed? What is driving the phenomenon and how can it be influenced? Is there any way of identifying the ontological status of medical conditions? Are there any real conditions that should be regarded as genuinely medical? Are all conditions of human life such as eating, sleeping, and worrying susceptible to being medicalized? Since medicalization is also conceived as a loss of control reducing freedom of choice and aggravating vulnerability this raises the question: How can the principle of respect for individual autonomy apply when medicalization is taking place? Although medicalization assumes that health is the supreme value in contemporary society, many people believe values such as autonomy,

H. ten Have and M. do C. Patrão Neves, *Dictionary of Global Bioethics*,
https://doi.org/10.1007/978-3-030-54161-3_350

dignity, and solidarity are more important. Moreover, medicalization has the power to transform social problems into individual challenges and pathologies such that they can be addressed with medication (an example is obesity where the focus is on individual responsibility rather than the socioeconomic conditions under which unhealthy food is provided).

Mental Health

The concept of health can be approached from a naturalistic or a normativistic position. Naturalistic perspectives articulate that health is objective, empirical, value free, and (like disease) observable in the natural world. An example is the biostatistical theory of Christopher Boorse that argues health is the normal state in which an organism and its organs function and in so doing contribute to its survival and reproduction and that what is normal can be statistically determined. Normativistic perspectives regard health as a value-laden concept and what is healthy depends on the goals, projects, and aspirations of an organism in a wide variety of situations—not on its natural state. Mental health can be defined negatively as the absence of mental illness and positively as subjective well-being, perceived self-efficacy, autonomy, competence, and self-actualization of someone's intellectual and emotional potential. Definitions of mental health vary across cultures. The WHO in its 2001 report on mental health concluded that it is nearly impossible to define mental health comprehensively from a cross-cultural perspective. However, it is generally agreed that mental health encompasses much more than simply lacking mental disorders. Such a conclusion endorses the normativistic perspective in that it articulates there are three sides to mental well-being: emotional, social, and psychological. In other words, different societies and cultures define and interpret what is mentally healthy and what interventions are appropriate in different ways. Although mental health is a value-laden and socially constructed concept, it is today frequently approached from a naturalistic perspective in which normal brain functioning can be studied with imaging technologies and neuroscientific methods. Such interventions are focused on the biology and anatomy of the brain. Although trepanation and lobotomy used to be employed to improve mental health in the past, a range of pharmaceutical drugs are available today to enhance mental health. The principal ethical issues raised by mental health are the lack of general understanding of what a mental condition is and how problems with mental health can lead to social consequences such as discrimination and stereotyping.

H. ten Have and M. do C. Patrão Neves, *Dictionary of Global Bioethics*,
https://doi.org/10.1007/978-3-030-54161-3_351

Mental Illness

The concept of mental illness like that of mental health is contested in the ongoing philosophical debate about the mind and mental states and raises questions concerning the interactions between mind and body. Illnesses of the mind affect the entirety of a person—not just his or her biological or psychological side. Understanding mental illness requires paying attention to subjective experiences of sufferers in the context of their life history and their relationships with other people. The concept can be distinguished from that of mental diseases in that the latter refers to disorders and dysfunctions of the brain. Psychiatric classifications usually define the criteria that need to be used for specific diseases to be diagnosed and treated. However, since such classifications are often based on subjective symptoms the criteria used and the boundaries between disorders are unclear. Mental illnesses and their diagnosis have significant ethical consequences such as being misused by removing people from social life and institutionalizing them, leading to stigmatization, and behaviors regarded as abnormal in societies and cultures being interpreted as illness. Ever since the seventeenth century mental illness has become increasingly medicalized as borne out by those considered insane and those demonstrating aberrant behavior being isolated from society in specific institutions. Using illness as a way of controlling inappropriate behaviors has reinforced the idea that the concept is socially constructed and therefore questionable. Thomas Szasz argued that mental illness is little more than a metaphor. Such a view instigated the antipsychiatry movement in the 1960s and 1970s, which argued that any diagnosis of mental illness was no more than an expression of unequal power relations between doctors and patients. Although many sufferers of mental illness today reject the view that mental illness does not exist since they experience real emotional pain and disability, many physicians continue to apply the medical model, focus on physical abnormalities in the brain, and apply a range of drugs in the hope of addressing symptoms. Other physicians take a broader approach involving psychotherapeutic and behavioral interventions. By applying such a biopsychosocial model they not only take into account biological but also psychological and sociocultural factors and in so doing are able to recognize the interrelations between mind and body.

H. ten Have and M. do C. Patrão Neves, *Dictionary of Global Bioethics*,
https://doi.org/10.1007/978-3-030-54161-3_352

Mercy

Mercy is synonymous with pity (expressing condescension) and with clemency (expressing forgiveness) in an asymmetric relationship between someone who has failed and someone who has the power and benevolence to pardon. Indeed, mercy expresses gratuitous forgiveness that goes beyond any attempt at justifying the fault committed or finding merit in the guilty individual. The act of mercy depends entirely on the person bestowing it and is dictated by feelings—not by reason—and harks back to Ancient Greece when Eleos was the personification—*daimon* (divinity)—of mercy, pity, clemency, and compassion and counterpart to the Roman goddess Clementia. Although mercy has always had a religious connotation when used to articulate the asymmetry between who forgives and who is forgiven, the concept of mercy is often used in a secular context as borne out by the well-known expression "mercy killing" in which the person who is suffering calls for euthanasia to finally bring to an end to his or her pain. Most groups, organizations, and associations who refer to mercy (e.g., mercy health) are religious and regard their actions (frequently addressing the needs of the poor) as guided by Christian values and principles.

H. ten Have and M. do C. Patrão Neves, *Dictionary of Global Bioethics*,
https://doi.org/10.1007/978-3-030-54161-3_353

Migration

Mobility is one of the characteristics of globalization and manifests itself in the movement of people from one place to another either within a country or more often from one country to another. The International Organization for Migration (IOM) was established in 1951 and was integrated in the United Nations as a specialized agency in 2006. It defines migration as "the movement of persons away from their place of usual residence, either across an international border or within a State." The IOM estimates there are currently 244 million international migrants (3.3% of the world's population) and believes the percentage is increasing (in 2000 it was 2.3%). Although many people migrate by choice, others are forced to migrate as borne out by the ever growing number of refugees, asylum seekers, and internally displaced persons. The IOM definition of migrants emphasizes that everybody moving away from his or her habitual place of residence is a migrant regardless of the person's legal status, the causes of movement, the length of stay, or whether the movement is voluntary or not. Migration can only be resolved by taking a comprehensive approach. The New York Declaration on Refugees and Migrants adopted in 2016 by UN member states reaffirmed the commitment to international human rights legislation to protect the safety, dignity, human rights, and fundamental freedoms of all migrants. It also acknowledged the positive contributions made by migrants to sustainable and inclusive development. States committed themselves to "ensure a people-centered, sensitive, humane, dignified, gender-responsive and prompt reception for all persons arriving in our countries, and particularly those in large movements, whether refugees or migrants. We will also ensure full respect and protection for their human rights and fundamental freedoms." In March 2017 the UN Secretary-General appointed a special representative for international migration tasked with following up the declaration. Migration poses significant bioethical problems such as providing migrants with healthcare. They are vulnerable in having limited or no access to healthcare (especially if they are undocumented workers or threatened with deportation), often having no family or social network to support them, and lacking health insurance. Another ethical challenge is the migration of healthcare workers (brain drain and care drain). Although migration is usually voluntary in this case, it often exacerbates inequalities in healthcare.

H. ten Have and M. do C. Patrão Neves, *Dictionary of Global Bioethics*,
https://doi.org/10.1007/978-3-030-54161-3_354

Military Ethics (*See* War)

Military ethics refers to ethical standards or behavior considered right, appropriate, and desirable in the military setting. It is a branch of applied ethics primarily practiced in military academies and universities. Although the bioethical principles of autonomy, beneficence, nonmaleficence, and justice apply in civilian ethics, in military ethics such principles often seem compromised because in the military setting there is the obligation to obey orders. A major ethical issue in military ethics is the ethics of war itself in which theories about when it is appropriate to go to war and what are the limits to warfare are discussed. The military setting also raises a number of specific issues such as the requirement to follow orders. This can put medical professionals in contradictory and unacceptably awkward positions such as being required to follow all legal orders, having to follow the ethical principles of their medical profession (they might be involved in interrogation, torture, and forced feeding of prisoners), being asked to certify that prisoners are fit for interrogation or torture, having to treat injuries as a result of such interrogation or torture, and having to force-feed prisoners on hunger strike. Bioethical issues also arise regarding soldiers. Although medical personnel in the military setting have limited control over medical decisions as a result of superior officers determining medical care and treatment required, in this setting medical information is usually not confidential. Moreover, military personnel are in a vulnerable situation since they often feel unable to refuse to participate in medical experiments as borne out by experimental vaccines first being tested on the military and new enhancement technologies (such as brain–machine interfaces) being tested on soldiers. Although the United States provides special healthcare services for its military, personnel cannot sue for medical negligence since this is assumed to undermine order and discipline in the military.

H. ten Have and M. do C. Patrão Neves, *Dictionary of Global Bioethics*,
https://doi.org/10.1007/978-3-030-54161-3_355

Minimalist Ethics

Minimalist ethics (a.k.a. the ethics of minima) refers to a philosophical perspective that advocates the need to identify minimum shared values in multicultural and pluralistic societies and thereby find the best possible common ground for people to live together peacefully. Utilitarian philosopher Jeremy Bentham was first to take such an approach and considered common law should consist of the fewest possible compulsory moral norms. Such an idea became ever more popular in utilitarian philosophy and was taken up by philosopher Stuart Mill who made use of the minimalist ethic and reduced moral life to the principle that people can live their lives as they wish so long as no harm is done to others. Other philosophers investigated minimalistic ethics and tried to identify a single moral principle that could ground normative ethics such as treating others as you would want them to treat you or doing no harm. Although such principles are procedural, secular, and focus on formal relationships, they are not substantial or contentful when it comes to sharing a vision of good living, which is sometimes religious in nature. Critics of minimalist ethics consider it insufficient for life to be good or for community life to be strong. Along with philosopher Daniel Callahan, they argue that such an approach remains too individualistic and formal since it is based on autonomy—not adequate for hard times when communities must work together for the common good. Yet another philosopher Tristram Engelhardt Jr. distinguished between moral friends and moral strangers and found a way of addressing such critics: for people to coexist peacefully and respectfully in the society of moral strangers in which we live we need some form of minimalist ethics. For life to be good according to the vision of the community to which we belong we need a contentful notion of what constitutes a good life—not alternatives, but different levels of engagement in society.

H. ten Have and M. do C. Patrão Neves, *Dictionary of Global Bioethics*,
https://doi.org/10.1007/978-3-030-54161-3_356

Mismanagement

Managing something incompetently, badly, or wrongly is called mismanagement. There are many forms of mismanagement the best known being financial mismanagement. An example is the bankruptcy in 2008 of Lehman Brothers, the fourth largest investment bank in the United States as a consequence of unwise investments in house-related assets. Other forms include document and information mismanagement (in which appropriate procedures for processing and protecting private information have not been applied), and disease mismanagement (in which diseases are not diagnosed or treated according to professional standards). Although mismanagement can be the result of incompetence, negligence, corruption, and conflicts of interests, a distinction is always made between mistakes (when decisions or failures of managers are often involuntary and down to unexpected factors) and deliberate actions (when decisions are intentional and thus constitute willful mismanagement). Managers have been known to set out deliberately to destroy or bankrupt a company or entity for their own benefit (usually financial). Although mismanagement in healthcare is a bioethical problem that should be avoided at all costs, it is a challenge for institutions and organizations to put in place measures to prevent and address it. Moreover, it is an issue of personal ethics that emphasizes the need for individual managers to have integrity, honesty, and competence.

H. ten Have and M. do C. Patrão Neves, *Dictionary of Global Bioethics*,
https://doi.org/10.1007/978-3-030-54161-3_357

Mistakes, Medical

Medical mistakes or errors are more common than people might think. Ever since publication of the 1998 report To Err Is Human: Building a Safer Health System by the Institute of Medicine (IOM) in the United States people have suddenly become aware that medical errors are a significant cause of death. The report estimates that 98,000 deaths a year occur in the United States as the result of preventable medical errors. The situation has not improved with recent estimates indicating that medical errors result in 210,000 to 440,000 deaths annually worldwide turning errors into the third leading cause of death after heart disease and cancer. The WHO estimates from European data that medical errors occur in 8% to 12% of all hospitalizations and that medication errors are common. A study in the United Kingdom in 2018 indicated that an estimated 237 million medical errors occur in the National Health Service every year. Although avoidable adverse drug reactions cause hundreds of deaths, about three quarters result in no harm to UK patients. Medical errors are preventable irrespective of whether the patient is harmed or not and take various forms such as misdiagnoses, improper transfusions, surgical injuries, falls, burns, pressure ulcers, and mistaken patient identities. They are more frequent and have serious consequences in intensive care units, operating rooms, and emergency departments. The growing focus on the number and types of errors has promoted increasing concern for patient safety. Although errors as such are not ethically problematic (occurring as they always will in complicated and ambiguous medical settings), those that cause harm and could have been prevented are. Important issues are the reporting of errors to improve quality of care and the disclosure of errors to patients. The only way to reduce medical errors and promote patient safety is to bring about a change in culture. The WHO declared patient safety a global health priority, organized global ministerial summits on this topic, and established September 17 as World Patient Safety Day. Surveys on patient safety culture in the United States are conducted by the Agency for Healthcare Research with the objective of reducing error rates. For a safety culture to be efficient it needs to acknowledge those activities of an organization that are high risk, create a blame-free environment in which individuals feel free to report errors, encourage collaboration between disciplines to seek solutions, and commit organizational resources to address safety concerns.

H. ten Have and M. do C. Patrão Neves, *Dictionary of Global Bioethics*,
https://doi.org/10.1007/978-3-030-54161-3_358

Moral Distress

Moral distress refers to an ethical dilemma that crops up in healthcare between doing the right thing in a given situation and not being duly authorized to act accordingly. It was first described by Andrew Jameton in 1984 as "knowing what to do in an ethical situation, but not being allowed to do it" and is particularly important in nursing practice (nurses often do not feel free to be moral because of constraints affecting their practice). The most obvious examples are found in clinical situations such as continuing life support in hopeless situations as a result of failures in communication with the family of patients and/or with the healthcare team. Although moral distress is more often caused by conflicts with other healthcare professionals, it can also be provoked by internal (religious) and external (institutional) constraints. Moral distress generates feelings of powerlessness, anxiety, anger, and frustration and contributes nothing to good practice or nurses' well-being. Critics of the concept of moral distress consider it demonstrates an epistemological arrogance, moral righteousness, and little more than a strategy to give nurses greater autonomy within the healthcare team. Moral distress is best addressed by taking an intermediate position and investing in dialogue, sharing concerns, and engaging in a team approach.

H. ten Have and M. do C. Patrão Neves, *Dictionary of Global Bioethics*,
https://doi.org/10.1007/978-3-030-54161-3_359

Moral Diversity (*See* Diversity)

Moral diversity is an empirical reality in democratic societies where all people are free to have their own beliefs, values, and principles. Such diversity has its roots in different religions, cultures, ideologies, and upbringings. Although moral diversity has been said to lead to divisiveness and conflict, this is not necessarily true. People with different moral values can still recognize a minimum level of common ethics essential to coexisting peacefully and respectfully with others, while acknowledging the importance of people being free to follow their own paths in life. Although moral diversity has been presented as a strong argument for moral relativism, empirical or descriptive moral relativism does not suppress transcultural moral values envisioned as an ethical minimum. The focus of life within moral diversity is on communication and consensus building. However, moral diversity has also been considered highly beneficial for individuals and communities at the theoretical and the practical level. At the theoretical level, moral diversity extends common sensitivity to wider realities, broadens the context for moral thought and practice, makes it more flexible and thoughtful, and contributes to strengthening old arguments and developing new ones by inspiring bright and honest debates. At the practical level, moral diversity creates new values such as tolerance, non-discrimination, and non-stigmatization and explores new forms of personal relationships and social cooperation.

H. ten Have and M. do C. Patrão Neves, *Dictionary of Global Bioethics*,
https://doi.org/10.1007/978-3-030-54161-3_360

Moral Entrepreneur

The concept of moral entrepreneur is specifically used in theories that explain how norms are implemented in everyday practice. Changes and improvements in ethical and human rights practices are more often the result of efforts made by grassroots movements than those of global institutions. Although institutions do create norms, their implementation is decentralized and domestic. Networks of people from civil society take up a cause (e.g., access to healthcare), organize themselves, engage in the struggle, and connect with similar movements and networks in other countries. Although they develop a mode of practice around the right to health, there is more to this than simply applying this right since activities need to be adapted to local circumstances and values and practices are constructed through collective labor within a specific context. Implementation or application of global principles and values is therefore a form of domestication in that they need to be transformed and internalized into domestic systems and local contexts. Since this is usually done by NGOs and individuals (not governments), the implementation of global ethical frameworks is much more than simply applying principles and complying with frameworks in practical settings in that it requires sustained work such as building convergence through localization; dialogue, debate, and interaction; and getting local people involved. Such work is not only practical but also involves discourse (with persuasion and learning being basic instruments to change values at the global level) and demonstrates the power of ideas. In such a context the role of so-called moral entrepreneurs can clearly be seen as people promoting specific principles in the international system, mobilizing support, and inspiring others to encapsulate such principles in local value systems. Moral entrepreneurs can be prominent individuals such as Henri Dunant who established the International Committee of the Red Cross in 1863, Eleanor Roosevelt who was a driving force in the adoption of the Universal Declaration of Human Rights, Czech dissident and later President Vaclav Havel, and Kofi Annan when he was UN Secretary-General; epistemic communities; or NGOs all promoting the universality of human rights and humanitarian intervention. Moral authority and norm building are not the exclusive remit of individuals since government bodies such as the National Commission for the Protection of Human Subjects of Biomedical and Behavioral Research have done so in bioethics. In 1978 its Belmont Report initiated collective action in the field of research ethics by promoting principles, mobilizing support, and inspiring legislation and regulation of practices (at first in the United States). The normative approach advocated was broadened such that it could be accepted more readily. Such practices were promoted in the 1990s by the Council for International Organizations of Medical Sciences (CIOMS) that acted as a global normative entrepreneur and in global bioethics by Doctors Without Borders (Médecins Sans

H. ten Have and M. do C. Patrão Neves, *Dictionary of Global Bioethics*,
https://doi.org/10.1007/978-3-030-54161-3_361

Frontières). Although the award money Doctors Without Borders received from the Nobel Peace Prize in 1999 was used to start the Campaign for Access to Essential Medicines to increase access to medication in developing countries, the same entrepreneurial role has been performed by professional organizations such as the Transplantation Society and the International Society of Nephrology who convened a summit meeting in 2008 at which the Declaration of Istanbul on Organ Trafficking and Transplant Tourism was adopted.

Moral Expertise

Moral expertise is a concept that questions whether experts specializing in moral theory in public policy debates add any value and whether professionalizing so-called moral experts in specific social sectors such as biomedical ethics has any relevance. In addition to these two questions an issue has arisen mainly in North America about whether there is such a thing as moral expertise. As a result of moral diversity and even moral relativism (i.e., the multiple views concerning beliefs and values), some consider it impossible to assign any kind of expertise to this field. Moreover, they go on to say a so-called moral expert would not be able to objectively and accurately explain human behavior and decision-making processes, let alone predict any outcome as a consequence of them. Philosophers argue that it is not reasonable or productive to impose a single model of development common to hard sciences on other fields of thought and practice such as morality. Furthermore, they argue moral expertise does not consist in acritically enunciating all particular moral convictions nor does it entail favoring one point of view to the detriment of all others and in so doing jeopardize any expert intervention. Moral expertise not only requires spatiotemporal knowledge of the main moral doctrines, but also (and above all) of ever new situations that question the right way to act. Although this requires thorough analysis, critical thought, and sound reasoning, sometimes the major input can be the capacity to put good questions or to reveal subtle consequences of the decisions under consideration. Sound justifications matter as does acknowledging distant consequences. Moral experts have sat for decades on relevant advisory bodies (especially for life sciences) and the way in which they work together has proven to be essential. They seem able to see the wider picture that concrete problems involve; they facilitate discussion using rational, logical, and coherent arguments; and have the imagination to come up with new solutions. Although this kind of work should be paid much like any other work, this is not common practice in many countries (especially in Europe). Professionalizing moral expertise would require assigning moral experts a clear role within institutions, making them full time, and paying them accordingly. Although moral experts are not yet perceived as a necessity by society, it should be kept in mind that all professions are validated by society (i.e., social activities become professions insofar as society acknowledges their utility for the common good). As moral expertise becomes more relevant and useful the more likely it will become a profession. However, currently it remains a controversial concept.

H. ten Have and M. do C. Patrão Neves, *Dictionary of Global Bioethics*,
https://doi.org/10.1007/978-3-030-54161-3_362

Moral Hazard

The concept of moral hazard refers to a situation in which someone profits from taking risks that will affect other people, does not personally suffer from taking such risks, and has little if any incentive to guard against risks. In such a situation people tend to continue to do what benefits them the most regardless of any moral considerations or trying to do what is right. Although such a concept was first used in the insurance industry to cover the risk of making bad decisions, it has also been employed in business and economy. Despite moral hazard also finding an equivalent use in healthcare, such services as insurance, business, and particularly healthcare would work much better in striving toward the common good, if there was no such thing as moral hazard. When it comes to healthcare (in particular), there are those who consider making people aware of the costs involved in their treatment could help society to have better services as borne out by several initiatives taken by national healthcare services to present patients with information about such costs. Although patients do not have a bill to pay in universally free healthcare services, presenting them with such information is supposed to lead them to consider whether their visit to the emergency department was justified, to get them to better understand the difficulties such national healthcare services struggle with, and to help them understand the importance of avoiding unnecessary demand. The concept of moral hazard entails the idea that people who are not exposed to risks behave differently than those who have to deal with the consequences of such decisions and actions with the former tending to be careless and the latter cautious.

H. ten Have and M. do C. Patrão Neves, *Dictionary of Global Bioethics*,
https://doi.org/10.1007/978-3-030-54161-3_363

Moral Relativism

Moral relativism refers to the idea that there is no universal or absolute set of moral principles since they are culturally embedded and change according to the beliefs or circumstances of different peoples. Moral relativism has its roots in Ancient Greece with the pre-Socratic philosopher Protagoras who is attributed with saying "man is the measure of all things," which Plato interpreted as meaning there is no objective truth. Although moral relativism has always been advocated throughout history, it never prevailed since it was generally viewed as an interpretation that could be contested. It was only in modern times that other thinkers adopted a relativist perspective such as the French philosopher Montaigne who wrote: "I study myself more than any other subject. That is my metaphysics, that is my physics." Nevertheless, it was the decline in moral universalism in the nineteenth century that gave way to different approaches to moral relativism. Moral relativism is today understood from three major perspectives: descriptive moral relativism (a.k.a. cultural relativism) argues that all moral values are culturally defined and thus change from culture to culture (critics answer this by arguing that the same moral value can be differently expressed by different cultures but remain the same transcultural moral value); meta-ethical moral relativism denies the possibility of any objective ground for preferring one set of values over another (critics answer this by arguing that moral relativism is unsustainable from the meta-ethical and epistemic perspective and that if all moral theories are relative then so is moral relativism); and normative moral relativism considers societies should accept each other's moral values because all values are worth the same in the absence of universal moral values (critics answer this by arguing that if all values are worth the same then none would be worth anything and lead to nihilism in which there is a void of values). Although a pure relativist approach to morality is difficult to maintain from a theoretical point of view and simply convenient from a practical perspective, what is important is questioning absolutist views of morality and refraining from a priori universalism. Moral relativism when considered alongside the acknowledged need for a universal horizon for moral rules has centered the focus of moral debate on fostering dialogue and tolerance.

H. ten Have and M. do C. Patrão Neves, *Dictionary of Global Bioethics*,
https://doi.org/10.1007/978-3-030-54161-3_364

Moral Residue

Moral residue refers to the feelings experienced by healthcare professionals (mostly nurses) in a distressing situation that they felt was not satisfactorily resolved. When facing a moral dilemma (i.e., a conflict between two obligations), the professional involved will still fail to accomplish one of the obligations whichever one was chosen and as a result suffer feelings of remorse and even guilt. Moreover, since moral distress is often the result of repetitive rather than one-off events the professional ends up believing his or her moral values are being systematically overridden thus compromising their moral identity and leading to moral residue. Another understandable reaction would be for a professional to step back from making ethical decisions. However, such an option is perceived as failing the professional mission and enhancing moral residue because of the crescendo effect in getting other professionals to make decisions rather than the professional involved. Moral residue is an emotional experience or state characterized by feelings of anxiety, frustration, and depression that strongly contributes to burnout and compromises the health and practice of professionals.

H. ten Have and M. do C. Patrão Neves, *Dictionary of Global Bioethics*,
https://doi.org/10.1007/978-3-030-54161-3_365

Moral Status

Moral status refers to a value recognized as belonging or ascribed to a particular entity according to which its protection is a moral and often legal obligation. Considered more broadly, the grounds for moral status always refer to the interests of the entity in question. For example, there may be overall interest in protecting the entity or such an entity may have interests of its own that ought to be protected. The most prominent case concerning moral status in the field of bioethics refers to the embryo. Those who defend the sanctity of human life consider that the embryo (and the zygote) has a moral status and should be fully protected. Those who share a gradualist view of the moral value of prenatal human life debate whether it is relevant to attribute moral status to the developmental stage and feel that moral worth and legal obligation can only apply to protecting a child after birth. The controversy has remained moot for decades as a result of contradictory arguments ranging from the biological, philosophical, religious, through to the social and political. A more recent issue about moral status concerns animals and the environment. Although animal rights advocates disagree on which animals deserve moral status, they agree on recognizing such a status for all sentient beings (perceiving and responding to sensations). Moral status has also been attributed to the environment in that it is in the interest of everybody to protect. Critics of attributing moral status to animals or the environment argue that, while there is a moral obligation to protect and preserve both, they should not be ascribed moral status since that is the exclusive reserve of human beings. They argue that having moral status means belonging to a moral community in which individuals behave morally—something that animals and the environment cannot do. Although the basis of moral status remains highly controversial as does identifying who or what are entitled to it, this is the reason some bioethicists question whether such a concept has any use whatsoever.

H. ten Have and M. do C. Patrão Neves, *Dictionary of Global Bioethics*,
https://doi.org/10.1007/978-3-030-54161-3_366

Moral Theories (*See* Deontology; Moral Theory)

Moral theories refer to a set of normative rules formulated to guide human action, to do good and avoid bad, to do what is right and reject what is wrong, and rationally justifying such rules. Throughout history there have been many different moral theories (i.e., coherent systems of concepts, methodologies, and arguments to justify different courses of action). It is generally accepted that three systems of moral theories stand out as being the most influential: the teleological, the deontological, and the procedural. Teleological moral theories focus on identifying ends or good outcomes that action should accomplish and by which it can be judged as good (or bad). The word "teleology" derives etymologically from the Greek *telos* (end). For any action to be deemed moral depends not only on the good that its end (outcome) brings but also on bringing about such an outcome (i.e., it is consequentialist). Teleological moral theories are descriptive theories first outlined by Aristotle in the third century BC. A good example of a contemporary teleological moral theory is utilitarianism. Deontological moral theories focus on defining the principles of action or duties. The word "deontological" derives etymologically from the Greek *deon* (duty). Moral actions should be duty oriented and considered morally right or wrong according to how they comply with duties regardless of consequences. Deontological moral theories are prescriptive theories first outlined by Kant in the eighteenth century. A good example of a contemporary deontological moral theory is liberalism. Procedural moral theories recognize the difficulty pluralistic societies have in defining a common solid concept of good and in identifying a universal concept of duty. Such theories refrain from considering previously established ends or principles and focus on finding reasonable ways to build consensus. In the twentieth century Jürgen Habermas and Karl-Otto Apel were first to establish a procedural moral theory. Such theories are today used alongside other moral theories in efforts to reach common ground and broad social consensus on how to implement what is deemed to be good and on how to formulate what are deemed to be duties. There are many other moral theories that have been proposed over time (and still proposed today) in attempts at presenting a comprehensive view of moral actions, widening the base of acceptance, and at enhancing widespread implementation of such a view. Nevertheless, they are all based on one of these three moral systems.

H. ten Have and M. do C. Patrão Neves, *Dictionary of Global Bioethics*,
https://doi.org/10.1007/978-3-030-54161-3_367

Moral Universalism

Moral universalism refers to the idea of a (absolute) moral truth and a single pattern (a.k.a. universal rule) of action acknowledged as good or right by all and applied to all. It dominated moral history until the nineteenth century when the moral universals it founded lost their consistency as a result of the decline of metaphysics. Moral universalism today is still accepted and lived by religious communities where it presents itself as moral absolutism (i.e., the theory that some actions are good or evil regardless of the context). Philosophy classifies moral universalism today as a meta-ethical position that advocates some moral values and principles are universally applicable regardless of particular circumstances and thus closely related to moral realism (i.e., the theory that moral truths do exist regardless of subjective opinions). From a broad perspective moral universalism can be seen as a position that argues a universal horizon is needed for each and every moral rule to attest to their validity. Although moral rules today are not effectively universal (they never really were), they are only legitimately formulated if they can be seen as universal. A simple example would be to consider the moral rule "do not lie." Although it is a moral rule, it is not universal because there are people who consider that under specific circumstances it would be preferable to lie. However, if everybody in the world acknowledged and accepted such a moral rule, then we could all live together harmoniously. Therefore "do not lie" has a universal horizon. In contrast, if the moral rule was "lie whenever it can benefit you," then it could never be universal because no one would believe or trust anyone else and social life would be impossible. The need for morality to have a universal horizon is clearly stressed in the Universal Declaration of Human Rights (1948), which presented a conventional moral code and a common moral base on which to build peaceful human relationships.

H. ten Have and M. do C. Patrão Neves, *Dictionary of Global Bioethics*,
https://doi.org/10.1007/978-3-030-54161-3_368

Multiculturalism

Multiculturalism refers to the coexistence of a multiplicity of different human cultures in the world and of cultural, ethnic, religious, or other groups within one and the same society. People gradually became aware of multiculturalism as they traveled to new parts of the world and encountered cultures who had different ways of living, traditions and customs, ways of thinking, and different values. The development of structuralism (a field of cultural anthropology) in the 1960s and 1970s helped people realize that cultures cannot be said to be superior or inferior despite it being possible to hierarchize civilizations according to their level of development. They are just different and as such ought to be respected. Although multiculturalism is often represented as morally improving a society, there are a couple of ethical issues that arise and need to be properly addressed. The first ethical issue is that multiculturalism is often more committed to emphasizing differences between cultures and more focused on describing their particularities and specific identities than on looking at what they have in common from a transcultural perspective; that a multicultural approach is politically more interested in features that differentiate peoples or groups than being committed to integrating such groups in the dominant culture; and that the very fact of integration can also be perceived as a lack of respect for cultural identity (although integration can happen and preserve cultural identity at the same time, this issue remains very complex and controversial in many societies). A second ethical issue raised by multiculturalism is the general idea that cultures are self-justifiable and morally neutral (in short, what is culturally justifiable is morally acceptable); that such an idea is strongly stressed by the principle of respect for cultural diversity formulated in various international documents; and that the acritical interpretation of this principle and the absolutism of its practice led to unacceptable cases contravening human rights theory such as female genital cutting (a cultural and religious tradition violating children's rights). In short, cultural diversity can only be respected as long as it does not violate human rights.

H. ten Have and M. do C. Patrão Neves, *Dictionary of Global Bioethics*,
https://doi.org/10.1007/978-3-030-54161-3_369

Nanoethics

The development of nanotechnologies in many different scientific fields and social activities raises many different ethical issues and regulatory challenges that need to be addressed at an early stage to prevent adverse impacts and to profit from potential benefits. Nanoethics is the field of ethics applied to nanotechnology that promotes ethical reflection on and addresses the implications of science and technology at the nanoscale. It provides a unified and integrated forum for informed and rational discussion of ethical and social concerns about technologies that converge at the nanoscale. Although most nanoethical issues are common to other emerging technologies, they differ in affecting things at the nanoscale such as security (nanotechnologies can penetrate information and communication systems like national defense systems that were once beyond reach and hamper them) and privacy (invading personal lives, controlling them, and affecting individual autonomy). Another ethical issue related to new technologies is their cost and how this prevents wide access to their benefits and potentially increases discrimination and injustice. The unaffordable cost of new technologies is likely to be a transitional stage and prices should decrease the more successful such technologies prove to be and as a result of economies of scale. In addition to these and other recurring themes in bioethics there are two particularly important issues in nanoethics: human enhancement and environmental impact. Although the application of nanotechnology to healthcare (especially nanopharma) is promising, its use blurs the distinction between disease and enhancement. Questions about injustice, discrimination, and even the creation of a superior race of humans or a new humanoid species (transhumanism) would then gain strength. When it comes to environmental issues, it is important to say that despite nanotechnology being proven useful in fighting global warming, land degradation, and polluted water, it produces environmental pollution of its own. This has been an important counterargument to those who argue new technologies should be able to overcome problems caused by old technologies. A specific ethical issue raised by nanotechnology is the possibility of nanorobots or nanostructures made of DNA self-replicating. Eric Drexler's sci-fi novel *Engines of Creation* (1986) led to many fearing such self-replicating nanorobots would be out of control and would create a hypothetical extreme situation called ecophagy (a.k.a. the gray goo theory) where they would consume all biomass and destroy all life on Earth. Taking into account the wide variety of fields that effective and foreseen applications of nanotechnologies might cover, the transverse ethical principle would be to preserve humanity as a universal end (not a means to achieve particular goals) and to proceed with caution.

H. ten Have and M. do C. Patrão Neves, *Dictionary of Global Bioethics*,
https://doi.org/10.1007/978-3-030-54161-3_370

Nanomedicine

Although nanotechnology has been used in many different fields with remarkable success, it is particularly useful in medicine where it is applied to prevent, diagnose, and treat health problems. Therefore it is justifiably considered either a field of nanotechnology or a branch of medicine concerned with the use of nanotechnology depending on how it is viewed. Nanomedicine can be summed up as the application of nanotechnology to healthcare. Bawa and Johnson (2007) wrote "nanomedicine, defined as the monitoring, control, construction, repair, defense, and improvement of human biologic systems via engineered nanodevices and nanomaterials" has the potential to bring major advances to medicine, change medical science and procedures, and contribute to high levels of efficacy (especially in cancer) and other major diseases (cardiovascular, neurodegenerative, musculoskeletal, and inflammatory). Nanodevices can be used in the prevention and early detection of health problems to monitor blood pressure, internal chemistry (glucose monitoring) of the body, pulse, brain wave activity, and other functions by sending warnings of imbalances through transmitters. They can also be used for healing purposes as a result of major achievements being made in drug or hormone delivery systems (e.g., artificial antibodies, artificial white- and red-blood cells, and antiviral nanorobots) aimed especially at patients suffering chronic imbalances or deficiencies. Nanotechnology also enables tailoring drugs at the molecular level to make them more effective and reduce side effects as borne out by many products being approved for the market such as defibrillators, pacemakers, biochips, insulin pumps, nebulizers, and needleless injectors (many others are currently undergoing clinical trials). Although nanomedicine is indeed a growing field as evidenced by major achievements like nanorobots (i.e., miniature devices acting as surgeons but faster and more accurate) sent into a patient's arteries to clear away blockages or into a patient's cells to repair or replace damaged structures and damaged genes and to alter or replace DNA (miniature devices acting as genetic engineers), it is nevertheless very expensive as a consequence. In addition to the advances made by nanomedicine in today's technology, there are other factors raising the costs such as the little interest shown by big pharmaceutical firms in the emerging nanomedical market, which is mainly driven by start-ups and small- and medium-sized enterprises (SMEs) that have significantly higher per-unit costs. Nevertheless, the number of patents (especially in the field of nano-based drug delivery) is constantly increasing. However, significantly growing healthcare costs do not create a friendly environment for the development and application of nanomedicine. In short, nanomedicine is by and large currently inaccessible to patients, not covered by insurance policies, and practiced experimentally.

H. ten Have and M. do C. Patrão Neves, *Dictionary of Global Bioethics*,
https://doi.org/10.1007/978-3-030-54161-3_371

Nanotechnology

Nanotechnology (a.k.a. nanotech) refers to the controlled manipulation of materials at the atomic or molecular scale termed the nanoscale (i.e., less than 100 nm where 1 inch corresponds to 25,400,000 nm and the diameter of a human hair is 100,000 nm) brought about after about 2 decades of basic nanoscience research. Although the distinctive trait of this technology is its size, there are other characteristics such as high strength, light weight, strong control of the light spectrum, and great chemical reactivity that contribute to nanotech being applied to an ever broader range of scientific fields such as chemistry, biology, physics, materials science, and engineering and to being applied in many different research and practice fields such as medicine (nanomedicine), the environment, information technology, and industry. Although wider applications of nanotech may well be in industry (construction materials and nanomachining of nanowires and nanorods), it will also apply to homeland security, transportation, and energy. This is the reason nanotechnologies (nanoscale technologies) are often referred to in the plural. Although the term nanotechnology was first coined by Norio Taniguchi in 1974, the idea that scientists might be able to manipulate and control individual atoms and molecules was first presented by the physicist Richard Feynman as early as 1959. However, nanotech only really began in 1981 when individual atoms could be seen for the first time as a result of the development of the scanning tunneling microscope (STM) and the atomic force microscope (AFM). Nevertheless, nanoscale materials have been used for centuries such as in the stained-glass windows of medieval churches. Although the potential of nanotech is today huge and growing, it does raise problems particularly regarding security (e.g., creation of atomic weapons and so-called smart bullets), privacy (e.g., creation of practically undetectable recording devices), and most importantly of all at the environmental level creating new toxins and pollutants (nanopollution).

H. ten Have and M. do C. Patrão Neves, *Dictionary of Global Bioethics*,
https://doi.org/10.1007/978-3-030-54161-3_372

Narrative Ethics

Although narrative ethics is a specific approach to ethical problems within the healthcare setting that focuses on narratives from the patient, it also involves others such as family and healthcare professionals. It grants an ethical status to narratives as a process to elucidate personal experiences of disease, pain, and suffering and thus contributes to a better understanding of the clinical situation and to better clinical decisions. Narrative ethics was first used to refer to the importance of narratives to understand participants' points of view in the decision-making process (Kathryn Montgomery). It is considered a methodology to enhance the deliberation process in which the creativity of participants is highly favored when it comes to finding new ways of problem solving (Rita Charon). Therefore narrative ethics is descriptive—not prescriptive (normative). Although it was formulated as a reaction to principlism in the 1990s, it can be complementary to the principlist approach. It was not until the end of the twentieth century that narrative ethics became philosophically more sound with the input of phenomenology (the science of phenomena as distinct from the nature of being) and hermeneutics (interpretation of what is meaningful to people) thus avoiding criticisms of subjectivism and relativism. It is through someone's own narrative that personal identity is reconstituted, the meaning of life is discovered, and suffering is integrated in his or her life (strongly inspired by Paul Ricoeur's thought). Narrative is a form of reasoning that should be integrated and implemented in clinical encounters (Howard Brody and Richard Zaner). Narrative ethics also refers today to using literary narratives to contribute to ethical deliberations by making use of concepts, reasoning and argumentation, and possible outcomes from literary works to formulate categories and paradigms as borne out by literature today being a methodological tool for bioethics teaching. The narrative approach to bioethics values empathetic listening and story sharing, enhances dialogue, promotes the enrichment of experiences, cherishes the unique circumstances of people's lives, and above all encourages participants to be creative in coming up with better solutions to concrete situations than those considered from a general normative point of view. Issues with the narrative approach concern uncertainty about the right procedure, the outcome, and the participant's capacity to communicate. This is the reason narrative ethics cannot be applied in many clinical situations such as emergencies and when incompetent patients are involved. Moreover, it takes up a lot of quality time that healthcare professionals can scarcely afford. Although narrative ethics is actively pursued by nurses within the healthcare setting as a result of the empathy it presupposes, it has also been appraised (positively) by feminist ethics as a result of reconstructing personal identity by interpretating the life of someone.

H. ten Have and M. do C. Patrão Neves, *Dictionary of Global Bioethics*,
https://doi.org/10.1007/978-3-030-54161-3_373

Natural Law

Natural law refers to a moral (legal and political) theory that considers human behavior should follow and comply with the nature of human beings and the nature of the world such that they are objectively and universally determined. Although it does not refer to the law of nature (which falls within the realm of sciences—not morality), such a distinction has not always been pertinent throughout history. Going back to the pre-Socratics, Heraclites considered the law that rules the cosmos (i.e., the world's natural dynamics) is one and the same law that should guide the lives of human beings. The idea of a single *logos* (reason) commanding the world and all living beings is particularly common to physiologists in pre-classical Greece. Later, Plato and Aristotle refer to the Greek *ergon* (work) meaning it is the function or task of humans to establish their own moral behavior according to what they are, their nature, and how they perfect human nature. In short, humans should become who they really are (i.e. what is their nature). Greek naturalism was developed by Christian philosophers such as Thomas Aquinas who argued God created the world and all living beings and that the natural law by which they develop is therefore necessarily good and the natural law by which human beings develop and provides them with reason is particularly important to guiding them in the right direction. In short, natural law is based on human nature, excludes all other authoritative interventions (as opposed to positive law), and human reason is the best way to proceed. The major criticism of natural law moral theory is that the existence of human nature has been challenged in recent times by existentialists. If there is no human nature and humans create and recreate themselves constantly during their lives, then natural law loses its foundation. Moreover, even if there is human nature, it cannot be normative because human will and freedom can always overcome it. Natural law morality was first developed bioethically by John Finnis in *Natural Law and Natural Rights* in 1980 in which seven equally valuable basic goods are presented: life, health, knowledge, play, friendship, religion, and aesthetic experience. Although such intrinsically valuable and universal goods are ends in themselves (not means to achieve something else), they should all be fulfilled together and in every situation in what constitutes morality. Broadly, biomedical ethics calls for people to respect and promote the core values rooted in human nature and in so doing comply with human dignity.

H. ten Have and M. do C. Patrão Neves, *Dictionary of Global Bioethics*,
https://doi.org/10.1007/978-3-030-54161-3_374

Nature versus Nurture

An ancient debate in philosophy, psychology, and science is whether human behavior is primarily determined by biological or environmental factors. The idea that the human constitution is primarily determined by nature dates back to Hippocratic writings that argued human behavior is determined by body fluids (blood, phlegm, and yellow and black bile) and thus nature. Later on philosophers such as Rousseau argued that the primary determinants are environmental (education, culture, and social conditions) and thus nurture. The implications for healthcare are different in that the focus on nature is associated with biological, physical, and genetic interventions, while the focus on nurture directs attention at environmental changes such as diet, lifestyle, and environmental pollution. Framing the debate in such terms (i.e., nature versus nurture) is attributed to Francis Galton in 1869. A controversial issue in the debate has been whether human intelligence is genetically determined or socially acquired through education? It is commonly accepted today that human behavior is a combination of nature and nurture. However, scientific disciplines often take a one-sided approach and continue to emphasize one category of determinants over the other such as genetic determinism articulating the primary role of genes and behaviorism focusing on the impact of environmental factors. Genetics today stresses the interaction between heredity and environment as borne out for example by height being not only determined by specific genes inherited in families but also requiring a healthy environment and adequate food for children to prosper.

H. ten Have and M. do C. Patrão Neves, *Dictionary of Global Bioethics*,
https://doi.org/10.1007/978-3-030-54161-3_375

Neoliberalism (*See* Globalization)

Neoliberalism is the dominant ideology of globalization. It is a conglomerate of ideas focused on promoting the free market such as competition, privatization, deregulation, reduction of public expenditure, tax reform, and protection of property rights. According to such an ideology globalization is basically liberalization and will foster individual liberty and human well-being when global markets are free. According to neoliberalism the role of the state should be limited to creating the institutional framework necessary to secure free trade and private property rights; the power of governments should be used to deregulate and remove constraints and social policies that curtail the flow of commodities; the state should withdraw from social provision and protection since that impedes the proper functioning of markets; public utilities, social welfare provisions, and public institutions should all be privatized; and every domain of human life should be open for market transactions such that individual citizens are free to choose what they want. Neoliberalism is clearly much more than an economic theory in that it presents a worldview in which the notion of market is not just economic. Neoliberalism was often used as a utopian vision, especially by political scientist Friedrich von Hayek, philosopher Ayn Rand, and economist Milton Friedman who used the word "market" as a metaphor for the organization of social life since they believed the market was the most fair and efficient form of social organization and the only framework within which individual liberty and freedom could flourish. They further believed the market was a self-regulating force that would provide the room necessary for self-actualizing individuals to be productive and creative to the benefit of all; there would be no need for protection or regulation since everything would be transformed into a commodity or service and transacted in a market where competition is the core virtue; and all individuals would be responsible and accountable for their own actions including their healthcare. Therefore the free market as argued by von Hayek, Rand, and Friedman is a recipe to solve the problems of the world. David Harvey in his history of neoliberalism defines it as follows: "a theory of political economic practices that proposes that human well-being can best be advanced by liberating individual entrepreneurial freedoms and skills within an institutional framework characterized by strong private property rights, free markets, and free trade …" Healthcare has not escaped the grasp of neoliberal policies as borne out by neoliberals believing that business will flourish in a climate of competitiveness and efficiency; that genetic tests and preventive and therapeutic interventions are consumables; that medical research thrives when operating in a global market; and that only then will healthcare be able to provide individuals with a range of choices concerning drugs and interventions that will be broader than ever before. However, the negative effects of neoliberalism are increasingly being recognized as

H. ten Have and M. do C. Patrão Neves, *Dictionary of Global Bioethics*,
https://doi.org/10.1007/978-3-030-54161-3_376

evidenced by increasing numbers of people becoming vulnerable worldwide; social inequalities increasing; and basic services such as public healthcare deteriorating in many countries. It has been shown that only the top 1% of people have benefited and they are the ones driving the process. For most populations existence has become more precarious and fragile, jobs have been lost or made temporary, social security has broken down, insecurity has increased, the environment has been neglected and degraded, and globalization has been blamed for growing injustices, social disintegration, and progressive exclusion. According to the market thinking of neoliberal globalization *Homo economicus* can be defined as rational self-interested individuals motivated by minimizing costs and maximizing gains for themselves, having only their own self-interest at heart, only relating to others through market exchanges, and made up of individuals who make choices rather than the community. Therefore market logic separates economic activities from social relationships such that competition rather than cooperation is the preferred mode of social interaction. The pervasive influences of neoliberal policies over the past few decades have created transformations that transcend rising inequality, poverty, and government debt and have led to people today being confronted by irreversible destruction and degradation; vast areas of land and water dead or polluted for centuries; tropical and subtropical forests rapidly disappearing; and fisheries depleted. Simultaneously, the numbers of displaced people are increasing, rates of incarceration rising, and growing populations even in richer countries being deprived not only of jobs but also basic services.

Neonatology (*See* Pediatrics)

Neonatology is a specialized field of medicine focused on the care and treatment of newborn babies and is considered a subspecialty of pediatrics in most countries. Patients include preterm babies born prematurely, term babies with serious conditions acquired during pregnancy or labor, and babies with congenital malformations and genetic conditions. Neonatal intensive care units (NICUs) treat severely ill neonates. A major ethical issue in neonatology relates to resuscitation and advanced life support and arises when a baby is very premature, when the severity of congenital malformations cannot be determined in utero, and when they turn out to be extremely severe after birth. Another ethical challenge relates to withholding and withdrawing treatment and is a frequent reason for death in the NICU. The decision to stop treatment is complicated since neonates are incompetent and decisions have to be made by others. One reason to withhold treatment is futility and another is quality of life after discharge from the unit. Such decisions are controversial because assessments made by physicians and parents can differ. Yet another ethical issue relates to neonatal euthanasia with some ethicists arguing it is justified since neonates cannot yet be regarded as persons and others emphasizing how suffering will impact future quality of life. Neonatal euthanasia is only tolerated in countries that have legalized euthanasia. A common ethical challenge that arises in the NICU is disagreement within the neonatal team or between physicians and parents. Although the guiding principle here is the best interests of the child, views may differ about what this implies. Clinical ethics committees and consultants play a role in attempting to solve such disagreements.

H. ten Have and M. do C. Patrão Neves, *Dictionary of Global Bioethics*,
https://doi.org/10.1007/978-3-030-54161-3_377

Neuroethics

Neuroethics is a new interdisciplinary field that studies the ethics of recording, monitoring, and intervening in the human brain. The term was first used by Jean-Pierre Changeux in the 1990s. It combines knowledge from clinical brain sciences, psychology, law, and moral philosophy. While neurosciences have expanded possibilities to study and intervene in the brain, neuroethics focuses on what should be done with the techniques and drugs that are available to alter the brain. Interventions in the human brain have long been carried out such as trepanation of the skull, electrical stimulation of the brain for patients with epilepsy, and lobotomy to treat mental disorders. Things changed in the 1950s when psychotropic drugs became available and there was no longer any need for psychosurgery. Magnetic resonance imaging has recently facilitated safer and more precise intracranial surgery and deep-brain stimulation is used today to influence dysfunctional brain circuits. Neurosurgery, drugs, and imaging raise a number of ethical questions many of which are traditional such as balancing benefits and harms and questioning whether patients and research subjects are capable of giving proper informed consent. However, neuroethics is different in that it raises new ethical challenges such as whether brain imaging truly identifies correlations between abnormalities in brain regions and certain behavior and whether anatomical and functional abnormalities determine behavior and undermine behavior control. Such claims have been used to excuse individuals from criminal responsibility. Although brain imaging may also detect incidental findings (e.g., aneurysms), it is unclear what should be done. Another ethical issue is cognitive enhancement such as whether interventions in the brain improve attention and concentration (Ritalin is a drug that helps people perform better in cognitive activities). Yet other ethical issues concern disorders of consciousness (especially patients in a vegetative state and in a minimally conscious state); the potential of interventions in the brain not only to determine the degree of consciousness but also recover some brain functions; and the potential of brain–computer interfaces to help patients with locked-in syndrome (not a state of unconsciousness but total paralysis) communicate their wishes.

H. ten Have and M. do C. Patrão Neves, *Dictionary of Global Bioethics*,
https://doi.org/10.1007/978-3-030-54161-3_378

Neurotechnology (*See* Neuroethics)

Advanced techniques and technologies are readily available today to assess, access, and examine the structures and functions of the brain. In 2013 the European Union initiated the Human Brain Project and the United States initiated Brain Research through Advancing Innovative Neurotechnologies (BRAIN). Such initiatives aim to revolutionize understanding of the human brain through assessment technologies based on different types of neuroimaging. Interventional technologies use neurotropic and psychotropic agents and drugs, neuromodulatory devices, neural stimulators, neural cells, tissue and genetic implants, and interface systems. Although many of these technologies are under development and their application in clinical settings is limited, they raise many ethical issues. Such issues are related to interpreting and validating knowledge produced by neuroscience and neurotechnology before it can be applied in human beings and to the potential use and misuse of neurotechnological interventions. Typical of the questions they raise are: Could such interventions be used to change cognitive and emotional states such that thoughts and feelings would be affected in the future? Could they be used for, say, the moral improvement of human beings?

H. ten Have and M. do C. Patrão Neves, *Dictionary of Global Bioethics*,
https://doi.org/10.1007/978-3-030-54161-3_379

Non-governmental Organizations

A non-governmental organization (NGO) is a voluntary association of individuals or groups (citizen based) that is legally constituted, registered with a central government, cooperates with governments, and dedicated to social causes such as human rights, health, education, environment, culture, and religion. Although NGOs are nonprofit and usually funded by governments, they maintain their independence through private donations from different corporations, grants, sales of goods and services, and membership fees. The first NGO to be created (even before such a designation was established) was the Anti-Slavery Society in 1839. However, it was not until after the Second World War that NGOs started to appear. In 1950 the UN Economic Social Council differentiated between the NGO participation rights of intergovernmental specialized agencies and those for international private organizations, issued such rights one by one, and codified them according to Resolution 288 X (B) of February 27, 1950 in which NGOs were defined as "any international organization that is not founded by an international treaty." Since that time minor amendments have been made such as deletion of the word "international" since it is no longer a requirement. The last revision took place in 1996. The World Bank identifies two broad groups of NGOs: the operational group centered on the development of projects and the advocacy group centered on the promotion of particular causes. Patient organizations are traditionally considered advocacy NGOs. In addition to NGOs relating to social movements, civil society, and campaigning for common goals, they frequently involve themselves in important areas of social intervention that do not receive the due attention of official institutions (especially in developing countries) and work with extremely vulnerable people. NGOs play an important role in global clinical assistance and health research by adopting a holistic approach that involves health education and improving the social determinants of health. NGOs have a splendid track record in the defense of biodiversity and environmental protection and raising awareness worldwide of the need for global responsibility toward the planet. A few NGOs are dedicated to bioethics. Although the work done by NGOs has received a lot of credit, they have also been involved in some ethical controversies such as conflicts of interests related to sources of financial support (the power donors gain within the organization) or services provided (payment of environmental certification for those who can afford it but not for all who deserve it); using

H. ten Have and M. do C. Patrão Neves, *Dictionary of Global Bioethics*,
https://doi.org/10.1007/978-3-030-54161-3_380

aggressive methods driven by an absolutist mission to disseminate their goals and effectively promote them rather than being open to dialogue or consensus building; (unwilling) complicity or cooperation with dictatorships while proceeding with their tasks; and the feeling of impunity they have as a result of not having codes of ethics or conduct. The number of NGOs in recent decades has greatly multiplied and has led to them competing among themselves. Getting people and institutions to engage in global missions is perhaps their highest achievement.

Nursing Ethics

Nursing ethics is a branch of applied ethics that examines the actions taken and decisions made by nurses. Although ethical issues gradually grew as nursing as a profession developed, it did not emerge as a special field of ethics until 1950. Florence Nightingale deemed by many as the founder of the nursing profession articulated that nursing should be guided by ethical standards such as equal care for all patients, the right of patients to receive care from competent nurses, and the priority patient interests should have over obedience to medical orders. Much like the Hippocratic Oath taken by doctors, nursing schools developed the Florence Nightingale Pledge. The first code of ethics for nurses was adopted in 1950 by the American Nurses Association. In 1953 the International Council of Nurses (ICN) launched the ICN Code of Ethics for Nurses. The code serves as a worldwide guide for nurses and has been revised several times. The Code is last revised in 2012 and is continuously reviewed. The preamble to the code states: "Inherent in nursing is a respect for human rights, including cultural rights, the right to life and choice, to dignity and to be treated with respect. Nursing care is respectful of and unrestricted by considerations of age, colour, creed, culture, disability or illness, gender, sexual orientation, nationality, politics, race or social status." Although the primary professional responsibility nurses have is to people requiring nursing care, they should also demonstrate professional values such as respectfulness, responsiveness, compassion, trustworthiness, and integrity. It is argued that the code should put stronger emphasis on equity and social justice, respect for the natural environment, and the role of technology, digitalized communication, and artificial intelligence and should also articulate the advocacy role nurses play as partners in decision-making with patients, families, and healthcare providers. The major issue that arises in developing and applying nursing ethics is differentiating it from medical ethics. While medicine is focused on cure and healing, nursing is more directed at care and health. Although such differences are less articulated today with doctors more focused on empathy and person-centered care and nurses more involved with technology and information systems, the ideal behind nursing is to treat patients rather than diseases. Different orientations such as these have led to the argument that nursing has different values and ethics than medicine such as nursing being more longer term than medical interventions and relationships and interactions with patients being more important such that trust can be built. Although the nurse in Anglo-Saxon countries is regarded as the go-to person to protect the interests of the patient and if necessary to act as his or her advocate in matters medical, in Europe greater emphasis is put on nurses promoting healthcare teamwork.

H. ten Have and M. do C. Patrão Neves, *Dictionary of Global Bioethics*,
https://doi.org/10.1007/978-3-030-54161-3_381

Occupational Safety

The focus of occupational safety is safe and healthy working conditions. Work can expose people to risks such as dangerous substances, radiation, infectious diseases, and accidents. Since 1950 the International Labour Organization (ILO) and the WHO have been running a joint program to promote occupational health in which three important objectives are emphasized such as maintaining and promoting workers' health and their capacity to work; improving the working environment and making work more conducive to safety and health; and developing work organizations and putting in place working cultures such that they support health and safety at work and promote a positive social climate. Occupational safety is not only concerned with physical risks but also with psychosocial risks, work-related stress, and non-communicable diseases such as cancers and respiratory diseases. During the Industrial Revolution in the nineteenth century concerns arose about occupational safety (especially children working in factories and workers in the mining industry). As a consequence of national governments increasingly introducing legislation and regulation to protect the health of workers and to guarantee their safety, employers in most countries are legally required to provide employees with a clean and safe working environment in which the focus is primarily on prevention of hazards and risks. It is estimated that currently (2019) worldwide more than 374 million people are injured or made ill every year through work-related accidents. Such a number is likely to increase in the future as a result of challenges posed by new technologies (digitization, robotics, and nanotechnology) and by new materials that have unmeasured health hazards and the potential to affect psychosocial health. Although demographic change is important since young workers have higher occupational injury rates and older workers need adaptive practices and equipment to enhance safety, climate change is more so in that it may lead to air pollution, heat stress, and new diseases. Although changes in the way work is organized may allow more people to enter the labor force, it may also lead to psychosocial problems (insecurity, compromised privacy, and insufficient time to rest). It is estimated today that 36% of the global workforce work excessive hours (more than 48 h per week). To promote awareness of safe, healthy, and decent working conditions in 2003 the ILO declared April 28 of each year as World Day for Safety and Health at Work.

H. ten Have and M. do C. Patrão Neves, *Dictionary of Global Bioethics*,
https://doi.org/10.1007/978-3-030-54161-3_382

Occupational Therapy

Occupational therapy (OT) is defined as "therapy based on engagement in meaningful activities of daily life (such as self-care skills, education, work, or social interaction), especially to enable or encourage participation in such activities despite impairments or limitations in physical or mental functioning" (*Merriam-Webster Dictionary*). The aim of OT is to enable persons with disabilities (occupational therapists prefer to speak of persons with disabilities rather than disabled people in an effort to avoid stigmatization) to participate in everyday life activities such as hemiplegic patients being taught to compensate loss of motor ability and in so doing increase their own independence. OT takes a broad view of occupation in that it not only relates to work and jobs but also to meaningful daily activities such as the ability of persons with disabilities independently to take a bath, get dressed, or cook a meal. Treatment often involves physical, emotional, and cognitive interventions combined with modifying the environment in which persons with disabilities live. Specific ethical issues arising in occupational therapy are related to the vulnerability of persons with disabilities many of whom have disabilities brought about by accidents, injuries, war, and disease. Such language emphasizes that disability is a different way of being rather than a defective condition that needs to be rectified. Specific codes of ethics have been developed for occupational therapy as borne out by the code developed by the American Occupational Therapy Association in 2015. The code consists of a number of principles. The first principle is beneficence (OT should be concerned with the well-being of the recipients of its services), the second principle is non-maleficence (OT should take reasonable precautions to avoid imposing or inflicting harm), the third principle requires respect not only for recipients and their surrogates but also for their rights, the fourth principle is to maintain high standards of competence, and yet other principles underline veracity (providing accurate information about OT services) and fidelity (treating colleagues and other professionals with fairness, discretion, and integrity).

H. ten Have and M. do C. Patrão Neves, *Dictionary of Global Bioethics*,
https://doi.org/10.1007/978-3-030-54161-3_383

Open Access

Open access refers to making academic and scientific work such as papers, journals, conference communications, technical reports, and theses freely available on the internet. The main objective is to make research results accessible to the entire scientific community. Were it not for open access, researchers and the institutions they work for would have little option but to buy the journals they need to keep updated in their respective fields—something that some institutions and researchers find unaffordable. The open access strategy allows scientific research to be widely disseminated and should prevent similar investigations being replicated by other teams elsewhere in the world thus saving money that can then be invested in new projects and help investigations dedicated to such projects to progress. All research builds on the results (positive and negative) of former research and in this way increases efficiency. Another objective of open access is to promote wider dialogue within the scientific community and in so doing contribute to better science. Open access publication is usually done in one of two ways such as getting published in open access journals that cost nothing (the number of such journals have multiplied lately and a community-curated list of them is available from the Directory of Open Access Journals) and self-archiving in an open access repository that involves authors publishing their work in traditional or open access journals and depositing them in disciplinary repositories (thematic) or institutional ones developed in universities (i.e., the Directory of Open Access Repositories) as well. Regrettably, the open access initiative is not working as originally intended since getting published as an open access work requires the author or the research center to pay the publisher. Although some of the research budget is already allocated to pay for publication, those who cannot afford to pay get published in a closed access regime as a kind of second league publication or do not get published at all regardless of the pertinence of the text produced. This means to get published the first criterion is being able to pay for open access and the merit of the work is relegated to being the second. All this while publishers are being paid by authors and still selling their publications to institutions.

H. ten Have and M. do C. Patrão Neves, *Dictionary of Global Bioethics*,
https://doi.org/10.1007/978-3-030-54161-3_384

Organ Donation (*See* Donation, Organs)

See for this entry under Donation.

H. ten Have and M. do C. Patrão Neves, *Dictionary of Global Bioethics*,
https://doi.org/10.1007/978-3-030-54161-3_385

Organ Trade (*See* Trafficking; Declaration of Istanbul)

The organ trade involves the selling and buying of human organs and reducing them to goods or merchandise subject to laws of the market such as supply and demand. It is the overwhelming need for life-saving organs combined with their scarcity that trigger and foster the organ trade. Although organs are usually sold in exchange for money, there are other kinds of payment such as goods or favors. In Western countries the organ trade is forbidden and there are strict rules to prevent it. Nevertheless, there is a black market for organs and other body parts. The situation is made much worse by Westerners in need of an organ traveling to developing and underdeveloped countries where there is no legislation against the organ trade or where such a practice is tolerated. The organ trade is a very profitable business and frequently included as part of a tourist package with a stay in a five-star hotel and a visit to a hospital for transplantation (transplant tourism). The poorest people in these countries are the sources of organs. They often sell a kidney to get some money to pay debts or to improve their living standards for a while thus making a mockery of it being termed a trade since it is little more than organ trafficking. Organ trafficking stoops to even greater depths such as kidnapping people, forcing them to have surgery, and removing the requisite organ (the kidnapped person is sometimes left alive). In other situations people (many of whom are children) are abducted and killed to harvest their organs (organ shortage is particularly acute in pediatrics). There have also been reports about the harvesting and selling of organs from prisoners condemned to the death penalty thus raising the suspicion that capital punishment is influenced by the demand for transplantation. The only way to stop the black market in the organ trade is for Western countries to pass legislation requiring all organs to be strictly traceable and physicians to be much more thorough about the source of the organ prior to transplantation. Although a global ban of the organ trade coupled with truly committed surveillance is what is needed to stop wealthy people exploiting vulnerable people in such a brutal way, all attempts at securing worldwide prohibition by international organizations have so far failed. In fact, the organ trade is not condemned by everybody since there are those who consider organ trafficking is what happens when the selling and buying of organs is prohibited. Although there will always be a shortage of

H. ten Have and M. do C. Patrão Neves, *Dictionary of Global Bioethics*,
https://doi.org/10.1007/978-3-030-54161-3_386

organs and what is worse people willing to buy the organ they need, there are those who argue that regulating this trade would result in current violations of the principles of autonomy and social justice no longer taking place. Incredibly, there are advocates of the organ trade who consider the principle of human dignity is not at stake and argue the sale of a body part is little different from prostitution in which the whole body is sold. Although they point out that prostitution is legal in many countries, they overlook the fact that it is the use of the body that is sold—not an unrecoverable part of it.

Organ Transplantation (*See* Transplantation Medicine)

See for this entry under Transplantation medicine.

H. ten Have and M. do C. Patrão Neves, *Dictionary of Global Bioethics*,
https://doi.org/10.1007/978-3-030-54161-3_387

Organizational Ethics (*See* Institutional Ethics)

The area of applied ethics examining the ethics of organizations is called organizational ethics. It studies the operating structure (horizontal or vertical) of organizations and their ethical behavior. The way organizations are organized influences relationships within the organization and the relationships or behavior they have with different stakeholders. Organizational ethics is different from business ethics in that the latter is focused on the relationship between ethics and commerce (many organizations are not businesses). Organizational ethics first emerged in the 1980s when organizational scandals involving corruption and fraud focused attention on ethical concerns. Although it is assumed organizations have a mission, core values, and principles that are manifested in their structure and behavior, it is not enough for individuals working for the organization to be ethically aware since it is the organization itself that should be regarded as a moral agent operating within an ethical climate and culture with integrity, responsibility, and trustworthiness. The culture of an organization is determined by power (control, loyalty, and stability), bureaucracy (productivity and efficiency), achievement and innovation (creativity and teamwork), and support (commitment and consensus). Although culture is highly dependent on organizational leadership (not to be confused with management or administration), other important values are fairness, compassion, integrity, honor, and responsibility. Other values have been identified as crucial to an ethical organization such as respect, customer focus, result orientation, risk taking, passion, and persistence. In addition to many organizations developing a professional code of conduct that has a positive influence on organizational culture, it is argued that every organization (implicit or explicit) should have a standard of conduct that communicates values, has acceptable criteria for decision-making, and has strict rules for behavior. Communicating values and guiding principles in a published code of conduct clarifies the significance of ethics. A code usually highlights the underlying values of the organization, its commitment to employees, the standards for doing business, and the relationships it has with wider society. Organizations realize that good conduct and integrity contribute to success and that key to developing a code of conduct is promoting a culture that encourages employees to internalize the principle of integrity, practice it, and help them do the right thing by allowing them to make appropriate decisions. The code of conduct should reflect the organization itself and there should be a commitment from the board of directors since ethical

H. ten Have and M. do C. Patrão Neves, *Dictionary of Global Bioethics*,
https://doi.org/10.1007/978-3-030-54161-3_388

responsibility ultimately lies with the board or its equivalent. Management's commitment to the code should be reflected in continuous awareness and promotion of the code and taking a wider approach to ethics and compliance. Although an awareness program should sustain interest in and commitment to the code, employees and other stakeholders should also be made aware of the consequences of not adhering to it.

Organoid

An organoid can be a mass of cells or tissue, a complex three-dimensional biological structure, or a miniature organ grown in vitro from pluripotent stem cells or primary human donor tissue that self-organizes and becomes functional as a real organ. It is already possible to produce human organoids similar to the retina, intestines, kidneys, pancreas, liver, and the inner ear. Organoids are currently considered a great resource for biomedical research to study different pathologies (cell behavior and tissue repair) and to test new molecules in a way that is accurate, easy to observe, and does not present clinical hazards or ethical problems (unless stem cells from human embryos are used). Furthermore, not only do organoids contribute to the diminished use of animals in research, they also open up great new opportunities for autologous transplantation (using the receiver's somatic stem cells) and personalized medicine. Although currently impossible, it is reasonable to foresee the regeneration of damaged human tissues or their substitution by organoids produced from cells of the receiver. Despite organoids rightfully being viewed as a highly beneficial clinical resource, new ethical problems have arisen as a result of the recent production of cerebral organoids. There are a couple of major concerns such as these mini-brains (which have matured in the same way as a few-month-old fetus) presenting neural connections and electrical activity, starting to bleed (vascularized human neural organoids), and begging the question whether or not they become sentient (sentient brain tissue in a petri dish); and transplanting such human neural organoids into animals (mice) where they are known to develop capillaries that penetrate into the animal's inner layers (a human brain in a mouse's body). Neural organoids today present one of the most challenging bioethical issues.

H. ten Have and M. do C. Patrão Neves, *Dictionary of Global Bioethics*,
https://doi.org/10.1007/978-3-030-54161-3_389

Outsourcing

Outsourcing is the practice of hiring an outside party to produce goods or provide services that used to be produced or provided by the company involved. It is usually done to reduce the costs of labor. Outsourcing can be done by domestic companies as well as foreign ones. If a specific function is relocated to another country, then it is called "offshoring." Although outsourcing in healthcare is on the increase in traditional support functions such as housekeeping, food, and laundry services, today information technology systems, medical billing, and clinical services such as anesthesia, dialysis, and emergency department staffing can also be added to the list. The argument is that such services do not depend on long-term relationships between physicians and patients. The Philippines is the top destination for healthcare outsourcing. Outsourcing is even more common in medical research as borne out by pharmaceutical companies increasingly outsourcing several stages of the clinical trial process and using contract research organizations (CROs) to recruit research subjects and analyze clinical trials. Moreover, medical devices and medication are increasingly produced in foreign countries. Outsourcing clearly has significant social effects such as leading to increasing unemployment in developed countries and loss of expertise in certain economic areas (e.g., manufacturing). However, less attention is paid to ethical issues such as: Who is responsible for malfeasance when goods and services are produced by foreign suppliers and distributed and traded by domestic companies?; What happens when the ethical rules of clinical research are violated or vulnerable populations exploited when trials are outsourced (while the sponsors may not be aware or attempt to hide such information)?; and Who should be assigned accountability when problems arise (e.g., when medication is contaminated or harmful)? Although the pharmaceutical industry is one of the most regulated in the world, outsourcing may undermine the global effectiveness of such regulations, especially when regions that have less stringent regulations are involved.

H. ten Have and M. do C. Patrão Neves, *Dictionary of Global Bioethics*,
https://doi.org/10.1007/978-3-030-54161-3_390

Ownership (*See* Patenting; Property Rights)

Ownership refers to having the exclusive right and control over some form of property. There can be different types of owners such as individuals, states, corporations, and organizations; different models of ownership such as private, public, collective, and common; and different types of property such as personal, immovable, and intellectual. Although ownership can apply to tangible objects such as computers and intangible ones such as intellectual property, the very act of possession does not necessarily mean ownership. Ownership is regarded as a bundle of rights that allow the owner to use and control an object. Although such rights allow the owner to possess, use, manage, destroy, consume, trade, bequeath, alienate, or profit from what is owned, the basic idea is that the owner can use the object involved as he or she desires without interference since ownership is considered a basic right. According to philosopher John Locke ownership is the result of labor such as when someone works with a natural resource (say, the land) the product can be appropriated as property (e.g., working and farming the land makes it his or her property). Although private property is characteristic of liberal and capitalist societies, a fundamental question in bioethics is what can be owned. Can the human body be regarded as property? Can human body parts and biological materials be owned? Such questions presuppose viewing the individual involved as the owner of his or her body. If the body is the property of an autonomous subject, then parts of the body can be alienated, traded, and commercialized as long as the subject voluntarily permits it. However, it is argued that the human body cannot be regarded as property since humans are embodied beings and as such the person cannot be separated from the body. Moreover, people are endowed with dignity that has to be respected such that the body cannot be transformed into a commodity (which denies respect). Discussions about property commonly crop up in bioethical discourse brought about by people donating cells, tissues, and other body materials to biobanks and raise the fundamental question: Who is then the owner? Although the practice of patenting implies corporations acquire exclusive ownership, this is known to impede innovation, scientific progress, and sharing of data and has led scientists and policy-makers to argue that it would be more beneficial for society and science if ownership was collective.

H. ten Have and M. do C. Patrão Neves, *Dictionary of Global Bioethics*,
https://doi.org/10.1007/978-3-030-54161-3_391

Pain

Pain is defined by the International Association for the Study of Pain as an unpleasant sensory and emotional experience associated with actual or potential tissue damage or described in terms of such damage. Pain in healthcare is often regarded as a symptom of an underlying condition, a signal from the nervous system that something may be wrong, and one of the most common reasons to consult a physician. Pain can have many different causes and manifestations. Chronic pain is especially prevalent as borne out by the WHO stating that one in 10 adults are diagnosed with chronic pain every year. In 2016 some 20% of US adults had chronic pain (approximately 64.6 million) and 8% of US adults (approximately 25.8 million) had high-impact chronic pain. Although pain impacts everyday activities, well-being, and quality of life, it has long not received the priority it deserves making pain management a recent concern. Bioethically, the major challenge is that chronic pain is undertreated. The WHO estimates that almost 80% of the world population has no or insufficient access to treatment for pain despite inexpensive and effective pain relief medicines existing. The reasons for not using them may be cultural, societal, religious, or political. Since biomedicine is often focused on the disease and pathological lesions, pain is considered a by-product, not given priority, subjective, and poorly understood. Furthermore, it is regarded as a symptom the relief of which has no magic bullets—not as a disease. A significant reason for undertreatment is that pain medication is frequently unavailable, effective drugs such as opioids are addictive, and doctors are afraid or reluctant to prescribe them as borne out by the current opioid epidemic in the United States leading to restrictions on prescribing opioids. A recent survey found that 23% of patients reporting pain said they were no longer receiving such prescriptions, 47% said they were on a lower dose, and as a result 84% reported having more pain and a reduced quality of life. This prompted the WHO to initiate a campaign to eliminate barriers to effective pain treatment and to point out that access to pain treatment is a human right. Although international human rights law makes it clear that countries have to provide pain treatment medications as part of their core obligations under the right to health, the growing emphasis on the human rights dimension of pain management advocates the need for a paradigm shift in medical practice in which ethical and legal obligations founded on patient rights are adhered to. In developed countries (mainly in medicalized societies) increasing intolerance to pain has led today to an overuse of drugs and all the negative effects that accompany it.

H. ten Have and M. do C. Patrão Neves, *Dictionary of Global Bioethics*,
https://doi.org/10.1007/978-3-030-54161-3_392

Palliative Care (*See* Hospice; Palliative Sedation)

Palliative care was defined by the WHO in 1990 as "the active total care of patients whose disease is not responsive to curative treatment." A broader and now commonly used definition is the one proposed by the WHO in 2002: "Palliative care is an approach that improves the quality of life of patients and their families facing the problems associated with life-threatening illness, through the prevention and relief of suffering by means of early identification and impeccable assessment and treatment of pain and other problems, physical, psychosocial and spiritual." The new definition demonstrates the wider scope of palliative care in that it is not only relevant for patients in the last stages of care but also for all those in the course of any chronic illness. Despite pain relief being important, it should also be remembered there are other patient needs to be addressed. Although different names are used (comfort care, terminal care, palliative medicine, supportive care, hospice care), palliative care is now the family name. Palliative care is related to a specific set of moral notions such as quality of life, human dignity, acceptance of human mortality, and total care and is characterized in five common ways. First, accepting that palliative care is a goal of healthcare such as when the medical and technological possibilities to cure a disease have become futile or prevention is not possible and another type of care is required (a common mistake is to present palliative care as a form of withholding treatment when there is no longer anything that can be done). Second, associating palliative care with specific virtues such as accepting that the person is suffering from an incurable disease; that his or her life is coming to an end sooner rather than later; that dying is a natural event and a normal process of life; that those involved in palliative care will not shy away from discussing death with the patient and will always show respect; and that nothing will be done to prolong dying or to hasten death. Third, accepting that palliative care not only involves patients but relatives, requires partnership and shared decision-making, and puts patients and relatives at the center of care. Fourth, accepting that palliative care is total care that should be used to help the patient face fundamental questions concerning life and death; human mortality; the meaning of life and death; fears, anxieties, and worries over loved ones left behind; and spiritual and existential concerns (palliative care is therefore essentially holistic, broad-spectrum care in which attention is given to spiritual issues). Finally, palliative care can also be characterized in other ways such as meeting the needs of patients, adopting an interdisciplinary approach (in which teamwork makes all the difference), building on all-round expertise, and involving as many stakeholders as possible. Three main moral problems in palliative care arise more frequently than in other care settings. The first concerns the different ways it is used, the different terminologies used to describe it (e.g., terminal sedation or deep sedation), and the vagueness of the concept despite the goal of

H. ten Have and M. do C. Patrão Neves, *Dictionary of Global Bioethics*,
https://doi.org/10.1007/978-3-030-54161-3_393

sedation clearly being to relieve refractory suffering by reducing consciousness. The second concerns euthanasia and the way in which palliative care from its very outset has always been interpreted as opposed to euthanasia, that definitions of palliative care often point out that death is a natural process, that nothing should be done to hasten or postpone it, and that better palliative care can prevent requests for euthanasia (it is more frequently argued today that euthanasia and palliative care complement each other rather than being mutually exclusive). The third concerns research and whether it is morally right for palliative care to be improved through systematic research despite research with severely ill and dying patients being ethically problematic and people with life-threatening illnesses constituting a vulnerable population with whom research cannot be morally justified. In addition to these three main moral problems there is another concerning the question of informed consent such as whether it is possible for people in the last stages of life and suffering from life-threatening illness to make free choices.

Palliative Sedation (*See* Palliative Care)

Sedation is employed in palliative care in varying ways and has resulted in different terminology being used such as terminal sedation and deep sedation and making the concept ever vaguer. Since the goal of sedation is to relieve refractory suffering by reducing consciousness, it is therefore an option of last resort in which symptoms can no longer be adequately treated or suffering relieved without compromising the consciousness of the patient. However, sedation should be guided by proportionality such that consciousness is reduced just enough to relieve suffering; otherwise it will result in shortening the life of the patient particularly when sedation is combined with withholding or withdrawing artificial food and fluids. This is the reason some opponents have identified palliative sedation with slow euthanasia. In an attempt to clarify matters it is proposed to define palliative sedation as "The intentional administration of sedative drugs in dosages and combinations required to reduce the consciousness of a terminal patient as much as necessary to adequately relieve one or more refractory symptoms." The emphasis in this definition on goal and intention (reflected by proportionality) demarcates palliative sedation from euthanasia (when the goal is to end the life of the patient). However, moral controversies remain surrounding the appropriate indicators needed for palliative sedation such as refractory physical symptoms (e.g., delirium, dyspnea, and pain), existential suffering, and when it is difficult to determine whether the suffering is refractory or not. Another controversy concerns the possible life-shortening effects of sedation. Although studies suggest that such effects are not proven, when the administration of artificial fluids is withdrawn the patient will certainly die and this leads to debate about whether sedation should be combined with the decision to withhold or withdraw artificial food and fluids. A third controversy is the decision-making process. Although the decision to sedate the patient is discussed with the patient and the family and only taken after explicit consent in most cases, such a decision is very difficult and harrowing for caregivers and family and has led to a number of sedative practices developing in contemporary palliative care that vary in terms of level of sedation (mild, intermediate, and deep), duration (intermittent or continuous), and rapidity (sudden or proportional). The prevalence of palliative sedation administered to patients varies across countries (e.g., from 2.5% in Denmark to 16.5%

H. ten Have and M. do C. Patrão Neves, *Dictionary of Global Bioethics*,
https://doi.org/10.1007/978-3-030-54161-3_394

in the United Kingdom). Deep continuous sedation is particularly controversial in that drugs are administered to keep the patient deeply sedated or in a coma until death and depriving him or her of artificial nutrition or hydration. Since the principle of proportionality is clearly not applied and sedation is not an option of last resort the explicit intention in many cases is to hasten death. There are countries such as Belgium and the Netherlands where continuous deep sedation is increasingly deliberately used in the last stage of life such that it can be labeled as palliative sedation, avoid having to be reported as euthanasia, and get around the paradigmatic case initially introduced to make a moral distinction between euthanasia and sedation.

Pandemics

A pandemic is an outbreak of disease that occurs over a wide geographic area and affects an exceptionally high proportion of the population. The WHO defines it as the worldwide spread of a new disease such as avian flu (2004), swine flu (2009), and coronavirus (2019). The term epidemic is used to refer to a sudden increase in the number of cases of a disease above what is normally expected. The term pandemic refers to an epidemic that has spread across several countries or continents and affected a large number of people. The worst pandemics include the Spanish flu in 1918–1919 that killed an estimated 20–50 million people and of course COVID-19 that as of December 12, 2020 had killed an estimated 1,699,285 people worldwide. Pandemics are by definition global phenomena and pose ethical challenges relating to individual liberty, avoiding harm, protecting the public, public health, solidarity, and trust as borne out by the need to balance individual autonomy against the interests of the public and the community during pandemics and to introduce infection control measures restricting personal freedom such as the freedom of movement. However, human rights need to be respected and states are obliged to implement the right to health and provide healthcare services whenever possible. Another ethical challenge concerns the obligations healthcare workers have during a pandemic (despite the risks they are exposed to) and whether they are part of the moral obligations they assume when entering the profession. Although patients during pandemics are people in need of help, they are also vectors (i.e., a source of risks). Furthermore, pandemics exacerbate existing health disparities in which the most vulnerable populations are often severely affected, the distribution of resources is challenging, and vulnerable populations are in need of prioritization. Another argument is that medical resources (especially when they are scarce) should be given to those who will benefit the most. Although public health responses to pandemics today focus on proactive planning and preparing for future pandemics, they also require advance planning, communication, public involvement, and transparency. Ethical concerns during a pandemic include priority of access to healthcare resources as a result of increased demand and possible shortages, obligations of healthcare workers in light of risks to their own health, striking a balance between reducing the spread of disease by self-isolating and travel measures and protecting the right of individuals to freedom of movement, and developing vaccines to fight the diseases. If the first priority of pandemic management is to allow society to function as normally as possible, then those individuals who are essential to the provision of healthcare, public safety, and key aspects of society should receive priority in the distribution of vaccines, antivirals, and other scarce resources.

H. ten Have and M. do C. Patrão Neves, *Dictionary of Global Bioethics*,
https://doi.org/10.1007/978-3-030-54161-3_395

Patenting (*See* Ownership; Property Rights)

A patent is a license conferring a right on someone or some other party for a set period of time (generally 20 years) that prevents others from making, using, or selling an invention. In other words, it is a form of intellectual property giving exclusive control and possession to a particular individual or party and preventing others from fabricating and marketing the same invention. The World Intellectual Property Organization (WIPO) describes a patent as "an exclusive right granted for an invention, which is a product or a process that provides, in general, a new way of doing something, or offers a new technical solution to a problem." To get a patent the party involved needs to disclose technical information about the invention to the public in a patent application. The underlying assumption behind patenting is that patents encourage the development of innovations and new technologies in every field by rewarding ideas. As alluded to above, intellectual property protection is granted for a limited period (generally 20 years) from the filing date of the application. Patenting in healthcare has surged since the 1980s after the Bayh–Dole Act in the United States in 1980 allowing university patents to be granted on results from publicly funded biomedical research (leading to the number of new biotech companies booming) and the US Supreme Court deciding (in Diamond vs. Chakrabarty) that a bioengineered microbe was patentable and hence life-forms could be owned (leading in 1988 to patenting the first genetically modified animal—the Harvard Oncomouse). In 1995 the Agreement on Trade-Related Aspects of Intellectual Property Rights (TRIPS) expanded the American system of property rights and patenting to cover the rest of the world. Since the 1980s there has been a proliferation of patent claims not only for new medications but also for genetic resources, traditional knowledge, biological organisms, and many life-forms. Although the expansion of intellectual property rights to cover many domains of contemporary life has only recently taken place (especially in life sciences and medicine), it has not only become a major component in globalization but it is being increasingly criticized too. Patenting raises a number of ethical issues such as the claim of ownership over life-forms being unacceptable in many cultures and religions (who robustly argue life-forms cannot be the property of human beings), its association with biopiracy, and the moral legitimacy of collecting biological or genetic materials to make commercial products (such that what has long been common knowledge is suddenly transformed into private property). Other issues are intellectual property rights reinforcing global inequality since patent activity is concentrated in a relatively small number of companies in the West (especially clear in biotechnology where 90% of all patents on life-forms are held by Western companies) and patenting making access to medication more difficult and expensive. The current regime of patent rules gives inventor companies a

H. ten Have and M. do C. Patrão Neves, *Dictionary of Global Bioethics*,
https://doi.org/10.1007/978-3-030-54161-3_396

temporary monopoly on their inventions, makes new drugs too expensive for many populations (particularly in the developing world), and deprives poor people of treatment for diseases that can be cured such as malaria, tuberculosis, and pneumonia. Consequently, millions of people die each year from curable medical conditions and has led to people like Thomas Pogge arguing that the global intellectual property regime not only reinforces inequalities but is itself the result of an unfair global institutional order. The World Trade Organization (WTO) and TRIPS came about as a result of coordinated pressure applied by Western countries and international businesses (i.e., the owners of property rights) on developing countries. When these two organizations were being set up, countries were not fairly represented, full information was not shared, no democratic bargaining took place (other than political and economic threats and coercion), the public was not involved, all negotiations were behind closed doors, and other countries were pressured to comply through bilateral trade agreements (the WTO has a dispute settlement system that can apply punitive measures if countries violate their rules). However, it is not only the process that is unfair but the system itself. It is stacked against non-Western countries as borne out by globalization of intellectual property primarily benefiting Western countries, the international legal context being created by the owners of intellectual property and heavily weighted in their interest, the intellectual property rights regime illustrating the lack of fairness in global health and healthcare, and the trade system favoring developed countries. By requiring all fields of technology to be patentable and patenting pharmaceutical products across the world TRIPS has burdened many developing countries with the need to establish legislation and bureaucratic systems. Although many countries lack adequate infrastructure and are unable to take advantage of intellectual property protection, there is little evidence that such protection would promote innovation in developing countries. What is more, even if innovation was promoted the focus would remain on the needs of developed countries as the most attractive market. Finally, developing countries are further disadvantaged since TRIPS makes it very difficult to use generic medication as an alternative to costly imported patented drugs.

Paternalism

The word "paternalism" derives etymologically from the Latin *paternus* (of a father, fatherly). It designates a doctrine or theory that allows interfering with someone's autonomy when trying to promote his or her well-being or to prevent him or her from coming to harm (much like the attitude a father adopts toward his children). Paternalism was philosophically addressed by John Stuart Mill in his work *On Liberty* in 1859. Within a particular sociopolitical context, Mill rejected paternalism and defined it as the state interfering unacceptably in the life of a citizen regardless of the justifications put forward for the simple reason it would restrict the liberty of citizens. Since then paternalism has always been condemned (most strongly in liberalism) as an undue constraint on individual freedom. Paternalism fell under the purview of bioethics theory and practice as a result of liberal and libertarian philosophy that shaped Anglo-American bioethics. It refers to the physician–patient relationship —not to the relationship between citizens and state. Paternalism uses the traditional model in which the physician knows best how to treat a patient as a result of possessing the necessary medical knowledge (skills and means). Such a presupposition is rooted in the Hippocratic tradition and the hegemony of the principle of beneficence both of which are intrinsically linked to paternalism such that the physician is like a father to his or her patients, knows best what his or her children/patients need, always tries to provide it, and acts at all times on their behalf for their well-being. Such a doctrine was challenged by the principle of autonomy in biomedical ethics in that the patient's right to make his or her own decision as to the best medical course of action contradicted the physician's paternalism. It was widely accepted that autonomy was unquestionably superior to paternalism and led to all paternalistic attitudes being strongly condemned. This Anglo-American perspective was never fully accepted in Southern European countries where it was considered that such a monolithic orientation left too much to patients and their families to decide and led to a feeling of abandonment, whereas they wanted the comfort of asking their doctor questions like "What do you think should be done?" rather than making decisions themselves. Such a daily healthcare reality prompted revising the theory by making a distinction between hard and soft paternalism (a.k.a. strong and weak paternalism). This led to soft paternalism being accepted under certain conditions such as when the patient's autonomy is not endangered. The general idea behind the soft form of paternalism is that it is ethically acceptable and sometimes even desirable to

H. ten Have and M. do C. Patrão Neves, *Dictionary of Global Bioethics*,
https://doi.org/10.1007/978-3-030-54161-3_397

interfere with the autonomy of patients when helping them choose the best way to fulfill their wishes or when evaluating whether the patient is competent and capable of making the decision. In any event any interference by the physician with the patient's autonomy must always unequivocally be presented as advice and the final decision left to the competent patient.

Patient Organizations

Patient organizations started with small gatherings of patients who wanted to share experiences about their own health problems such as addictions and diseases and wanted to put in place some form of reciprocal support on the basis of mutual understanding and reinforcement. The first organized meetings were those of Alcoholics Anonymous in the United States before the Second World War. Later the patient rights movement led to such meetings also being organized for other health problems and the number of patients, family members, and caregivers involved growing. They organized themselves in associations to strengthen their claim to being able to help in the provision of care. Most common and serious diseases today have their very own patient organization (e.g., cancer, diabetes, and cardiovascular disease). Patient organizations contribute a great deal to improving care for patients suffering from the specific diseases they represent by facilitating access to new-generation drugs and cutting edge biotechnologies, by lowering the prices of care and drugs, by providing support for patients and families through sharing experiences, by counseling on treatment options, by promoting health education, and by raising public awareness of disease symptoms and risk factors. More recently patient groups have become involved in scientific and therapeutic activism as a result of research funding and diversifying their outreach activities by setting up websites and blogs. Patient organizations provide important services to millions of people and have proven to be particularly significant when it comes to rare diseases, mental illnesses, and degenerative diseases all of which leave patients and families feeling extremely vulnerable. Such organizations are good examples of the recognition given today to active citizenship being important in discussions about public health policies. Although patient organizations have contributed much to the quality of healthcare services and the promotion of health within society, they have also been the subject of criticism. There are two main aspects that have been criticized. The first refers to the strong power imbalance in relation to governments since many of the larger associations originated as lobbies and as such are closely tied to governments. They lobby mostly to increase funding for treatments and research and to get governments to put in place adequate legislation to protect and benefit the patients and families that the association represents. This has resulted in smaller associations with a few patients who have other diseases not being given the

H. ten Have and M. do C. Patrão Neves, *Dictionary of Global Bioethics*,
https://doi.org/10.1007/978-3-030-54161-3_398

governmental support they really need. Patient organizations only advocate for the disease they represent regardless of the needs of other patients. The second main aspect refers to the uncomfortably close relations that many patient organizations have with the pharmaceutical industry whose interests are strongly promoted by their lobbying activities. Since such organizations are frequently financed by the pharmaceutical industry, conflict of interest will always be a very important ethical issue for them.

Patient Rights

The claim to patient rights first arose when the human rights movement (specifying groups that are particularly vulnerable) and the humanization of healthcare movement (strengthening patient autonomy) converged. Patient rights have been used to establish the principles and rules for the ethical relationship between the patient and the healthcare professional and between the healthcare institution and the state. The first document establishing patient rights was produced in 1973 by the American Hospital Association's House of Delegates and called the Patient's Bill of Rights (approved in 1992). It became very influential in the development of similar documents in other parts of the world. In 1981 the World Medical Association approved the Declaration of Lisbon on the Rights of the Patient (a worldwide statement later amended in 1995, editorially revised in 2005, and reaffirmed in 2015). The rights established were the right to medical care of good quality (without discrimination, in accordance with the best interests of patients, and quality guaranteed), the right to freedom of choice (freely choosing the physician and hospital and the right to ask for a second opinion), the right to self-determination (to give or withhold consent for any diagnostic procedure or therapy), the right to refuse to participate in research or the teaching of medicine, the right to informed consent even if the patient is unconscious or legally incompetent (informed consent must be obtained whenever possible), the right to information (full information about his/her health status including the medical facts about his/her condition and the right not to be informed on his/her explicit request), the right to confidentiality (all identifiable information about a patient's health status must be kept confidential), the right to health education (physicians have an obligation to participate actively in educational efforts), the right to dignity (entitlement to humane terminal care), and the right to religious assistance (to receive or decline spiritual and moral comfort). Such rights are crucial to respecting the patient's dignity and as such the declaration represents a cornerstone of the fundamental rights of patients. However, such rights are not recognized worldwide. Indeed, patients' rights can vary not only in different cultures and healthcare systems, but also in different models of the patient–physician relationship ranging from paternalistic models (where intervention from the physician is predominant and behavior of the patient is more passive) to informative models (where the presence of the physician is more discrete and the patient is the decision maker). Nevertheless, despite patient rights being formally established in 1981 and adopted by most countries and most healthcare institutions, they remain as works in progress or goals to be achieved rather than a reality (even in some developed countries). Patient organizations play an important role in the implementation of

H. ten Have and M. do C. Patrão Neves, *Dictionary of Global Bioethics*,
https://doi.org/10.1007/978-3-030-54161-3_399

patient rights by getting patients to understand their rights and getting healthcare professionals to respect them. It is important to stress that the most basic right of the patient is universal access to healthcare—something that is not a reality worldwide.

Pediatrics (*See* Neonatology; Children and Ethics)

Pediatrics is a medical specialty involving the medical care of infants, children, and adolescents. The first medical textbook on diseases of children was published in 1764 and the first pediatric hospital was established in Paris in 1802. Pediatrics recognizes the important biological, physiological, and psychological differences between adults and children and was first instituted as a medical specialty in the 1930s with the founding of the American Pediatric Society and the American Academy of Pediatrics. Pediatrics now has several subspecialties such as neonatology, pediatric cardiology, pediatric nephrology, and pediatric oncology. It is still recommended that pediatric care be extended to people until they reach 21 years of age.

H. ten Have and M. do C. Patrão Neves, *Dictionary of Global Bioethics*,
https://doi.org/10.1007/978-3-030-54161-3_400

Persistent Vegetative State

Persistent vegetative state (PVS) was first described in 1972 and refers to a clinical condition in which there is a lack of self-awareness (complete unawareness of the self) despite the patient having sleep–wake cycles, other basic reflexes (such as blinking or withdrawing the hand when it is squeezed hard), being capable of spontaneous breathing, and maintaining complete or partial hypothalamic and brainstem autonomic functions. PVS is a chronic neurological disorder of consciousness in which patients in a vegetative state are unable to connect with anything or anyone and show no signs of experiencing emotions or cognitive function. Such a clinical condition is the result of modern medical technology being capable of suspending the process of death but incapable of fully recovering people who have suffered severe and widespread damage to the brain (sometimes during cardiopulmonary arrest followed by resuscitation). The Karen Quinlan case in the 1970s brought the issue to the attention of the general public, the Nancy Cruzan case in the 1980s broadened the debate, and the Terri Schiavo case in 2005 showed the debate is still ongoing. A PVS diagnosis can only be pronounced a month after the event that caused the clinical condition since distinguishing between PVS and minimally conscious states (MCS) is not always straightforward despite current research in neurosciences trying to determine the degree of consciousness of such patients. Although PVS patients have a shortened lifespan and their recovery is extremely unlikely, some patients do recover consciousness, but there are always significant cognitive and physical difficulties. PVS raises the difficult question about the kind of care such patients should be provided. Healthcare provision depends heavily on evaluating on a case-by-case basis the level of mental functioning, sentience, and the possibility of recovery of the patient. The consensus is that all invasive or disproportionate actions should be rejected. The most pressing issues are about withdrawing life support including artificial nutrition and hydration. There are a number of questions that need addressing: Are nutrition and hydration basic needs or therapeutic resources? If the only goal of artificial nutrition and hydration is to prolong life can it be considered a therapeutic measure and therefore futile and withdrawn? Should the decision about the care due to a PVS patient be taken by the medical team (i.e., according to good clinical practices) or by a surrogate (i.e., according to the patient's wishes or the surrogate's)? What happens when family members do not share the same opinion? Although different solutions have been found for different cases, the patient's best interests have to be considered together with the principles of autonomy and beneficence. Nevertheless, the principle of justice in terms of the consumption of futile healthcare resources has also to be balanced against the other ethical principles.

H. ten Have and M. do C. Patrão Neves, *Dictionary of Global Bioethics*,
https://doi.org/10.1007/978-3-030-54161-3_401

Personalism

Personalism is a philosophical doctrine centered on the value of a person considered as a being in a relationship (thus transcending the concept of a human being) and within a community (opposed to the concept of an individual). It developed from an existential orientation that was Christian in nature and had a theological foundation and a transcendent horizon in the search for the meaning of life. Such a philosophical movement emerged in the early twentieth century as a result of papers written by Emmanuel Mounier (who points out the concept was first used by the North American poet Walt Whitman in 1867) in the journal *Esprit* (1932). Although Mounier was called the great disseminator of personalism, its roots are identifiable in the history of philosophy. Personalism was first used as a reaction to Kantian rationalism, Hegelian idealism, Schelling's pantheism, Marx's materialism, and Nietzsche's nihilism and has been presented as an alternative to dominant philosophical schools since the eighteenth century. However, personalism is more than a reactive theory in that it is also an affirmative one by supporting Maine de Biran's appraisal of psychological intimacy, Max Scheler's central place for persons, or Karl Jaspers' Christian existentialism. The centrality of the person expresses two main ideas such as personal dignity (i.e., the inherent, unique, and unconditional value of persons that separates them from all other non-person beings) and interpersonal relationships (the essential and ineradicable context of persons who never exist in isolation and are not independent of others, but live in a community). Personalism is a form of communitarian anthropology and a moral doctrine. Personalist bioethics was developed by the Italian bioethicist and Catholic cardinal Elio Sgreccia mostly in his work *Personalist Bioethics: Foundations and Applications* (2012) as a reaction to North American mainstream bioethics. Sgreccia's book goes beyond a religious perspective in being a strictly rational reflection based on the dignity of the person and focused on the intimate meaning of human existence. Broadly speaking, personalism within biomedical ethics advocates patient-centered care that adopts a holistic approach and takes on board the spiritual dimension of the person, whereas personalism within environmental ethics advocates that the supremacy of the person does not allow the human and non-human community to which persons belong to be neglected.

H. ten Have and M. do C. Patrão Neves, *Dictionary of Global Bioethics*,
https://doi.org/10.1007/978-3-030-54161-3_402

Personalized Medicine

Personalized medicine is a form of precision medicine (the two terms largely overlap). It is based on the unique genetic or molecular profile of each patient and provides customized treatment specifically tailored to characteristics of that particular patient (genetic, environmental, and lifestyle). The best examples of personalized medicine are found in cancer (breast cancer and melanoma) and in cardiovascular diseases where it makes use of specific information about the patient's condition and targeted therapies. Personalized medicine involves classifying patients into subpopulations with a special susceptibility to a particular disease or a different response to a specific treatment and tailoring new drugs and medical devices to address such specificities. Such new drugs or medical devices are not unique to the patient but to individual characteristics of each patient. This is the reason some professionals prefer using the term precision medicine instead of personalized medicine to avoid misunderstandings. Personalized medicine is more efficient, less invasive, and has fewer harmful side effects. Although it is more beneficial than regular medicine and can even be less costly than a trial-and-error approach, current prices of gene sequencing still make it very expensive as a result of it being mostly accessible at the experimental stage and to just for few people in developed countries. Restricted accessibility and thus injustice are two of the major ethical problems related to personalized medicine. In addition to the usual ethical problems with precision medicine, there are others that are intensified and aggravated in personalized medicine including the amount of health data produced, their ownership, and dual use; transverse ethical issues such as privacy and confidentiality, stigmatization and discrimination; informed consent requiring patients to understand the risks and benefits of participating in this innovative biomedical research; and genetic tests becoming part of routine healthcare and bringing with them all the ethical issues characteristic of this field one of which is the disclosure of sensitive information about an individual simultaneously affecting other family members. An additional ethical issue is misleading advertising in which commercial products are purported as being designed for each individual's needs, which they are not. Believed by many to be the future of medicine, personalized medicine will continue to grow since it improves health outcomes and lowers costs.

H. ten Have and M. do C. Patrão Neves, *Dictionary of Global Bioethics*,
https://doi.org/10.1007/978-3-030-54161-3_403

Pharmacogenomics

Pharmacogenomics is a new field of research and clinical assistance that combines pharmacology (the science of drugs) and genomics (the science of genes and their functioning). It refers to the study of genes that determine drug behavior and the influence such genes have on someone's response to a particular drug. Therefore pharmacogenomics contributes to precision and personalized medicine both of which are branches of clinical research and assistance. Although it is a growing field whose use is today quite limited, it should lead to higher efficiency and safety in the diagnosis and treatment of various diseases. Although the term pharmacogenetics was first coined in 1959 by Friedrich Vogel, it has been used interchangeably with pharmacogenomics introduced in the 1990s with the development of genome sciences. Study of the role genetic variation plays in drug response or adverse reactions to drugs started in the 1950s and continued to be developed up to the 1990s when the Human Genome Project began. Although this project decisively contributed to the development of pharmacogenomics, it was not until the completion of both the Human Genome Project and the International HapMap Project on the identification and cataloging of genetic similarities and differences in humans that pharmacogenomics truly started to develop. Pharmacogenomics is expected to be highly beneficial in the future as a consequence of its uses increasing (e.g., tailoring drugs to treat a wide range of health problems such as cardiovascular disease, cancer, Alzheimer disease, and HIV/AIDS). The ethical issues raised are similar to those of precision and personalized medicine and are stronger upstream (high cost, low availability, additional financial stress for governments and insurances, false hopes, and use and misuse and storage of genetic information obtained for pharmacogenetic analysis) than downstream where benefits are higher than potential risks.

H. ten Have and M. do C. Patrão Neves, *Dictionary of Global Bioethics*,
https://doi.org/10.1007/978-3-030-54161-3_404

Pharmacy Ethics

Pharmacy ethics refers to the ethical standards (behavior considered right, appropriate, and desirable) for the practice of pharmacy. The primary role of pharmacists in healthcare is to ensure that medicines are used in the safest and most effective manner. However, pharmacists are unusual in that they occupy a space between healthcare and business in that they provide medicines prescribed by healthcare providers and at the same time run a business with the purpose of making a profit from the sale of medication. Such an intermediary position brings about certain ethical challenges because pharmacists have many responsibilities such as moral obligations to patients (making sure patients receive optimal benefit from their medication since the well-being and best interests of the patient come first); respecting the self-autonomy of patients and ensuring the information provided to patients is understood; having a duty to act with honesty and integrity and to avoid discriminatory practices and unfair competition; the way they act toward other health professionals; respecting the values of colleagues and consulting them when uncertain as to what is best for the patient; the role they have toward society in seeking justice in the way health resources are distributed; and for promoting, preserving, and improving the health of the population. Balancing the principles of healthcare with business is precarious as borne out by the dilemma that arises when pharmacists recommend lower cost generic medication knowing that it will be to the benefit of patients and society but that it will lower profitability. This is the reason pharmacies also sell non-medical products such as food supplements, cosmetics, and slimming substances. Although tobacco used to be sold by pharmacies, this has now been outlawed in most countries. Since pharmacies have both professional and commercial sides this has led to codes of ethics giving priority to the professional side of pharmacies and some countries like Germany having detailed regulations that ensure the professional side of pharmacies is not infringed by commercial concerns. It should be noted that pharmacy advertising is prohibited in most countries.

H. ten Have and M. do C. Patrão Neves, *Dictionary of Global Bioethics*,
https://doi.org/10.1007/978-3-030-54161-3_405

Placebo

The word "placebo" derives etymologically from the Latin *placere* (to please). It has been used in medicine since the eighteenth century to refer to a substance originally sugar or flour tablets or just saline (today any kind of fake treatment such as a pill or a shot) given more to please than to benefit the patient. Although the placebo has no active pharmaceutical ingredient (API), its positive effect has been recorded ever since the early twentieth century and has come to be known as the placebo effect. This refers to patient responses that are mostly positive (improvement of symptoms) but also sometimes negative (side effects) to the placebo administered. Placebos have been used in clinical care when there is no known treatment or when doctors believe the patient is a hypochondriac. However, its principal use is in clinical trials as a control to compare the effects of non-intervention to the effects of a treatment under trial. A placebo-controlled, double-blind, randomized clinical trial is the gold standard of clinical research. The use of placebo controls in randomized clinical trials is controversial in that it deprives patients of treatment. The use of placebo controls today is highly restricted and only permitted when there is no proven effective treatment for the condition under study, its use does not entail any harm, and patients are fully aware they may be in a placebo group. In 2013 the Declaration of Helsinki was revised in Fortaleza (Brazil) where placebo use was one of the most important issues of debate as a result of many South American countries strongly opposing it. The approved (revised) text is "The benefits, risks, burdens and effectiveness of a new intervention must be tested against those of the best proven intervention(s), except in the following circumstances: where no proven intervention exists, the use of placebo, or no intervention, is acceptable; or where for compelling and scientifically sound methodological reasons the use of any intervention less effective than the best proven one, the use of placebo, or no intervention is necessary to determine the efficacy or safety of an intervention and the patients who receive any intervention less effective than the best proven one, placebo, or no intervention will not be subject to additional risks of serious or irreversible harm as a result of not receiving the best proven intervention. Extreme care must be taken to avoid abuse of this option." More recently there has been research into the use of placebos and the placebo effect in laboratory and clinical settings (outside blind trials) in which people know they are receiving a placebo instead of a real treatment. Surprisingly, the placebo continues to register positive psychobiological effects and just goes to show that it can be an effective treatment.

H. ten Have and M. do C. Patrão Neves, *Dictionary of Global Bioethics*,
https://doi.org/10.1007/978-3-030-54161-3_406

Plagiarism

The word "plagiarism" derives etymologically from the Latin *plagium* (the action of stealing). In ancient Rome it referred to the stealing or kidnapping of slaves (or free people to be enslaved) and was metaphorically used by the Roman poet Marcus Valerius Martialis (first century) to refer to literary theft of his work. In ancient Greek *plagios* (oblique, crossed, felonious) referred to something that had been deviated (interpreted as taken from someone and given to someone else). Thus plagiarism refers to the criminal action of taking something that belongs to someone else. The word became more common in the seventeenth century and gradually started to refer specifically to somebody stealing the intellectual work or ideas of someone else and reproducing them as his or her own without proper acknowledgement of the original author. Plagiarism has been broken down into a number of categories to assess the severity of the wrongdoing such as complete or reckless, intentional or unintentional, and with or without consent and can lead to criminal and/or disciplinary charges. Paradoxical as it may seem, plagiarism has today become much more frequent and received much more criticism than ever before. In the wake of the digital revolution and easy access to the seemingly infinite amount of intellectual work available on the internet, plagiarism has become so common that people seem to be unaware that freely using someone else's work as their own is wrong despite strong reaction against such use and institutions, scholars, and professionals taking action to prevent it through education, surveillance, and penalties. Plagiarism is condemned today by all parties (individuals and institutions) involved in the production of intellectual work such as science, humanities, and arts. When it comes to scientific research, plagiarism is part of the well-known trilogy that characterizes research misconduct: fabrication, falsification, and plagiarism.

H. ten Have and M. do C. Patrão Neves, *Dictionary of Global Bioethics*,
https://doi.org/10.1007/978-3-030-54161-3_407

Pluralism

The concept of pluralism in philosophy refers to the statement or conviction that reality is not constituted by a single substance (monism) or two fundamental substances (dualism), but by a diversity of elements that should be understood together in their own reciprocal dynamism. Therefore pluralism is not a synonym of diversity because it entails a holistic view of the various elements considered. Such a concept is used in many different fields and is particularly relevant in politics and morality. Moral pluralism refers to the view that there is a plurality or a multiplicity of values, principles, rules, and virtues guiding human actions all of which deserve due consideration. Such a contemporary position is the result of the decline in moral universals during the nineteenth century. Indeed, in the absence of a moral authority recognized by all the adherence to moral rules becomes an individual affair that generates and fosters pluralism. Pluralism can be beneficial for individuals and societies by favoring different paths that individuals can take to flourish, that are more adequate for their needs, and that enrich social life by providing them with different theoretical perspectives and practical solutions. However, pluralism can also have negative effects since it leads to individualism or relativism both of which weaken morality. Pluralism is as beneficial as it is permeable (i.e., it presents itself as a forum for honest and transparent dialogue animated by collective search for the best decision for all).

H. ten Have and M. do C. Patrão Neves, *Dictionary of Global Bioethics*,
https://doi.org/10.1007/978-3-030-54161-3_408

Pollution

The word "pollution" derives etymologically from the Latin *pollŭo* (to soil, to dirty, to make impure, to infect with a disease). Pollution generally refers to contamination of the environment in which substances that are harmful or have poisonous effects are present or have been introduced into the environment. Different types of pollution can be distinguished such as water, air, soil, thermal, radioactive, noise, and light. Although all forms of pollution impact human health, when it comes to public health the most important are air pollution and water pollution. In 2005 the WHO warned that 23% of the global disease burden and 24% of all deaths can be attributed to environmental factors such as air pollution and unsafe drinking water, that air pollution is associated with 6.5 million deaths globally each year most of which occur in India and China, that the main source of air pollution is energy production (especially the emissions of coal power plants), that approximately 11,800 new cases of chronic bronchitis and 21,000 hospital admissions in the European Union are the result of such power plants, that air pollution is a global problem since toxic substances affect people even in countries where such power plants have been closed, that addressing air pollution demands global cooperation and taking into account differences in stages of development and economic growth, and that evidence that air pollution has a negative impact on global health is growing such that more countries are engaged in phasing out coal-fired power plants. The UN Climate Change Conference agreement reached in Paris in 2015 states that coal as a source of energy should be abandoned. Water pollution is similar in that the contamination of water is usually the result of human activities such as discharging chemicals, microorganisms, plastics, and waste into seas, rivers, lakes, and reservoirs. About 80% of all wastewater is dumped (mostly untreated) into the environment. The pollution of drinkable freshwater is a special public health problem. According to the WHO (2015) 663 million people lack reliable sources of drinking water, contaminated water kills more than 840,000 people each year (as a result of diarrhea-related diseases), and more than 800 children under 5 years of age die every day from diarrhea linked to unsafe water, lack of sanitation, and poor hygiene. There are many examples of problems arising at local levels such as the infamous lead poisoning of water in the city of Flint, Michigan (USA) in 2014 when the authorities decided for economic reasons to switch the water supply to the river and overlooked the need to treat the water since it would flow through the old lead water pipes of people's homes. This resulted in more than 100,000 residents being exposed to high lead levels and more than 5% of all children

H. ten Have and M. do C. Patrão Neves, *Dictionary of Global Bioethics*,
https://doi.org/10.1007/978-3-030-54161-3_409

having elevated levels of lead in their blood. When it comes to global bioethics it is widely accepted that air pollution and water pollution are the major health problems and ethical challenges that not only cannot be denied but will also demand more powerful global governance.

Population Ethics

Population ethics is an area of applied ethics concerned with the ethical problems caused by human activities aimed at controlling who is born and how people are born in the future. Its focus is sometimes on the size and sometimes on the composition of present populations or future populations. Depending on the normative view taken, population ethics can have practical implications relating to such issues as the allocation of resources for family planning and prenatal care and the provision of aid to developing countries. These are highly controversial issues because developed countries often only provide assistance if it can be linked with the developing country imposing laws legalizing contraception and abortion (as a contraceptive measure). This runs the risk of neighboring countries withholding aid because such measures often conflict with their own cultural environment. In 1798 Thomas Malthus predicted a catastrophe would take place since the growth of populations was exponential while that of the food supply was linear. By arguing population growth should be controlled, his ideas led to a movement advocating human population planning named neo-Malthusianism. Much later, concerns with overpopulation and associated environmental degradation and depletion of resources revived such ideas in the 1960s and 1970s and regarded population growth as a time bomb. A recent example of neo-Malthusianism is the one-child policy in China (1979–2015) aimed at controlling the size of the population. Such a policy was highly controversial since it was enforced and left little room for individual choices. A major part of the population refused to accept the policy and the Chinese government were obliged to modify it in the 1980s. Although the modified policy allowed parents in rural areas to have a second child if the first child was female, ethnic minorities were excluded. Another example is David Benatar's antinatalism that argues human beings should stop having children, human life is bad, it is better never to be born, bringing people into the world is little more than harming them, and procreation is therefore wrong. Such utilitarian approaches in population ethics have been characterized as excessively pessimistic and even inhuman, especially since the global food supply is sufficient to nourish the world population (the main problem lies in the way available food is distributed).

H. ten Have and M. do C. Patrão Neves, *Dictionary of Global Bioethics*,
https://doi.org/10.1007/978-3-030-54161-3_410

Poverty

Poor people can be defined as people unable to meet their basic needs. Looked at from an economic perspective such people lack the financial resources to cover their needs for food, safe drinking water, shelter, clothing, education, and healthcare. Poverty is measured by the number of people who fall below a certain level of income called the poverty line. In 1990 the poverty line was first determined at USD1 a day and has gradually increased since then to reach USD1.90 in 2015. Although the poverty line is a measure of absolute poverty, it means much more than lack of income. According to the United Nations it relates to a situation in which basic human needs are missing and essential services cannot be accessed. Absolute poverty differs from relative poverty in that the latter depends on social conditions where people cannot meet the living standards of others in the same time and place. Relative poverty varies between countries and is the result of income inequalities between them. The WHO report that 1.2 billion people in the world live in absolute poverty and that poverty should be a global health concern because it is associated with diseases, malnutrition, infant mortality, low life expectancy, and violence. This has led to international organizations such as the WHO and World Bank having poverty reduction strategies. The first of the 17 UN Sustainable Development Goals is to end poverty in all its forms everywhere by 2030 (the same target was given to the other goals). The main ethical challenge is finding the best way to eliminate poverty. Although almost all people will agree that poverty is unacceptable, they are unsure of their obligations to poor people (if any). Some argue there is no duty to assist the poor and that what is needed is beneficence and charity. The libertarian perspective argues that nothing is owed to the poor and that poverty is unfortunate —not unjust. Charity is criticized because human beings already have a duty to help the most needy and vulnerable, as well as moral obligations to provide assistance. Peter Singer argued that absolute poverty is bad and that when it can be prevented it should be done without hesitation. Another perspective such as that of Thomas Pogge argues that obligations are institutional since it is global institutional arrangements that are responsible for global poverty and inequality, that citizens and governments of wealthy societies have created such arrangements, that by doing done so have given no thought to preventing poverty, and that they have a duty to compensate for and rectify this situation by creating fairer institutions. When it comes to global bioethics, fortunately but belatedly this last position is receiving growing support.

H. ten Have and M. do C. Patrão Neves, *Dictionary of Global Bioethics*,
https://doi.org/10.1007/978-3-030-54161-3_411

Pre-approval Access (*See* Compassionate Use; Right to Try)

Access to investigational medicines prior to approval is called "pre-approval access" (a.k.a. expanded use). When the emphasis is on the moral principle of beneficence, the same practice is called "compassionate use"; and when the emphasis is on individual rights and the moral principle of personal autonomy, the current term is "right to try." In all cases the primary goal is treatment of the patient rather than the collection of data or scientific investigation. Although the preferred option for patients to get access to investigational drugs is to participate in clinical trials, most pharmaceutical companies today have pre-approval access policies. On behalf of individual patients or certain groups of patients with similar characteristics doctors can request access to therapies that are not yet approved for sale by the FDA or other regulatory agencies. Although each case is regarded as unique and considered individually, it is recognized that there might be extraordinary circumstances where access to such investigational drugs outside a clinical trial is appropriate. A request can be considered if three basic criteria are met: the patient suffers from a serious or life-threatening illness or condition; there are no other suitable treatment options; and the patient does not qualify for ongoing clinical trials.

H. ten Have and M. do C. Patrão Neves, *Dictionary of Global Bioethics*,
https://doi.org/10.1007/978-3-030-54161-3_412

Precautionary Principle

Although the precautionary principle lacks a consensual rigorous definition, it is commonly understood as the duty to refrain from acting whenever doing so has potentially serious and irreversible consequences that are not yet accurately determined (taking the via negativa approach as applied to the production of genetically modified organisms) and the obligation to take appropriate measures to avoid potentially severe and irreparable risks in the absence of scientific certainty (taking the via positiva approach as applied to biological control and its effects in ecosystems). The precautionary principle originates from the German *Vorsorgeprinzip*, which can be broken down into its constituent parts of *Sorge* (care, concern) and *Vorsorge* (precaution, alertness, foresight, precaution). The precautionary principle refers to the need to envision consequences and take care to prevent them; it entered common use in the late 1960s or early 1970s. Theoretically developed by the German philosopher Hans Jonas in *The Imperative of Responsibility: In Search of an Ethics for the Technological Age* (1979) in which the need for such a principle was argued as necessary as a result of the overwhelming cumulative development of modern technology extending its consequences to a space without borders and a time without limits and potentially causing irreversible yet unpredictable harm. Instead of taking risks humankind should adopt all possible measures to prevent harm and should refrain from acting without sound knowledge of its effects (be they short or long term). The two main domains to which Jonas applies the precautionary principle are environmental and genetic engineering since they both endanger the survival of humanity as such. The precautionary principle has not played a significant role in the application of biotechnologies to human beings but has been extraordinarily successful in its application to environmental issues. In 1992 it became part of the UN Rio Declaration on Environment and Development: "where there are threats of serious or irreversible damage, the lack of full scientific certainty shall now be used as a reason for the postponing cost-effective measures to prevent environmental degradation." In the same year (1992) it became part of the EU Treaty of Maastricht and has been used ever since in a number of public realms such as public health and governance (currently). The principle has also been criticized as obstructing progress and failing to propose specific measures.

H. ten Have and M. do C. Patrão Neves, *Dictionary of Global Bioethics*,
https://doi.org/10.1007/978-3-030-54161-3_413

Precision Medicine

Precision medicine refers to medicine that is accurate (or precise). It is an innovative approach to patient care based on the genetic or molecular profile or on the individual characteristics (personalized medicine) of patients or groups of patients such as ethnic groups who are more susceptible to a particular disease and do not frequently respond to standard medications. The aim of such an approach is to provide patients with specific, targeted therapies and to optimize the efficiency of treatment. Precision medicine is a recent field of biomedical research that is growing as a result of the beneficial results it provides. It increases positive outcomes, reduces negative side effects, and is better able than other forms of medicine to recover the health of patients fully and quickly. The main problem with precision medicine is cost because sequencing large amounts of DNA is very expensive as are subsequent drugs that have to be developed to target the genetic or molecular characteristics (pharmacogenomics) of patients or groups of patients. Although there are no ethical problems with precision medicine from the health point of view, there are some relating to the costs involved such as private investment in research being mainly oriented at diseases that are prevalent in populations and groups that can afford it (restricting its availability to developed countries), discrimination in healthcare increasing, global injustice intensifying, financial pressure on national health care services (already struggling to keep up) becoming greater, and the burden on private insurance companies getting ever larger. However, a significant drop in the costs of DNA sequencing is already happening and should make precision medicine more accessible and thus fairer.

H. ten Have and M. do C. Patrão Neves, *Dictionary of Global Bioethics*,
https://doi.org/10.1007/978-3-030-54161-3_414

Predictive Medicine

Although the idea that medicine could have a predictive side is not new, predictive medicine as a subspecialty in healthcare is quite recent. It reflects the growing recognition of the importance of prevention in medicine and of the new means acquired by medicine for effective and efficient prevention. Predictive medicine not only contributes to promoting and maintaining someone's health, it also reduces the health costs for the state by optimizing services. Moreover, predictive medicine benefits today from important advances in genetics as borne out by the increasing number of genetic tests or molecular biomarkers whose accuracy is ever growing. Although predictive medicine emerged in the field of genetics, it was important to venture beyond such a field since most serious diseases are multifactorial and even genetic disorders are influenced by other health factors such as diet, exercise, and stress. A lot of effort has gone into developing biomarkers that can also be used for more common diseases. Biomarker testing can assess the likelihood of someone developing a specific disease (such as cancer), its early detection (screening), its identification (diagnosis), and can be used to plan the best treatment. Evaluating someone's individual risk for some specific disease can facilitate preventive intervention. Predictive medicine is highly efficient at developing the predictive side of medicine. The objective of predictive medicine is to collect as much data about individual patients as possible, analyze them, and be able to predict a person's individual risk for certain diseases. It is also here that the most significant ethical issues arise such as the nature of the data collected, the process employed to collect them, and their different possible uses. There are also issues regarding informed consent, privacy, stigmatization, discrimination, and ownership. Patient data are currently collected from many different sources such as simple blood tests or even personal health-monitoring devices unbeknown to the person in question. Big data collected worldwide help to predict the flu virus for the next season and are used to predict epidemic outbreaks like Ebola. Important health benefits are achieved at the expense of individual privacy and very often have no advantages for the population that provided (involuntarily) the data. In conclusion, a balance must be reached between expected benefits and foreseen losses.

H. ten Have and M. do C. Patrão Neves, *Dictionary of Global Bioethics*,
https://doi.org/10.1007/978-3-030-54161-3_415

Prenatal Genetic Screening

Prenatal genetic screening refers to the analysis of genetic material of a fetus to identify the presence, absence or modification of a particular DNA sequence, gene or chromosome. It is a type of medical test, performed during pregnancy, harmless to the mother or the fetus, to check the health of a baby before birth, although it cannot identify all possible inherited disorders and birth defects. It is particularly useful to identify certain genetic disorders, such as Down syndrome, cystic fibrosis, Tay-Sachs disease, or sickle cell anemia. It should only be performed upon the recommendation of a geneticist and when there is an increased risk that the baby will have a genetic or chromosomal disorder, due to a family history of genetic disorders, or the couple's ethnic background, or other risk factors such as the woman's age. A sound justification for prenatal genetic screening is an ethical requirement. In case of testing positive, it is to the parents to make decisions about the pregnancy. Some analysts consider that prenatal genetic screening should only be provided to women or couples who will abort if the fetus tests positive. Others, however, consider that testing can also reduce anxiety related to uncertainty, contribute to make informed decisions and, eventually, to help the parents to prepare to welcome a disabled child (to gather more information about what to expect) if they decide to carry on with the pregnancy. The goals of testing and respective consequences are perhaps the major ethical issues raised by prenatal genetic screening. The moral status of the fetus related to abortion is also implicated. At the same time, it should be realized that genetic screenings can give false positive results and, in this case, should be followed by diagnostic tests (the more invasive chorionic villus sampling and amniocentesis) for a definitive diagnosis. In any case, it will not be beneficial to the fetus because there is no treatment available. Other ethical issues can include: the limiting of a woman's control over decisions about her pregnancy, social pressure to have only healthy children, creating a pattern of normality, and aggravating discrimination.

H. ten Have and M. do C. Patrão Neves, *Dictionary of Global Bioethics*,
https://doi.org/10.1007/978-3-030-54161-3_416

Prevention

Prevention is a traditional goal of medicine as borne out by Hippocratic writings pointing out the importance a healthy lifestyle has in preventing the occurrence of disease (prevention is better than cure is thus an ancient wisdom). Prevention in the context of healthcare covers a wide range of activities from protecting health to preventing diseases, handicaps, and disabilities. Health protection refers to systemic measures undertaken to prevent the health of populations being harmed such as ensuring food quality meets relevant standards and prohibiting the release of toxic substances into the air. Health promotion refers to systemic measures undertaken to influence individual behavior and lifestyle such as campaigns to reduce smoking and alcohol use. In short, disease prevention can be defined as targeting activities at preventing health problems. There are three types of prevention termed primary prevention, secondary prevention, and tertiary prevention. Primary prevention (e.g., vaccination) aims to prevent the emergence of a health problem. Secondary prevention (e.g., population screening) aims to identify and treat a health problem as soon as it crops up to reduce or eliminate its impact. Tertiary prevention (e.g., treatment of glaucoma) aims at counteracting the progression of health problems to prevent them from becoming chronic. Ethical concerns with prevention are related to three specific features of preventive medicine. The first is whether putting the focus on healthy people or at least people who do not feel ill (yet) can be justified when they have not presented with a complaint or requested seeing a doctor (as is the case in curative medicine). The second is whether putting the focus on groups and populations rather than individual patients (e.g., babies, children, or pregnant women) means such groups may have a higher risk for potential health problems and thus preventive activities are medicalizing people with health complaints. The third is whether putting the focus on the effects of preventive programs means they are unobservable by individual participants, whereas the results of curative interventions are noticeable by patients themselves mostly within a short timeframe. Although preventive activities call for individual efforts to be made that are often beneficial in the longer run, even when successful the effect will not even be experienced since the potential disease will not emerge. Although preventive programs are mostly undertaken at the population level where mortality and morbidity rates are low, individuals are invited to participate since prevention will promote the good of others and the community. Termed the prevention paradox such a preventive measure is beneficial for the community but has less to offer to each participating individual. A fourth characteristic of prevention is that many interventions take place outside the field of healthcare such as governments requiring drivers to wear seat belts (a.k.a. safety belts) in automobiles. Although individual freedom can rightly be restricted for the sake of health, there are ethical concerns with prevention

H. ten Have and M. do C. Patrão Neves, *Dictionary of Global Bioethics*,
https://doi.org/10.1007/978-3-030-54161-3_417

such as balancing individual autonomy, the interests of other people, and the interests of society. The moral justification of preventive activities can be looked at in two diametrically opposite ways: the advantages and the disadvantages. The advantages of prevention are reduced mortality and morbidity; death and suffering that can be avoided do not occur; quality of life, well-being, and control over life can be enhanced; and healthcare expenses can be greatly reduced. The disadvantages of prevention are the psychological harm individuals may experience (by creating anxiety and worries about test outcomes), the medicalization of society, and increased healthcare costs (by identifying more cases of disease). There is evidence to show that preventing disease (e.g., acute heart failure and cancer) may lead to diseases later in life (e.g., Alzheimer disease). Although the good accomplished by prevention (better health and longer life) is therefore relative, people will always of course value living and health over disease and dying. Another issue concerns the means used to bring about the advantages (the good) of preventive medicine. Prevention is often a collective good as borne out by individuals having limited abilities to protect themselves against epidemics or pandemics such as COVID-19 and protection requiring governmental interference. Although individual and collective interests are often at odds, individual autonomy needs to be limited such as obliging an infectious tuberculosis patient to stay at home or in an institution to safeguard the public interest. The question revolves around the degree of pressure (or coercion) that should be allowed. An example is the current debate about childhood vaccination. In many countries vaccination is voluntary and information is provided to persuade parents to have their children vaccinated. However, when growing groups of parents do not follow the advice (often as a result of misleading information propagated on social media) and therefore endanger other people, the real question is how this can be overcome and whether some sort of coercion is necessary.

Principlism

Principlism is the name given to theoretical models of bioethics structured by giving voice to and founding ethical principles that have evolved as a result of the many different ways such principles have been applied to concrete cases. Principles are general, formal, and abstract statements that ground and regulate an action and are expressed by obligations (prohibitions or permissions). The term "principlism" initially had a pejorative meaning when it was first used as a theoretical model of bioethics proposed by Tom Beauchamp and James Childress in *Principles of Biomedical Ethics* (1979). The first two of the four prima facie principles they presented were beneficence and nonmaleficence rooted in the Hippocratic tradition and teleological in nature (requiring the accomplishment of good). The second two were autonomy and justice embedded in liberal philosophy and deontological in nature (requiring respect regardless of the consequences). These four principles led to the derivation of four major rules (confidentiality and fidelity as well as veracity and privacy, respectively). Although rules are less general, formal, and abstract than principles, they are more contentful and specific and therefore closer to the cases they have been devised to solve. Such principles and rules were applied deductively to concrete cases in the early editions of the work, which aroused strong criticism and even gave a pejorative meaning to principlism. In subsequent editions of the work (to date there have been seven revised editions) the proposed method has been further developed and refined. Instead of a mere top-down application model (from principles to cases), the authors propose an integrated model in which principles and rules are specified, their abstract character is reduced, and the relative weight and strength of principles and rules within the broader framework of a common morality (top-down) are balanced. When a case is resolved and the outcome is not deemed satisfactory according to the perception of common morality, then the rules and principles are reconsidered and further refined. Since 1979 several other theoretical models of bioethics have appeared. Although many claim to be rooted in different traditions and have developed different methodologies, most of them are still structured by principles. In fact, not only is Beauchamp and Childress's four principles approach the most successful bioethical theory, principlism is also the dominant approach taken in global bioethics as borne out by all major bioethical international documents being principle based such as the Convention on Human Rights and Biomedicine and the Universal Declaration on Bioethics and Human Rights.

H. ten Have and M. do C. Patrão Neves, *Dictionary of Global Bioethics*,
https://doi.org/10.1007/978-3-030-54161-3_418

Prisoners

As a result of being deprived of their liberty and sometimes being the focus of medical research, prisoners are regarded as a vulnerable population. Although more than 10 million people are incarcerated in penal institutions worldwide, imprisonment levels vary across countries. The United States has the highest number of prisoners in the world with an incarceration rate in 2019 of 737 per 100,000 persons (followed by Russia with 615 and the Ukraine with 350), which is in stark contrast with Portugal with 120 and Norway with 66. Although prisoners have lost many rights (their needs are met by the state), they have not lost the right to health. The provision of adequate healthcare in prisons has been helped by the United Nations issuing several documents stipulating standard minimum rules for the treatment of prisoners such as the Standard Minimum Rules for the Treatment of Prisoners (a.k.a. the Nelson Mandela Rules) that had been adopted as early as 1955. Since prisons often fail to address the care needs of prisoners, healthcare providers often have to face a number of ethical issues such as the difficulty of providing care in the best interest of inmates in a correctional setting; the limited resources available within prisons for adequate care; confidentiality (prisoners have the same right to confidentiality as everyone else but in a prison setting this is sometimes difficult to maintain); and dual loyalty (although physicians have commitments to patients and legal authorities, they are asked to assess the mental health and danger prisoners pose to other inmates or staff and to determine whether they should be held in solitary confinement). Dual loyalty is a special problem in countries where torture is legal and physicians are asked to assess the health of prisoners and their capacity to tolerate such a barbaric practice. Many medical bodies argue that physicians should never participate in any activity that violates human rights. Special problems arise when prisoners are on hunger strike and physicians are asked to force-feed them. The World Medical Association's Declaration of Malta on Hunger Strikes (last revised in 2017) points out that physicians should respect individual autonomy and, having assessed the wishes of a hunger-striker, should not forcibly provide treatment that the prisoner refuses: "Applying, instructing or assisting forced feeding contrary to an informed and voluntary refusal is unjustifiable." As pointed out earlier, prisoners are regarded as a vulnerable population. However, it is now accepted that their ability to make informed and voluntary decisions is compromised when it comes to participating in research. Additional regulations have been stipulated: studies should involve minimal risk and inconvenience; should focus on the situation of prisoners (e.g., on the causes, effects, and processes of incarceration and criminal behavior); and should focus on conditions that affect prisoners as a class (e.g., research into hepatitis, which is more prevalent in prisons).

H. ten Have and M. do C. Patrão Neves, *Dictionary of Global Bioethics*,
https://doi.org/10.1007/978-3-030-54161-3_419

Privacy (*See*: Confidentiality)

Privacy refers to the right to keep confidential, secret, or private anything that concerns someone's own private life. Privacy as a concept has a long history and the word "private" has been specifically used since the times of Aristotle as the opposite of "public." The public sphere refers to political activity and getting together with fellow citizens, whereas the private sphere concerns domestic life and relationships with family members. The opposition between private and public is probably more pertinent today than it has ever been. The word "private" derives etymologically from the Latin *privatus* (private, withdrawn from public life, apart from other people; privacy, seclusion, freedom from intrusion). Despite its remote ancestry, privacy has become particularly important in the philosophical, political, and legal realms and in the common language of modern times. Although cultures value it in different ways and criticisms are made about its relevance, the consensus is that it is a valuable concept at the personal level (most people agree that privacy is important to building personal identity) and at the social level (most contexts in which social activity takes place benefit from there being a private side to it). Privacy essentially refers to limiting access to such matters as data, the body, and relationships and should be ethically established, politically implemented, and legally protected. In the second half of the 20th century, the human rights movement used its influence to get privacy recognized as a patient right in the Lisbon Declaration of 1981. The declaration specifically states that the only person who can disclose clinical information about himself or herself is the (competent) patient. Indeed, it is the patient's right to privacy that brought about the healthcare professional's duty of confidentiality. If health professionals fail to keep the information they gather in their professional relationships with patients confidential, the patient's own privacy is jeopardized. Privacy and confidentiality are two sides of the same coin in that they are based on respect for people and human dignity and bounded together by trust. However, privacy is strongly challenged by today's new technologies of information and communication through the use of social networks by individuals and the ease of access to digitally stored data despite the fact that barriers are continuously being raised. Lack of privacy leads to feelings of insecurity at the individual level and increases the chances for discrimination, stigmatization, and injustice at the social level.

H. ten Have and M. do C. Patrão Neves, *Dictionary of Global Bioethics*,
https://doi.org/10.1007/978-3-030-54161-3_420

Professional Ethics

Professional ethics refers to specific moral standards that regulate the practice of a particular profession. The history of professional ethics can be traced back to Ancient Greece when it was already well established among liberal professionals working in medicine and law. Even back then it had a set of rules that had to be followed by all professionals such as the Hippocratic Corpus (that regulated medical practice) and the Hippocratic Oath (a professional ethics code). Moreover, the principle of beneficence and the rule of confidentiality had already been formally established even at that time. Most professions can be characterized by requiring their members to have a high level of education and by getting their members well represented on regulatory bodies such that they have a great deal of control over the standards of practice. Although professional ethics has become symbolic of the quality standards of a profession, it also responds to society's expectations regarding professional behavior and standards. Professional ethics is self-regulated (moral standards are established by the very professionals to whom they apply in a sort of closed shop), minimalist (the norms are not very demanding because all professionals need to be able to follow them), and corporative by defending the profession and responding to social needs, interests, and expectations (social recognition). Although the moral standards approved for each profession are drawn up by its own members and brought together in respective codes of ethics, all professionals have to comply with the rules established on pain of facing disciplinary proceedings and losing the right to practice.

H. ten Have and M. do C. Patrão Neves, *Dictionary of Global Bioethics*,
https://doi.org/10.1007/978-3-030-54161-3_421

Professionalism

Professionalism refers to complying with the goals and procedures established for a profession and responding to social needs, interests, and expectations (social recognition). Indeed, professions that do not respond to social recognition become obsolete and disappear as borne out by many professions vanishing over time and new social activities becoming professions when they are viewed by people as significantly contributing to the well-being and development of society. As ever increasing numbers of people dedicated themselves to that activity and made a good living from it, this allowed many of their members the luxury of sitting on different bodies that represented them such as professional orders, associations, and unions. Such representative professional organizations formally established good practices according to the mission and social responsibility of the profession. A code of ethics (defining professional standards) is always testimony to the maturity of a profession and reinforces its commitment to society. The first ethical requirement for professionalism is sound and constantly updated knowledge (theoretical and scientific competency) and is mostly acquired through reading journal papers and attending courses and conferences (education). The second ethical requirement is know-how (skills) and is mostly acquired through experience (training). However, these two requirements on their own are not enough to assure professionalism since ethical sensitivity (good judgment) is also required. Although these three requirements are paramount to whatever a professional does in his or her practice, professionalism itself is the most basic ethical requirement in professional practice.

H. ten Have and M. do C. Patrão Neves, *Dictionary of Global Bioethics*,
https://doi.org/10.1007/978-3-030-54161-3_422

Property Rights (*See* Ownership; Patenting)

Property rights refer to the ownership of resources and their use. Although property is defined by the state and enforced by law, property rights are usually attributed to individuals or to groups of people. Property rights differ from other kind of rights in that they are transferable and the owner can sell them, exchange them, or give them away. Property can also be transferred to the next generation as inheritance. There is a distinction between ownership and possession since owners do not always use their property and sometimes rent it instead to others who are then in possession of the property. Ownership involves power as borne out by the owners of capital exerting varying degrees of control over the lives of people working for them. Nevertheless, human beings cannot be the object of property since property rights only apply to things that have no rights of their own (e.g., land). The two types of property are private and public. Private property is characteristic of contemporary societies where it can be individual or collective. While private property is owned by an individual, collective property is owned by a collective entity such as a private corporation. Public or common property is used for the common good and belongs to the community—not to individuals or groups of individuals. Private property rights are the basis on which the capitalist economic system rests. John Locke argued that the right to private property is the result of human beings laboring and extending their personality into the products they create. He further argued that if somebody turns barren land into farmland that can be worked, then such land becomes his or her property; that property rights cannot be infringed by anyone or any government; that human beings will not have the incentive to work or to be creative and take risks if such rights are not protected; and that rewarding individuals for their work is what contributes to the progress of civilization. However, private property rights have been criticized since they have been said to make humans greedy (leading to their moral degradation); create inequality in that property is scarce and usually owned by those in the best position to acquire more property (increasing the gap between the haves and have-nots); and the owners of property such as land, manufacturing businesses, and information systems have too great an influence on politics (encouraging exploitation, domination, and corruption). This has led many to argue that the private property system is morally harmful since it prevents many from living a fulfilling life and results in the unjust distribution of wealth. When it comes to healthcare, the ethical discussion focuses on intellectual property rights (particularly patents) since property rights today are attributed to inventions such as new medications and even new life-forms created in the laboratory. Property rights are criticized from the global bioethics perspective because they reinforce global inequality; patents in medicine and biotechnology are owned by a limited number of companies

H. ten Have and M. do C. Patrão Neves, *Dictionary of Global Bioethics*,
https://doi.org/10.1007/978-3-030-54161-3_423

in the West (making medicines difficult to access in developing countries); the current intellectual property rights regime enshrined in law by the World Trade Organization and the agreement on Trade-Related Aspects of Intellectual Property Rights is the outcome of unfair pressure applied by Western countries and international businesses; and that property rights often take precedence over human rights such as the right to health.

Proportionality

Proportionality refers to the correspondence certain characteristics or variables have with each other or the balance between two elements. Proportionality within bioethics refers to the moral principle that an action should be no more severe than what is required (e.g., a punishment should be no harsher than the wrongdoing that led to it). Historically, the principle of proportionality can be traced back to Aristotle and his theory of distributive justice ("equals should be treated equally and unequals unequally"). This theory states that attributions should be proportional or correspond to differences between people. Although such a perspective became important in legal affairs, it was mainly within Roman Catholic moral theology (specifically Aquinas) that the principle of proportionality became preponderant. By stating that the reason to act should correspond to the moral value or disvalue that results from an action a relationship between means and ends is established. This led to proportionate reason becoming one of the conditions for the principle of double effect (a.k.a. the doctrine of double effect) that helped assure good effects outweigh bad consequences. In the 17th century Hugo Grotius developed the principle of proportionality in the debate over what constitutes a just war and went on to highlight that a proportional response to aggression was one of its most important requirements. In the 1960s the principle of proportionality was introduced by the German Constitutional Court as a guideline for public authorities. In the 1970s the principle was also applied to animal protection in scientific research as a result of European legislation stipulating that animal experimentation is acceptable as long as the research serves a higher purpose. In fact, the same argument (based on the principle of proportionality) is also used for research on embryos. This argument was first presented by Alan Gewirth in *Reason and Morality* (1978). He argued that the right not to be killed derives from the level of agency of the being concerned and that before the being (e.g., a fetus) acquires this capacity its rights depend on how close it is to acquiring such capacities. The same argument has been applied to mentally deficient people who lack full-fledged generic rights, but nevertheless have rights that correspond to their abilities. The notion of proportionality is now used for ethical review such that a proportional ethical review would be an assessment that takes into account the level of risk involved in the research project under consideration (e.g., research whose risk level is low should receive relative or minimal ethical scrutiny). Informally, such a perspective is often applied by different ethical bodies. Formally, it requires a concrete presentation of the potential risks, which is not always easy or possible to do.

H. ten Have and M. do C. Patrão Neves, *Dictionary of Global Bioethics*,
https://doi.org/10.1007/978-3-030-54161-3_424

Proteomics

Proteomics is a branch of molecular biology and an interdisciplinary field dedicated to the large-scale study and characterization of proteomes (i.e., the complete set of proteins produced in an organism or biological system encoded by its particular genome). The word "proteomics" first appeared in 1997 and derives from the word "proteome" (protein + genome) first coined by Mark Wilkins in 1994 to refer to a body of proteins and the word "technics" from the Greek *techne* (skill). Proteomics can thus be defined as the application of molecular techniques to the study of proteins. Genomics (an interdisciplinary biological area of biology focused on the structure, function, evolution, mapping, and editing of genomes) led naturally to proteomics as it developed. Proteomics provides significant biological information about the functioning of cells (i.e., by studying genes) that no other field of science can. Since proteins are responsible for the phenotypes of cells and the mechanisms of some diseases, aging, and effects of the environment require much more than genomics can deliver, many areas of research benefit from proteomics with its special focus on drug discovery, pharmacogenomics, and the identification of biomarkers. Although proteomics has great scientific potential and shows a lot of promise in the clinical setting, it is important to stress that it raises ethical problems concerning ownership and intellectual property, storage and privacy, and misuse of samples and data. Although such ethical issues are not specific to proteomics, they are enhanced in this field and thus additional ethical and legal safeguards need to be in place as a result of the ever increasing amount of information gathered and its ever increasing sensitivity.

H. ten Have and M. do C. Patrão Neves, *Dictionary of Global Bioethics*,
https://doi.org/10.1007/978-3-030-54161-3_425

Psychiatry Ethics

Psychiatric ethics is a branch of applied ethics that examines the actions and decisions of psychiatrists. There are those who argue psychiatry as a medical specialty is unique since it is concerned with people with behavioral and mental illnesses who are often stigmatized and subjected to socioeconomic disadvantages and discrimination. Therefore psychiatric ethics needs to go beyond the ethical standards of general medicine because patients are vulnerable, patient information is sensitive, and there is a need for special protection and confidentiality. Although psychiatric assistance depends on establishing good relations with patients and is crucial to patients feeling trusted, psychiatrists at the same time need to maintain a certain professional distance to provide the best advice in the interests of patients. A special problem is the way governments abuse psychiatry for political purposes as borne out by psychiatrists in the former Soviet Union being used to diagnose political opponents and dissidents with something they called "sluggish schizophrenia" in attempts to get them incarcerated in mental institutions; and in China where members of Falun Gong (a religious practice combining Buddhism and Taoism) are classified as mentally ill and thus removed from social life. A number of professional organizations have issued codes of conduct such as the American Psychiatric Association who published the first edition of *The Principles of Medical Ethics* with guidelines for psychiatric practice in 1973 (last updated in 2013); and the World Psychiatric Association (WPA) who first published its code of global ethics in 1977 as the Declaration of Hawaii and revised it in 1996 as the Madrid Declaration on Ethical Standards for Psychiatric Practice (it was updated in 2011 and supplemented by the WPA Ethics Policy in 2014). Such initiatives came about as a result of a number of countries misusing psychiatry in the 1970s. The WPA code of conduct articulates the ethical principles of respect for autonomy and beneficence and emphasizes the importance of informed consent, confidentiality, fair and equal treatment, and social justice when dealing with mentally ill people: "Psychiatrists serve patients by providing the best therapy available consistent with accepted scientific knowledge and ethical principles. Psychiatrists should devise therapeutic interventions that are least restrictive to the freedom of the patient and seek advice in areas of their work about which they do not have primary expertise. While doing so, psychiatrists should be aware of and concerned with the equitable allocation of health resources." The code also addresses specific issues such as euthanasia, torture, the death penalty, sex selection (i.e., attempts at controlling the sex of offspring to bring about the desired sex), organ transplantation, information propagated by the media, and conflicts of interest in relations with industry. The code points out the responsibility psychiatrists have to protecting the human rights of patients and to preventing

H. ten Have and M. do C. Patrão Neves, *Dictionary of Global Bioethics*,
https://doi.org/10.1007/978-3-030-54161-3_426

stigmatization and furthermore provides rules for forensic evaluation and compulsory intervention that emphasize the importance of psychiatrists not misusing their professional expertise and not recommending compulsory intervention in the absence of mental disorders.

Psychosurgery

Psychosurgery refers to neurosurgical procedures to alter the thoughts, emotions, personality, or behavior of people. Although sporadically practiced in earlier times, psychosurgery became common as a treatment of people with psychiatric illness in the early 19th century. Procedures such as leucotomy and lobotomy flourished in the 1930s and 1940s (introduced by the Portuguese neurologist Egas Moniz) and involved alcohol being injected into the frontal lobe to cause necrosis. The procedure was later modified when a special instrument called a leucotome (resembling an ice pick) was inserted into the brain through the nose in order to severe the neural pathways in the frontal lobe of the brain. Although Moniz was awarded the Nobel Prize for physiology and medicine in 1949, the procedure was condemned in the 1950s not only as unethical but also as associated with negative outcomes (alternatives became available with psychotropic medication). The procedure became even more problematic when people became aware that it was not only used for mental illness but also for violent prisoners. Psychosurgery has now been refined by stereotactic surgery and brain imaging technology both of which make it possible to target areas of ablative lesioning. Contemporary psychosurgery involves inserting electrodes into the brain to stimulate specific areas to mitigate symptoms of a specific mental illness. Rather than ablation the aim is the inhibition of neural activity. Deep brain stimulation (DBS) first introduced in the 1990s is now increasingly used. The basic assumptions of these psychosurgical interventions have not changed in that mental illness continues to be regarded as the result of neurological dysfunction (defects in the brain that can be repaired or eliminated). Although psychosurgery has significantly changed, it is still used as a last resort when other treatments have not been successful (primarily because DBS is not destructive and generally reversible).

H. ten Have and M. do C. Patrão Neves, *Dictionary of Global Bioethics*,
https://doi.org/10.1007/978-3-030-54161-3_427

Public Health

Public health is defined by the WHO as the science and art of promoting health, preventing disease, and prolonging life through organized efforts made by society. It can be broken down into three elements the first of which is its focus on the health and quality of life of populations (the reason it is called public); the second is it employs a broad notion of health that includes lifestyles, living conditions, and environmental and socioeconomic determinants of health and healthcare; the third is it adopts collective approaches. Public health does not look at health problems from an individual perspective but intervenes in the context and circumstances of groups of individuals. The growing interest today in public health is related to the processes of globalization that clearly demonstrate that disease, health, and care in all parts of the world are interconnected. Most of the world's population lives in poverty and this is associated with higher morbidity, lower life expectancy, and healthcare that is by and large inaccessible. Healthcare has a number of new challenges to address as a direct consequence of globalization. One challenge of globalization is pandemics as new infectious diseases rapidly spread throughout the globe such as avian flu (which starts in poultry in Asian countries and then infects people who take it to other countries) and COVID-19 (which likely emerged in a wet market in the Chinese province of Wuhan and then rapidly spread across the entire globe and left millions of people infected in its wake). Another challenge of globalization is food safety. Most food today is produced in factories that are often located in other countries. Although food production is controlled and regulated in some countries, this is not the case in many other countries. Moreover, practices can differ between countries as borne out by some countries allowing growth hormones and antibiotics to be used in cattle breeding (more than half of all antibiotics used in the world are applied in bioindustry). An example of a food safety issue was the bovine spongiform encephalopathy (BSE) epidemic in the United Kingdom in 1996, which was caused by prions and spread as a result of the meat and bones of dead animals being ground into food for cattle. Human beings were infected as a result of consuming prion-diseased beef and contracted Creutzfeldt–Jakob disease (a serious neurological disorder). This resulted in countries worldwide banning the import of British beef (Russia has only recently lifted its ban). A further challenge of globalization is biosecurity in that pathogens and toxins developed in labs to address health problems can escape or be deliberately produced as biological weapons. Yet another challenge of globalization relates to humanitarian aid. The increasing numbers of disasters have increased the role played by humanitarian organizations such as Doctors without Borders. Healthcare in some developing countries is almost completely sustained by NGOs. Although the ethical motive to

H. ten Have and M. do C. Patrão Neves, *Dictionary of Global Bioethics*,
https://doi.org/10.1007/978-3-030-54161-3_428

assist is difficult to criticize, new ethical issues have emerged here such as triage. When a disaster produces many victims, choices have to be made to prioritize the recipients of first aid. Another issue concerns whether the focus of assistance should be on immediate life-saving efforts or longer term perspectives (such that the health infrastructure will be strengthened and thus help people cope with future emergences).

Publication Ethics

Publication ethics refers to the ethical standards that apply to the publication of scholarly work aiming to promote the highest quality of scientific work and to guarantee genuine authorship. Publication is the last stage in research projects that have often entailed a long process and heavy investment (time, energy, and a variety of resources). Furthermore, it is important to point out that published works and only published works are recognized as scientific production. Although publication is therefore the major (the ultimate) goal (publish or perish) of scientific research, the number of citations a publication gets and similar respective metrics are arguably more important. Indeed, measurement of the impact of published research (the number of citations of a journal article, and author) in the field of study is decisive when applying for grants or specific positions. Although the need to get published has reinforced the publishing industry, publishers are today heavily charging researchers to publish their work on their open access websites. The amounts charged vary according to the journal's index factor and many researchers cannot afford to get published in higher ranked journals. Such a situation does not of course necessarily contribute to better science or to disseminating science more widely. Moreover, such a situation has made it possible to identify publication practices that for different reasons are considered ethically unacceptable or even scientific misconduct. Although most practices considered scientific fraud have been banned by research integrity codes such as FFP (fabrication, falsification, and plagiarism), many other practices are said to be questionable. The Organization for Economic Cooperation and Development (OECD) refers to a number of questionable practices such as research practice misconduct, data-related misconduct, personal misconduct, financial and other misconduct, and publication-related misconduct. A number of concerns directly relate to the publication process such as supervisors taking advantage of a student's work; center directors demanding authorship for all the work produced by the center (despite maybe not being involved); peer reviewers profiting from the information gathered in their reviewing task or delaying another's publication for their own advantage; multiple submissions, redundant publications, duplicate publications, and self-citation; and institutions possibly being complicit in improper behavior in the publication process because they too benefit from a high publication score (thus facilitating access to research funds and profitable research partnerships). The international Committee on Publication Ethics (COPE) consists of editors, publishers, authors, and other interested parties and has established a code of conduct and best practice guidelines for authors and editors when faced with publication misconduct.

H. ten Have and M. do C. Patrão Neves, *Dictionary of Global Bioethics*,
https://doi.org/10.1007/978-3-030-54161-3_429

Quality of Care

The WHO defines quality of care as "the extent to which health care services provided to individuals and patient populations improve desired health outcomes." Although quality of care can only be brought about if it is safe, effective, timely, efficient, equitable, and people centered, it is nevertheless regarded as key to the human right to health. Quality of care used to focus on the structure of the healthcare system, the process employed to provide care, and the outcomes of the care provided. It built on a model developed by Avedis Donabedian (the father of quality assurance) that distinguished between structure, process, and outcome. Another related approach to quality of care distinguishes between three perspectives to define, assess, and improve quality of care. The user perspective emphasizes satisfaction and outcome; focuses on expectations and preferences (patient demands); takes into account such factors as support, information, and choice. The provider perspective emphasizes professional standards and guidelines; focuses on needs as defined by professionals; takes into account such factors as effectiveness, accessibility, continuity, technical competence, and personal attention. The policy perspective emphasizes the system and organization of care; focuses on the context in which care is delivered; takes into account such factors as availability, efficiency, and organization. Striking a balance between these three perspectives results in effective quality of care. The Donabedian Model of care allows construction of a more detailed framework of relevant factors each of which can be refined using the three perspectives of users, providers, and organizations (policy-makers in universities, hospitals, and professional bodies) emphasizing respectively perception, performance, and effectiveness. The user perspective calls for clinical methods to aim at interaction, communication, and deliberation; the provider perspective articulates integration; and the policy perspective emphasizes accountability. Once the relevant factors of quality of care are identified steps can be taken to enhance such care and various activities can then be undertaken such as development of standards, guidelines, and arrangements for quality of care; documenting and assessing care practices; analyzing discrepancies between what is desired and what is achieved; assuring and improving quality of care (i.e., improving the balance between quality desired and quality achieved). Quality of care and ethical principles go hand in hand. Although patients expect effective quality of care and institutions often brag that they provide the best care possible, the various factors involved in quality of care frequently call for trade-offs to be made between goals and values. The effectiveness of maximizing outcomes is sometimes not compatible with patient-centered care and sometimes fails to satisfy patient preferences because access to care can be impeded by the structure of the healthcare system such that safe and timely care is not provided and the standards of care cannot be applied in a generalized manner

H. ten Have and M. do C. Patrão Neves, *Dictionary of Global Bioethics*,
https://doi.org/10.1007/978-3-030-54161-3_430

and hence need adapting to the specific circumstances of each patient. Quality improvement methods therefore require ethical analysis. In addition to everyday healthcare often not complying with the ideals of quality of care there are serious problems such as medical errors, unnecessary surgery, inappropriate use of medication, inadequate prevention, avoidable deaths, and long delays in the provision of care. Although such deficiencies make improving quality of care an imperative, the measures taken to do this should be applied within an ethics of care framework such that they do not treat the patient unfairly, waste scarce resources, or cause harm.

Quality of Life (*See* Life, Quality of; QALY)

Quality of life has become a common factor in discussions today about healthcare and bioethics. The WHO defines it as "an individual's perception of their position in life in the context of the culture and value systems in which they live and in relation to their goals, expectations, standards and concerns. It is a broad ranging concept affected in a complex way by the person's physical health, psychological state, personal beliefs, social relationships and their relationship to salient features of their environment." Quality of life is a controversial notion in that there is no agreement about what it means, who is qualified to evaluate it, and what such an evaluation entails. Quality of life is a broad notion used in various contexts such as an ethical principle to assist decision-making about whether to withhold or withdraw medical treatment and as a criterion to determine the competence of a patient to request euthanasia or choose assisted suicide. Several attempts have been made to quantify or measure quality of life at different levels such as quality of life in general (WHOQOL 100), quality of health (Sickness Impact Profile/SIP 136, SIP 68), and quality of health within a specific chronic disease (diabetes DQOL). Although attempts have been made to determine quality of life by applying different metrics such as the Quality of Life Scale (QOLS) first defined in the 1970s by American psychologist John Flanagan, there are many other instruments that do the same such as the quality-adjusted life year (QALY) model that measures the effect a specific clinical intervention has on the patient's quality of life for the rest of his or her life. However, quality of life is deemed to be a subjective concept that can only be invoked by the self—not by someone else such as a health professional unless it is possible to reduce or eliminate the subjectivity and present it as fundamentally objective. Although efforts to measure quality of life empirically determine what quality of life is, as a guidance for medical decision-making it often operates as a normative concept that identifies the goal of intervention. Empirical measurement of quality of life can thus provide helpful information, but cannot in itself provide normative guidance since what "ought to be" cannot be derived from what "is."

H. ten Have and M. do C. Patrão Neves, *Dictionary of Global Bioethics*,
https://doi.org/10.1007/978-3-030-54161-3_431

QALY (*See* Quality of Life)

QALY stands for quality-adjusted life year and is a unit of measurement used to determine a patient's health-related quality of life and the length of life expected to be lived and at the same time to assess the utility of a specific clinical treatment for a concrete disease in that patient. By primarily valuing health outcomes (health gains) it is a means of economic evaluation (economic worth) that is frequently used to analyze the cost-effectiveness of health outcomes compared with the economic costs associated with different possible clinical interventions (medical economics). QALYs are thus clearly linked to utilitarian moral theory that ultimately justifies the morality of an action by the overall good it produces. The QALY model measures the good each proposed clinical procedure is likely to produce, is thus helpful when it comes to making decisions, and gave rise to a series of tools to assess the health states of patients such as the Health Index in the United States and the EuroQol Five-Dimensional Questionnaire (EQ-5D) in Europe. The QALY model has been greeted as an important tool to provide objective and impartial grounds for decision-making processes, contributes to optimizing the management of healthcare facilities, and has been particularly useful (by evaluating medical futility) in selecting patients for transplantation or admission to intensive care units. However, the QALY model has been criticized because it is based on ambiguous notions such as health (whether health should be defined broadly following the WHO definition or narrowly as happens within everyday practice) and the objective assessment of what is subjective by nature. Although quality of life is subjective, it raises a number of questions: Who determines what constitutes quality of life? What criteria should be used? Can these criteria be equally applied to different patients? How does the methodology of assessment influence the result of the evaluation? The ever growing economization of healthcare is a serious ethical concern in that overconcentrating on the economic perspective tends to blur other fundamental values in the field of health. Although the QALY model standardizes care that needs to be personalized for it to be humane and is thus a helpful tool in healthcare, it should not be the only criterion or the decisive one in any decision.

H. ten Have and M. do C. Patrão Neves, *Dictionary of Global Bioethics*,
https://doi.org/10.1007/978-3-030-54161-3_432

Refugees

More than 44,000 people are forcibly displaced each day because of war, violence, and persecution. Included in this number today is a new and recent category of displaced people called climate refugees. Worldwide there are 68.5 million displaced people of whom 40 million are internally displaced, 25.4 million are refugees, and 3.1 million are asylum seekers. Such a high level of displacement has never previously been recorded (a good way of looking at it is that someone is forcibly displaced almost every 2 seconds). The Convention Relating to the Status of Refugees adopted by the United Nations in 1951 defines a refugee as "somebody who is forced to leave home because of a well-founded fear of being persecuted for reasons of race, religion, nationality, membership of a particular social group or political opinion, is outside the country of his nationality and is unable or, owing to such fear, is unwilling to avail himself of the protection of that country; or who, not having a nationality and being outside the country of his former habitual residence as a result of such events, is unable or, owing to such fear, is unwilling to return to it." The convention goes on to define the rights of refugees (the right to work, to housing and education, to freedom of movement and religion) and the responsibilities of states. A fundamental principle is non-refoulement that stipulates refugees should not be returned to a country where they face serious threats to their life or freedom. Refugees are considered a special category within migration and displacement and a clear distinction is made between refugees and migrants. The latter are people who have left their country for a number of reasons such as poverty or lack of work—not because of persecution. Most displaced people are internally displaced because of civil war or violence. Healthcare is problematic when it comes to displaced people as a result of medical professionals often being involved in evaluating asylum claims and then having to balance the needs of individual persons with formal immigration policies; bioethical dilemmas arising when displaced populations need treatment (refugee camps are frequently sources of diseases that may develop into a threat to the general population); resources being limited when disasters occur and people are displaced; and priorities needing to be identified and triage applied. Although the refugee crisis is a global problem, it has not received much attention bioethically. This is strange because whenever people are in need and it is a matter of life or death the ethical thing to do is to accept refugees (especially when the risks and sacrifices of accepting countries are often low). History has shown that countries by and large benefit enormously from migration and when countries with the capacity to help do not assist, other countries that do not have such a capacity are obliged to take on an unfair burden. Refugees are often stigmatized today as potential terrorists and criminals in Western nations that

H. ten Have and M. do C. Patrão Neves, *Dictionary of Global Bioethics*,
https://doi.org/10.1007/978-3-030-54161-3_433

are directly or indirectly responsible for making people refugees as a result of such nations' belligerent activities. Although many people undertake risky and dangerous journeys and thousands of people drown in the Mediterranean, search and rescue missions have been scaled down. However, it should be kept in mind that rescuing people is a moral duty. Since many governments have decided not to honor the moral duty they have to rescuing people NGO vessels are now playing a growing role in doing so.

Regenerative Medicine

Regenerative medicine aims at replacing, engineering, and regenerating human cells, tissues, and organs to restore normal functioning. It is a new branch of medicine that has emerged during the last two decades and shows great promise in healing, repairing, and replacing damaged tissues and failing organ systems. Regenerative medicine is an interdisciplinary field that combines biotechnologies such as tissue engineering, stem cells, molecular biology, and bioprinting with the objective of developing therapies that can replace tissues that have been damaged by disease or degraded by ageing. One approach is to engineer tissues and organs to stimulate the repair mechanisms of the body to heal previously irreparable tissues and organs. Another approach is to grow tissues and organs in the laboratory and implant them in the body. Such an approach is expected to solve the problem of organ scarcity for transplantation and that of organ rejection since these newly regenerated cells are derived from the tissues of the patients themselves. Regenerative medicine also hopes to produce artificial organs. Some regenerative therapies such as for wound healing and orthopedics are approved by the FDA and commercially available. An example is Carticel, a biologic that uses autologous chondrocytes to treat focal articular cartilage defects. The cells are harvested from the patient's cartilage, expanded, and then implanted at the site of injury. Regenerative medicine has shown a great deal of promise in the treatment of chronic diseases and the deficiencies of ageing. Many new devices are currently in the stage of preclinical and clinical research. Specialized journals and research institutes have been established such as the Center for Regenerative Medicine at the National Institutes of Health in the United States (in 2011). The ethical issues of regenerative medicine are similar to those arising in biotechnology in general. Special attention is given to ethics issues of stem cell research when human embryos are used. Most of the approaches of regenerative medicine, such as stem cell therapy, gene therapy and tissue engineering are still in the promissory stage. Although they will be classified as experimental treatments focused on small cohorts of patient and on rare diseases with unknown long-term risks and potentially high costs (once they have been fully explored), they will probably not be used in routine medical practice. Particular ethical problems arise in connection to enhancement. Protagonist of enhancement frequently argue that renewal and rejuvenation of tissues and organs will not only be useful for treating diseases or injuries but also to eliminate the problems of ageing such that regenerative medicine will be a powerful force of life extension.

H. ten Have and M. do C. Patrão Neves, *Dictionary of Global Bioethics*,
https://doi.org/10.1007/978-3-030-54161-3_434

Regulation (EU) on Clinical Trials

The European Union regulated clinical trials long before the most recent regulation entered into force (2014) as borne out by issuing three earlier directives: Council Directive 65/65/EEC of 26 January 1965 on getting member states to approximate the provisions laid down by law, regulation, or administrative action relating to proprietary medicinal products (labeling, marketing, pharmaceutical products); Council Directive 75/318/EEC of 20 May 1975 on getting member states to approximate laws relating to analytical, pharmacotoxicological, and clinical standards and protocols in the testing of proprietary medicinal products; and Directive 2001/20/EC of 4 April 2001 of the European Parliament and of the Council on getting member states to approximate the laws, regulations, and administrative provisions relating to the implementation of good clinical practice in the conduct of clinical trials on medicinal products for human use. These three directives are totally coherent with each other and chronologically highlight the extent to which the drive to regulate clinical trials has intensified when it comes to safety of and respect for ethical principles throughout the European Union. From the legal perspective (i.e., according to European Treaties), directives have to be transposed by all European member states into their own legal system. However, such a transposition process allows member states to introduce changes they consider suitable or justifiable from a national perspective. Therefore the text adopted by each member state can present some significant differences and as a result clinical trials are not equally regulated in all member states. In 2012 the European Commission launched the Clinical Trial Regulation (Regulation (EU) No. 536/2014), which entered into force in 2014. The principal objective was to ensure the same ruling applied to clinical trials throughout the European Union thus harmonizing assessment and supervisory processes. Indeed, regulations forbid member states from making changes during adoption processes. Moreover, the European Commission wanted to speed up the process of approval of clinical trials; guarantee there would be a single process concerning clinical trials valid for all member states; cut the red tape; and reduce administrative costs. Such measures should increase the efficiency of clinical trials, avoid unnecessary duplications, and foster biomedical research and innovation in the European Union. In short, the goal is to create an environment not only favorable but also attractive to conducting clinical trials in the European Union. Such an environment would profit from the possible availability of next-generation drugs and cutting edge medical devices in addition to all the direct and indirect financial revenues associated with clinical trials. The EU Regulation on Clinical Trials relies on the Clinical Trials Information System (CTIS) as

H. ten Have and M. do C. Patrão Neves, *Dictionary of Global Bioethics*,
https://doi.org/10.1007/978-3-030-54161-3_435

the single entry point for submitting clinical trial information. The European Medicines Agency (EMA) will set up and maintain the information system in collaboration with member states and the European Commission and make information stored in the system publicly available and subject to transparency rules. Although the regulation was adopted and entered into force in 2014 (as mentioned above), the CTIS is not yet fully functional. The regulation becomes applicable 6 months after the European Commission publishes notice of its confirmation.

Rehabilitation

The WHO describes rehabilitation as "a set of interventions needed when a person is experiencing or is likely to experience limitations in everyday functioning due to ageing or a health condition, including chronic diseases or disorders, injuries or traumas." The word "rehabilitation" derives etymologically from the Latin *re* (again) and *habitare* (to inhabit, to dwell, to live). Its general connotation is to restore something that has fallen into disrepair. The term is used in a number of settings such as addiction (when someone learns to live without addictive substances), criminal law (when convicted offenders learn to be reintegrated in society), environmental policies (when forests are restored and urban slums are turned into homes), and media (when the reputation of someone is restored). Rehabilitation is an interdisciplinary approach in healthcare in which physicians, nurses, and physical, occupational, and speech therapists work together to reduce disability, improve quality of life, get individuals to participate in daily life, and encourage functional independence. Although the damage done by disease or trauma often cannot be undone, the individual can be restored to as much health, functioning, and well-being as possible. As a result of non-communicable diseases and ageing proliferating worldwide, the WHO argues the importance of fulfilling the ever growing need for rehabilitation. However, in many countries there is a shortage of trained professionals to provide rehabilitation services and such a global need goes unmet. Resource-rich countries are not free from ethical conflicts relating to concerns about resource allocation and disagreements about medical or institutional practice, rights of patients, and payment issues. Other ethical issues relate to the decision-making capacity of patients. Although the autonomy of patients should be respected, in some cases decision-making capacity can be compromised such as happens with elderly or mentally ill patents and patients with head injuries. Rehabilitation is process based, takes time, and thus requires regular evaluation of the competence of patients. However, it should be kept in mind that rehabilitation cannot achieve optimal outcomes without the active participation of patients.

H. ten Have and M. do C. Patrão Neves, *Dictionary of Global Bioethics*,
https://doi.org/10.1007/978-3-030-54161-3_436

Religion and Bioethics (*See* Bioethics and Religion)

When bioethics emerged as a new discipline in the 1970s, it was strongly connected to religion. In the United States the major religious influence came from Protestant theologians, whereas in Europe Catholic theologians were the most influential in the development of bioethics. All major religions pay a lot of attention to ethical matters and moral conduct. Theologians were among the founding fathers of the discipline of bioethics. However, when bioethics developed into a discipline of applied ethics, theological and religious discourse gradually became marginalized. Although religions differ regarding the amount of morality they espouse, they agree on the nature of morality. They regard ethics as a matter of truth transcending human communities and universally binding on all people. However, religious traditions use a number of sources when making ethical judgments such as sacred texts that reveal the moral order (the Torah or the Bible), institutions and traditions (the hadith in Islam provides moral guidance), human reason (rationality is a resource given by God to achieve moral judgments that are right), natural order (natural law is built into the structures of creation), and religious experience (a source of moral values). Religious traditions have struggled to identify common values especially when confronted by globalization. In 1993 approximately 200 leaders from more than 40 religious and spiritual traditions gathered together at a Parliament of the World's Religions (there have been a few such gatherings—the first in 1893) and signed a statement titled Towards a Global Ethics. This statement was drafted by German theologian Hans Küng and declared that all traditions share common values such as respect for life, solidarity, tolerance, non-violence, and equal rights. The document emphasizes the importance of showing what the world religions have in common rather than pointing out how they differ. It affirms that ethics should always take priority over the market economy. Since the principles identified in the statement offer guidelines for global bioethics the document can be regarded as a precursor of the Universal Declaration on Bioethics and Human Rights.

H. ten Have and M. do C. Patrão Neves, *Dictionary of Global Bioethics*,
https://doi.org/10.1007/978-3-030-54161-3_437

Reproductive Autonomy

Reproductive autonomy refers to the right women have to full control over their bodies (they bear the children) and particularly for reproductive purposes (i.e., control over all aspects of reproduction). This entails society acknowledging a number of rights women have such as contraception (whether and when to use contraception and which contraceptive method to use), abortion (whether and when to continue or discontinue a pregnancy), reproductive technologies (if needed), sterilization (if wanted), and all procedures in family planning. Although reproductive autonomy directly benefits women, it also empowers them, allows them to take charge of their own lives, benefits society as a whole, promotes general well-being, and contributes to physically and psychologically balancing life within the family and the community. Reproductive autonomy can also be claimed by men. If a man decides not to become father, then it is his responsibility to ensure his partner does not become pregnant. However, if he does decide to become a father, then he will have to rely on a woman for her oocytes and womb irrespective of whether reproductive technologies are used or not. Although reproductive autonomy is recognized in the West, it is often not in developing countries for a number of reasons such as the status of women being subordinate to men, the rule of law (forbidding some of the procedures that guarantee autonomy), the level of education (lack of awareness of rights or of confidence in making informed decisions), and poverty (no or little access to family planning). Although methods have been proposed to measure the reproductive autonomy of women, some of these reasons still apply in developed countries despite reproductive autonomy levels being generally higher. In the West there are other problems that arise as borne out by the Hastings Center running a couple of projects about them. The first wanted "to understand why fertile patients were more likely than other women in the United States to give birth to twins, triplets." The conclusion was that it was a result of the autonomy of women who did not mind having multiple births because it increased their chances of getting pregnant. The second project wanted to understand why routine prenatal testing did not respect the reproductive autonomy of women. The conclusion was that often there was not enough information about the testing itself or of possible consequences that had previously been provided. The main argument against absolute reproductive autonomy concerns the rights of the zygote, the embryo, or the fetus. Should the rights of children-to-be be allowed to impose responsibilities on parents-to-be and should they be allowed to restrict autonomy? If so, then protection of human life from conception to birth will forbid abortion and at the same time encourage family planning.

H. ten Have and M. do C. Patrão Neves, *Dictionary of Global Bioethics*,
https://doi.org/10.1007/978-3-030-54161-3_438

Reproductive Ethics

Reproductive ethics is a broad field covering all moral questions related to the medicalization of human reproduction. The four major chronological events that raise ethical questions here are pre-generation, generation, pregnancy, and birth. The most relevant ethical issues raised at pre-generation (before conception) are family planning and all situations that prevent family planning from being freely and easily available to everybody. There are many issues that occur during generation (after conception) such as contraception and abortion (access and methods); infertility and assisted reproductive technologies (artificial insemination, in vitro fertilization, intracytoplasmic sperm injection, gestational surrogacy, and commodification of women); cryopreservation, gamete and embryo banks, surplus embryos (synthetic gametes, donation or adoption of embryos); preimplantation genetic diagnosis; sex selection of embryos or fetuses (designer babies). Each of these topics has ethical problems of its own. Most of the ethical issues that arise during pregnancy are related to the impact maternal behavior (smoking or taking drugs) has on the health of the fetus. The ethical issues that characterize birth vary from prematurity to viability. The introduction of artificial procedures to biotechnologically manipulate conception and birth (the most natural and universal human events) raised many questions from the outset. Bioethics contributes to clarifying what is really at stake in each situation by presenting different arguments and comparing them and has attempted to reach a dynamic consensus that respects human dignity (as a personal value) and social justice (as a communitarian value).

H. ten Have and M. do C. Patrão Neves, *Dictionary of Global Bioethics*,
https://doi.org/10.1007/978-3-030-54161-3_439

Research (*See* Clinical Research; Research Ethics)

Research refers to the detailed, meticulous, systematic, and thorough study of a specific object and has the aim of discovering new information or confirming a hypothesis or interpretation by describing, explaining, predicting, and controlling the object observed. There are numerous fields of research ranging from the humanities, arts, and social sciences (soft sciences) to biology, physics, and mathematics (hard sciences). This has led to research methods varying according to the field under consideration and the goal of such research. Research can be deductive (reasoning starts with a general statement or hypothesis and examines whether a specific logical conclusion can be reached) or inductive (reasoning makes broad generalizations from specific observations). Research can be qualitative (observational, gathering non-numerical data, understanding problems by conducting inquiries) or quantitative (gathering numerical data and discovering patterns). Research can also be categorized as basic (main goal is to develop knowledge), applied (focused on concrete challenges that require a solution), and problem oriented (addressing specific problems). Research procedures and methods historically adopted a narrow approach to experimental research in the early 19th century as borne out by their following and strengthening a positivist orientation that believed in the importance of distinguishing between facts and values. It was believed that research strictly referred to facts and that values should be discarded as jeopardizing the objectivity of science. Such an estrangement between facts and values and between scientific research and ethical thought was considered an imperative back then and self-regulation of researchers was considered enough. However, some major events throughout the 20th century showed that self-regulation was not enough to prevent abuses against humanity. Although dropping atomic bombs on Hiroshima and Nagasaki was considered by many a major step forward for physics and believed to have shortened the war, it was considered a crime against humanity by others. Although the biomedical knowledge gained by Nazi doctors experimenting on Holocaust victims may have built some perverted sense of a scientific advance, it was a dreadful abuse of humanity. Ever since those horrific days different strategies have been adopted to bring about hetero-regulation (i.e., regulation by others) of research in which funding is closely evaluated and ethically assessed. More recently the importance of research integrity has become paramount with the development and specification of integrity rules and making it a legal requirement that they are complied with. When people mistakenly consider research integrity as synonymous with research ethics and sufficient for the ethical surveillance of scientific research, then it falls once again into the ever deepening "rabbit hole" that self-regulation of research represents.

H. ten Have and M. do C. Patrão Neves, *Dictionary of Global Bioethics*,
https://doi.org/10.1007/978-3-030-54161-3_440

Research Ethics, Animal (*See* Animal Research)

Animals have been used in scientific research since ancient times. Guinea pigs have long been used in anatomical and physiological studies with scant regard given to their interests. Such a situation persisted until the 20th century when things started to change (albeit reticently) after publication of *The Principles of Humane Experimental Technique* by William Russel and Rex Burch in 1959. Since then strict rules have been developed determining whether animals should be used in biomedical research. Such rules forbid some types of research and painful procedures as borne out by some species being protected and excluded from research, the number of animals involved in research decreasing, and concerns about animal well-being and the assistance they are given after use in the labs increasing. Although the tendency today is to improve the protection and well-being of animals and to reduce animal experimentation, it still persists. The use of animals in biomedical research has benefited the development of biomedical sciences and biotechnologies that have been shown to be essential to life-saving discoveries and to have contributed much to finding causes, diagnoses, and treatments for many diseases. Animals share most of their DNA with humans (e.g., mice share more than 98% with humans), suffer from many of the same diseases that affect humans (e.g., cancer, diabetes, and heart disease), and are very useful in studying the evolution of diseases as a result of animals having a shorter lifespan. Antibiotics, vaccines, and organ transplantation are all first studied in animals. Some diseases involve complex physiological processes (e.g., asthma and cystic fibrosis) that need to be studied in a live animal. Such studies led to the use of animal models for human pathologies and to the production of transgenic animals and more recently hybrid animals or chimeras. The main question today is whether animal experimentation should be banned totally, especially since today there are sophisticated computer systems (capable of reproducing the evolution of human diseases), mathematical models, and tissues and cell cultures at the organic level. Although such alternatives are increasingly used (lowering expenditure and time spent on research) and live animals decreasingly used, for the time being a total ban is unlikely because animal experimentation is vital to lowering the risks for humans during clinical trials (like those trying to find vaccines to fight COVID-19).

H. ten Have and M. do C. Patrão Neves, *Dictionary of Global Bioethics*,
https://doi.org/10.1007/978-3-030-54161-3_441

Research Ethics, Clinical Research

Clinical research is an important part of biomedical research. Its aims include the production of new knowledge and innovation and gathering evidence in support of the safety and effectiveness of medical products and devices. It is part of the general goal of biomedical sciences to better understand, prevent, diagnose, and treat illness and does so by translating basic research into new health treatments (testing the theory behind different possible applications). Therefore clinical research (applying new knowledge and innovation to healthcare) differs from basic research (developing new theories and knowledge) and from translational research (connecting different areas of research with the aim of maximizing research by making its application more effective). However, despite the differences they are all important and complementary. The two major types of clinical research are interventional studies or clinical trials that test new drugs or treatments for a disease; and observational or natural history studies (a.k.a. cohort studies or epidemiological studies) that collect health information with the aim of better understanding a disease and its development. The very fact that both types are carried out on humans (healthy or sick) makes them highly sensitive. Ethical concerns focus primarily on the safety of the people involved and on issues related to their privacy such as restricting access to personal information and using personal data. Other concerns relate to the quality of the research project and the competence of the researchers. From a global perspective the main concern relates to countries complying with the same ethical standards regardless of where the research is undertaken or of the population involved.

H. ten Have and M. do C. Patrão Neves, *Dictionary of Global Bioethics*,
https://doi.org/10.1007/978-3-030-54161-3_442

Research Ethics, Data Sharing

The sharing of research data has long been an aim of institutions and publishers and has led to the development of new models of scientific research called data-sharing policies based on the ethical obligation to make data freely available or as unrestricted as possible. Such models were first proposed and implemented by public funding institutions and then increasingly by publishers that considered research data a public good and wanted to bring about the ethical values of transparency, openness, and reproducibility of research and as a consequence reinforce public trust in scientific research. Data sharing has significant advantages since it results in better research (enabling data to be verified and better scrutinized and knowledge to be more accurate and sound), faster and more effective advances (avoiding failure and redundant lines of research), cost reduction (optimizing investment), and networking enhancement (facilitating innovation). In short, it maximizes resources (human and financial) applied to scientific research and benefits society. Data sharing can include raw and processed data, methods and protocols, materials and software. Although data-sharing policies should always be responsibly implemented and take into account legal, regulatory, and commercial constraints such as authorship and patenting rights (data ownership), the amount of data shared today is still low. A likely reason is the lack of adequate regulation of such policies (equally suitable to different types of research and to different kinds of data). Another reason is the reluctance of researchers to disclose findings before they are absolutely sure of their accuracy and comfortable with the different uses to which they may be subjected.

H. ten Have and M. do C. Patrão Neves, *Dictionary of Global Bioethics*,
https://doi.org/10.1007/978-3-030-54161-3_443

Research Ethics, Embryo

Embryo research refers to the study of human embryos (2 weeks after fertilization up to the 8th week). However, embryo research is only carried out up to the 14th day (a convention based on the first stage of embryonic development when the embryo has naturally completed its implantation into the uterus and splitting into twins is no longer possible). Although scientists knew that research on embryos invariably killed them, embryo research started with the development of assisted reproduction technologies (ARTs), specifically in vitro fertilization (IVF) in the 1970s. The very fact of artificially generating or producing embryos outside the human body in the lab made them readily accessible to research. The first ARTs aimed at enhancing their success by always producing more embryos than could be transferred to the womb of the woman in question and in so doing generating spare embryos. Such embryos were either discarded (immediately or when indefinite cryopreservation was not an option) or used in scientific research (adoption was not an option at that time). In some countries the principle of human dignity prevailed where it was deemed more respectful to let the embryo die naturally than to use embryonic human life as a mere object (as happens in scientific research). In other countries the utilitarian argument prevailed where it was deemed the impossibility to cryopreserve embryos indefinitely could lead to producing a higher good in which the use or embryos in scientific research could improve ARTs, develop knowledge about different human diseases, and discover new treatments rather than letting them die. Even in countries espousing the utilitarian approach embryo research was initially only performed on spare embryos. Although mostly aimed at improving IVF techniques, better understanding embryonic development, and increasing knowledge about inheritable disorders, the high expectations raised by embryo research led to the intensive use of human embryos. However, some scientific projects required precise manipulation of embryos at a very early stage as a result of embryos containing pluripotent stem cells that could be used in a wide range of biomedical research projects and of the promise they show in curing many diseases and injuries (birth defects and genetic disorders). It was with this in mind that scientists proposed producing human embryos specifically for research. This meant stretching the utilitarian principle to the point where it was acceptable that human life at an early stage could be produced solely for research purposes when it could demonstrate the overall benefits it would provide for healthcare. Although most countries accept experimentation on spare embryos, they forbid the production of human embryos solely for research purposes. Nevertheless, some countries buy embryonic stem cell lines in foreign countries for their research projects.

H. ten Have and M. do C. Patrão Neves, *Dictionary of Global Bioethics*,
https://doi.org/10.1007/978-3-030-54161-3_444

Research Ethics, Integrity (*See* Integrity)

Research ethics refers to the (theoretical) principles and norms (guidelines) that regulate (the practice of) research. It is a field of applied ethics that involves human beings and living beings generally. Moreover, it is a sensitive field as a result of the strength its impact has and the power it has to change people, societies, and the world. Although this led to researchers always trying to adopt the best scientific and ethical practices, research has nevertheless been traditionally regulated by researchers themselves. Such a situation changed in the aftermath of the Second World War when it became clear that despite the impressive scientific achievements of physics and biology, they also produced human tragedies as a consequence of using atomic bombs and medical experimentation. It was no longer possible to rely on self-regulation and societies started to establish hetero-regulation (i.e., regulation by others) of research and did so by ethical review. As research became ever more broad, diverse, accurate, effective, deep, and transformative in the impacts produced, it then took on an economic, financial, and political nature. Such a context led to pressure being applied to researchers to present good research data and resulted in shortcuts being taken that endangered the accuracy and soundness of research. Moreover, research teams became larger and included experts from around the world with different academic and professional backgrounds. This new reality diluted individual responsibility by allowing institutions and countries rather than researchers to dictate research paths that did not always coincide with more thorough approaches. This increased the risk of research deviating from best practices and led the research community to make codes of conduct to regain the trust of citizens and articulate ethical norms of research such as honesty, reliability, objectivity, impartiality, independence, integrity, accountability, openness, accessibility, and fairness. These are referred to as the norms of research integrity in which integrity is currently understood both as an ethical principle and as the core of research ethics (guidelines for ethical research). Although this led to integrity often being treated as synonymous with ethics, the scope of integrity is significantly narrower than that of ethics. The most important point to stress is that ethics applies to research, while research integrity applies to research conducted to the highest possible standards, to quality research, and to better science.

H. ten Have and M. do C. Patrão Neves, *Dictionary of Global Bioethics*,
https://doi.org/10.1007/978-3-030-54161-3_445

Research Ethics, Interspecies (*See* Chimera)

Interspecies research (ISR) refers to the production of a new being called an interspecies chimera in biomedical research. It can also refer to interspecies collaborative research (a.k.a. interspecies communication research) in social sciences research and ethology research. Both approaches challenge the traditional boundaries between human and non-human animals by crossing the human species boundary. Interspecies collaborative research has the objective of understanding how different species interact with one another and with humans. Ethical concerns relate to the methods used and the consequences such research has for animal life. The tendency today is to carry out such research on animals in their natural environment (in the wild), avoid the need for a controlled setting such as captivity, and only undertake non-invasive research or research designed to primarily benefit animals. Although the creation of interspecies chimeras (human–animal and animal–animal combinations) has long been a narrative in the mythologies of many cultures and received by people with awe and fear, recent scientific advances today have increased the chances of such beings becoming a reality by being able to create interspecies transgenics, chimeras, hybrids, and xenografts. In 1988 scientists at Harvard patented the Oncomouse after successfully transferring a human cancer gene into a mouse. Although the initial goal was to study the evolution of cancer and try different biomedical approaches to cure it, researchers soon realized that interspecies chimeras could be used as a unique research model for the treatment of chronic and end-stage disease (a major goal of interspecies biomedical research). Scientific advances led to other objectives (mostly in the field of transplantation) such as the production of human tissues and organs in animals (xenotransplantation). The major ethical issues raised by interspecies biomedical research is the denaturalization (interfering with the integrity of nature, distorting it, and violating the precautionary principle), the objectification (reducing life to an inert passive reality under the control of human beings thus disrespecting the intrinsic value of life), and the instrumentalization (disposing of life as an instrument totally subordinated to the intentions of human beings thus raising questions about social responsibility) of life overall. In conclusion, extreme caution should be taken when undertaking ISR because the very identity of humans is at stake. Moreover, the well-being of animals should be paramount and such research should only be pursued when the principles of beneficence and freedom of research can be met.

H. ten Have and M. do C. Patrão Neves, *Dictionary of Global Bioethics*,
https://doi.org/10.1007/978-3-030-54161-3_446

Research Ethics, Research Ethics Committees (*See* Institutional Review Boards)

The historical origin of research ethics committees (RECs) can be traced back to the 1970s when it became increasingly clear that biomedical research needed some form of monitoring. Such committees expanded worldwide, took on a more transdisciplinary and pluralistic nature, and their conclusions were deemed binding when they evaluated scientific research projects involving humans (particularly in clinical trials) as borne out by clinical trials today only being executed after approval by a REC. The reason for implementing RECs worldwide was to prevent some countries being excluded from international biomedical research and from direct and indirect revenues (better healthcare, capacity building, economics). The original institutional structure or role played by RECs was to guarantee that research complies with the best scientific and ethical standards and that human research participants had their rights and well-being protected. Over the years (mostly in the 21st century) RECs have expanded into other fields of research such as soft sciences and observational studies. Although ethical issues may be different in these other fields, the social requirement of sound research and the need to protect the rights of participants remain paramount. Moreover, in most countries RECs are no longer limited to healthcare or to research institutions (as institutional review boards are), have become mandatory in universities where research is wide-ranging, and most importantly operate today on a regional or national basis. Although the advantage of having RECs worldwide is that research evaluation adheres to similar requirements, gains consistency, and increases its legitimacy, it is also important to have local RECs that can monitor ongoing studies. As the number of RECs get ever larger this raises the important issue of the competence of its members. Without the necessary skills, members turn their attention from the important matter of ethical evaluation to legal matters such as informed consent leading to ethical evaluation being relegated to little more than an administrative procedure.

H. ten Have and M. do C. Patrão Neves, *Dictionary of Global Bioethics*,
https://doi.org/10.1007/978-3-030-54161-3_447

Research Policy

Scientific research and innovation today are widely recognized in Western countries as critical to the development of societies and to the well-being and quality of life of people as borne out by the majority of developed countries increasing their investment in science and innovation. Although research policies have long strongly favored biological sciences (particularly biomedical sciences), more recently greater attention has been paid to the digitization of society. Although investment in research is highly beneficial overall, it is important to develop policies that prevent potential unintended consequences while enhancing the benefits. There are a number of major challenges that research policies have to address. The first concerns the fields in which investment is made and bringing about research development that is balanced. Research policies most often focus on technological areas, experimental sciences, and quantitative studies (where impacts are faster and measurable) and do so at the expense of other domains of research thus contributing to knowledge and culture being developed in an unbalanced way and reducing critical thinking (particularly needed in the current era of information). The second challenge is the intergenerational gap and the need for an inclusive policy. Although younger generations are into research and love the speed at which it proceeds, this alienates older generations leaving them behind and worsening social injustice. The third challenge for research policies revolves around whether international relations should be dictated by competition or by cooperation. Competition leads to brain drains and the impoverishment of countries such that other countries can reap the benefits. Cooperation can benefit all countries by sharing resources between each of them thus avoiding a breakdown in international justice.

H. ten Have and M. do C. Patrão Neves, *Dictionary of Global Bioethics*,
https://doi.org/10.1007/978-3-030-54161-3_448

Resource Allocation

Resource allocation refers to assigning available resources for various uses and is a practice commonly applied in economics, management, and strategic planning. Resources are often scarce leading to not all projected goals being accomplished and priorities needing to be set. Resource allocation is a means of improving efficiency and using resources in the best possible way to accomplish certain ends. Resource allocation requires tasks, goals, activities, and available resources needed to reach them to be planned, specified, and divided up. Moreover, it requires continuous monitoring of the implementation of activities and utilization of resources over time. The allocation of healthcare resources has been a growing concern in recent decades in that increasing costs and growing demand have made resources ever more limited and in that healthcare resources have not been distributed in a just and equitable fashion. Although this normative challenge is based on the individual and social value of health, health is not a commodity that can be simply allocated since resource allocation needs to be ethically justified. Although there is no consensus on the best way to allocate scarce resources, a distinction is commonly made between priority setting at the macrolevel and rationing at the microlevel. At the macrolevel policy-making decisions are made about budgets available for healthcare and its various sectors and such decisions determine the resources available at the microlevel of patient care. Therefore priorities at the policy level need to be assigned when resources are limited. Several indicators are used to set priorities such as estimated health benefits, quality-adjusted life years, and cost-effectiveness as borne out by the Committee on Choices in Health Care in the Netherlands proposing similar priority principles in 1992 such as necessity (medical benefit), effectiveness, efficiency, and individual responsibility. Such principles should be applied in this order such that services can be prioritized and non-essential services eliminated from the basic healthcare package. Although rationing also refers to resource allocation, it does so at the microlevel. While priority setting ranks categories of services from essential to non-essential, rationing means that potentially beneficial treatments are withheld from some individuals as borne out by many hospitals providing a given number of surgical interventions each year but having exhausted their budgets in the final months of the year. Rationing is ethically problematic since it involves unfair treatment of individual patients and leaves the burden of patient selection to healthcare providers. Although it is argued that rationing is unavoidable, macrolevel decisions will always be ethically preferable since they involve distributing public goods and thus such decisions need to be taken at the societal level.

H. ten Have and M. do C. Patrão Neves, *Dictionary of Global Bioethics*,
https://doi.org/10.1007/978-3-030-54161-3_449

Respect for Autonomy (*See* Autonomy)

Respect for autonomy is a principle of bioethics (i.e., an ethical obligation in the life sciences) that acknowledges individual autonomy and the duty to respect, comply, and act accordingly with it. It was introduced by Tom Beauchamp and James Childress in their classic work *Principles of Biomedical Ethics* first published in 1979. At the time the authors referred to the "principle of autonomy" rather than the "principle of respect for people" used in the very same year in the *Belmont Report: Ethical Principles and Guidelines for the Protection of Human Subjects of Research.* The "principle of respect for people" required acknowledging the autonomy or self-determination people have to make their own choices, something that should never be restricted unless its consequences affected the autonomy of others. It was not until 1989 and publication of the third edition of *Principles of Biomedical Ethics* that the authors changed the "principle of autonomy" to the "principle of respect for autonomy." They underlined the fact that there is a difference between "being autonomous and choosing autonomously" and "being respected as an autonomous agent." The latter is more demanding in that it requires not only recognizing someone's capacity to act autonomously, but also taking actions to enable him or her to do so. Although respect for autonomy empowers the person, it also makes the other responsible for the practice of autonomy by fostering adequate conditions for decision making such as disclosing information. Autonomy can only be correctly understood and practiced in bioethics under this well-developed concept. It is important to keep in mind that respect for autonomy requires individual autonomy to be previously evaluated and only then is there a strict obligation to comply with the principle. Respect for autonomy does not require reasons or agreement since individual beliefs or convictions (not commonly shared) can be put forward to ground autonomy. However, since autonomy is not a condition that can be confirmed or denied, this suggests that a person can be autonomous and not autonomous at the same time. In fact, the same person can be autonomous for some decisions at a certain time and not autonomous for other issues at the same time. An autonomous person previously affected by an emotional event will take autonomous decisions in every aspect of his or her life, except when such decisions are associated with that traumatic event. Put another way, someone who is claustrophobic and usually autonomous will not be autonomous when stuck in an elevator.

H. ten Have and M. do C. Patrão Neves, *Dictionary of Global Bioethics*,
https://doi.org/10.1007/978-3-030-54161-3_450

Responsibility, Collective

Collective responsibility refers to the duty of individuals as members of a collective or a community (organization, association, group) to take responsibility for decisions taken by the collective relating to its actions or its failure to take action. Someone who belongs to a collective, ignores controversial decisions taken by it, tolerates collective options that are unfair, accepts collective choices that are offensive, and has the power to intervene and stop them but decides not to do so is responsible for the collective decision. Collective responsibility applies to all crucial collective choices (such as the death penalty, euthanasia, and surrogate motherhood) that someone could intervene in and stop but decides not to do so for some reason. Human beings are social beings and as part of a collective they share the responsibility for collective decisions unless they disagreed with them and made a genuine effort to change them. Otherwise, they are complicit in such actions of the collective and any claim of passivity in the taking of collective decisions does not excuse them from responsibility. Neutrality is never an option since the option not to take a position is in itself already a position and an option.

H. ten Have and M. do C. Patrão Neves, *Dictionary of Global Bioethics*,
https://doi.org/10.1007/978-3-030-54161-3_451

Responsibility, Concept

The word "responsibility" derives etymologically from the Latin *respondere* (to offer something in return, to reply, to answer). Although responsibility is commonly conceived today as similar to accountability and liability, it is different. Accountability refers to individuals or institutions acknowledging and accepting something for which they are answerable (i.e., accountable); responsibility refers to their being answerable for something that may not be within their remit; and liability refers to a legal responsibility. Since responsibility can be applied to all domains of human activity other than just the legal one, the concept of responsibility is broader than that of accountability or liability. Generally speaking, responsibility refers to the psychological capacity or the legal and moral obligation of answering to someone for something within an individual's or an institution's initiative or power to control or manage. Responsibility thus requires consciousness (i.e., awareness of and information about a situation) and freedom (i.e., exercising free will and acting freely).

H. ten Have and M. do C. Patrão Neves, *Dictionary of Global Bioethics*,
https://doi.org/10.1007/978-3-030-54161-3_452

Responsibility, Corporate

Corporate responsibility refers to the duties corporations, organizations, and associations (groups of people organized around a common goal and acting as a single entity) have toward individuals and groups affected directly or indirectly by their decisions and activities. Regardless of the field in which they operate, corporations are set up such that they work in a social context that can be officially recognized by society. Although they work on behalf of themselves or on behalf of their members (i.e., pursuing their own mission and rationale), they should nevertheless be aware of their impact on other groups or on society as a whole and ensure that what benefits they produce do not harm others as borne out by patient associations taking corporate responsibility within bioethics and each one being created to defend the specific interests of patients suffering from a particular health condition. Their primary concern is enhancing the provision of the best treatment, care, conditions of life for their members, patients, and families. Such associations undertake their corporate responsibility for their members (e.g., satisfying specific needs of patients suffering particular health conditions) without negatively affecting the work of other similar associations (e.g., without asking for special privileges).

H. ten Have and M. do C. Patrão Neves, *Dictionary of Global Bioethics*,
https://doi.org/10.1007/978-3-030-54161-3_453

Responsibility, General

Responsibility refers to the duty someone has to answer for what he or she has brought about. It has a long history that can be traced back to Ancient Greece. Plato and Aristotle used the Greek *aitios* (authorship, cause) to relate the actions someone takes and the effects they have to that individual taking responsibility for them in some way. However, the word "responsibility" has only become systematically used since the eighteenth century (specifically within the legal and political realm where it meant liability and accountability, respectively). It was not until the second half of the twentieth century that responsibility became an important philosophical topic and took on a moral dimension mostly as a result of the work of Emmanuel Levinas and Hans Jonas. Responsibility moved from the narrow legal domain to the broader moral realm and was given wider significance in that it referred not only to the obligation to answer for what had been done but also to the obligation to answer for what ought to have been done when the agent had the means and power to do so. Therefore, its meaning shifted from exclusive consideration of the past to contemplation of the future and from a duty toward other human beings to a duty toward other interests including those of other living beings and nature in general (environment). In short, it moved from interpersonal responsibility to global responsibility. Such reasoning argues responsibility is proportional to the level of power someone has such that the more power he or she has the more responsible he or she should be. Responsibility is also recognized as an obligation within bioethics as borne out in the Universal Declaration on Bioethics and Human Rights (2005).

H. ten Have and M. do C. Patrão Neves, *Dictionary of Global Bioethics*,
https://doi.org/10.1007/978-3-030-54161-3_454

Responsibility, Individual

Individual or personal responsibility refers to the duty everyone has to answer for what they should or should not do and for what they do or do not do. People are responsible legally and morally for their actions and their omissions (actions they fail to take). In fact, omissions can lead to legal prosecution such as when an individual has the power to help someone else but fails to do so. A good example would be a physician not stopping at a road accident despite seeing that someone is injured and knowing it is his or her responsibility (moral and legal) to help because he or she has the knowledge and skills needed. Responsibility (the duty to answer) is equivalent to possessing the power to answer. Since responsibility is based on the possession of power, it extends beyond people in the immediate vicinity and beyond direct causes. The further individual power extends, the wider individual responsibility becomes. Ultimately, it will include humanity, next generations, all living beings and their habitats such as ecosystems and the planet (referring to indirect effects that are distant in space and time). The more information people have about the consequences of their actions and omissions, the more conscious they will be about their duties, the freer they will be when it comes to decision-making, and the more responsible they will become. The dialectic relation between consciousness, liberty, and responsibility is indissoluble and the three implicate themselves reciprocally. Such a relation led to individual responsibility being contemplated together with autonomy in the Universal Declaration of Bioethics and Human Rights (2005). Article 5 refers to the principle of autonomy and individual responsibility and states: "The autonomy of persons to make decisions, while taking responsibility for those decisions and respecting the autonomy of others, is to be respected. For persons who are not capable of exercising autonomy, special measures are to be taken to protect their rights and interests." This text is interesting and rather bold by connecting responsibility to autonomy because it moves away from the Anglo-American hegemony of autonomy and gets closer to other perspectives such as the European, which considers autonomy entails responsibility thus stressing the importance of individual responsibility within bioethics. However, this is a problematic bioethical issue particularly when considered within the clinical framework. Although there is consensus that individuals are responsible for their own health, a number of questions arise: How far should this responsibility go? What kind of responsibility is at stake? Can it be decided that an alcohol addict with liver failure will not receive a new liver? The answer to the last question is simple because exclusion from the transplantation list is decided on the basis of cost-effectiveness of the organ—not as a consequence of irresponsible behavior. Although people who live a reckless life or have an unhealthy lifestyle are indeed responsible for their health condition, they are only morally so—

H. ten Have and M. do C. Patrão Neves, *Dictionary of Global Bioethics*,
https://doi.org/10.1007/978-3-030-54161-3_455

not legally. This means that there will be no legal consequences when it comes to health problems. Even if their behavior is the direct cause of the health problem, they will still receive treatment just like any other person. A similar case would be someone suffering from a genetic condition or a hereditary disease for which he or she is not responsible. In the rare event that individual responsibility has to be established in healthcare it is always done on a case-by-case basis.

Responsibility, Social

Social responsibility refers to the duty of everyone—be they individuals or collectives (public or private)—to answer for what they do, should or should not do, and do not do but should do toward the society to which they belong. All social entities (mostly businesses and enterprises) owe their existence to society and provide society with the services it needs. Their existence is thus justifiable insofar as they answer positively to the needs, expectations, and desires of citizens and communities. Social entities progressively acknowledge their social responsibility by developing appropriate measures in a number of fields such as social protection (banning the import of goods produced by child labor), environmental preservation (adopting clean energies), and animal well-being (refusing to work on experiments with animals). Social responsibility is also practiced by supporting community projects in, say, arts or sports and granting scholarships. Social entities that have the means to implement such actions contribute to the overall well-being and development of the community to which they belong and thus benefit all. The social responsibility social entities have as a result of implementing such actions is not a legal obligation but a moral one in that they are voluntary and encouraged by citizens and communities. The more citizens and communities are empowered, the more they press social entities to develop social responsibility programs that directly contribute to the common good. Social responsibility also applies to promoting adequate conditions for human health to be maintained and improved as clearly stated in the Universal Declaration of Bioethics and Human Rights (2005). Article 14 refers to the principle of social responsibility and health: "progress in science and technology should advance: (a) access to quality health care and essential medicines, especially for the health of women and children, because health is essential to life itself and must be considered to be a social and human good; (b) access to adequate nutrition and water; (c) improvement of living conditions and the environment; (d) elimination of the marginalization and the exclusion of persons on the basis of any grounds; (e) reduction of poverty and illiteracy." Therefore health is not just the responsibility of the individual. It is also a social responsibility as a result of the recognition that the increasing weight of external factors on human health such as environmental disasters seriously affects human health years after the event.

H. ten Have and M. do C. Patrão Neves, *Dictionary of Global Bioethics*,
https://doi.org/10.1007/978-3-030-54161-3_456

Resuscitation (including DNR Orders)

Cardio-pulmonary resuscitation (CPR) is often shortened to "resuscitation" and refers to emergency procedures to maintain circulation when the heart has stopped beating and the lungs are struggling to work. It is often enough to restart the heart by applying chest compressions and to restore lung function by artificial ventilation thus maintaining circulatory flow and oxygenation during cardiac arrest. The word "revival" is used for "resuscitation" in Latin countries. Resuscitation techniques were developed in 1960 in response to the discovery by Peter Safar and James Elam of mouth-to-mouth resuscitation in 1956. In 1960 the American Heart Association set up a course on CPR for doctors that combined mouth-to-mouth resuscitation with chest compression (conceived by Kouwenhoven, Safar, and Jude). In 1972 mass citizen training in CPR was organized in the United States. Since then CPR has evolved and mechanical devices have been developed. The possibility to restore heart and lung function and bring people back from the dead was truly amazing. Although CPR started to be used indiscriminately for cardio-pulmonary arrest, the survival rates (especially neurological outcomes) were poor for such patients. The limited success of CPR led to a new clinical condition in which there was physical survival but emotional death (suppressing any capacity to connect or to form a relationship) of the same person. Although it successfully halted the process of death, it failed to recover life completely and led to new, severe clinical conditions such as persistent vegetative state, coma, physical disabilities, and mental disability. Such serious and undesirable situations came about because of the difficulty in determining precisely when resuscitation should be terminated (it was often extended for too long). The cause of cardiac arrest, the patient's age, and the time spent during the attempt at resuscitation are the most important factors in determining how long the process should last. Achieving a knowledgeable and wise balance between the principle of beneficence and non-maleficence is of paramount importance. Such factors were important in determining whether resuscitation maneuvers should not be used in clinical conditions that advised against it (specifically regarding terminally ill patients). Do Not Resuscitate (DNR) orders apply when a patient decides he or she no longer wishes to prolong his or her life. DNR orders may also be issued by the attending physician who may consider resuscitation poor clinical practice in the

H. ten Have and M. do C. Patrão Neves, *Dictionary of Global Bioethics*,
https://doi.org/10.1007/978-3-030-54161-3_457

particular situation at hand. This raises a couple of ethical issues such as who gives the order (question of autonomy) and under what circumstances. A third related issue is whether non-resuscitation orders can be classified as euthanasia (active or passive). Indeed, DNR orders do not fit into the precise definition of euthanasia since the goal of non-resuscitation orders is neither to artificially prolong a life where there is no expectation of recovery nor to hasten death at one and the same time. In short, DNR orders can be proportionate and beneficial or disproportionate and maleficent depending on the clinical situation.

Right to Die

The right to die refers to the claim (based on autonomy) that individuals should be entitled to choose to end their lives under specific clinical conditions such as imminent death, terminal illness, and when they are suffering physically or psychologically. The right to die in the clinical setting entails the direct or indirect assistance of a physician or healthcare professional to guarantee death is effective and painless (in contrast to suicide that may be unsuccessful and inflict severe pain). The right to die can thus be materialized by assisted suicide or euthanasia and can be brought about by refusing therapy or by rejecting artificial nutrition and hydration. Legal formulation of the right to die in the end-of-life context came about as a result of increasingly aggressive therapeutic practices in the 1970s and 1980s (note that only a few countries have legalized the right to die). The Karen Quinlan case in 1976 was the first to establish a legal precedent on the right to die and demonstrated to the public at large exactly what the right to die was. It was legally grounded on the principle of autonomy and on ownership of one's body. Consequently, people should be free to dispose of their bodies as they see fit including at end of life. Contrary to common public perception, euthanasia cannot be required solely under the principle of autonomy as it requires the intervention of a health professional and even the provision of public services to supervise, organize, and assist with euthanasia. However, it is argued that states that give the right to die to individuals should also fulfill the obligation to provide the material conditions necessary for citizens to carry out such a right. It is also argued that the right to die conflicts with the right to life, which is legally protected. Nevertheless, legal initiatives in the United States and elsewhere have established the right to die. The specific circumstances under which this right is recognized are established by law and can be narrowly applied (terminally ill and competent patients only) or more broadly applied (patients who are suffering and at the request of others wanting to bring the suffering to an end). Many associations and NGOs are involved in the right-to-die movement and can provide information fully compliant with national laws about the right to die.

H. ten Have and M. do C. Patrão Neves, *Dictionary of Global Bioethics*,
https://doi.org/10.1007/978-3-030-54161-3_458

Right to Health

The right to health refers to everyone enjoying the highest attainable standard of physical and mental health. This right is not new in that it was already mentioned as a fundamental right in the constitution of the WHO in 1946. The 1948 Universal Declaration of Human Rights also mentioned health as part of the right to an adequate standard of living (Article 25). The right to health was again recognized as a human right in the 1966 International Covenant on Economic, Social and Cultural Rights. Although numerous international human rights documents have recognized this right since then, it was not until recently that it became influential (probably as a result of the strong criticism it received since it was first formulated because its broad concept was perceived more as an ideal than a goal). Since 2000 this has changed in that the normative content of the right to health and the obligations of states and other actors have been clarified and specified. Since 2002 independent experts, called special rapporteurs on the right to health, have been appointed by the Human Rights Council. They publish annual reports, examine national situations, and deal with individual complaints about violations of the right. Citizens and NGOs can use the growing body of legal interpretations and evidence they have built up to put pressure on governments to improve access to treatment and care. In 2000 the UN Economic and Social Council clarified that the right to health is not the same as the right to be healthy. The right should be interpreted as extending not only to timely and appropriate healthcare but also to the underlying determinants of health such as access to safe potable water; good sanitation; adequate supplies of safe food, nutrition, and housing; healthy occupational and environmental conditions; and access to health-related education and information (including that on sexual and reproductive health). Other determinants of health considered are the individual genome (natural and intrinsic), lifestyle (voluntary and extrinsic), and their interaction. Therefore the right to health is different from the right to healthcare. The right to health consists of freedoms (from non-consensual medical treatment or forced sterilization) and entitlements (the prevention, treatment and control of diseases; access to essential medicines; maternal, child, and reproductive health; equal and timely access to basic health services; and provision of health-related education and information). All services, goods, and facilities provided must be available, accessible, acceptable, and of good quality. Since states have the primary obligation to protect and promote human rights, this implies they have the obligation to implement the right to health. Although it is unreasonable to expect this to be done at once, the right to health is subject to progressive realization in which the right is brought about little by little and in accord with the resources

H. ten Have and M. do C. Patrão Neves, *Dictionary of Global Bioethics*,
https://doi.org/10.1007/978-3-030-54161-3_459

available. In addition to states having to show they are making every possible effort to better protect this right, they have other obligations too such as protecting human rights. Although states should ensure that non-state parties do not infringe on human rights, they need to pass legislation to ensure that equal access to healthcare is provided for such parties.

Right to Try (*See* Compassionate Use; Pre-approval Access)

Right to try is the name given to legislation in the United States that allows terminally ill patients to use experimental drugs that are not yet approved by the FDA. Although 40 US states have today introduced right-to-try laws, such legislation was unanimously adopted at the federal level and signed into law on May 30, 2018. Right-to-try legislation stipulates that people with life-threatening diseases have the right to use unapproved experimental drugs because they do not have the time to wait for new drugs to be tested and approved. It has the further advantage of making medical tourism unnecessary in a lot of cases such that patients no longer have to travel to China or other countries for innovative untested treatments. The only requirement such legislation insists on is that any new medication has been cleared in a Phase 1 clinical trial and thus has undergone preliminary safety testing. Once cleared, patients can find a doctor willing to try the therapy, doctors can ask a drug company for permission to test the treatment, and patients can try the treatment. The two driving forces behind right-to-try legislation are the robust discourse (especially in the United States) surrounding hope and miracles (nearly 80% of Americans believe in miracles) and the politics of deregulation. The Goldwater Institute (a conservative thinktank promoting limited government, economic freedom, and individual liberty) was instrumental in getting right-to-try legislation passed into law. The institute's aim is to reduce and eliminate the role of federal institutions (such as the FDA) as borne out by advocating the right to try since 2014 at state levels and putting forward a model bill. Criticism of the right-to-try law argues that it is deceptive, since patients are given the right to ask while pharmaceutical companies are not obliged to provide the medication requested. The only thing that the law accomplishes is that the FDA is no longer involved in access to new drugs, and such legislation is in any case unnecessary since the FDA already has several ways to expedite access to experimental treatment: expanded access (since 1987); fast-track designation (since 1988); priority review and accelerated approval (since 1992); designation as a breakthrough therapy (since 2012); and programs introduced especially for stem cell therapies such as regenerative medicine and advanced therapy (since 2016). Moreover, FDA processing is fast as borne out by the FDA receiving over 1,000 expanded access applications per year over the last 10 years, such requests being reviewed within hours or days (more than 99% approved), and the application process taking 45 min. Therefore the argument that FDA review impedes access to new medication is fallacious. There are other criticisms of right-to-try legislation: it is risky and dangerous since it is based on the mistaken assumption that experimental drugs in clinical trials are safe and hold the promise to produce miracles whereas in reality clinical experiments are often not successful (roughly 70% of early-phase clinical trials of drugs fail and over 50% at

H. ten Have and M. do C. Patrão Neves, *Dictionary of Global Bioethics*,
https://doi.org/10.1007/978-3-030-54161-3_460

Phase 3), it may undermine the science that is needed to investigate medication and determine its safety and efficacy, by expanding access it may delay generation of data to make evidence-based decisions about approval, it may deter enrolment in clinical trials, and it undermines public regulation designed to protect patients and the general public. Oversight came about and agencies such as the FDA were set up as a result of a long series of scandals in research ethics, continuous abuse and fraud in scientific investigations and publications. The system of review board and ethics committees called institutional review boards (IRBs) that has been created to prevent such scandals and abuses is now by passed.

Risk

Risk refers to the possibility of loss, injury, adverse outcomes, or unwelcome circumstances. The International Organization for Standardization defined risks as the effect of uncertainty on objectives. Uncertainties may be events that may or may not happen and may be caused by ambiguity or lack of information. Businesses face various types of risks. An example is Kodak once dominant in the photography market deciding not to develop a digital camera invented by its own engineers because it was regarded as a strategic risk to its core business. Health interventions without risks do not exist and continuous assessment is consequently required to determine which risks are acceptable and under what circumstances. Various disciplines are involved in studying risks such as administrative studies, business continuity, security and risk management, risk management and organizational continuity, investment and financial risk, and risk management for banking all of which assume that risk is not merely a quantitative notion, but has to do both with potential harm and with fairness. Therefore risk perception is an important issue. Although experts may argue that the risks of a new biotechnology are negligible, public perception may well be different. Risk ethics argues that risk should attend to contextual factors as well as formal scientific methods because the identification of risks is not a simple rational affair as it involves emotions. The two types of uncertainties that can be distinguished in bioethics are ethical uncertainties depending on cultural differences (e.g., regarding what kind of information needs to be provided in a research project) and epistemic uncertainties (e.g., whether all options for decisions have been considered and whether the consequences of decisions are known). In both cases irrespective of the decisions made risk is always involved.

H. ten Have and M. do C. Patrão Neves, *Dictionary of Global Bioethics*,
https://doi.org/10.1007/978-3-030-54161-3_461

Robotics

Robotics refers to the study of robots (i.e., machines designed, constructed, and used by humans to perform tasks either under human supervision or working by themselves as autonomous entities). Robotics is the branch of technology (in the field of engineering and computer sciences) that creates programmable machines. In 1954 George Devol, who later worked with Joe Engelberger (the father of modern robotics), designed the first programmable robotic arm called Unimate, which in 1962 became the first industrial robot at General Motors. Since then robots have been used in industry (mostly) to carry out high-precision and rapid work, perform repetitive tasks, or work in environments hazardous for humankind. Robotics and artificial intelligence today have converged, opened up the way to automation (robotics process automation), and brought about the so-called "fourth industrial revolution" that is changing the way we live and work. Advances in robotics (especially autonomous robots) raise many ethical issues and regulatory challenges that have to be addressed to prevent undesirable outcomes. The major current fear is unemployment as a result of replacing humans with robots (everything a robot is able to do, it does better, faster, and cheaper than humans). The major future fear is the empowerment of robots allowing them to overcome human ends and project their own goals. It is here that robot ethics took on its present form, which promotes ethical reflection on the social implications of robotic technology and especially in domains where social robots interact with humans. The field covered by robot ethics widens every day and ranges from elder care and medical robotics, to robots for various search and rescue missions including military robots, to waitresses in many different services such as restaurants or for entertainment purposes, through to sex robots and robotic pets. This has led to the feared rise of machines, creation of the first humanoid robots, growing general acceptance of their presence in society, ever more human tasks being taken over, and maybe making them indistinguishable from humans. In 2017 Sophia became the first robot to receive citizenship (in Saudi Arabia). In the same year the European Parliament proposed granting electronic personhood to the most advanced machines to ensure their rights and responsibilities within a set of regulations governing the use and creation of artificial intelligence. This was strongly criticized by prominent experts (more than 150 from 14 countries) in medicine, robotics, artificial intelligence, and ethics who argue it is too early to consider robots as persons and point out that doing so might lead to manufacturers wiping their hands of responsibility for creating such humanoids.

H. ten Have and M. do C. Patrão Neves, *Dictionary of Global Bioethics*,
https://doi.org/10.1007/978-3-030-54161-3_462

Safety (*See* Biosafety)

Safety refers to the condition in which someone is protected from (potential) harm, danger, risk, or injury. It is a wider notion than biosafety (which focuses on the handling and containment of infectious microorganisms and hazardous biological materials). Biosafety is basically restricted to laboratories, while safety has much wider applications as borne out by seat belts fitted to automobiles or homes made resistant to earthquakes. Safety is relative in that not all possible risks can be prevented. Safety in healthcare primarily applies to the use of new technologies, medicines, and treatment. The WHO defines patient safety as "the prevention of errors and adverse effects to patients associated with health care." As a result of adverse events and medical errors being relatively frequent, patient safety is thus a crucial element of quality of care since patients need to be able to trust that the care they receive is safe. Harm should be avoided in any healthcare delivery system by preventing errors, learning from errors, and building a culture of safety that involves professionals, patients, and organizations. Major safety concerns in healthcare are patient data and protecting such data against loss and unauthorized use (especially when stored online). The issues raised here are how can data safety best be maintained and how can states and commercial entities be stopped from abusing data. Medical data are personal information and patients want to keep such data private and confidential. Although it was traditionally the duty of healthcare providers to keep data safe, today data safety is beyond their control since many actors (ministries, insurance companies, drug stores, and pharmaceutical industries) not only wanting to access patient data but also in many cases being able to do so. Health data breaches are relatively frequent and often the result of hospital administrations getting hacked. This raises the ethical problem of who should have access to patient data and imposes an obligation on healthcare institutions to protect the safety of patient data as much as possible.

H. ten Have and M. do C. Patrão Neves, *Dictionary of Global Bioethics*,
https://doi.org/10.1007/978-3-030-54161-3_463

SARS

Severe acute respiratory syndrome (SARS) is an illness caused by an airborne coronavirus and was first identified in 2003. Coronaviruses are thought to be animal viruses (bats are the primary candidate) that jump from hosts to other animals. The first infection of humans (animal-to-human transmission) took place in the province of Guangdong (China) in November 2002. The virus is part of a family of viruses that have existed for centuries and caused a number of diseases from the simple cold to Middle East respiratory syndrome (MERS). Transmission from person to person occurs through respiratory secretions carried in small droplets of saliva coughed or sneezed into the air by an infected person mainly during the second week of the illness. SARS is highly contagious, potentially life threatening, has influenza-like symptoms such as fever, headache, malaise, body aches (myalgia), diarrhea, shivering, dry cough, hypoxia, and usually pneumonia, which requires intensive care. Accurate and timely diagnosis is very difficult when people are asymptomatic. There is currently no cure, standard treatment, or vaccine for SARS (vaccines for COVID-19 first became available in late 2020). SARS-CoV was the first widespread epidemic of the twenty-first century having affected 26 countries, infecting more than 8,000 people, causing 774 deaths (1 in 10 of the people infected), and particularly affecting people over the age of 65 (half the deaths). In 2019 there was a new SARS outbreak involving a novel coronavirus. SARS-CoV-2 is the virus that caused coronavirus disease 2019 (COVID-19), spread worldwide, and turned into a pandemic. There are a number of ethical issues raised by SARS that are common to other epidemics such as the autonomy and social responsibility of citizens; the quality of communication between states and their citizens about following rules, lockdowns, and quarantines; and solidarity among states.

H. ten Have and M. do C. Patrão Neves, *Dictionary of Global Bioethics*,
https://doi.org/10.1007/978-3-030-54161-3_464

Science Ethics

Science ethics can refer strictly to the ethics of science (a.k.a. the ethics in science) or broadly to the relationship between science and ethics. The former (ethics of science) aims to establish a set of good practices within the field of scientific research concerning the behavior of scientists and the procedures of science according to which both ought to be acceptable to common morality and to guarantee the accuracy and soundness of scientific knowledge. There is a tendency here to restrict the ethics of science to scientific integrity by looking at the goals, procedures, and social impact of science and at the same time the conduct of all stakeholders (individual and collective). The latter (relationship between science and ethics) can be broken down into three major stages. The first views ethics as limiting science and playing a repressive function and as an authority capable of imposing limits to scientific advances and technological innovation mainly as a result of the general social fear of all things new. This is the stage at which science tries to escape ethical scrutiny and ethics tries to catch up with the fast pace of science inevitably failing and resulting in science and ethics confronting each other. The second views ethics as playing a normative function regarding science by proposing guidelines for good practices and guaranteeing scientific knowledge and technological advances contribute to overall well-being and to common good. This is the stage at which cooperation between science and ethics is fostered, society participates in the destiny of science, citizens' trust in science increases, and the social responsibility of sciences is enhanced. However, there is a temptation at this stage to convert ethics into merely a legal, administrative, or bureaucratic procedure. This is the reason there is a third stage in the relationship between science and ethics. This third stage views ethics as playing an educational role and raising the capacity for decision-making beyond the simple exteriority of submission to the law or compliance with the guidelines. Although this becomes particularly relevant when there is no established rule for the situation in question, an ethical decision must nevertheless be made. It should be kept in mind that ethics does not simply interact with science at the last stage of a process (imposing limits in a conflicting relationship) or simply surveil the development of sciences (issuing guidelines in a complementary relationship), it is active from the very beginning of the scientific endeavor in an indissoluble relationship that acknowledges that both ethics and science are expressions of the human spirit.

H. ten Have and M. do C. Patrão Neves, *Dictionary of Global Bioethics*,
https://doi.org/10.1007/978-3-030-54161-3_465

Scientific Misconduct

Scientific misconduct is usually strictly defined as the "fabrication, falsification, or plagiarism in proposing, performing, or reviewing research, or in reporting research results." (The Office of Research Integrity). The European Code of Conduct for Research Integrity (2017): "Fabrication is making up results and recording them as if they were real; Falsification is manipulating research materials, equipment or processes or changing, omitting or suppressing data or results without justification; Plagiarism is using other people's work and ideas without giving proper credit to the original source, thus violating the rights of the original author(s) to their intellectual outputs." Although the fabrication, falsification, plagiarism (FFP) categorization is widely used for the three most important forms of violation of research integrity, it is nevertheless important to distinguish between them. Although fabrication and falsification undermine scientific production, disseminate fake results that might be used to ground further scientific research, tampering with it, and entail severe consequences for scientific advances, in contrast plagiarism does not really affect scientific knowledge—just authorship. Since plagiarism harms researchers and fabrication and falsification damage research (science, knowledge), they are all unacceptable in the field of science. Although they are sometimes designated scientific fraud, in a broader (perhaps weaker) sense scientific misconduct already includes scientific fraud (i.e., FFP) and questionable practices. The Organization for Economic Cooperation and Development (OECD) considers research practice misconduct, data-related misconduct, publication-related misconduct, personal misconduct, and financial and other misconduct as questionable practices. By explicitly referring to such misconduct as questionable practices the OECD is acknowledging that scientific misconduct is not restricted to FFP and that there is a lot of ambiguity in identifying and condemning other behaviors that might be candidates for scientific misconduct.

H. ten Have and M. do C. Patrão Neves, *Dictionary of Global Bioethics*,
https://doi.org/10.1007/978-3-030-54161-3_466

Sexual Ethics

Sexual ethics is the area of applied ethics focused on human sexuality and sexual behavior. Sex has long been culturally and historically the subject of moral deliberation, regulation, and often legislation. Although it is primarily associated with human procreation, it is also a powerful force in human relations. Attempts at understanding sexual relationships have done so by looking at such relationships from a number of different perspectives. The pessimistic perspective concedes that sex is necessary for procreation, while sex for any other purposes simply reveals the animal nature of human beings in which the sexual act objectifies people, uses them for personal gratification, is driven by desires and emotions, and therefore threatens human reason when it comes to behavior. Although the pessimistic perspective argues sexual activity is only morally permissible in the context of marriage, an intermediate perspective highlights the association between love and sex in which sexual activity is complementary to a loving relationship. The optimistic perspective emphasizes the positive value sexual activity has in human relationships as a bonding mechanism contributing to the quality of intimate relationships. It further points out that sex provides pleasure for its own sake and therefore calls into question the so-called moral link between sex and marriage. Such a view has led to greater tolerance toward various forms of sexual behavior in many countries in recent decades as borne out by the acceptance and decriminalization of forms of sexuality that used to be regarded as perversion such as homosexuality, transsexuality, swing (a.k.a. group sex), and polygamy. Most people today accept that sex is more than a mechanism for human procreation and understand that sex and intimate relations are fundamental needs of human beings.

H. ten Have and M. do C. Patrão Neves, *Dictionary of Global Bioethics*,
https://doi.org/10.1007/978-3-030-54161-3_467

Slippery Slope

The slippery slope (a course of action that will likely lead to something disastrous) is an argument used in bioethics and many other areas. It is an intuitively convincing and logically complex concept in which an initial small step or course of action leads to a cascade of related events culminating in harmful, disastrous, and unacceptable consequences. A good example is the argument that allowing euthanasia on the basis of a voluntary request will lead to termination of human life without request. If A is allowed to happen, then B will eventually happen. Therefore A should not happen. The slippery slope argument comprises a number of steps such as the action considered, the sequence of actions following the initial action, an area of indeterminacy where the agent is losing control over what is happening, and finally a catastrophic outcome. As soon as the first step is taken the agent cannot prevent the subsequent steps and will necessarily be confronted with a negative outcome in which unpredicted and undesirable consequences appear that can no longer be controlled. Slippery slope arguments can be broken down into empirical and logical versions. The empirical version identifies causal processes leading from doing A to ultimately doing B. Although the assumption is that the ethos of society can change, moral experience can also change because of social processes thus making acceptable what once was unacceptable. There are two logical versions of the slippery slope argument. The first states there is no conceptual difference between A and B, what justifies A will also apply to B, and accepting A will imply accepting B. The second logical version is that there is a difference between A and B, but there are no relevant differences between intermediary actions between A and B such that allowing A will ultimately imply accepting B. However, sometimes there is a gray zone that may make it possible draw a line (e.g., in legislation) such that the slippery slope can be avoided.

H. ten Have and M. do C. Patrão Neves, *Dictionary of Global Bioethics*,
https://doi.org/10.1007/978-3-030-54161-3_468

Social Ethics

Social ethics differs from individual ethics in being concerned with the relationship individuals have with others and with society. A crucial notion is the common good to which humans as social beings and members of communities should subscribe. Therefore social ethics focuses on the principles and guidelines that influence the welfare of societies and communities. Globalization has led to social ethics receiving increasing attention. An example is the discourse on vulnerability that is linked to social and political concerns and society being construed as more than the aggregate of equal and autonomous individuals. When interconnectedness and dependence result in making people vulnerable it is important to find out what has brought this about, something that can only be ascertained from the perspective of justice and good society—not that of the individual. Although vulnerability is a notion of social ethics rather than individual ethics, the same considerations apply to the discourse on human rights because such rights apply to groups, agencies, organizations, and movements around the world—not just individuals. Human rights are public goods and as such express a commitment to social ethics. The right to health emphasizes that human rights evoke a societal perspective in which there is an obligation to act for the common good—not just for the rights of individuals to be free from government interference. Although the emphasis in healthcare was initially on individual rights (particularly patient rights), attention has now shifted to improving social conditions and the social determinants of health (i.e., social and economic human rights). The fundamental ideas in social ethics are interconnectedness and relationality. Since someone's identity and well-being depend on society, individuals have the obligation and the responsibility to contribute to social development and the common good. Such a perspective of social ethics is expressed in several principles of UNESCO's Universal Declaration on Bioethics and Human Rights (2005): equality, justice and equity (Article 10), non-discrimination and non-stigmatization (Article 11), respect for cultural diversity and pluralism (Article 12), solidarity and cooperation (Article 13), social responsibility and health (Article 14), sharing of benefits (Article 15), protecting future generations (Article 16), and protection of the environment, the biosphere and biodiversity (Article 17).

H. ten Have and M. do C. Patrão Neves, *Dictionary of Global Bioethics*,
https://doi.org/10.1007/978-3-030-54161-3_469

Social Media

Social media refers to the means people can use to interact online and share digital contents with one another. Indeed, the creation and constant development of information and communication technologies have given citizens easy access to websites and applications such as forums, microblogging, networking, social bookmarking, social curation, and wikis and have given them the opportunity to participate in social networking by producing and sharing a wide variety of content (text, pictures, music, and videos). The variety of online communication channels today is wide and ever increasing as borne out by the likes of Facebook (social networking), Twitter (microblogging), Google+ (project networking), LinkedIn (business networking), Reddit (social news website), and Pinterest (social curation website). Social media is based on computer-mediated communication (CMC) that now relates to the creation of new communicative communities and digital communities bringing together people who share the same interests and goals (previously it related to exchanging and/or disseminating information or data in general). Social media plays a major part in the lives of people today and does so in many ways from helping families and friends keep in contact, to fighting loneliness, through to launching businesses, selling products, and promoting brands. It is a powerful tool for education, political campaigners, and civic movements. In short, social media connects people from all over the world and all cultures in terms of information, education, understanding, tolerance, participation, and civic activism. However, there is a negative side to all of this brought about by the abuse of social media such as the deliberate defamation of people and institutions, the violation of privacy and exposure of intimate life, cyberbullying, the spread of hate movements, violence, terrorism, and the constitution of communities of pedophiles or segregationists (racial). Matters are made worse by some people substituting physical interaction with virtual chatting and others becoming addicted to online activities such as gaming. This negative side has led to growing numbers of people today disconnecting from social media in an attempt to avoid hate mail. They consider social media has become an amoral world where individuals can say or do what they want without any kind of boundaries, without respect for human rights, or without even the most basic common sense. The major challenge that social media faces today is its safe and secure use and finding ways to restrict contents circulating online to those who respect human rights and fulfill democratic values.

H. ten Have and M. do C. Patrão Neves, *Dictionary of Global Bioethics*,
https://doi.org/10.1007/978-3-030-54161-3_470

Social Work

The International Federation of Social Workers defines social work as "a practice-based profession and an academic discipline that promotes social change and development, social cohesion, and the empowerment and liberation of people." It is concerned with individuals, families, groups, and communities and has the goal of enhancing social functioning and well-being. Social work as a discipline developed in the late nineteenth century primarily in North America and Europe. It is an interdisciplinary field in that it covers such areas as psychology, sociology, economics, politics, and law. The task of social workers in healthcare is to address poverty, mental illness, drug abuse, homelessness, elderly care, disabilities, crime, discrimination, conflicts, and abuse. Working at the interface of individual care and social settings means social workers are confronted with many ethical issues focused on social justice, human rights, and cultural diversity. Even though they often work as advocates and stakeholders for patients and families and as community organizers, they are frequently still part of the interdisciplinary healthcare team in addition to being members of ethics committees. Professional organizations have developed codes of conduct for social work such as that of the National Association of Social Workers in the United States. Its code of conduct identifies core values such as service, human dignity, social justice, competence, integrity, and the importance of human relationships. Ethical standards and responsibilities are defined on the basis of such values. The International Federation of Social Workers has produced arguably the most relevant ethical guidelines for social workers the most recent being the Global Social Work Statement of Ethical Principles and the Statement of Ethical Principles and Professional Integrity.

H. ten Have and M. do C. Patrão Neves, *Dictionary of Global Bioethics*,
https://doi.org/10.1007/978-3-030-54161-3_471

Solidarity

The word "solidarity" derives etymologically from the Latin *solidus* (solid entity) and refers to an entire three-dimensional body that is both consistent and sound. Its conceptual origin derives from the French *solidaire* (solidarity) used in the fifteenth century within the legal system in situations in which each person was held responsible for the actions of people overall. It was only later (in the eighteenth century) that Diderot used solidarity with the meaning it has today: to join a cause. Therefore solidarity is the link or the bond that ties humans together in a whole where everyone responds for everybody else thus making people stronger. Solidarity should be considered an onto-anthropological fact and a social and ethical virtue and principle. Indeed, there is no disputing the fact that all individuals are interdependent and live as interconnected beings in human societies. Although this reality or fact was initially translated into a virtue, a trait of character, a voluntary willingness to help others, it has currently become an ethical principle or obligation: to respond for others on their behalf. The principle of solidarity requires acknowledging that the interdependence and the ethical obligation of people to strengthen human relationships is natural. Broadly speaking, solidarity expresses mutual dependence that corresponds to reciprocal responsibility. Therefore solidarity is a two-way system that, on the one hand, helps and shares with those most in need and, on the other, expects them to recover and join with those who have helped and shared with them to bring about a more egalitarian situation with the aim of making the common good stronger. Article 13 of the Universal Declaration on Bioethics and Human Rights (2005) establishes the principle of solidarity and cooperation as "solidarity among human beings and international cooperation towards that end are to be encouraged." Although the establishment of solidarity as a bioethical principle was important, its definition was too narrow and vague to be as useful as it could and should be. This was the reason the Nuffield Council on Bioethics acknowledged the importance of solidarity in bioethics and the lack of a precise and operational definition of the concept, published a work on its definition (2011), and established three levels of solidarity. The first was the interpersonal level at which solidarity consists of "manifestations of the willingness to carry costs to assist others with whom a person recognizes sameness or similarity in at least one relevant

H. ten Have and M. do C. Patrão Neves, *Dictionary of Global Bioethics*,
https://doi.org/10.1007/978-3-030-54161-3_472

respect." The second was the group practice level at which solidarity is described "as manifestations of a collective commitment to carry costs to assist others (who are all linked by means of a shared situation or cause)." The third was at the level of contractual legal relationships at which solidarity is described as "welfare state and social welfare arrangements, but also contracts between different private actors and international declarations or treaties." Solidarity became ever more important as bioethics moved from individual to societal considerations and as bioethical issues increasingly became global.

Spirituality

The word "spirituality" derives etymologically from the Latin *spiritus* (breath, breath of life). Although *spiritus* originally designated a vital principle (i.e., that which makes beings alive, the principle of life), the notion of spirit later designated the non-material part of the human being and started to be interpreted as the seat of emotions, feelings, the psyche, the character, the soul, and eventually the immortal soul thus gaining also a religious connotation. Nevertheless, spirituality is much broader than religion (which entails faith in a superior being such as a divinity like God) as borne out by religious people being spiritual, but not all spiritual people being religious. Spirituality refers today to a non-physical and non-psychic human dimension that is intangible, to a human reality that is not objectifiable and cannot be reduced to an object or to something concrete or material that is accessible to the senses. Although religion is an expression of spirituality, so are artistic production, scientific research, and analytical and critical thought. Recognition of the importance of spirituality in healthcare is recent and came about as a result of a shift in focus from disease to the person and as a result of acknowledging that the same pathology experienced by different people can also evolve differently. Patient-centered care requires attention be paid not only to someone's physical and psychological dimensions, but also to taking a truly holistic approach that considers all human dimensions and their unicity (which is greater than the sum of the parts). Therefore spirituality can be considered an important part of someone's well-being and as such requires assistance within the healthcare setting. Spiritual assistance is more frequently needed in palliative care where it provides support in the dying process. It can also be extended to family members helping them to cope with the situation. The first step in providing spiritual assistance in the healthcare setting usually involves either integrating a religious minister in the institution or facilitating access to religious leaders outside the institution. However, there are difficulties here in that spiritual care does not coincide with religious assistance and that there are non-religious patients who need and could benefit from spiritual assistance. Therefore it is important to acknowledge and meet other patients' spiritual needs by setting up affective emotional gatherings and creating an environment suitable to meditation.

H. ten Have and M. do C. Patrão Neves, *Dictionary of Global Bioethics*,
https://doi.org/10.1007/978-3-030-54161-3_473

Sports (*See* Doping)

Bioethical issues regarding sport are primarily related to doping. The scope of doping is wider today since technologies can be used to enhance sporting performance. For example the case of Oscar Pistorius focused attention to the use of biomechanical prostheses. He was a professional sprinter (convicted of murder in 2015) who had both his feet amputated when he was 11 months old as a result of a congenital defect and was born without fibulae. Although he became Paralympic champion in 2006, he was criticized because his specially developed artificial limbs gave him an advantage over athletes with natural feet and legs. Since the early 2000s, genetics has been used in sports (gene doping). Although healthcare is increasingly used for non-therapeutic purposes to enhance the performance of athletes, the paradox is that the same sports organization is often involved in testing and creating drugs to enhance performance and in pursuing methods that can be used for their detection.

H. ten Have and M. do C. Patrão Neves, *Dictionary of Global Bioethics*,
https://doi.org/10.1007/978-3-030-54161-3_474

Standards of Care (*See* Double Standards)

Standards of care are used to define appropriate treatments and preventive activities in clinical healthcare irrespective of whether they are new or established. They are usually determined by the medical profession and promulgated through guidelines based on scientific evidence. Standards not only prescribe what minimally competent physicians should do but also play a role in claims of medical malpractice (standards define what should be expected under normal circumstances and what is customarily done to avoid harm). Although standards do not aim at perfect medical practice or guarantee particular outcomes, they do ensure that medical intervention is as safe as possible and that the procedures followed meet the standard of expertise reached by the profession. Although standards are usually defined in clinical practice guidelines, a major controversy about standards of care emerged in global bioethics in the 1990s. In a number of developing countries antiretroviral treatment for pregnant women was tested against a placebo ignoring that the standard of care in developed countries was treatment with zidovudine. Some researchers argued that following this standard was too expensive for developing countries, that such a standard could not be applied in such countries, and that testing against a placebo was the only option available. This case led to intensive debate in which it was argued that it was an example of double standards and ethical imperialism. References were made to the Declaration of Helsinki that emphasizes every patient should be assured of receiving the best proven diagnostic and therapeutic method. Other researchers argued that standards of care are relative and depend on the local social and cultural context. The controversy led to efforts being made to change the Declaration of Helsinki such that it referred to the best available treatment. In 2000 the declaration changed its wording to the best current treatment: "the benefits, risks, burdens and effectiveness of a new method should be tested against those of the best current prophylactic, diagnostic, and therapeutic methods." Dissatisfied with this change the FDA decided in 2008 to abandon the Declaration of Helsinki as an ethical guideline and now considers that trials outside the United States should comply with the Good Clinical Practices of the International Conference of Harmonization (ICH-GCP).

H. ten Have and M. do C. Patrão Neves, *Dictionary of Global Bioethics*,
https://doi.org/10.1007/978-3-030-54161-3_475

Stem Cells, Adult

Stem cells refer to adult or somatic stem cells obtained from adult body tissues. They are found in the body from the earliest days (3–4 days) of embryonic development and endure throughout the lifecycle, although they are rare, dispersed, and difficult to identify. Adult stem cells are generally restricted to tissues found in the brain, bone marrow, peripheral blood and blood vessels, spinal cord, dental pulp, skeletal muscles, liver, pancreas, cornea, retina, and the epithelia of the skin. Nevertheless, they also reside in most organs. At the beginning of stem cell research scientists focused their attention on embryonic stem cells because their potential to differentiate was higher. However, there is now evidence to show that adult stem cells not only differentiate based on their tissue of origin, but can also differentiate to become other cell types as well. Moreover, despite adult stem cells being multipotent (or unipotent) they have proven to be safer than pluripotent stem cells and can be of broader use in both clinical trials and clinical practice. Bone marrow transplantation and the possibility to postpone or even overcome the need for organ transplantation are two major areas in which adult stem cell–based therapy has proven successful. Even unipotent cells have vast therapeutic potential in the treatment of injuries and diseases. Apart from the general ethical problems raised by stem cell–based research and therapy, adult stem cells solve more problems by becoming an effective alternative to embryonic stems cells. Furthermore, the source and the procedure used to harvest them do not raise specific concerns and their use paves the way for personalized medicine. In short, they are potentially highly beneficial.

H. ten Have and M. do C. Patrão Neves, *Dictionary of Global Bioethics*,
https://doi.org/10.1007/978-3-030-54161-3_476

Stem Cells, Embryonic

The earliest source of stem cells in humans are embryos at the blastocyst stage (a hollow ball of about 150–200 cells that is 4–5 days old) of their development. Such cells are identified and collected mainly from surplus embryos discarded after in vitro fertilization (IVF). The procedure requires destruction of the blastocyst. However, embryonic stem cells (ESCs) are immortal in culture and can be grown relatively easily and maintained in permanent culture, frozen and thawed, and transported between laboratories once derived from a pluripotent cell collected from the inner cell mass of the blastocyst. Although the derivation of mouse ESCs was first reported in 1981, it was only in 1998 that the first human ESC lines were derived and reported. Today there are many human ESC lines in the world that are widely shared among different research groups. Since they cannot generate another embryo, ESCs are pluripotent—not totipotent (i.e., they have the potential to develop into any cell type of the adult organism). Such pluripotent cells begin to differentiate after the blastocyst stage when the embryo implants in the uterus. Being pluripotent, ESCs are less specialized than other stem cells, consequently more suitable for a wider range of research projects, and thus preferred over adult ones. ESCs have been highly controversial ever since their discovery because of the need to destroy the embryo (blastocyst) to harvest them. Destruction of a human life at the embryonic stage is considered by many unacceptable regardless of any derived beneficence. Many others argue that researchers are using surplus embryos that would die in any event since they are discarded after IVF and that such embryos would at least be of some use before their destruction. Nevertheless, some consider that it is more respectful of the dignity of human life to immediately destroy surplus embryos rather than using them. Some countries such as the United Kingdom were encouraged by the high therapeutic hopes of stem cell research and proposed creating human embryos specifically for research. Although this intensified the criticism of using embryonic stem cells, at the same time other countries such as Germany (opposed to embryo destruction for stem cell harvesting) started to buy human ESC lines from other countries' labs to foster their own research. Although ethical problems raised at the beginning of stem cell–based research still prevail, as a result of reprogramming techniques it is now possible to have pluripotent stem cells that do not involve destroying a human blastocyst.

H. ten Have and M. do C. Patrão Neves, *Dictionary of Global Bioethics*,
https://doi.org/10.1007/978-3-030-54161-3_477

Stem Cells, General

Stem cells refer to undifferentiated cells (i.e., yet to have a specific role) that can turn into almost any cell an organism needs in its lifecycle. Although the word "stem" generally refers to a central, supportive, or main section of something from which other parts can develop and grow, in biology stem cells refer to precursor or progenitor cells that have not only the potential for unlimited or long-term self-renewal (they can make identical copies of themselves for long periods of time) but also the capacity to generate multiple mature differentiated cell types (with characteristic morphologies and specialized functions). All stem cells share these two characteristics. Although the two main sources of stem cells are adult body tissues and embryos, today it is also possible to obtain stem cells from other (somatic) cells using genetic reprogramming techniques. The potency of stem cells (their differentiation capability) can vary depending on the source from which they are harvested. They can be totipotent (i.e., capable of developing into a complete organism). The only totipotent cell is the zygote, a single diploid cell formed by the fusion of an egg cell and a sperm cell. Stem cells present in the inner layer of the blastocyst are pluripotent (i.e., capable of developing into all cell types of an adult organism). Stem cells generated experimentally using reprogramming factors are also pluripotent (induced pluripotent stem cells). Adult stem cells are multipotent (i.e., they have the potential to differentiate into multiple connective tissues). There are also unipotent stem cells found in adult tissues that have the capacity to differentiate into only one type of cell. As a result of being able to renew themselves and generate multiple cell types, stem cells have huge potential in different fields of research (particularly in tissue regeneration, drug screening, and organogenesis). Stem cell–based therapy has already been successful in bone marrow transplantation and helped many leukemia patients. In addition to tissue regeneration (repairing wounds and tissue damage in people after an illness or injury) being probably the most important current use of stem cells, many other stem cell–based therapies are presently undergoing clinical trials. There seem to be no limits to the types of diseases and disabilities that could benefit from this research such as autoimmune diseases, macular degeneration, multiple sclerosis, heart disease, diabetes, and Parkinson disease. Another important benefit of such research is the toleration of transplanted organs. Major

H. ten Have and M. do C. Patrão Neves, *Dictionary of Global Bioethics*,
https://doi.org/10.1007/978-3-030-54161-3_478

ethical issues related to the use of stem cells today are high therapeutic expectations after too short a period of research increasing the number of terminally ill and other desperate patients scrambling to get enrolled in clinical trials. The vulnerability of these patients (and their families) has to be taken into account and researchers should restrain their enthusiasm when proposing an innovative stem cell–based therapy. Another increasingly important ethical issue is the use of stem cells not only to address the signs and symptoms of old age, but also to reverse the ageing process and even to try and conquer immortality. Stem cell interventions that bring about some form of rejuvenation under the guise of being considered therapeutic exploit people's emotions and financial resources and challenge the very nature of life and mortality.

Stem Cells, Induced Pluripotent

Induced pluripotent stem cells (iPSCs) refer to a type of stem cell directly generated using reprogramming techniques from multipotent adult stem cells. Therefore iPSCs gained a lower level of differentiation and could present themselves as pluripotent. Although the only pluripotent cells in nature are embryonic stem cells, researchers acquired the knowledge, skills, and the power necessary to reinvert the natural differentiation and specialization process of a cell and thus artificially obtain a pluripotent cell from a multipotent cell. Such an achievement provided researchers with pluripotent stem cells free of the ethical problems raised by embryonic stem cells. Indeed, it is reasonable to argue that it was the ethical problems related to the destruction of blastocysts for pluripotent stem cell harvesting that led to researchers in 2006 discovering reprogramming techniques.

H. ten Have and M. do C. Patrão Neves, *Dictionary of Global Bioethics*,
https://doi.org/10.1007/978-3-030-54161-3_479

Stewardship

Stewardship refers to the care and management of resources and goods that is expected to be done in a responsible manner and stewards refer to the people entrusted with such care and management. It should be kept in mind that stewards are guardians or trustees. Although they do not own what is entrusted to them such as resources and goods, they are accountable for how such resources and goods are managed. The concept can be traced back to Ancient Greece and the Bible. The basic idea back then was that nature and creation were not the product of human efforts and that human beings were merely custodians who had to administer nature and creation with due care and who were answerable for how they cared for it. Four principles are often attributed to stewardship. The first is the principle of ownership in which resources and goods are not our property but we are managers and administrators acting on behalf of the owners (biblical stewardship assumes that God owns everything and that stewardship expresses our obedience to God by committing ourselves to administering everything placed under our control). The second is the principle of responsibility in which nothing belongs to us but we are responsible for how we deal with it. The third is the principle of accountability in which we have to explain how we are managing the possessions of others and are called to give an account of our management. The fourth is the principle of reward in which careful and responsible management is expected to deliver future benefits and rewards. Although stewardship has theological and religious roots, it is not necessarily a religious notion. Secular notions emphasize responsible use as balancing the interests of society, future generations. and other species. The notion has significant ethical implications especially for sustainable development, climate change, and biodiversity preservation. Stewardship is often connected to the precautionary principle in that stewards as custodians of threatened resources accept such a principle since it aims to avoid serious or irreversible damage.

H. ten Have and M. do C. Patrão Neves, *Dictionary of Global Bioethics*,
https://doi.org/10.1007/978-3-030-54161-3_480

Stigmatization

Stigma refers to a mark of disgrace associated with a particular event, quality, or individual and brands that person as less worthy than others. In ancient times a stigma was a mark branded into a slave. Stigmatizing someone today means labeling him or her negatively usually as an expression of disapproval and disrespect. With the exception of Catholicism where stigmata are perceived as marks of holiness since they refer to the crucifixion wounds of Jesus the concept has an overall negative meaning. Although stigma is a mark of shame, disgrace, or discredit and stigmatizing someone makes it easier to treat the individual concerned with disrespect by naming, shaming, and punishing him or her, there is no definitive list of stigmatizing practices. However, labeling people in such a way has been historically used for foreigners or people of other races, colors, religions, and social classes and has led to members of such groups being stereotyped always in a derogatory and insulting way that suggests they are dangerous, criminal, and even evil. Although discrimination is illegal and strictly prohibited under international human rights law, stigmatization is more a social concern but nevertheless firmly based on the same fundamental ethical principles of equality, justice, and human dignity. When it comes to bioethics, stigmatization was first addressed in the International Declaration on Human Genetic Data adopted by UNESCO in 2003. Article 7a states that "Every effort should be made to ensure that human genetic data and human proteomic data are not used for purposes that discriminate in a way that is intended to infringe, or has the effect of infringing, human rights, fundamental freedoms or human dignity of an individual or for purposes that lead to the stigmatization of an individual, a family, a group or communities." The focus here was explicitly on stigmatization on the basis of someone's genetic constitution. The Universal Declaration on Bioethics and Human Rights (UDBHR) has now expanded this to cover the many other causes of stigmatization. Article 11 of the declaration states that "No individual or group should be discriminated against or stigmatized on any grounds in violation of human dignity, human rights and fundamental freedoms."

H. ten Have and M. do C. Patrão Neves, *Dictionary of Global Bioethics*,
https://doi.org/10.1007/978-3-030-54161-3_481

Strikes

Strikes refer to a collective form of protest that is used when employers or governments impose a condition or a set of conditions on workers or populations, respectively, that are felt to be unfair and unjust. They usually involve the withdrawal of labor or a refusal to work and are organized by a body of employees (e.g., a union) aimed at gaining concessions from their employer or getting employers to change the conditions of work. A number of types of strikes can be distinguished such as general strikes (widespread collective action by many unions and mass refusal to work due to disputes about wages or conditions of work), capital strikes (refusal to invest or reinvest by those who own and control wealth and resources), essential strikes (strikes of essential workers such as physicians and nurses), and hunger strikes. Strikes are usually actions of last resort by certain populations who feel disadvantaged, who feel their rights have been violated, and who feel they have tried everything to resolve the situation but have failed to reach an agreement. Essential strikes are problematic in that they affect areas of society that are essential for the provision of basic human needs (e.g., strikes of police, public health professionals, physicians, emergency services, and water distribution). It is sometimes argued that workers in such areas have no right to strike, that the consequences of strikes are socially disruptive, and that they can be harmful for individuals who cannot escape the consequences. However, if a right to strike is acknowledged, then it should at least be balanced by taking the basic needs of citizens into consideration and strictly regulated (e.g., by mandatory arbitration). This is particularly true of essential strikes, which should pursue the least burdensome and least harmful action. However, the less strikes affect the population the weaker they are and the less successful they become. Hunger strikes refer to individuals or groups refusing to eat or drink who want to protest against what they see as injustice, improper treatment, conditions of imprisonment, etc. as borne out by the hunger strikes of Mohandas Gandhi. Although, like other types of strikes, the aim is to bring about change, they differ from other types of strikes in that the impact of the strike affects no one other than the hunger-strikers themselves. They hope that their suffering will induce feelings of shame, guilt, and sympathy not only in those at whom the strike is directed but also in people watching and following their strike more widely. There can also be mass hunger strikes such as the strike in the US detention camps at Guantanamo Bay. The authorities responded by force-feeding the prisoners (assisted by medical personnel). Force-feeding of prisoners has been prohibited since 1975 in the Declaration of Tokyo of the World Medical Association and is regarded as a form of torture.

H. ten Have and M. do C. Patrão Neves, *Dictionary of Global Bioethics*,
https://doi.org/10.1007/978-3-030-54161-3_482

Subsidiarity

Subsidiarity refers to organizing or managing matters at a more decentralized level than at a larger and more complex level. The principle of subsidiarity applies to all human institutions including the state, is a general principle in EU law, and is defined in Article 5 of the Treaty on European Union (1992): "Under the principle of subsidiarity, in areas which do not fall within its exclusive competence, the Union shall act only if and in so far as the objectives of the proposed action cannot be sufficiently achieved by the Member States, either at central level or at regional and local level, but can rather, by reason of the scale or effects of the proposed action, be better achieved at Union level." The aim is to make sure that decisions are taken as close to the citizen level as possible and if that cannot be done then it should be done at the national, regional, or local level before action is taken at the EU level. Subsidiarity as an organizing principle came about as a result of Catholic social teaching and was first formulated in 1891 by Pope Leo XIII in his encyclical Rerum Novarum and later in 1931 by Pope Pius XI in his encyclical Quadragesimo Anno. By identifying a middle way between capitalism and communism, it articulates that human beings must consider not only their own advantage but also the common good. It argues that when the role of the state has become encompassing and authoritarian it should be limited or the ancient principle of subsidiarity should be applied: "Just as it is gravely wrong to take from individuals what they can accomplish by their own initiative and industry and give it to the community, so also it is an injustice and at the same time a grave evil and disturbance of right order to assign to a greater and higher association what lesser and subordinate organizations can do. For every social activity ought of its very nature to furnish help to the members of the body social, and never destroy and absorb them." If such a principle is ignored, then the dignity of humans will often be violated, governments will try to intervene and solve all manner of problems, and society itself (communities, families) will have failed in their own responsibilities. Back in 1931 Pope Pius XI was concerned that a situation would emerge in which there were only states and individuals—no intermediary communities, institutions, or levels. Although the aim of the principle of subsidiarity is to ensure that society is well ordered and directed toward the common good, this requires states, individuals, institutions, civil organizations, and churches to work together in civil society. However, much ethical deliberation will be needed to determine when the state should intervene and when it should not. Although conditions today differ widely from those at the time of Pope Pius XI's encyclical and neoliberal policies and market ideology have become dominant, it should be kept in mind that subsidiarity is not an argument for reducing state activities and interventions. It is a two-sided coin in which the state has the responsibility not only to respect and promote various levels of society,

H. ten Have and M. do C. Patrão Neves, *Dictionary of Global Bioethics*,
https://doi.org/10.1007/978-3-030-54161-3_483

but also the right and responsibility to intervene to protect the poor and the vulnerable. In short, the goal of subsidiarity is not smaller government but to promote and safeguard the common good. When it comes to bioethics, the principle of subsidiarity refers to ensuring new technologies are only used for the purpose intended (e.g., reproductive technologies should only be used to treat infertility—not to satisfy individual desires).

Substance Abuse (*See* Addiction)

Substance abuse refers to the harmful or hazardous use of addictive substances such as alcohol, tobacco, and illicit drugs. Using psychoactive substances can lead to dependence syndrome, which presents as a cluster of cognitive, physiological, and psychological symptoms that develop after repeated use and are associated with a strong desire to take the substance in question and continue to use it despite the harmful consequences and deleterious impacts doing so has on other activities. Psychotropic substances such as opium, cannabis, coca, and alcohol have long been used by humans for a variety of reasons ranging from medical, religious, and recreational through to ritual. Abuse and misuse were the reasons restrictions were put on their use and some cultures prohibiting them. The ethical issues of substance abuse have to do with the weighing of benefits and harms that should be done at both the individual level and the public health level. Although most countries have put in place barriers to substance abuse (e.g., minimum age limits for people to buy alcohol and restricting areas for people to buy and smoke tobacco) and support systems to help addicts (e.g., decriminalizing illicit drug consumers and freely distributing methadone to them), there is also the major issue of individual autonomy (especially when it comes to addictive substances). Individual autonomy raises questions such as how can it be argued that an addict is making free choices when he or she risks inflicting chronic self-damage and what are the goals of intervention. Although the traditional goal was abstinence, today this is replaced by more modest goals such as patient survival, health improvement, quality of life improvement, and harm reduction in which treatment is focused on the individual needs of the patient and can result in putting the patient on opioid maintenance therapy or in providing him or her with clean needles. Much emphasis is placed on harm reduction approaches that attempt to improve the health of chronic addicts who are unable or unwilling to discontinue substance abuse. Although such approaches are criticized (especially in countries taking a primarily legal approach) since they prolong dependence, make substance abuse acceptable or at least tolerable for young people, and remove the incentives for treatment, such fears are not substantiated by research. Nevertheless, such approaches do not

H. ten Have and M. do C. Patrão Neves, *Dictionary of Global Bioethics*,
https://doi.org/10.1007/978-3-030-54161-3_484

remove the need to reduce or eliminate addictive behavior through rehabilitation programs. Another ethical issue concerns coercive care. Non-voluntary confinement and treatment regimes are applied to enforce abstinence in many countries as borne out by compulsory treatment options for drug dependence ranging from drug detention facilities, short-term and long-term inpatient treatment, and group-based outpatient treatment through to prison-based treatment. However, although systematic evaluation of the effectiveness of compulsory drug treatment is limited, whenever studies are done they seem to suggest harm is done and benefits are unproven. Non-compulsory treatment modalities should therefore be prioritized given the potential for human rights abuses within compulsory treatment settings.

Suffering

Suffering refers to the pain or anguish someone is undergoing. Since pain is objective and suffering is subjective, suffering (being subjective) is not the same for everyone even though the cause may be the same and thus can be unpredictable. Although suffering is triggered by an external event, its reality is interior to the person as borne out by the same event being interpreted and felt differently according to the personal context, not all people suffering in the same way or to the same extent, and not all people being capable of overcoming or recovering equally from it. The diversity inherent in suffering varies according to whether it is physical, psychological, social, or spiritual each of which evoke a different and adapted response by others (family, friends, and professionals). Suffering has been interpreted throughout history as being the consequence of, say, destiny, an imperfect human nature, the need for punishment, expiation (penance), or the need for redemption. Although such consequences are no longer given any credit, back then they had serious effects on those suffering. The rise of modern medicine has led to care encompassing not only the disease (cause), but also the pain (symptom) and the suffering (the way disease is experienced), even when suffering does not have an objective cause. Although suffering is not tolerated in our contemporary hedonist society, medical science is tasked with eliminating it, and the person suffering is expected to hide it, suffering is nevertheless a natural expression of our emotional life.

H. ten Have and M. do C. Patrão Neves, *Dictionary of Global Bioethics*,
https://doi.org/10.1007/978-3-030-54161-3_485

Suicide

In September 2019 the WHO reported that someone commits suicide every 40 s somewhere in the world and that 800,000 people end their lives each year. For young people aged between 15 and 29 suicide is the second commonest cause of death after traffic accidents. Suicide is defined as the action of killing oneself intentionally. However, suicidal behavior does not always lead to death. Suicide might be attempted but fails. Intentional means that death is the result of a free decision in the absence of coercion or external restraints. Suicide has been discussed as a moral issue since antiquity in all cultures. In debates discussing whether or not suicide is morally acceptable several arguments are used such as autonomy. Although individuals are free to decide to end their lives, whether people in a suicidal state of mind are really free is a moot point since they are likely desperate and see no other solution. Another argument is that suicide may be related to mental illness in which case the decision is not a rational decision of an autonomous individual but of someone who in fact needs help and medical assistance. The autonomy argument is related to the property argument and raises important questions: Do our lives belong to us or to a more powerful entity? If the latter, then is it a gift that cannot be rejected? There are a couple of versions of the sanctity of life argument. One states that life is given and has an inherent value (if we cannot give life to ourselves, then we cannot end it). Another version elaborated by Kant is that we have a moral duty of self-preservation (rationality and autonomy mean we cannot eradicate self-preservation by killing ourselves). A third type of argument refers to the social dimension of human life. It argues that since human beings exist in communities and societies, suicide can be justified if it leads to the survival of other people.

H. ten Have and M. do C. Patrão Neves, *Dictionary of Global Bioethics*,
https://doi.org/10.1007/978-3-030-54161-3_486

Surgery

Surgery refers to the branch of medical practice that treats injuries, diseases, and deformities by physically removing, repairing, or readjusting organs and tissues and usually involves cutting into the body. It is one of the oldest branches of medicine and is usually undertaken to resolve acute injuries and illnesses rather than chronic diseases. Although it is often regarded as a technical discipline, its practitioners are primarily physicians responsible for the well-being of patients. Surgery can be broken down into a number of subspecialties such as transplantation, cosmetic surgery, and pediatric surgery. Ethical issues affect some of these subspecialties more than others. Although surgery at the outset of Western medicine in the nineteenth century up until recently could not really be considered a medical expertise since it was carried out by people whose skill lied elsewhere such as dentists, today it is arguably the most prestigious medical form of expertise. Recently interest in surgical ethics has been growing. A major ethical issue concerns informed consent and the importance of it being obtained prior to any intervention. This entails explaining the benefits, risks, and alternatives in such a way that takes into account the vulnerability of patients while at the same time pointing out that the risk inherent in surgery can never be completely predicted. Another ethical issue concerns responsibility. Although surgeons are of course committed to the welfare of the patient, they are not simply administering a medication but directly involved in the intervention whose outcome depends on their skills, acumen, experience, and attention. A further ethical concern relates to the way in which the benefits and risks of intervention are evaluated since there is a real possibility of harm. This is the reason surgeons are not allowed to perform interventions requested by patients. A final ethical concern relates to innovation. Although surgery can only progress and procedures improve through research, many novel interventions are first applied in the surgical theater. Unlike new medication where testing and evaluation are thoroughly undertaken, new surgical interventions are left for surgeons to apply in medical practice without being regulated. However, there is a world of difference between performing innovative surgery for the benefit of a patient and doing surgical research. Surgical research involves the surgeon creating a research protocol and following the same procedures for review and oversight as happens with other types of medical research.

H. ten Have and M. do C. Patrão Neves, *Dictionary of Global Bioethics*,
https://doi.org/10.1007/978-3-030-54161-3_487

Surrogate Decision-Making

Surrogate decision-making refers to authorizing a third person (a proxy) to make decisions about the healthcare treatment provided by a professional to a patient who is unable to express his or her wishes as a result of being, say, a minor, mentally incapacitated, senile, or no longer self-aware. Such a proxy can be a close family member biologically (parents and children), legally (spouses), or affectively (a close family member or friend). The proxy can perform this role informally or when legally designated formally. Legal designation covers a number of variables such as the proxy being nominated by the patient when the latter was still autonomous, by the proxy receiving durable medical power of attorney (a document nominating an agent to take medical decisions on behalf of someone when that person is no longer able to do it himself or herself), or by the proxy receiving a court order. All such variables depend on what is foreseen by national law. Although proxy decisions have been strongly contested in a number of clinical situations, more frequently they refer to patients in a persistent vegetative state, to severely disabled persons (e.g., tetraplegic), or to children affected by incurable diseases. When there is no consensus about what to do among family members and when there is disagreement between family members and healthcare professionals, the case is brought to court. The only way to decrease the occurrence of such situations or reduce the level of conflict among informal proxies would be to get every adult autonomous person to designate a surrogate or a second who can be called if the first cannot be reached (durable medical power of attorney). The advantage of people having a legally designated proxy is knowing that their wishes will be respected even in a state of incompetence. The disadvantages are emotional proxies impairing a rational decision, lack of knowledge about what the patient would really want in a situation that was never discussed between the proxy and the patient, or even selfish interests on the part of the proxy. In any event legal recognition of a surrogate decision-maker is always based on the principle of autonomy or an extension of it.

H. ten Have and M. do C. Patrão Neves, *Dictionary of Global Bioethics*,
https://doi.org/10.1007/978-3-030-54161-3_488

Surrogate Motherhood

Surrogate motherhood (a.k.a. gestational surrogacy or surrogacy) refers to a process in which a woman becomes pregnant on behalf of someone else and gives birth to that person's child. Surrogacy can either be biological (when the oocyte is not from the gestational mother but from the woman who will raise the child and who is indeed the child's biological mother) or legal (when irrespective of whether there is a biological tie there is a formal contract attributing maternal rights and duties to another woman other than the gestational woman). Cases can be commercial in which surrogacy is paid by whoever contracts the service or altruistic, solidary, and compassionate in which surrogacy is free of charge or at most covers the costs of the surrogate. Surrogacy became possible as a result of advances in assisted reproductive technologies (ARTs) that facilitate the production of a human embryo outside the human body in a laboratory by combining different biological materials such as sperm, oocyte, and uterus from different persons. Although surrogacy is one of a number of reproductive technologies, it is specifically used to address female infertility due to aplasia of the uterus. Surrogacy is not only employed today for clinical reasons but also for social purposes (as happens with other ARTs) such as to enable gay couples or single men to have children biologically related to them. Gestational surrogacy has raised complex ethical problems since the beginning such as what happens if the child born has a disability and no one wants it or what happens when the surrogate mother becomes emotionally attached to the child and does not want to give it away. Although surrogacy has become big business in countries where it is permitted, it is expensive and has led to surrogacy tourism. The countries that tolerate surrogacy and have no legal restrictions to it are developing countries with poor populations who are happy to bear another person's child for money (less expensively than in developed countries where it is legal). Such tourism has strongly grown despite largely being considered a form of exploitation of the most vulnerable by rich people. Countries where surrogacy is legal have mostly opted for the altruistic model. However, it should be kept in mind that the only situation in which surrogacy is totally free and credible is when the mother of the mother-to-be (generally in a high-risk pregnancy) or the sister of the mother-to-be (or other family members) gives birth. Nevertheless, it should also be kept in mind that familial surrogacy presents different but no less serious problems such as psychological pressure to become a surrogate, mixed emotions, a plurality of long-term affective issues (toward the parents-to-be but mainly toward the child), and subjecting the child to tension or even emotional blackmail throughout his or her family life. Most countries that have laws on ART make their position on surrogacy very clear and generally forbid such a practice. Although this is the case in most European countries, there are some countries in Europe and worldwide that have

H. ten Have and M. do C. Patrão Neves, *Dictionary of Global Bioethics*,
https://doi.org/10.1007/978-3-030-54161-3_489

approved legislation allowing altruistic surrogacy. When this is the case the law is more or less restrictive and clearly determines who is allowed to access surrogacy be they nationals or foreigners, heterosexual couples or homosexual couples, single men or single women and further determines whether surrogacy can be used as a subsidiary principle to overcome infertility considered a disease, a social resource, or a reproductive alternative (e.g., for a single man).

Sustainability

Sustainability refers to meeting the needs of the present generation without compromising the ability of future generations to meet their needs and was famously defined as such in 1987 by the World Commission on Environment and Development. In addition to the commission's report identifying the three pillars of sustainability as environmental protection, economic development, and social equity, more recently culture was identified as a fourth pillar. Sustainability is closely related to conservation, resilience, and stewardship. For example, environmental sustainability and stewardship include the protection and preservation of natural resources and resilience as the capacity to endure stress while functioning and adapting to changing circumstances. The commission's report marked the start of various global initiatives culminating in the adoption of the UN's Sustainable Development Goals in 2015. In healthcare the term "sustainability" is used in a very different context since it refers to the capacity to sustain programs that provide care and treatment and the very use of the word implies that healthcare does little environmental damage, produces little waste, and uses little energy. The notion of sustainability as put forward by the commission is in fact a compromise between sociopolitical development, economy, and ecology. However, the interactions between these three spheres are very inequal. Western countries have seen enormous development and economic growth at the expense of environmental degradation not only of their own countries but also of others across the world. Developing countries argue that to grow and develop they should not be bound by environmental limits, that the issue of global inequality should be addressed in another way that involves developed countries being asked to make sacrifices, and that any compromise between sociopolitical development, economy, and ecology would be unfair. The notion of sustainability in the approach taken by the commission is weak in that it rests on the assumption that the sociopolitical, economic, and ecological spheres are separate with their own logic and values but that they can be integrated. However, in reality economic growth and social development are often given priority, sustainability is strongly anthropocentric with human beings and their needs taking precedence, and the basic assumption is that everything in nature has a value that can be used by humans for their and only their benefit. Therefore a stronger notion of sustainability will be necessary: one that requires a drastic change in patterns of production and consumption, one that respects the intrinsic value of nature regardless of any benefits for humans, and one that assumes the sociopolitical and economic spheres are embedded in the ecological sphere—not separated from it. Ethical concerns about the debate on sustainability relate to the moral obligations the current generation has to future generations and to whether the moral interests of non-human species and the natural environment should be taken into account.

H. ten Have and M. do C. Patrão Neves, *Dictionary of Global Bioethics*,
https://doi.org/10.1007/978-3-030-54161-3_490

Synthetic Biology

Synthetic biology refers to the discipline that combines biology and engineering to design and fabricate biological devices, systems, and components that do not currently exist in the natural world such as green chemicals for agricultural waste, new vaccines synthesized in the laboratory, and engineered organisms functioning as biosensors. Although such novel organisms (e.g., vaccines, biofuels, environmental cleansers, and better crops) have been developed for human purposes, the engineering of biology is the result of the revolutionary development in life sciences beginning with recombinant DNA in the 1970s. Synthetic biology raises a number of ethical concerns one of which is related to intellectual property rights and patenting. Should new life-forms created through synthetic biology be patentable (patents might limit access and hinder scientific progress)? Can life-forms be owned and if so by whom? Another concern relates to potential benefits and harms. Although new technological developments tend to increase disparities, synthetic biology has the potential to help address the global burden of disease by focusing on, say, neglected diseases despite often driven nevertheless by market considerations. Yet another concern is the artificial creation of life or new life-forms as borne out by many critics long arguing scientists should not "play God." Finally, there are concerns about the biosafety and biosecurity of synthetic biology since nothing is known about what has been put in place (if anything) to stop new life-forms from escaping the laboratory, to stop them from being intentionally released into the environment, or to control them just in case they do escape. Although synthetic life sciences have beneficial uses, they hold the potential to cause serious harm and thus oversight and regulation are necessary. It is also argued that the precautionary principle must be applied before new synthetic organisms are fabricated.

H. ten Have and M. do C. Patrão Neves, *Dictionary of Global Bioethics*,
https://doi.org/10.1007/978-3-030-54161-3_491

Technology Assessment

Technology assessment (TA) refers to the systematic study of the effects introduction, extension, or modification of a technology might have on society (especially those that are unintended, indirect, and delayed). The Office of Technology Assessment in the United States has provided the following definition: "Medical technology assessment is, in a narrow sense, the evaluation or testing of a medical technology for safety and efficacy. In a broader sense, it is a process of policy research that examines the short- and long-term consequences of individual medical technologies." New technologies are usually introduced and applied in healthcare as soon as they become available without evaluating potential negative effects, positive results, safety, or efficacy (in stark contrast to what happens with new medication). The new discipline of TA was developed because of the potential effects new technologies have on healthcare such as cost increases and social effects. The introduction of new technologies such as stem cell research or artificial reproductive technologies raises significant ethical issues. For example, new methods of assisted reproduction have introduced hitherto unknown relationships based on genetic parenthood. In many countries TA has become a component of healthcare planning and strategy, which has the aim of producing relevant information to help policy-makers make decisions and to clarify ethical issues such that they can be considered in advance. The evaluation of new technologies is thus not only a scientific approach that provides factual information about medical, social, economic, and ethical consequences, but also a meta-technology that can function as an early warning system. The assumption here is that technologies have a lifecycle in which a technology emerges (prior to adoption), becomes a new technology (is adopted), then an accepted technology (in general use), and finally an obsolete technology (taken out of use). Evaluation at the right phase of the lifecycle can assist careful introduction and prudent use. However, another assumption is that information and decision-making are connected: the more and better the knowledge the better the decisions that will be made. However, in practice technical rationality often does not prevail and the acceptance of new technology is many times not the result of rational balancing of information. An important consideration to keep in mind is that TA will enable ethical analysis; the development of healthcare should not be driven by the mere fact that new technologies become available and are there to be applied. The evaluation that has to be made is whether they *should* be applied at all. In such an evaluation the important moral criteria are desirability and permissibility—not mere availability (the problem being that most TA studies are focused on economic assessment rather than ethical evaluation).

H. ten Have and M. do C. Patrão Neves, *Dictionary of Global Bioethics*,
https://doi.org/10.1007/978-3-030-54161-3_492

Telecare

Telecare refers primarily to the provision of remote care to people such as the elderly and disabled who have difficulties in getting to conventional care settings. Telecare is particularly important for people living in remote places such as islands and sparsely populated areas who would otherwise not have access to healthcare (it is widely used in less developed countries). It is often used when there is a need for a second opinion or the advice of a consultant and at times of pandemics such as COVID-19 when people are sheltering at home and disadvised from visiting hospitals. Providing care at a distance enables people to stay at home and be as independent as possible. Although telecare is usually provided by telephone, more sophisticated technology such as sensors and fall detectors alert care providers when help is needed and allow them to monitor patients. Telecare raises a number of ethical concerns such as the risk of its use to replace face-to-face care to reduce costs, shifting responsibility to patients, and introducing new forms of dependence. Although such technology can be used to increase self-awareness, it implies a shift in relations and responsibilities. It is important to keep in mind that telecare is not a solution to care problems and should only be introduced by involving, consulting, and engaging care users. Furthermore, careful reflection should be focused on which problems telecare should be used to address. Although telecare is vitally important in emergencies, it cannot be used as a remedy to combat the problems of ageing. Rather than simply monitoring the elderly, telecare can be used to enhance social support. However, this raises issues of privacy and confidentiality about who should have access to personal data and sensor data especially when considering the capacities of the elderly to live alone. Another ethical concern is that telecare will change the home environment in that the home will be constantly monitored and people who want to stay at home as long as possible will be continuously observed thus making them more vulnerable than they already are.

H. ten Have and M. do C. Patrão Neves, *Dictionary of Global Bioethics*,
https://doi.org/10.1007/978-3-030-54161-3_493

Testing, Genetic

Genetic testing refers to analyzing genetic material to identify the presence, absence, or modification of a particular DNA sequence, gene, or chromosome to determine the possibility of developing or passing on a genetic disorder common in someone's family or in his or her ethnic group. It is a type of medical test that (unlike genetic screening) should accurately determine the presence or absence of the condition being tested for. Genetic tests are more invasive than genetic screening and can entail side effects (particularly in pregnant women). There are many genetic tests (more than 1,000 at the time of writing and more are being developed) that can be sorted in some major types (disregarding the distinction between screening and testing) such as newborn screening, diagnostic testing, carrier testing, prenatal testing, preimplantation testing, predictive and presymptomatic testing, and forensic testing each of which raises specific ethical issues. Although genetic testing has benefits in eliminating fear and uncertainty surrounding someone's health, in confirming a disorder that can be clinically treated, and thus in empowering people to take control of their lives, it also has limitations and risks. It is important not only to balance potential benefits and risks before making the decision to test, but also to evaluate the emotional state of the person prior to the test (feelings of guilt, anger, anxiety, or depression are not uncommon). Although genetic testing is voluntary in almost all countries, it is still subject to ethical requirements in that it can only be undertaken on the recommendation and in the presence of a geneticist or genetic counselor (only geneticists or counselors should interpret the results and provide genetic counseling).

H. ten Have and M. do C. Patrão Neves, *Dictionary of Global Bioethics*,
https://doi.org/10.1007/978-3-030-54161-3_494

Testing, Premarital

Premarital testing refers to analyzing the genetic material of a couple who are planning to get married to identify whether either of them carries a copy of a gene mutation (carrier testing), which when present in two copies causes a genetic disorder (that might run in both their families). It is particularly useful in identifying certain genetic disorders such as cystic fibrosis, sickle cell anemia, and thalassemia and infectious diseases such as hepatitis B, hepatitis C, and HIV/AIDS. The main goal of premarital testing is to avoid the birth of disabled children, promote a healthy future generation, and protect the health of one or other partner. Premarital testing is beneficial in that it prevents the transmission of infectious diseases and avoids the ethical problems related to prenatal genetic screening (essentially abortion). However, it raises ethical problems of its own. The major one threatens a couple's autonomy when premarital testing becomes a public health policy recommended by national governments or even made mandatory as happens in some countries such as Saudi Arabia. Such cases raise many ethical questions: Would couples be allowed to get married when there is a risk for transmitting a genetic disease to their offspring? What happens when the pressure from society is too great for them to get married? Should genetic information only be used to counsel them by, say, recommending reproductive technologies over natural means of reproduction? What happens when a couple do not follow genetic counseling and give birth to a baby with a genetic disorder? Would that mean they have to take full moral or legal responsibility for the child and lose any social and financial support? Israel has a policy "for screening tests in the population aimed at identifying couples [according to their ethnic and geographic origin] at risk of bearing children with severe genetic diseases" such as cystic fibrosis and fragile X syndrome (the latter affecting a small proportion of the Jewish population). Although premarital genetic screening has been criticized for being made mandatory for thalassemia, sickle cell anemia, hepatitis B, and sexually transmitted diseases in Dubai and Saudi Arabia, the results of this policy show that the prevalence of β-thalassemia and sickle cell anemia has steadily decreased. Nevertheless, the ethical challenge of providing premarital testing and counseling that do not interfere with the autonomy of individuals remains. Several countries today have chosen to bypass such a challenge by putting in place premarital screening programs that ascertain the likelihood of one partner transmitting specific diseases to the other or to his or her children and by providing couples with options to plan a healthy family.

H. ten Have and M. do C. Patrão Neves, *Dictionary of Global Bioethics*,
https://doi.org/10.1007/978-3-030-54161-3_495

Torture

Torture refers to inflicting pain on someone to extract information or a confession and has long been commonly practiced in human history. For example, doctors during the Renaissance doctors were required to attend when someone was tortured and asked to certify whether he or she was fit to undergo torture. The 1984 Convention against Torture and Other Cruel, Inhuman or Degrading Treatment or Punishment defines torture as "any act by which severe pain or suffering, whether physical or mental, is intentionally inflicted on a person for such purposes as obtaining from him or a third person information or a confession, punishing him for an act he or a third person has committed or is suspected of having committed, or intimidating or coercing him or a third person, or for any reason based on discrimination of any kind, when such pain or suffering is inflicted by or at the instigation of or with the consent or acquiescence of a public official or other person acting in an official capacity." Therefore torture has both an intention (to inflict pain or suffering) and an aim (to gain information or extract a confession). Torture can be broken down into a number of categories: physical torture in which pain is inflicted on someone (the capacity of humans to invent new ways of torture beggars belief); psychological torture in which circumstances are created to make someone feel so anxious that he or she will provide information; and torture by proxy, for example by rendition that involves transferring, say, suspected terrorists to countries where torture is common practice and getting others to carry out the dirty work. The ethical debate on torture has been revived because of the War on Terror and falls into one of three categories: the first is that torture should be allowed if the information gained can save lives (putting the public interest first); the second is that torture should be allowed only in extraordinary circumstances (being extremely rare means torture does not risk becoming institutionalized); the third is that torture is deontological, never justifiable, and an immoral act that calls into question the institutions of liberal democracy.

H. ten Have and M. do C. Patrão Neves, *Dictionary of Global Bioethics*,
https://doi.org/10.1007/978-3-030-54161-3_496

Traditional Medicine

According to the WHO traditional medicine (TM) can be defined as the sum total of the knowledge, skills, and practices based on theories, beliefs, and experiences indigenous to different cultures (whether explicable or not) used in the maintenance of health and in the prevention, diagnosis, improvement, or treatment of physical and mental illness. TM refers to the wide variety of health practices, approaches, knowledge, and beliefs that incorporate not only medicine based on plants, animals, and minerals but also spiritual therapies and manual techniques, all of which are applied to treat, diagnose, and prevent illnesses and maintain well-being. Although TM is sometimes used to contrast Western scientific medicine with that of indigenous peoples, every country and society (Western or not) has its TM (a.k.a. alternative medicine or complementary medicine). TM is a part of traditional knowledge (a.k.a. indigenous or local knowledge) that consists of the mature long-standing traditions and practices of regional, indigenous, or local communities. Most indigenous people have traditional songs, stories, legends, dreams, methods, and practices that they use to endow specific human attributes to traditional knowledge as sometimes borne out by artifacts handed down from father to son or mother to daughter. Generally, there is no real separation between secular and sacred knowledge and practice in indigenous knowledge systems—they are one and the same. Moreover, knowledge in virtually all such systems is transmitted directly from individual to individual. Traditional knowledge can be characterized as being practical common sense based on teachings and experiences passed on from generation to generation; covering knowledge of the environment and relationships between things; being holistic and rooted in the spiritual health, culture, and language of people; and representing a way of life that combines the physical with the spiritual. TM is becoming increasingly used. In countries such as Ghana up to 80% of the population uses it for primary healthcare. In China traditional herbal preparations account for 30%–50% of total medicinal consumption, and 90% of the population of Germany have used complementary medicine at least once. Although doctors in Western countries often view the use of TM as a threat to effective treatment traditionally handled by standard care, nevertheless people in the West still go to alternative practitioners. It is widely argued that TM should be considered complementary to scientific medicine and never a substitute for it. The WHO launched its comprehensive TM strategy in 2002 and the WHO Congress on Traditional Medicine adopted the Beijing Declaration in 2008 that states "the knowledge of traditional medicine, treatments and practices should be respected, preserved, promoted and communicated widely and appropriately based on the circumstances in each country." It also underlines the responsibility governments have to the health of their populations and to formulating national policies,

H. ten Have and M. do C. Patrão Neves, *Dictionary of Global Bioethics*,
https://doi.org/10.1007/978-3-030-54161-3_497

regulations, and standards as part of comprehensive national health systems to ensure appropriate, safe, and effective use of TM. The purpose of the WHO strategy is not only to develop national policies but also to evaluate and regulate TM practices. Although more evidence needs to be amassed regarding the safety, efficacy, and quality of TM products and practices, when TM therapies deemed safe and effective are widely available this may well improve healthcare in many countries irrespective of whether they are in the West or elsewhere. The final aim is to integrate traditional medicine with mainstream medicine and healthcare.

Trafficking

Trafficking refers to trading human beings to exploit them. Violence, deception, or coercion are used to recruit, harbor (i.e., receive and hold people), transport, and exploit people by forcing them to work against their will. Such exploitation includes forced prostitution, forced labor, forced marriage, servitude, and even organ removal and is regarded as modern-day slavery. Traffickers use force, fraud, or coercion to lure their victims and look for people who are especially vulnerable such as those experiencing economic hardship, falling outside social safety nets, and victims of natural disasters or political instability. Matters are made worse because many trafficked people do not identify as victims or ask for help. More than 21 million individuals are today believed to have been trafficked around the world about 68% of whom are subject to forced labor and 22% to forced sexual exploitation. Human trafficking is believed to generate profits of USD32 billion every year. It is the third largest form of trafficking behind illegal drug trafficking and arms trafficking and is not restricted to developing countries as indicated by 4,460 cases of human trafficking being reported in the United States in 2017. Healthcare professionals are often in a unique position to identify and intervene on behalf of victims. Actions have been undertaken to stop human trafficking such as raising awareness of the phenomenon by making January 11 Human Trafficking Awareness Day (since 2010) and by countries in 2003 adopting the Protocol to Prevent, Suppress, and Punish Trafficking in Persons, Especially Women and Children (a.k.a. Trafficking Protocol). Adopted as a supplement to the UN Convention against Transnational Organized Crime the protocol defines people trafficking as: "the recruitment, transportation, transfer, harboring or receipt of persons, by means of the threat or use of force or other forms of coercion, of abduction, of fraud, of deception, of the abuse of power or of a position of vulnerability or of the giving or receiving of payments or benefits to achieve the consent of a person having control over another person, for the purpose of exploitation." The protocol makes it clear that exploiting vulnerable people is a crime irrespective of whether those exploited give consent. The protocol encourages countries to take measures to prevent and combat trafficking and assist victims. Although some countries have taken action such as the United States by introducing the Trafficking Victims Protection Act in 2000, other countries such as India (where more than 200,000 children are trafficked each year) have found it hard to effect legislation because corruption, lack of training, and shortage of resources hinder putting effective antitrafficking programs in place.

H. ten Have and M. do C. Patrão Neves, *Dictionary of Global Bioethics*,
https://doi.org/10.1007/978-3-030-54161-3_498

Transhumanism (*See* Enhancement; Transplantation; Genetic Engineering)

Transhumanism refers to a philosophical movement that believes the human race can evolve beyond its current physical and mental limitations such as ageing, susceptibility to disease, and proneness to disabilities by means of science and technology. It also calls for strong investment in new biotechnologies to overcome so-called human weaknesses such as short height, average intelligence, and poor memory. The major fields that have contributed to transhumanism and strengthened transhumanist aspirations are transplantation medicine brought about by bioengineering, bionics, and genetic engineering; genome editing and its capacity to alter the human genome; and cognitive sciences and emerging information technologies that have brought brain–computer interfaces and the possibility of merging humans and machines to create cyborgs ever nearer. Transhumanism as a philosophy looks so far beyond therapy that transhumanists consider it a human enhancement tool where the goal is not to repair human bodies or balance brain functions but to equip humans with what is needed to make them stronger, more powerful, and more resistant. Pursuing such a path should lead to human limitations being progressively overcome and give rise to an artificial new humanoid reality (or species) that many have called "transhuman." Transhumanism argues that humankind should be allowed to develop as much as possible, transgress natural limits, and be pursued by everybody. Transhumanism raises two major ethical issues. The first relates to fairness and the growth of inequity in the world as borne out by current biotechnologies only being accessible to wealthier population and not being equally available to all who need or desire them. Although today's biotechnologies will be less expensive in the future and will become available to more people, they will hardly be affordable for the majority of the world's population and in any event there will always be new and widely inaccessible technologies. Therefore the transhumanism movement will increase inequalities worldwide. The second issue relates to whether transhumanism should be allowed to develop at all because of its potential to make humankind as currently known go extinct—let alone allowing the rise of a new humanoid species defined by how well it performs at all sorts of activities, how powerful it is, and how capable it is of subjugating normal humans. Although initiatives to protect humankind and the human genome have been approved worldwide such as UNESCO's Universal Declaration on the Human Genome and Human Rights (1997), the declaration's scope is narrow and may not be able to face up to the challenges of today's emerging technologies.

H. ten Have and M. do C. Patrão Neves, *Dictionary of Global Bioethics*,
https://doi.org/10.1007/978-3-030-54161-3_499

Transplantation Medicine

Transplantation medicine refers to a relatively recent form of medicine that started in the second half of the twentieth century after the first successful experimental kidney transplant in 1954. This new medical field became well known as a result of the discovery of histocompatibility (a.k.a. tissue compatibility) and the development of immune-suppressing medication to overcome organ rejection. However, the major achievement that brought transplantation to the attention of the public was the first heart transplant in 1967 by Christiaan Barnard. The public were impressed not only by the medical achievement of successfully overcoming the technical complexity of the surgery, but also by the fact that the heart (the symbolic center of human identity, individual identity, even of the soul) was involved. Such beliefs led to the first ethical problems raised by transplantation medicine being concerned with the identity of the receiver and the integrity of the donor. These were specifically anthropological questions: doubts were expressed whether transplant patients would preserve their identity or their personality or whether they would assimilate and express donor characteristics (this was a concern at the time Christiaan Barnard performed the first heart transplant). When transplants went from post-mortem donation to living donation, the question turned on its head and concerned the integrity of the donor—not the receiver: Would the donor remain the same after giving away an integral part of himself or herself? The field continued to evolve and in 1981 benefited from the discovery of cyclosporine used to prevent organ rejection after a transplant and acknowledged as being one of the most remarkable successes in the history of medicine. Today it is possible to transplant the cornea, heart, lung, trachea, liver, kidney, pancreas, skin, and vascular tissues. Moreover, transplants today take on a number of forms such as from a deceased donor or a living donor (direct donation when the donor chooses the recipient or less commonly indirect donation); paired kidney exchange (when A wants to give a kidney to B but is only a match for C, while D wants to give a kidney to C but is only a match for B—hence A gives to C and D to B); or domino transplantation (when A is not sick enough to receive an organ such as a liver from the normal donor pool but sick enough to need a new liver for a healthy life, while B is suffering from the genetic blood disease amyloidosis that destroys organs such as liver and is at the top of the list for a liver transplant—hence B will donate the liver to A who will not develop amyloidosis in the foreseeable future). Transplantation has become a victim of its own success. The number of people whose lives can be saved by a transplant is continuously growing while there are not enough organs available. Every year many die while on the waiting list for an organ as a consequence of the imbalance between the number of organs required and those available. Such an imbalance causes serious ethical problems such as the trafficking of organs, tissues, and body parts and all the attendant human dramas involved in human dignity

H. ten Have and M. do C. Patrão Neves, *Dictionary of Global Bioethics*,
https://doi.org/10.1007/978-3-030-54161-3_500

and social justice being violated. The imperative to increase the number of organs for transplantation is best achieved by optimizing health service organization (skilled teamwork, close communication, and efficient transport of the organ from professional identification of a possible donor through to finding a suitable recipient) and putting in place a model of informed consent that requires people to opt out. Informed consent is undoubtedly the core ethical issue in transplantation and can be done either by opting in (informed consent stricto sensu in which each citizen is requested to express his or her free will concerning organ donation and citizens who do not express themselves are considered non-donors) or by opting out (presumed consent in which each citizen is considered a donor unless he or she takes a position against it). In the latter case the role of the family has been questioned since the family cannot legally oppose the harvesting of organs and other body parts. Although most physicians act ethically and continue to ask the family's permission and risk losing a viable donor, practically it would be better for the physician to talk to the family detailing how the procedure works, justifying it, and assuring the family that due respect will be shown to their loved-one rather than asking their permission and risking losing life-saving organs. Transplantation medicine continues to evolve and currently different sources of human organs are being investigated such as animals (xenotransplantation), the growing of stem cells (bioprinting), and electromechanical devices (bionics).

Triage (*See* Emergency Medicine)

Triage refers to selecting people for treatment from a larger group of people all of whom are in need of treatment. Triage is used when events such as war, terrorist attacks, or disasters cause so many casualties that healthcare resources are too limited to treat everyone. The word "triage" derives etymologically from the French *trier* (to sort out) and was first introduced by French chief surgeon Dominique Larry during the Napoleonic Wars. He formulated three principles based on providing medical care according to systematic and rational standards. The first called for standards to be based on medical urgency, not related to military rank, nationality, or social status, and bias should be avoided. The second called for medical urgency to be focused on saving as many lives as possible. The third called for a fair system of triage to sustain trust among soldiers. The concept of military triage has now been widened to include mass emergencies when humanitarian aid is needed and situations where resources are scarce such as limited numbers of beds in an intensive care unit. Triage is used in life-saving situations in which patients are categorized according to the severity of their injuries and illnesses and treatment is prioritized based on the probability of survival and recovery. The ethical issues concerning triage are considerable and matters are made worse by there being no agreement on the ethical criteria to use and make triage decisions. Although the focus in ordinary triage is on the interests of individual patients, in extraordinary circumstances the focus is unclear as to whether it is on survival of the greatest number of people or on those most likely to survive. Even when principles are proposed to guide decisions to allocate scarce resources, they often present a checklist of moral points to take into account rather than the medical needs of those injured leaving some to doubt whether triage itself is ethically justified, whether triage in a disaster setting will save more lives, and whether triage may actually worsen outcomes. The concept of triage introduces a military and paternalistic perspective. Larry's focus on an egalitarian approach did not imply that available resources were focused on the interests of the individual person in need but on those of the war effort. This is further exemplified today in military directives such as NATO's *Handbook of Emergency War Surgery* (2009), which identifies three groups of patients in thermonuclear warfare: those with minimal wounds, those with wounds that are too extensive (they continue fighting or die), and those with relatively simple injuries that require immediate surgery. Only the latter will be evacuated and treated. Although the focus of triage here is salvage value (saving the greatest number of lives), the ultimate goal is return of the greatest possible number of soldiers to the front. The same military rationale underlies the concept of minimum qualifications for survival: saving individual lives is not important as such but only within a broader context. Triage was later looked at from a paternalistic perspective in which individual freedom and human

H. ten Have and M. do C. Patrão Neves, *Dictionary of Global Bioethics*,
https://doi.org/10.1007/978-3-030-54161-3_501

rights were restricted for the sake of the public good or the well-being of the population as a whole. Although basic values when triage is being considered are not determined democratically or by public deliberation but by authorities who take control of the events, such control can be legally regulated. Even though emergencies and disasters are declared by government bodies and addressed by unconventional legal responses, decisions nevertheless need to be made about what to legislate first. This led to the term legal triage being introduced to help construct a legal environment to facilitate legitimate public responses.

Trust

Trust refers to the personal feeling and social bond an individual feels toward someone else concerning the latter's truthfulness or reliability and establishes a specific bond between them that in the absence of trust does not really exist. Trust has long been used in a religious context (especially since medieval times) as the only right attitude to adopt before God. Although it entailed accepting dependence and vulnerability toward a being in whom you place your trust such as God, later political philosophy continued to express the idea that by trusting someone you transfer your power to that person. In moral philosophy (shaped by Kant) there is an absolute duty to keep promises and doing so is synonymous with trust. More recently, trust has been a recurring theme in the work of many philosophers (particularly Nicolai Hartmann, Thomas Scanlon, and Diego Gambetta) who look beyond religion at other moral attitudes that are based on trust such as loyalty. Trust is a delicate and very fragile relationship that once broken is very difficult or even impossible to restore. This is the reason it should be cherished. The only way to nurture it is to be truthful (at the intellectual level) and faithful (at the practical level). Trust is of paramount importance within the healthcare setting. If the patient does not trust the physician, then there is a risk of the patient not providing the information needed for an accurate diagnosis, not receiving inadequate therapy, and not following prescriptions. It is important not only for healthcare professionals to gain the trust of patients and their families but also for the latter to trust the healthcare team. In their *For the Patient's Good* (1988) Edmund Pellegrino and David Thomasma underlined the importance of trust in the physician–patient relationship when presenting their model of beneficence-in-trust as a counterpart to the hegemony of autonomy. Indeed, outside a trustful relationship care will not be efficient or personalized and may even be humane. The same applies to clinical research, health policies, other domains of global bioethics, and human relationships overall. Trust calls for absolute truth and fidelity no matter where it is found.

H. ten Have and M. do C. Patrão Neves, *Dictionary of Global Bioethics*,
https://doi.org/10.1007/978-3-030-54161-3_502

Truth Telling

Truth telling refers to communicating information that is factual (true). By doing so someone discloses what he or she knows as true to someone else who is in need of the information. Telling the truth is vitally important in the clinical setting. Deontologically, truth should always be told because otherwise, of course, there will be no trust both of which are essential for ethical relationships. When viewed from the deontological principle of autonomy, truth telling is considered a duty since autonomy requires knowing the truth. However, taking such a line of reasoning to its limit can lead to someone who does not want or is not ready to cope with a painful truth (especially when suddenly transmitted) experiencing negative feelings such as unacceptance, anger, and depression. Truth can be deeply harmful. Truth telling from a consequentialist perspective should always be considered along with any foreseen consequences. If they are negative, then truth should be withheld and possibly replaced by something called "mercy lying," which is a deliberate lie told for compassionate (mercy) reasons to avoid suffering. However, the capacity of someone to cope with the truth must be evaluated by a third party (not by the person himself) and this can result in hiding pertinent information from that person. Therefore mercy lying can be even more harmful than truth telling and is widely regarded as a source of distrust, disappointment, and frustration. There is some middle ground that adopts the best of both worlds by complying with the deontological duty and the precepts of common morality. This involves telling the truth and combining it with consideration of the expected impact following the consequentialist procedure. Truth in this middle ground should be told gradually allowing the person either to slowly assimilate the information bit by bit or to halt the process definitively or until a later time once the new reality has sunk in. There must be conditions here such that a person's right not to know is not jeopardized (except for contagious diseases). Ethically, when all things are considered the duty of truth telling should not be challenged despite being able to override the right not to know in certain circumstances. The focus should be on communicating and ensuring all healthcare professionals (especially physicians) receive specific training in techniques of communication.

H. ten Have and M. do C. Patrão Neves, *Dictionary of Global Bioethics*,
https://doi.org/10.1007/978-3-030-54161-3_503

Tuberculosis

Tuberculosis (TB) is a bacterial disease caused by *Mycobacterium tuberculosis* that most often affects the lungs and sometimes the kidney, spine, and brain. Not all persons infected become sick as borne out by some people developing latent TB (when symptoms may be mild for some time such that care is delayed and bacteria are transmitted in the meantime) while others have active TB. Active disease manifests itself through symptoms such as coughs (bringing up sputum and blood), chest pains, weakness, weight loss, fever, and night sweats. People with active TB can infect 10–15 other people through close contact over the course of a year as a result of the disease being spread through the air. It is estimated that 25% of the global population have latent TB and a 5%–15% lifetime risk of falling ill with active TB. The risks for people with compromised immune systems (e.g., HIV patients or malnourished people) falling ill are much higher. Approximately 45% of people with TB will die without proper treatment and nearly all HIV-positive people with TB die despite TB being curable and preventable. TB is among the top 10 causes of death in the world. In 2017, 1.6 million people died from TB, around 1 million children became ill with TB and 230,000 of them died of TB, while TB is the leading killer of HIV-positive people. A major problem is drug resistance brought about by using drugs inappropriately, prescribing them incorrectly, giving people poor-quality drugs, and patients stopping treatment prematurely all of which contribute to TB strains becoming resistant to anti-TB medicines. Multidrug-resistant tuberculosis (MDR-TB) is a form of TB caused by bacteria that do not respond to isoniazid and rifampicin, the two most powerful, first-line, anti-TB drugs. Although MDR-TB is treatable and curable using second-line drugs, it is still regarded a public health crisis and a health security threat. In 2017 the WHO estimated there were 558,000 new cases of TB that were resistant to rifampicin (the most effective first-line drug) and that 82% of these patients had MDR-TB. The MDR-TB burden largely falls on India, China, and the Russian Federation that together account for nearly half of global cases. The sad truth is that worldwide only 55% of MDR-TB patients are currently successfully treated, tuberculosis is making a comeback in many countries, and the number of new cases is growing. In 2016, there are 10.4 million new cases of TB and more than 1.7 million deaths. In 2017 the G20 Leaders' Summit in Hamburg decided to take action on TB declaring it a health priority. It made a commitment to stimulating more research and improving access to affordable drugs (the summit recognized access to treatment was another ethical problem in addition to drug resistance and acknowledged only one in 10 MDR-TB patients had access to treatment). There are many reasons people do not seek treatment such as poor infrastructure, limited medical resources, shortage of healthcare professionals, high drug prices (new drugs in particular), and treatment regimes with newer drugs such as

H. ten Have and M. do C. Patrão Neves, *Dictionary of Global Bioethics*,
https://doi.org/10.1007/978-3-030-54161-3_504

bedaquiline and delamanid costing much more than older treatment regimes (people who need at least 18 months of bedaquiline now pay more than USD2,000 which is a 50% price increase). The fundamental ethical problem here is that the economic conditions under which people live are so poor that assistance is needed for health to be restored and lives to be saved. TB raises another traditional ethical issue related to infectious diseases concerning the involuntary confinement of patients with open TB: How should individual freedom and public health be balanced?

Ubuntu Ethics

Ubuntu refers to a value system traditionally used in certain parts of Africa, comes from the Nguni Bantu word meaning "the quality of being human" or "humanity," and entails values such as respect for others, helpfulness, community, sharing, caring, trust, and unselfishness. Ubuntu is an ethical system founded on some basic social truths that normal humans cannot deny such as being social, living together, needing to cooperate, and pursuing the common good. This is often literally translated into English in aphorisms like "a person is a person through other persons" or "I am because we are." Persons here refer to ancestors and future generations. Ubuntu ethics is related to the ethical tradition of communitarianism a good example of which bioethically is the application of group consent. Although this does not replace individual consent, any participation of someone in clinical research is first extensively discussed by the community or tribe to which the individual belongs and is decided by group consensus. The reason ubuntu ethics was explicitly promoted in post-apartheid South Africa in the 1990s was because it was capable of getting people from all backgrounds to come together. Globally, ubuntu bioethics has been promoted as a way of countering hatred, intolerance, exploitation, poverty, and abuse of human rights.

H. ten Have and M. do C. Patrão Neves, *Dictionary of Global Bioethics*,
https://doi.org/10.1007/978-3-030-54161-3_505

Utilitarianism (*See* Consequentialism)

Utilitarianism refers to a philosophical theory that is particularly important in sociopolitical and moral thought and is characterized by the primacy of the consequentialist principle of utility in which good and bad are determined, respectively, by the degree of happiness (pleasure and the absence of pain) or unhappiness they produce for society as a whole—not just for the individual. The utilitarian goal is to maximize the good (welfare) and make it available to as many people as possible (principle of utility). Utilitarianism was introduced by Jeremy Bentham and later developed and systematized by John Stuart Mill (both considered classical utilitarians) who between them strongly shaped Anglophone morality from the eighteenth century right up to today. Its antecedents go back to Ancient Greek philosophy, Epicurus, and hedonists who identified good with pleasure. However, the direct precursors of utilitarianism are self-preservation (Hobbes argued this was the summum bonum), valuing emotions as natural phenomena human beings experience in relation to happiness or unhappiness (argued by Hume), empiricism, and naturalism. The major difference between Bentham's utilitarianism and its precursors was the attribution of a prescriptive dimension to the formerly descriptive dimension of the principle of utility (Hume felt the utility principle was an observational fact—not a rule of action) that then becomes normative. Stuart Mill stressed the importance of coherence between utilitarianism and liberalism that influenced contemporary morality and significantly shaped bioethics. He argued the path to happiness was singular and thus each individual had the right to pursue it freely. Later Henry Sidgwick rejected the former empiricism and naturalism of utilitarianism, disconnected it from any kind of metaphysical project, and was instrumental in building the transition between classic and contemporary utilitarianism. Although contemporary utilitarianism abandons hedonist influences and stresses the importance of moral rules, such rules are only justifiable if they create more utility and greater good than other possible rules.

Contemporary utilitarianism is strongly centered on the notion of justice and this is the reason it has been widely criticized (particularly by John Rawls). Achieving the greatest good for as many people as possible not only neglects the singularity of the person and his or her particular concrete situation, but runs the risk for not respecting that person's individual rights. Nevertheless, utilitarianism, as a society-centered endeavor, can be a useful approach in public health policies and in working out what is in the best interest of a particular group of patients such as those with communicable or genetic diseases.

Utilitarianism is often used as the basis for measures to protect animal interests and animal welfare.

Utilitarianism remains a significant influence in bioethical debates.

H. ten Have and M. do C. Patrão Neves, *Dictionary of Global Bioethics*,
https://doi.org/10.1007/978-3-030-54161-3_506

Vaccination

Vaccination refers to injecting someone with a microbe (or parts of a microbe) that has been-killed or weakened to stimulate the immune system against the microbe, to produce immunity against a disease, and to prevent disease (hopefully). A healthy immune system recognizes invading bacteria and viruses and produces antibodies to destroy or disable them. Vaccinations or immunizations ready the immune system such that it can prevent disease. Different types of vaccines are available such as life-attenuated vaccines (weakened germs) against measles, mumps, and rubella (MMR combined vaccine), smallpox, chickenpox, rotavirus, and yellow fever; inactivated vaccines (killed microbes) against hepatitis A, flu, polio, and rabies; subunit, recombinant, polysaccharide, and conjugate vaccines that use specific pieces of germs (e.g., protein or sugar) against hepatitis B, human papillomavirus, whooping cough, pneumococcal disease, meningococcal disease, and shingles; and toxoid vaccines that use a toxin produced by the germ that causes the disease against tetanus and diphtheria. Against Covid-19 new RNA vaccines have successfully been developed (genetically engineered virus RNA that provokes immune responses against the spikes of the coronavirus). Countries apply vaccination schedules that entail their populations receiving repeat vaccinations between birth and 18 years of age (e.g., the MMR vaccine is given at 12–15 months and repeated between 4 and 6 years later). Vaccination is regarded as one of the most effective health interventions since it has eradicated smallpox, almost eradicated poliomyelitis, and today prevents between 2 and 3 million deaths every year. Although its effectiveness and low costs have encouraged countries to make vaccinations mandatory (for example, in 11 European countries) the benefits and harms of vaccination are increasingly contested today. There is a growing antivaccination movement that has been listed by the WHO as a major global health threat (reluctance to get vaccinated has resulted in the incidence of measles rising). Although the idea that parents should have the opportunity to choose whether or not their children are vaccinated and which vaccines their children receive is morally right, it is important to stress that vaccination rates for highly infectious diseases such as measles should be above 90% to maintain herd immunity. Although harm related to vaccination can never be completely excluded, many false ideas about the harmfulness of vaccines have been circulated one of which was that the MMR vaccine caused autism. This was based on two publications in 1998 and 2002 by author Andrew Wakefield that had to be retracted because they were deemed to be fraudulent. Although a number of other studies disproved any relationship with autism, the idea that vaccines are associated with autism persists in popular discourse. The major ethical challenge facing vaccination is striking a balance between individual and collective interests. Although individual autonomy means individuals have the right to decide

H. ten Have and M. do C. Patrão Neves, *Dictionary of Global Bioethics*,
https://doi.org/10.1007/978-3-030-54161-3_507

whether or not to be vaccinated, many who take such a decision do so for selfish reasons because they believe they are protected by herd immunity and do not need to expose themselves to any risks. It is important to keep in mind that vaccination is a common good, that the greater the number of people vaccinated the better individuals are protected (even those who decided not to be vaccinated), and that the individual decision to be vaccinated not only protects the individual but also the community. When most people are immune and the disease cannot disseminate easily or lead to an outbreak, this is called "herd immunity." Although there have always been exemptions to vaccination (e.g., for religious reasons), there is a limit beyond which herd immunity will no longer provide sufficient indirect protection against infectious diseases.

Values

Values refer to the importance, worth, or usefulness of something (a reality) and are the criteria used to evaluate such a reality. There are so many kinds of values ranging from religious, cultural, esthetic, economic, moral, political, transcultural through to environmental that there is a science dedicated to it called axiology (a.k.a. the study of values or value theory). The word "axiology" derives etymologically from the Greek *axios* (value) and *logos* (science, study). Those values of ethical interest are by and large moral values and transcultural values since they relate to human actions. Moral values can be objective or subjective. Objective values are intrinsic to specific realities, constitutive of that reality in itself, and tend to be universal and permanent. The value belongs to something (the reality considered in itself). The classic example of an objective value is the person in that he or she has an absolute, equal, and unconditional value regardless of any particular features. The principle of human dignity is based on the objective value of the person. More recently nature has been recognized as possessing an objective value instead of its traditionally assigned utility value. The value of nature does not depend on the person who appreciates it because nature has a worth of its own regardless of whether it is admired by anyone or not. However, subjective values do depend on the person appreciating them and on that person's interests or tastes. In short, subjective value relies on someone evaluating something (most values are subjective). Global bioethics tries to identify, define, and reach a consensus on transcultural values considered universal and in so doing provide a basis for common or global moral reasoning.

H. ten Have and M. do C. Patrão Neves, *Dictionary of Global Bioethics*,
https://doi.org/10.1007/978-3-030-54161-3_508

Vegetarianism (*See* Animal Ethics; Animal Welfare; Zoocentrism)

Vegetarianism refers to the practice of voluntarily abstaining from eating animal flesh. Such a practice can be broken down into a number of categories such as semi-vegetarianism or flexitarianism (in which no meat is consumed although occasionally red meat or poultry are); vegetarianism strictu sensu (in which no animal protein including meat, fish, and fowl is consumed); and pure vegetarianism or veganism (in which no animals or their by-products such as milk, eggs, or honey are consumed). Other categories of vegetarianism have been established such as lactovegetarianism (in which red or white meat, fish, poultry, or eggs are not consumed but dairy products are); ovovegetarianism (in which red or white meat, fish, poultry, or dairy products are not consumed but eggs are); lacto-ovovegetarianism (in which red or white meat and fish are not consumed but dairy products and eggs are); pollotarianism (in which red meat, fish, or seafood are not consumed but poultry are); and pescatarianism (in which red or white meat and poultry are not consumed but fish and seafood are). Such products have long comprised the diet of human beings who have biologically evolved as omnivores and whose teeth and digestive system have adapted to eat a wide variety of food of both plant and animal origin. Therefore the health of human beings depends on having a balanced diet including animal and plant proteins. Nevertheless, there have always been food restrictions. In ancient times, Pythagoras rejected the consumption of meat because he believed in the cross-species transmigration of souls and Plutarch and Porphyry (also vegetarians) advocated the moral obligation to respect all life-forms. Most religions have food interdictions: Buddhism encourages vegetarianism; Hinduism prohibits beef and restricts other meats and fish; Eastern Orthodox Christianity restricts the consumption of meat and fish; Islam prohibits pork and some birds; and Judaism prohibits pork and shellfish. Although the origin of such food restrictions varies, they are by and large requirements to intensify spiritual life by ingesting a lighter diet and to preserve the cleanliness of the soul because eating animals is considered unclean (for different spiritual reasons). Such food restrictions today have rationally been established out of respect for other life-forms. People regard vegetarianism today as an ideological, philosophical, or ethical way of rejecting the consumption of sentient beings like animals, promoting animal welfare, and recognizing the intrinsic value of animal life and animal rights. Although the objective of intensive animal farming is to maximize production and minimize cost, this has led to livestock being reared, transported, and slaughtered under conditions in which they suffer, health problems are rife, and environmental impacts are serious. Refusing to eat meat is a way to protest against the dreadful living conditions imposed on animals. By refusing to be part of this system, they help the strategy calling for a decrease in the demand for animal flesh thus reducing its supply. The inherent value of animal life is enough on its own for

H. ten Have and M. do C. Patrão Neves, *Dictionary of Global Bioethics*,
https://doi.org/10.1007/978-3-030-54161-3_509

people to refrain from eating animals because it is recognized as a value in itself—not just the instrumental value it has for humans. Although vegetarianism has benefited from the development of nutritional sciences contributing to the design of healthy diets without animal proteins, it raises the ethical question of whether it is right for vegetarian parents to restrict the food their children eat, did not choose, and that could be harmful to their health.

Veterinary Ethics

Veterinary ethics refers to the ethical study of veterinary medicine. Veterinarians have to balance the demands of animal health, clients, industry, employers, society, business, and animal welfare. The five types of ethical obligations widely recognized in veterinary medicine are: first, obligations to the client who hires the veterinarian and usually owns the animal; second, obligations to peers and the profession; third, obligations to society in general such as protecting public health; fourth, obligations to the veterinarian himself or herself; and fifth, obligations to the animal. Professional ethics in veterinary medicine is articulated by professional organizations. Veterinarians in many countries take a Veterinary Oath when they enter the profession. The American Veterinary Medical Association (AVMA) promotes a code of ethics. The first principle in the code is: "A veterinarian shall be influenced only by the welfare of the patient, the needs of the client, the safety of the public, and the need to uphold the public trust vested in the veterinary profession, and shall avoid conflicts of interest or the appearance thereof." The fundamental issue in veterinary ethics is whether the owner of the animal or the animal itself is the primary responsibility of the veterinarian. There are major differences in the way such responsibility is viewed such as the owner may give priority to making economic gains by asking veterinarians to prescribe animal food containing hormones, vitamins, and antibiotics to stimulate growth (while from the perspective of animal welfare and public health these practices are harmful); the owner may request euthanasia as an alternative to paying for expensive treatments for the animal (veterinarians sometimes refuse euthanasia under these conditions); the veterinarian may be involved in the mass killing of animals when there are outbreaks of infectious diseases; the veterinarian may be involved in genetic selection and animal breeding practices to enhance food production (sometimes at the expense of animal welfare); and the veterinarian is often involved in the sterilization of pets (despite animal rights advocates considering it invasive of animal nature and unacceptable). Confronted with global challenges such as pandemics, it is widely argued that a One Health approach is needed. Interdisciplinary collaboration between physicians, nurses, veterinarians, and other health professionals is required to make sure that all aspects of healthcare of humans, animals, and the environment are taken into consideration (especially at the human–animal interface). It is no longer possible to separate human health from animal health or environmental health.

H. ten Have and M. do C. Patrão Neves, *Dictionary of Global Bioethics*,
https://doi.org/10.1007/978-3-030-54161-3_510

Violence

Violence refers to behavior in which physical force is deliberately used to hurt, damage, or kill someone or something. It has always been a pervasive part of human life. The WHO has been closely looking at violence and health since 2002 and defines violence as: "The intentional use of physical force or power, threatened or actual, against oneself, another person, or against a group or community, that either results in or has a high likelihood of resulting in injury, death, psychological harm, maldevelopment or deprivation." This definition emphasizes a number of unfortunate but very apparent truths: violence does not necessarily result in death or injury; violence can lead to psychological harm, deprivation, and maldevelopment; violence is easier to pursue against women, children, and the elderly; and violence is intentional. Violence is clearly distinguishable from unintended events (accidents) that result in injuries. However, the intent to use force does not necessarily mean that there was an intent to cause damage. For example, vigorously shaking a crying child with the intent to quieten it may cause brain damage—but has not the intent to cause injury. The WHO definition relating violence to the health or well-being of individuals recognizes that violence is culturally determined and that people may not perceive acts that harm others as violent. For example, in some backward cultures hitting your spouse may be regarded by some as an acceptable practice but they are violent acts with important health implications. Violence is usually broken down into three types such as self-inflicted violence (suicidal behavior and self-abuse); interpersonal violence (committed against family members such as spouses, children, and the elderly and community violence between unrelated individuals), and collective violence (social, political, and economic violence). Violence is regarded as a global public health problem and as one of the leading causes of death worldwide for people between 15 and 44 years of age. Although most of the violence in everyday life is invisible since it occurs in homes, workplaces, and medical and social institutions, in 2000 an estimated 1.6 million people worldwide died as a result of self-inflicted, interpersonal, or collective violence of which nearly one half were suicides, almost one third were homicides, and about one fifth were war related. Although violence has always occurred, it is today considered an unacceptable part of being human. A great deal of ethical, religious, and philosophical thinking has been put into efforts to reflect on the causes and prevention of violence and since the 1980s public health research has been analyzing the roots of violence to better understand how to prevent its occurrence. From the viewpoint of bioethics, the phenomenon of violence is rejected from various perspectives as disrespecting the dignity of human beings and the principle of personal autonomy, as being clearly in conflict with the principles of beneficence and non-maleficence, and as implying unfair treatment (by not implementing the principle of justice).

H. ten Have and M. do C. Patrão Neves, *Dictionary of Global Bioethics*,
https://doi.org/10.1007/978-3-030-54161-3_511

Virtue Ethics

Virtue ethics refers to a theoretical model of ethics that emphasizes the central role virtues play in moral life. Virtues are dispositions people have to act in a specific way. Although they do not express obligations, they are presented as highly beneficial to self-flourishment and achievement of the common good and therefore encouraged by education and example. Although virtue is central in this ethical theory, making use of practical reasoning is the second main feature of the theory (i.e., a rational deliberative process for decision-making). Taking into consideration the ends to be achieved, the means available, and the singularity of the concrete situation, a moral agent will dialectically balance them such that the best course of action is taken. Historically, virtue ethics has its roots in Ancient Greek moral philosophy where it was fundamentally developed by Aristotle (after Socrates and Plato) in his *Nicomachean Ethics*. Virtue (a.k.a. human excellence or moral character) is an excellent trait of character helping someone fulfill his or her own promise leading to self-accomplishment of potentialities, to self-perfection, to happiness (defined as fulfilling one's self) and thus contributing to practical wisdom and enhancing the importance of deliberation (i.e., balancing arguments applied to the specific situation at stake). The history of Western morality (and of Eastern spirituality since Mencius and Confucius) was defined as a form of virtue ethics right up to the eighteenth century and Kantian deontological morality. Since then right through to the late twentieth century virtues have by and large disappeared from ethical theories. Recently virtue ethics was revived by the Scottish philosopher Alasdair MacIntyre in his book *After Virtue* (1981). MacIntyre recovers Aristotelian virtue ethics and Thomistic metaphysics and ethics in an attempt to renew human agency by examining virtues and their importance to good community life. Virtue ethics morality was first developed in bioethics by Edmund Pellegrino and David Thomasma in *For the Patient's Good* (1988). This work was a reaction to the hegemony of autonomy in principlism and to the individualism of the human rights approach in biomedical ethics (deontology) and to utilitarianism (consequentialism). The authors wanted to stress the importance the moral character of healthcare professionals played in overcoming the limitations of normative ethics. In the absence of an adequate moral rule to guide healthcare professionals in a concrete situation, the authors argue people should rely on the moral character of such professionals to make the best decision. Even when there are rules it does not necessarily mean they will be applied unless, of course, people are dealing with professionals whose moral character impels them to respect such rules. Virtue ethics has been gaining recognition in recent decades so much so that it is currently one of the most prominent ethical doctrines. Virtues have now been associated and integrated in almost all models of applied ethics (notably in principlism). In short, moral life requires human virtues.

H. ten Have and M. do C. Patrão Neves, *Dictionary of Global Bioethics*,
https://doi.org/10.1007/978-3-030-54161-3_512

Virus Sharing

Virus sharing refers to the sharing of viruses for global pandemic preparedness, pandemic risk assessment, candidate vaccine virus development, updating of diagnostic reagents and test kits, and monitoring resistance to antiviral medicines. Benefit sharing is a fundamental principle of global bioethics. An example is the sharing of influenza virus countries first affected by influenza epidemics providing virus samples to an international network of influenza laboratories coordinated by the WHO Global Influenza Surveillance and Response System (GISRS). The purpose is to assess the risk of an influenza pandemic and to assist in preparedness measures such as the early fabrication of vaccines and antiviral medication. In 2007 the Indonesian government decided not to share avian flu virus samples with the WHO. Although countries like Indonesia where flu epidemics often start are usually the first to provide materials to prepare preventive and therapeutic responses in other countries, the exchange of virus samples for vaccines was at the time unfair and unequal. The virus sample is given by the WHO to private companies, the vaccine produced being patented by these companies, the countries that donated the sample having to buy the vaccine, and the benefits of international cooperation going to pharma enterprises that receive biological and genetic materials for free and to populations of developed countries wealthy enough to secure vaccines before the epidemic shows up within their borders. Although (as happened with the 2005 avian influenza outbreak) Indonesia had previously provided virus samples, it later discovered that the WHO had transferred the sample to a company that patented the resulting vaccine thus making the vaccine unaffordable for most countries. While believing the system of sharing was built on the principle of global solidarity, Indonesia realized in practice that it had been commercialized and turned (by the WHO itself) into a system that further increased health inequality. Matters were made worse when Indonesia was hit hard by the 2005 avian flu. Although it had provided the materials to make the vaccine, it was unable to buy the vaccine since wealthy countries had reserved and bought all the available stock. Such flagrant injustice and inequality led to the decision being made to stop virus sharing until balanced benefit sharing was accomplished (no vaccine, no virus). The Indonesian decision initiated a long and heated debate in which the opposing views of developing and developed countries were aired and resulted in an agreement being reached in May 2011 to ensure the equitable distribution of benefits (Pandemic Influenza Preparedness Framework). Countries were encouraged to share virus samples and recipients of the virus (pharma companies) were obliged to ensure benefits such as vaccines from virus and genetic material were being distributed in a fair and equitable way.

H. ten Have and M. do C. Patrão Neves, *Dictionary of Global Bioethics*,
https://doi.org/10.1007/978-3-030-54161-3_513

Vivisection (*See* Animal Ethics; Animal Research)

Vivisection refers to the practice of carrying out operations on live animals for the purpose of experimentation or scientific research. When vivisection is carried out on humans it is regarded today as torture. Animal experiments have long been used in the history of medicine and have been justified by the general but false assumption of Descartes that animals were unable to feel pain. However, even when it was generally acknowledged that animals feel pain, there were those who argued vivisection could still be considered justifiable for the greater good of humankind by amassing knowledge and developing skills to treat diseases that affect human beings. Animals have long been used to study physiology and pathology (particularly since the rise of experimental medicine in the nineteenth century). Although one of the founders of experimental medicine Claude Bernard (1813–1878) argued that vivisection was indispensable to physiological research, around the same time concerns emerged in societies about the treatment of animals. The philosopher Jeremy Bentham argued that animals were sentient beings and had interests that should be considered. In 1824 the Society for the Prevention of Cruelty to Animals was established in Great Britain and was instrumental in raising concerns about the suffering of laboratory animals and in the foundation of the first antivivisection organizations in the 1870s. The first acts prohibiting cruel treatment of animals were introduced in several countries in the 1830s. Although the word "vivisection" derives etymologically from the Latin *vivus* (alive) and *sectio* (cutting) and harks back to the second century AD and the Greek physician Galen, opponents of animal experiments increasingly used the term in the nineteenth century. The Cruelty to Animals Act became an act of parliament in 1876 in Great Britain and had the aim of regulating the use of live animals for experimental purposes. Research requiring vivisection is subjected in most countries today to ethical review by special animal research committees. Across Europe 11.5 million animals were used in experiments in 2011 most of which took place in the United Kingdom. In 2014 approximately 318,000 experiments on rabbits, 4,000 on cats, 11,000 on horses, 23,000 on dogs, and 8,898 on primates took place in Europe. Most animals used are mice, rats, flies, and worms. Although most animals are not used to test drugs and most animal experiments focus on basic research and producing genetically modified animals such as mice, more than 100 million animals are killed in laboratories each year. Moreover, even though the numbers refer to experiments with animals and not necessarily to vivisection, it is clear that dissection on living animals is common.

H. ten Have and M. do C. Patrão Neves, *Dictionary of Global Bioethics*,
https://doi.org/10.1007/978-3-030-54161-3_514

Vulnerability

The word "vulnerability" derives etymologically from the Latin *vulnus* or *vulneris* (wound) and can be defined as being exposed to the possibility of being attacked or harmed either physically or emotionally by someone or something. According to the philosophical and broad approach of Emmanuel Levinas and Hans Jonas, vulnerability is little more than a universal condition that is common to all living beings due to the simple fact of being mortal. In short, the mortal nature of all living beings makes them vulnerable. Vulnerability has a radical ontological (and anthropological) dimension of its own that cannot be eliminated. Ethically, cooperation and solidarity are the only adequate responses to vulnerability (i.e., being exposed to the possibility of being hurt). According to the bioethical and more specific approach taken today, vulnerability was used in its adjectival form (i.e., "vulnerable") to classify people or groups: first, in clinical research; then, as a way of excluding them from clinical trials (since the Belmont Report in 1979); and, later, in clinical practice as a way of recognizing people requiring special protection (since the Declaration of Helsinki revised in 1996). Vulnerability became relevant bioethically in the 1990s and was finally established as an ethical principle in 2005 by UNESCO's Universal Declaration on Bioethics and Human Rights. In Article 8 the declaration introduced the principle of respect for human vulnerability and personal integrity: "In applying and advancing scientific knowledge, medical practice and associated technologies, human vulnerability should be taken into account. Individuals and groups of special vulnerability should be protected and the personal integrity of such individuals respected." Despite being contested as a principle because it was unclear whether it entailed an obligation, there is broad consensus that it is an essential bioethical concept that could be employed to overcome the shortcomings of autonomy and to call for adequate care for all regardless of lack of clarity over their respective power to claim it. The concept of vulnerability is important personally and globally because it stresses how social and economic inequalities around the world deepen people's vulnerability and at the same time highlights the importance of global cooperation and solidarity to mitigate vulnerability. As soon as vulnerability is acknowledged there is always a need to take moral action.

H. ten Have and M. do C. Patrão Neves, *Dictionary of Global Bioethics*,
https://doi.org/10.1007/978-3-030-54161-3_515

War (*See* Military Ethics)

Throughout history war has been a subject of ethical debate raising such questions as: Can war be morally justified? What are the moral rules that apply to warfare? When is war justified? This debate has led to a substantial body of just war theory. The rules of war can be broken down into two moral types: rules that govern the initiation of a war (*jus ad bellum*; justice of war) and those that govern how wars should be fought (*jus in bello*; justice in war). Classical just war theory is based on Thomas Aquinas who formulated three conditions that have to be met for warfare to be justified: first, war should be formally declared by a legitimate authority; second, there should be just cause; and, third, war should be waged by just (i.e., moderate) means. It has been argued that these three conditions are not sufficient and that supplementary conditions should be formulated such as war should be the last resort; war is unjustified when there are non-belligerent means to end hostilities; the danger that war seeks to prevent must be real and certain—not speculative; war is only justified when there is a reasonable hope of success; and there must be the intention to achieve something good or avoid something evil (i.e., revenge or hatred cannot justify warfare). The third condition of Aquinas applies to ensuring just means are used in the way war is waged. What is permissible conduct in warfare? Three principles have been proposed: first, the principle of proportionality (the level of force and damage should be proportionate to the goals of the war); second, the principle of discrimination (determining legitimate targets of action and prohibiting attacks on non-combatants); and, third, the principle of double effect (distinguishing between killing in war and murder). Although warfare often results in the killing of innocent people, this is usually deemed an unintended effect but difficult to defend in atomic warfare. War is a bioethical problem that extends beyond people being killed: it is destructive to health and human life, inflicts long-lasting suffering on people, and the medical consequences of war continuing years after it has ended. Therefore when assessing the criteria used to justify war, medical considerations should be included and bioethics can play a part in examining the principle of proportionality and in considering likely medical harm in the short and long term.

H. ten Have and M. do C. Patrão Neves, *Dictionary of Global Bioethics*,
https://doi.org/10.1007/978-3-030-54161-3_516

Water

Water refers to an inorganic, odorless, tasteless, transparent, and almost colorless liquid that is the principal constituent not only of the hydrosphere but also of the fluids of all known living organisms and as such is essential for human health and survival. Until recently water was never regarded as problematic since it was abundantly provided by biodiversity, human beings in general had easy access to it, and most countries had little difficulty in providing it to their populations. However, today water has become a global problem as a consequence of pollution, contamination, scarcity, unequal access, and poor distribution. In 2015 the WHO estimated that 663 million people lacked access to safe drinking water and at the global level at least 1.8 billion people used a source of drinking water that was contaminated and could transmit diseases. Water is now increasingly regarded as a scarce resource and a serious political problem as borne out by disputes between countries that share common water sources such as rivers and over the building of dams. Ethical discourse on water and water policy has been greatly stimulated over the last two decades by UNESCO. Its Commission on the Ethics of Scientific Knowledge and Technology (COMEST) selected the ethics of freshwater use as one of its priorities and developed a set of guiding principles for the use of freshwater. The major ethical concerns with water are associated with security, common good, human rights, justice, and governance. Water security involves the quality of water and the wise use of water resources that goes beyond that needed to fulfill immediate human needs. The main challenge is to ensure as many people as possible have access to water. Water security involves discussions about the right to water in which people should have access to water, be provided with water, and be charged an affordable amount for its supply. Water is also regarded as a common good because it is provided by nature, can be used but not owned, is the ecological basis of all life, is indispensable to the survival of humankind, is the heritage of humankind, and therefore common property. It cannot be claimed by anyone as his or her property since it belongs to humankind and cannot be turned into a commercial product. Framing water as a human right has become an increasingly popular way of articulating ethical concerns about water. Contrary to food, clothing, housing, and medical care the issue of water was overlooked in early human rights documents. It has been argued that such an oversight is unsurprising since it is generally assumed that water is a necessary part of the preparation of food and its consumption. Without water two explicitly stated rights cannot be realized: the right to life (Article 3) and the right to a standard of living adequate for the health and well-being of a person and his or her family (Article 25). The general interpretation is that water despite not being mentioned is implicitly assumed. Justice is another ethical concern as borne out by the evident inequalities in water and water governance and major disparities in water use such as the average North American using

H. ten Have and M. do C. Patrão Neves, *Dictionary of Global Bioethics*,
https://doi.org/10.1007/978-3-030-54161-3_517

much more per day than people elsewhere, the availability of water differing substantially from region to region, and millions of people lacking safe water resources. Water justice means that a society can only be deemed just when all its citizens receive sufficient water to lead a healthy and dignified life. Such ethical considerations should have consequences for water governance. The Sustainable Development Goals (SDGs) adopted by the United Nations in 2015 present the priorities and actions that are absolutely necessary for all countries to take to overcome the global problems they will face in the next 15 years and to protect the planet and its people. Goal 6 (of 17 goals) is: "Ensure availability and sustainable management of water and sanitation for all."

Weapons (*See* Biological Weapons)

Weapons have long been the focus of ethical discourse since a long time. In 1096 Pope Urban II prohibited the use of crossbows introduced from China in 1096 and Pope Innocent II repeating the prohibition in 1139 both without any real effect. New weapons have long been regarded as inhuman and unfair such as the machine gun in 1884 (despite the damage such a weapon could do the inventor argued it would prevent human suffering), nuclear weapons in 1945 (giving rise to international action to prevent their use), and biological and chemical weapons. Ethical attention today is focused on biological and chemical weapons. The use of poison gas in the First World War led to countries agreeing the Geneva Protocol in 1925 and the first international law banning such weapons. In 1972 the Biological Weapons Convention was adopted prohibiting the development, production, stockpiling, retention, acquisition, and use of bacteriological agents and toxins. It was followed in 1993 by the adoption of the Chemical Weapons Convention banning the use, development, production, and stockpiling of chemical weapons. Although it is not fully clear why biological and chemical weapons receive special attention while many other weapons are just as horrendously destructive, one reason may be the impossibility or difficulty to control the effects such weapons have on their targets (biological weapons have been known to affect the perpetrators). Another reason is because innocent civilians are killed. Although there is high pressure from public opinion today for militaries to develop surgical precision in targeting such that civilians are avoided, it should be kept in mind that precision weapons are notoriously unreliable. Of all the weapons designed to harm people only biological and chemical weapons are banned. One argument in favor of such weapons is that they can bring warfare to an end more quickly because they inflict mass casualties and therefore ultimately save lives. Such an argument was used to justify gas warfare in the First World War. However, it will cause little surprise that such an argument is widely contested. Although states today are forbidden from using such weapons, it still occurs and the risk is run that terrorists and rogue states will get hold of them and use them —hence the need for agencies such as the Organisation for the Prohibition of Chemical Weapons to carry out continuous surveillance. One exception outlined in such conventions refers to the use of toxic chemicals against humans for law enforcement purposes such as when there are domestic riots. This raises the question of whether it is acceptable to use agents such as tear gas to control riots, especially in light of other sometimes more powerful chemical agents being used in some countries. Interpreting which chemicals can be used is therefore ambiguous.

H. ten Have and M. do C. Patrão Neves, *Dictionary of Global Bioethics*,
https://doi.org/10.1007/978-3-030-54161-3_518

Whistle-Blowing

Whistle-blowing refers to someone perceiving something as wrong within an organization and communicating that information to actors outside the organization (usually the media). A recent example is the case of Edward Snowden who in 2013 leaked classified information from the National Security Agency in the United States to show how they ran many secret surveillance programs and by whistle-blowers from the world of healthcare who report unsafe and incorrect surgical practices. Whistle-blowing is problematic because employers and other responsible parties within the organization generally assume that such communication should not have been made, that in-house procedures should have been followed instead, and that contracts of employment have been broken. Therefore whistle-blowing is a risky affair. As a result of whistle-blowers being fired or even prosecuted, many countries have introduced laws to protect them. Whistle-blowers can be characterized as people who disregard their own self-interest or benefit, make information public solely on behalf of the common good, gain no personal advantages out of doing so, never promote their own profile or reputation, and take risks because they have a higher goal in mind. It should be kept in mind that the ethical issues that crop up in whistle-blowing primarily concern the perceived wrongdoing and that not every issue justifies whistle-blowing. Moral concerns should only ever be raised when the public interest or the common good is endangered by certain practices such as scientific fraud, negligent treatment, discrimination, corruption, and harassment. Whistle-blowing often involves a pattern, practice, or procedure that reflects an immoral culture within the organization rather than a single incident. Moreover, there are ethical questions about whether unauthorized public disclosure is justified, whether disclosure does more harm than good, whether it will rectify or prevent the wrongdoing, whether it is done in a responsible manner, and whether it is done after exhausting the usual channels of communication within the organization.

H. ten Have and M. do C. Patrão Neves, *Dictionary of Global Bioethics*,
https://doi.org/10.1007/978-3-030-54161-3_519

Wrongful Birth

Wrongful birth refers to parents taking legal action against a healthcare professional or institution that has failed to warn them of the risks of conceiving and/or giving birth to a child suffering from a serious incurable disease or severe disabilities passed on by the parents. Wrongful birth presupposes that the life of such a child would hardly be worth living even though the disease was not fatal in the short term; that standard clinical knowledge, screening, and diagnostic resources could and should have provided the parents with information about the probabilities of having a seriously disabled child; and that family planning was accessible, efficient, and could indeed have prevented conception of the child. The wrongful birth argument is thus that the child should never have been conceived. The development of knowledge about human reproduction, contraception, embryology, and genetics coupled with advances in technological innovation in these fields has improved the capacity to foresee and control the outcomes of pregnancies in an unprecedented way. In short, the greater the power a healthcare professional has the greater the responsibility he or she will have to shoulder. The most common cases of wrongful birth concern the failure of healthcare professionals to inform parents that they carry a genetic disease and the likelihood of passing the disease on given the family history. Legally, wrongful birth implies negligence or malpractice and argues that the medical or healthcare institution has liability. The plaintiff (the parents) complain that the burden following the birth of the child falls upon them and they seek compensation for the harm done to their lives. Ethically, it is possible to argue that the absence of life can be preferable to life itself, especially when life has no potential for joy and flourishment. However, the same argument could be presented in the case of children who are unhappy or frustrated with their parents. The main ethical issue regarding wrongful birth concerns whether parents have been provided in a reasonable way with the necessary medical information to enable them to make an informed (and thus responsible) decision about their reproductive choices.

H. ten Have and M. do C. Patrão Neves, *Dictionary of Global Bioethics*,
https://doi.org/10.1007/978-3-030-54161-3_520

Wrongful Life

Wrongful life refers to parents or custodians taking legal action against a healthcare professional or institution that has failed to prevent the birth of a child suffering from a serious incurable disease or severe disability. Wrongful life presupposes that the life of such a child would hardly be worth living even though the disease was not fatal in the short term; that standard healthcare during pregnancy should have been able to detect and identify the medical problem; and that abortion was legal for the situation in question. The wrongful life argument is thus that the child should never have been born. The concept of wrongful life has only relatively recently been shown to have relevance in common law (the second half of the twentieth century). It was at this time that the means for clinical diagnosis of impaired fetal growth were developed and became highly accurate and that abortion became legal for a number of situations and for a longer period during pregnancy. Although the first wrongful life cases were brought in the 1960s and dealt with illegitimate births followed by wrongful life cases dealing with birth deformities, it was not until the 1980s that wrongful life cases were brought concerning the failure to detect malformations during pregnancy. This meant that long-established standards of care would have to be revisited. Legally, wrongful birth implies negligence or malpractice and argues that the medical or healthcare institution has liability. The plaintiff (the parents or custodians) complain that wrongful birth has resulted in the child suffering, being deprived of any chance of flourishing or leading a fulfilling life, and having its human dignity violated and seek compensation for the harm done. Ethically, it is difficult if not impossible to argue that lives are not worth living. However, it is argued there are circumstances in which the common notion of human dignity as an intrinsic, unconditional, and absolute value of human beings regardless of singular features or particular conditions can be denied. Moreover, well-founded and objective medical criteria and metrics on quality of life need to be presented to evaluate a life as not worth living. However, this is another instance of the slippery slope argument in which clinical criteria suddenly include social criteria and then maybe state criteria. All criteria put forward to evaluate whether a life is worth living have the effect of singling out humans according to their characteristics thus objectivizing and quantifying them and conflicting with the principle of human dignity.

H. ten Have and M. do C. Patrão Neves, *Dictionary of Global Bioethics*,
https://doi.org/10.1007/978-3-030-54161-3_521

Xenograft

The prefix "xeno" derives etymologically from the Greek *xeno* (foreigner) and a "graft" medically refers to a piece of living tissue surgically transplanted. Xenografts are thus foreign to the recipient in that they are harvested from an animal species different from that of the recipient. Specialized scientific knowledge and expertise in medical sciences are necessary for xenografts to take (succeed). The success of xenografts has been limited to a few living tissues harvested from non-human animals for human use (mostly pigs) such as heart valves and porcine islet xenografts for the treatment of diabetes. Although beneficial to humans, xenografts raise two significant ethical issues. Although the first concerns the antinatural procedure of combining biological elements from different species (i.e., cross-species) and the supposed risk of hybridization, such an argument is losing relevance due to the singularity and small size of the animal living tissue transplanted to humans. The second ethical issue relates to the conflict that arises between using animals to supply and satisfy human needs and the argument in favor of animal rights. Such an argument is gaining relevance as a result of the animal rights movement strengthening worldwide.

H. ten Have and M. do C. Patrão Neves, *Dictionary of Global Bioethics*,
https://doi.org/10.1007/978-3-030-54161-3_522

Xenotransplantation

The prefix "xeno" derives etymologically from the Greek *xeno* (foreigner). Xenotransplantation refers to the transplantation of cells, tissues, or organs between different species such as from animals to humans. Xenotransplantation first began in the early 1900s when animals such as pigs, goats, lambs, and monkeys were used to harvest vital organs to save the lives of human beings. Such experiments inexorably failed until scientists learned more about histocompatibility and the causes of organ rejection, which are even more complex between species than between human beings. Indeed, it was only after the first immunosuppressive drugs were identified in the early 1960s that transplantation between humans definitively moved from an experimental to a therapeutic phase (mostly in the 1980s). Human-to-human transplantation has always been characterized by a chronic scarcity of organs for transplantation and too high a number of deaths all over the world. Such a shortage of organs for transplantation encouraged scientists to look for different non-human biological sources of body parts in addition to those procured from cadavers or those donated by family members, friends, or even unrelated people. Should xenotransplantation succeed it would be a major contribution to overcoming the scarcity of vital non-regenerable organs by providing an unlimited supply of cells, tissues, and organs (xenografts). However, it has not been possible to fully overcome the two major difficulties of xenotransplantation: histocompatibility problems from non-humans to humans and cross-species infection. The implementation of new strategies to diminish immunogenicity after xenotransplantation has led to the procedure today pursuing two different paths. The first is at the therapeutic level and involves procuring animal body parts to transplant into human beings (mostly from pigs since they are the most suitable animal donors for humans) as borne out by bioprosthetic heart valves and insulin from pigs being already successfully used in humans. The second is at the experimental level and involves creating genetically engineered animals (transgenic animals) modified to express human complement-regulatory proteins (by pronuclear injection and later by nuclear transfer and cloning) to become organ donors. The production of human organs in non-human bodies (with modified functional human genes introduced into the donor animal's organs) should prevent rejection. The first major ethical issue raised by xenotransplantation was the experimental procedure itself as happened with by Baby Fae who was suffering from hypoplastic left heart syndrome (HLHS) and received the heart of a baboon in 1984 only to die 21 days later as a result of severe organ rejection. Although the reason given for cross-species transplantation was to gain some time for a suitable pediatric heart to be found, it was argued Baby Fae was used by physicians and researchers as a simple experimental resource. There was also vigorous criticism from animal rights advocates against the unnecessary death of the

H. ten Have and M. do C. Patrão Neves, *Dictionary of Global Bioethics*,
https://doi.org/10.1007/978-3-030-54161-3_523

baboon. Experimental procedures are heavily regulated today and the most pressing ethical issues concerning xenotransplantation relate to animal rights. It is frequently argued that animal lives have an intrinsic value of their own and should not be used to satisfy human needs. Nevertheless, it can be argued that xenotransplantation is much the same from an ethical point of view as breeding animals for food. Although the demand for organs will continue to grow, it is possible that xenotransplantation will lose pertinence in the not too distant future as a consequence of the development of new sources of organs without ethical problems (especially mechanical devices and autologous transplantation).

Zika

Zika virus refers to a disease caused by a virus primarily transmitted by mosquitoes from the genus *Aedes* that bite during the day. It is the same mosquito that transmits dengue, chikungunya, and yellow fever. Apart from bites zika virus is also transmitted from mother to fetus during pregnancy through sexual contact, transfusion of blood and blood products, and organ transplantation. Although most people infected with zika virus do not develop symptoms, those that do have them (after an incubation period of 3–14 days) find they last for 2–7 days, are mild, and consist of fever, rash, conjunctivitis, muscle and joint pain, malaise, and headache. There is no treatment for zika infection. Zika virus was first identified in humans in Uganda and Tanzania in 1952. Although sporadic outbreaks have been reported in Africa and Asia, it was not until 2007 that a major outbreak was reported in the Pacific as borne out by a major one taking place in French Polynesia in 2013. In 2015 zika virus infection was reported in Brazil. Later that year the infection was found to be associated with Guillain-Barré syndrome and microcephaly. Today 86 countries and territories have reported zika infections.Zika infection poses a number of dangers including increased risk of neurological complications such as Guillain-Barré syndrome, neuropathy, and myelitis; infection during pregnancy produces congenital malformations (particularly microcephaly); and the fact that no vaccine or treatment has been developed to deal with it. Ethical issues concerning zika relate to prevention the most concerning of which are the potential link between pregnant women infected with zika virus and the risk of their babies being born with microcephaly, infected patients often having no symptoms, pregnant women being advised to postpone travel to areas where zika transmission is ongoing, and lack of clarity when infection is detected about what advice should be given to the pregnant mother or couple.

H. ten Have and M. do C. Patrão Neves, *Dictionary of Global Bioethics*,
https://doi.org/10.1007/978-3-030-54161-3_524

Zoocentrism (*See* Animal Ethics; Anthropocentrism; Biocentrism; Ecocentrism)

The word "zoocentrism" derives etymologically from the Greek *zoon* (animal) and *kentron* (center). The Greek suffix *-ismós* (Latin *-ismus*) expresses the general scope of the word to which it is added. Zoocentrism refers to the different doctrines that consider animals as having moral worth, regard animals from a perspective centered on them and animal life, and include animals in the moral community. Such doctrines tend to adopt sentience as the criterion for the moral worth of animals and in so doing establish them as part of the moral community (from which non-sentient animals are excluded). An egalitarian perspective on animals would tend toward biocentrism. Historically, zoocentrism moved away from hegemonic anthropocentrism once sensitivity was recognized in many animals (thus narrowing the gap between different species). The work of utilitarian philosophers such as Jeremy Bentham in the eighteenth century led to zoocentrism emerging as a doctrine. Even today it is still through utilitarian philosophy that contemporary zoocentrism is evolving. It is doing so through two main orientations: animal liberation (defended by Peter Singer) and animal rights (defended by Tom Regan). Nevertheless, it is possible to track zoocentric ideas from the birth of humankind to current times in religious and cultural traditions that attribute significant importance to certain animals involved in the transmigration of souls across species barriers or the worship of sacred animals (cows, monkeys, and elephants) that still persist in regions of Asia or Africa. More recently, scientific studies (biology, ethology, psychology) have contributed to animal life in terms of sensitivity, practical reasoning, self-awareness, and culture being better understood and to recognition of the intrinsic value of animal life and the obligation to respect and protect it. Zoocentrism differs from anthropocentrism by denying human beings should be given superiority. It differs from biocentrism by implicitly establishing a hierarchy among life-forms that favors or values sentient animals (i.e., those closer to human beings). It differs from ecocentrism by taking a narrower view of our common home—planet Earth. Since global bioethics is not restricted to biomedical ethics, it should also include zoocentric perspectives.

H. ten Have and M. do C. Patrão Neves, *Dictionary of Global Bioethics*,
https://doi.org/10.1007/978-3-030-54161-3_525